Translating Physiological Tools to Augment Crop Breeding

Harohalli Masthigowda Mamrutha •
Gopalareddy Krishnappa • Rinki Khobra •
Gyanendra Singh • Gyanendra Pratap Singh
Editors

# Translating Physiological Tools to Augment Crop Breeding

Springer

*Editors*
Harohalli Masthigowda Mamrutha
ICAR-Indian Institute of Wheat and Barley Research
Karnal, Haryana, India

Gopalareddy Krishnappa
ICAR-Indian Institute of Wheat and Barley Research
Karnal, Haryana, India

Rinki Khobra
ICAR-Indian Institute of Wheat and Barley Research
Karnal, Haryana, India

Gyanendra Singh
ICAR-Indian Institute of Wheat and Barley Research
Karnal, Haryana, India

Gyanendra Pratap Singh
ICAR-Indian Institute of Wheat and Barley Research
Karnal, Haryana, India

ISBN 978-981-19-7500-4 ISBN 978-981-19-7498-4 (eBook)
https://doi.org/10.1007/978-981-19-7498-4



This Springer imprint is published by the registered company Springer Nature Singapore Pte Ltd.
The registered company address is: 152 Beach Road, #21-01/04 Gateway East, Singapore 189721, Singapore

# Preface

To ensure food security for the ever-growing population, agricultural production needs to be increased by 50% by 2050 with the dwindling natural resources. Recently, crop yields are stagnating due to narrow genetic base, climate change, biotic and abiotic stresses. To break the yield barriers under changing climatic scenario, physiological breeding plays a key role. This book covers different physiological traits, processes, tools, and their application in crop breeding, new and advanced field phenotyping tools used in several physiological processes, evidence and success stories of physiological breeding to complement crop breeding. The book mainly focuses on the practical utilization of physiological tools in crop breeding to accelerate genetic gain and up to date literature of physiological techniques for ready reference by researchers, scientists, and students.

The book comprises 21 chapters contributed by renowned national and international scientists working on different areas of crop physiology. Each chapter emphasizes specific trait/physiological process and its importance in crops, their phenotyping information, and how best it can be employed for crop improvement. It covers a wide range of physiological topics including advances in field phenotyping, application of stable isotopes, high-throughput phenomics, role of transpiration in physiological breeding, impact of high temperature stress, root phenotyping for improved resource use efficiency, cuticular wax and its application, advances in photosynthetic studies, root exudation, rhizosphere efficiency, leaf spectral reflectance, role of endophytic fungi, metabolomics, physiological breeding in hardy crops like millets, and role of next-generation sequencing and genetic mapping in trait identification. This book also covers futuristic research areas like artificial intelligence and machine learning for crop improvement.

Though basic knowledge on different physiological process is extensively available, information on their application and compilation is very meager. Hence, in this book an extensive attempt was made to compile all such applications of physiological tools along with their advanced research in crop improvement, which is the need of the hour for both academics and researchers for ready reference. This book will be of interest to teachers, researchers, climate change scientists, capacity builders, and policymakers. Also, the book serves as additional reading material for undergraduate, graduate, and postgraduate students of agriculture, physiology, botany, ecology, and environmental sciences and also for national and international agricultural scientists.

We are assertive that the present book will be a landmark, as this kind of compilation on physiological breeding is first of its kind and the chapters contain updated information as well as views conveyed by researchers based on their vast experience and expertise in physiological breeding. We thank all the eminent authors for their valuable contributions towards enhancing the quality of the book. Undoubtedly, ***Translating Physiological Tools to Augment Crop Breeding*** is a timely and much warranted book considering the significance of physiological breeding to achieve future food demand.

Karnal, Haryana, India

Harohalli Masthigowda Mamrutha
Gopalareddy Krishnappa
Rinki Khobra
Gyanendra Singh
Gyanendra Pratap Singh

# Contents

# About the Editors

**Harohalli Masthigowda Mamrutha, Ph.D.,** is presently serving as a Senior Scientist (Plant Physiology) at ICAR-Indian Institute of Wheat and Barley Research, Karnal, Haryana, India. She obtained her B.Sc., M.Sc., and Ph.D. degrees from the University of Agricultural Sciences, Bengaluru, India. She holds about 15 years of experience in R&D and has published around 45 research and review articles in peer-reviewed national and international journals, 25 book chapters, 3 technical bulletins, and more than 60 popular articles. She handled 12 research projects in various capacities including projects funded by USDA, DBT-BBSRC, CIMMYT, ACIAR, and ICAR and presented research findings in several national and international forums. Her research is largely focused on physiological trait-based breeding for improving photosynthesis, yield, and abiotic stress tolerance and use of CRISPR/Cas9 technology in wheat improvement. She has co-developed five high yielding wheat varieties and ten wheat genetic stocks for abiotic stress tolerance and for improving nutritional quality. She has been serving as an editor of *Plant Physiology Reports*. She also serves as an editorial board member of a journal and as reviewer of many international journals. She is committed to her professionalism and honored with several recognitions, including Prof. Mahatim Singh Memorial Award from Society for Advancement of Wheat & Barley Research (SAWBAR) India, RD Asana gold medal award, and JC Bose gold medal award from Indian Society of Plant Physiology and ICAR-IIWBR best scientist award. She is also the recipient of international fellowships from wheat initiative, Donald Danforth Plant Science Centre, USA, and IUBMB, France.

**Gopalareddy Krishnappa, Ph.D.,** Scientist, ICAR-Sugarcane Breeding Institute (ICAR-SBI), Coimbatore, Tamil Nadu, completed his bachelor's degree in 2006 and master's degree in 2008 from the University of Agricultural Sciences, Bangalore, and doctorate degree in Genetics from ICAR-Indian Agricultural Research Institute, New Delhi, in 2016. Dr. Reddy is an accomplished and dedicated agricultural researcher with a rich experience in wheat research over a decade. Dr. Reddy developed 11 wheat varieties and 26 wheat genetic stocks to benefit a multitude of stakeholders including farmers, consumers, researchers, and industry. His research is largely focused on the development of high yielding wheat varieties with improved quality attributes. Dr. Reddy utilized modern breeding tools like genomic selection, MARS, MAS, and GWAS to complement conventional breeding techniques to dissect complex traits in wheat. He published more than 30 research and review articles of national and international repute with high impact journals, 02 books, 15 book chapters, and 30 technical or popular articles. He has been teaching for more than 3 years to students of different state agricultural and private universities. Dr. Reddy was conferred with AIASA Young Scientist Award 2020, Best Worker 2020, NEEDEF Young Scientist Award 2021, and Dr. VS Mathur memorial award 2022.

**Rinki Khobra, Ph.D.,** presently serves as a scientist (Plant Physiology) at ICAR-Indian Institute of Wheat and Barley Research, under the aegis of Indian Council of Agricultural Research. She received her M.Sc. degree from Kurukshetra University, Kurukshetra, in Botany and Ph.D. from ICAR-Indian Agricultural Research Institute, New Delhi, India, in Plant Physiology. Her research is primarily focused on abiotic stress tolerance, mainly drought and lodging in wheat. She is also working on identification of plant traits contributing towards higher yield potential along with root system architectural studies. Dr. Khobra is associated with the development of one bread wheat variety and ten genetic stocks for various physiological traits. She published more than 15 research and review articles of national and international repute along with 1 bulletin, 7 book chapters, and 45 popular articles. Apart from research, Dr. Khobra is also involved in teaching Ph.D. scholars. She is a lifetime member of few professional societies.

**Gyanendra Singh, Ph.D.,** Director, ICAR-Indian Institute of Wheat and Barley Research, Karnal, India, graduated from Janta Vedic College in 1982 followed by an M.Sc. in 1984 and a Ph.D. in 1998. He re-defined and executed shuttle breeding methodology for wider adaptability, resistance, and yield in wheat. He devised a new double-digit appraisal scale for spot blotch disease severity in wheat, which is convenient and time saving especially while evaluating large breeding populations. He contributed to the development of 25 varieties, 50 genetic stocks, and 15 RIL populations in wheat to facilitate molecular studies. He published 160 research articles, 10 edited books, 08 bulletins, 19 book chapters, 05 review articles, 06 training manuals, and 100 popular/technical articles. Dr. Singh is elected fellow of Society for Advancement of Wheat and Barley Research and Indian Society of Genetics and Plant Breeding. Dr. Singh received USAID fellowship at Mississippi State University, USA, and certificate of achievements from United Sates Department of Agriculture (USDA) for training at the University of Florida. He is a member of the scientific advisory committee of the PPV&FRA, New Delhi. He received distinguished scientist award from SSDAT, best research team award of DWR, Karnal, in 2006, DWR best scientist, and VS Mathur award (2020).

**Gyanendra Pratap Singh, Ph.D.,** Director, ICAR - National Bureau of Plant Genetic Resources, New Delhi, has more than 27 years of experience in agriculture research including 10 years of teaching and more than 6 years in administration, which led to many milestones. Dr. Singh is instrumental in the development of 51 wheat (5 biofortified varieties) and 03 barley varieties and 01 potato variety benefitting farmers, consumers, and industries. He is the main force behind the development and rapid spread of improved mega-wheat varieties including DBW 187, and DBW 222, HD 2967, HD

3086, and DBW 303 as these were readily adopted by farmers. Dr. Singh largely focused on intrinsic research on heat and drought tolerance for wheat improvement in India and developed many climate-resilient wheat varieties. He is also the leader of cutting-edge technologies like Marker Assisted Recurrent Selection and Precision Phenotyping for heat and drought tolerance. He published more than 300 research articles of national and international repute with high impact journals, 20 books, 65 book chapters, 57 technical bulletins, 55 popular articles, and 05 Policy/Strategy papers. His strenuous efforts in advancing the research and development led to several accolades at national and international level by honoring with 17 prestigious awards. He is also a fellow of Indian National Science Academy (INSA), National Academy of Sciences India (NASI), National Academy of Agricultural Sciences (NAAS), Indian Society of Genetics and Plant Breeding (ISGPB), and Society for Advancement of Wheat and Barley Research (SAWBAR).

# Chapter 1
# Importance of Integrating Physiological Breeding to Augment Crop Breeding

**Harohalli Masthigowda Mamrutha, Rinki Khobra, Gopalareddy Krishnappa, Zeenat Wadhwa, Ratan Tiwari, Gyanendra Singh, and Gyanendra Pratap Singh**

**Abstract** Global food consumption is expected to surpass the crops genetic gain by 2050, and climate change threatens the breeding programmes by hampering yield advantages. The current yield gain achieved is mostly through conventional breeding, and there is an urgent need for exploring new breeding approaches for augmenting the food demand. Recently, the physiological breeding and specifically trait-based breeding approaches are more explored in different crop breeding programmes and shown significant cumulative effect on yield advantage under both control and stress conditions. This helped in having complete information of genetic resource and the capability to improve favourable alleles by phenomic screening of offspring. This chapter mainly discusses the importance and role of physiological breeding, success stories of physiological breeding and the methods and need for integrating physiological breeding with all conventional crop improvement programmes to achieve significant yield advantages.

**Keywords** Physiological breeding · Conventional breeding · Phenotyping · Trait-specific breeding · Crop improvement

H. M. Mamrutha (✉) · R. Khobra · Z. Wadhwa · R. Tiwari · G. Singh
ICAR-Indian Institute of Wheat and Barley Research, Karnal, Haryana, India
e-mail: mamrutha.M@icar.gov.in

G. Krishnappa
ICAR-Indian Institute of Wheat and Barley Research, Karnal, Haryana, India

ICAR-Sugarcane Breeding Institute, Coimbatore, Tamil Nadu, India

G. Pratap Singh
ICAR-Indian Institute of Wheat and Barley Research, Karnal, Haryana, India

ICAR-National Bureau of Plant Genetic Resources, New Delhi, India

 
H. M. Mamrutha et al. (eds.), *Translating Physiological Tools to Augment Crop Breeding*, https://doi.org/10.1007/978-981-19-7498-4_1

## 1.1 Introduction

Plant breeders and agricultural scientists of the twenty-first century are challenged to increase the yield potentials of crops to feed the growing world population. Climate change, the resultant stresses and increasing nutrient deficiencies are factors to be considered in designing modern plant breeding pipelines. An increasingly challenging crop environment and the rapid advances in genetic technologies call for better understanding of the physiological processes involved in achieving crop productivity, and their interaction with environment. Three new factors that can help to achieve this new models of improved plant processes and crop ideotypes are high-throughput phenotyping technologies that permit evaluation of complex trait expression on a breeding scale in realistic field environments, renewed focus on preserving and utilizing plant genetic resources and awareness on climate change for trait selection in improving new allelic diversity into existing gene pools (Boote and Sinclair 2006).

Annual genetic yield gains in cereals are currently in the range of 0.5–1% (Fischer et al. 2014), achieved through conventional approaches, mainly by unspecified recombination of genes of minor effect among elite germplasm, and the introduction of new genetic diversity often associated with disease resistance and grain quality (Braun et al. 2010; Eagles et al. 2014). Physiological breeding complements this approach by adding two main elements: knowledge of well-characterized genetic resources to design crossing strategies and the ability to enrich for favourable alleles through phenomic and genomic screening of progeny. This increases the probability of achieving cumulative gene action for yield compared to crossing physiologically uncharacterized lines. In practice, it differs from conventional breeding by considering a larger range of traits including genetically complex physiological characteristics (Reynolds et al. 2012; Richards et al. 2010).

## 1.2 History of Physiological Breeding

The history of breeding association with physiology dates back to the great development in crop and yield physiology in the 1960s and 1970s in the book *The Growth of Cereals and Grasses* (Milthorpe and Ivins 1965). The integrated sorghum physiology and breeding research group at the University of Nebraska in the early 1970s was the first real and significant contribution of crop physiology to sorghum and maize breeding in USA through Rockefeller Foundation (Hallauer 2007). At the same time, crop physiology and breeding research at Texas A&M University discovered the genetic and physiological basis to photoperiod and temperature effect on flowering in sorghum. This research supported in converting tropical sorghum from Africa and Asia into temperate types which were further used in hybrid sorghum breeding to achieve a higher level of yield and grain quality. Genes for osmotic adjustment and non-senescence were also isolated from these sorghum lines for inducing drought resistance and thus the integration of crop physiology into plant breeding was adopted in the breeding programs of most Consultative Group on International Agricultural Research (CGIAR) institutions.

The physiological basis of the Green revolution in the cereals was identified very early as increase in harvest index was observed from 20–30% to about 40–50%, depending on the crop. The yield components involved in this increase were also identified. Crop physiology led breeders to understand that yield formation in cereals is derived from an intricate balance between yield components development, source to sink communication, crop assimilation and assimilate transport—linked to crop phenology and plant architecture.

Later the crop physiologists developed the concept of use efficiency regarding how the plant uses its essential resources such as irradiance, water, or nutrients. Thus, we have water use efficiency (WUE), radiation use efficiency (RUE) and nutrient use efficiency (NUE), and it has been shown that breeding for increased efficiency can lead to greater productivity. Consequently, it has been proposed that any serious improvement in cereal yield potential beyond the common average 0.5–1.0% per annum must come from a genetic–physiological intervention in photosystem biochemistry and function.

## 1.3 Need and Importance of Physiological Breeding

Thorough understanding of plant physiology is needed to design the traits targeted by various breeding approaches. In breeding programmes, the use of physiological traits could be done either by direct selection approach or through surrogate traits. Numerous physiological studies identified plant traits, which are associated with plant adaptability to stress prone environments (Cattivelli et al. 2008). Early escape during drought stress is a well-documented trait which is commonly exploited by the breeding community to avoid significant yield losses. However, the recent advances in plant physiology, phenotyping and genomics have led to new insights in abiotic stress tolerance, thus providing breeders with more precise information and new tools to increase crop yield (Tuberosa and Salvi 2006). Plant physiology improves our basic understanding for the abiotic stress tolerance-related traits which improves selection efficiency. Physiology also provides the basic data to supplement genomics and molecular biology approaches for identifying the candidate genes and quantitative trait loci (QTLs) associated with these traits.

A team of physiologists and breeders developed a general model for drought stress adaptation that includes traits such as pre-anthesis growth, WUE, relatively cool canopy, transpiration efficiency (TE), WUE of spike photosynthesis and photoprotection, carbon isotope discrimination of leaves, antioxidant systems and anatomical traits (Reynolds et al. 2005). This model is intended to help breeders to choose crossing parents with different drought adaption mechanisms (Mir et al. 2012). Numerous other physiological breeding experiments have shown that breeding for enhanced TE, WUE and numerous root characteristics allows for lower transpiration, more water uptake from soil and comparatively higher output under stress circumstances. Though the lack of precise, rapid and economical phenotyping procedures ideal for large breeding populations limits the field of physiological breeding, the benefits are likely to be enormous once successful because it enhances the probability of additive gene effect in crossing.

It has been more than 20 years since the gap between breeders and physiologists regarding physiological comprehension in breeding programme was recognized. For any improvement in the crop, physiologists play an important role as they help to interpret target environments, evaluate the potential attributes to enhance yield potential or crop transition within these environments, select appropriate secondary traits that seem to be easy to measure, have such a high heritability and also have a significant correlation with crop productivity, and comprehend complex relationships among both traits and with the environment. Thus, crop physiology can establish key links between phenotypic traits and crop performance.

## 1.4 Diverse Pipeline Used in Physiological Breeding for Crop Improvement

1. **Designing of crop**: For developing improved plant is a unique component of physiological breeding wherever, novel trait combinations may be extensively examined across a range of target conditions in terms of their influence on yield by implementing new strategic crosses.
2. **Preparation of genetic resources**: Physiological analysis of genetic resources has the potential to widen the agricultural gene pool in a highly targeted manner. However, most breeders are cautious about crossing with exotic germplasm, and precedent often refers to prerequisites. Although extensive physiological and genetic dissection is not yet a standard technique for choosing parents among advanced breeding lines, advances in field phenotyping, in conjunction with high-throughput genotyping will uncover more potential lines for use in trait-based crossover. It is mainly used for the identification of genetic resource to cover allelic variation for crossing.
3. **Plant phenotyping**: New advancements in high-throughput field phenotyping have improved the efficacy of physiological breeding. It implies the use of non-invasive techniques, such as proximal/remote sensing of spectral reflectance from plant tissue. It primarily employs phenotyping techniques and experimental treatments to maximize the resolution of physiological trait expression in parents.
4. **Plant genotyping:** Plant breeding and physiology are tangle related with crop improvement, and genetic dissection of traits is enhanced by using modern tools like molecular breeding and modern biotechnology that mainly help in genetic dissection as well as in advanced gene-based selection of traits.
5. **Crossing and selection**: Using simple quantitative models to assess potentially complementary features in this strategic hybridization to achieve cumulative gene

action for yield, paired with high-throughput phenotyping and genotyping to select progeny.

6. **Assessment of genetic improvements:** Based on multilocation trial data, analysis of trait/allele combinations that generate environmentally resistant genetic advances (to build new crosses); informatics services for iterative refining of breeding techniques.

## 1.5 Future Prospects

Future progress of plant breeding requires selection of trait-specific plant phenotypes by employing high-throughput multi-tier screening phenotyping approaches to identify genotypes expressing the desired trait. A well-defined framework may be needed to understand the role that physiological phenotyping can play in breeding programs and also physiological breeding should be an integral part of crop improvement programmes to address the gap between breeders and physiologists and to get a significant yield jump.

## References

Boote KJ, Sinclair TR (2006) Crop physiology: significant discoveries and our changing perspective on research. Crop Sci 46(5):2270–2277

Braun H, Atlin G, Payne T (2010) Multi-location testing as a tool to identify plant response to global climate change. In: Reynolds MP (ed) Climate change and crop production. CABI, Wallingford, pp 115–138

Cattivelli L, Rizza F, Badeck FW, Mazzucotelli E, Mastrangelo AM, Francia E, Marè C, Tondelli A, Stanca AM (2008) Drought tolerance improvement in crop plants: an integrated view from breeding to genomics. Field Crop Res 105(1–2):1–4

Eagles HA, Cane K, Trevaskis B, Vallance N, Eastwood RF, Gororo NN, Kuchel H, Martin PJ (2014) Ppd1, Vrn1, ALMT1 and Rht genes and their effects on grain yield in lower rainfall environments in southern Australia. Crop Pasture Sci 65:159

Fischer RA, Byerlee D, Edmeades GO (2014) Crop yields and food security: will yield increases continue to feed the world? ACIAR monograph 158. Australian Centre for International Agricultural Research, Canberra

Hallauer AR (2007) History, contribution, and future of quantitative genetics in plant breeding: lessons from maize. Crop Sci 47:S4–S19

Milthorpe FL, Ivins JD (1966) The growth of cereals and grasses. Butterworths, London. (Proceedings of 12th Easter School in Agricultural Sciences, University of Nottingham, 1965)

Mir RR, Zaman-Allah M, Sreenivasulu N et al (2012) Integrated genomics, physiology and breeding approaches for improving drought tolerance in crops. Theor Appl Genet 125:625–645

Reynolds MP, Mujeeb-Kazi A, Sawkins M (2005) Prospects for utilising plant-adaptive mechanisms to improve wheat and other crops in drought- and salinity-prone environments. Ann Appl Biol 146:239–259

Reynolds M, Foulkes J, Furbank R, Griffiths S, King J, Murchie E, Parry M, Slafer G (2012) Achieving yield gains in wheat. Plant Cell Environ 35:1799–1823

Richards RA, Rebetzke GJ, Watt M, Condon AG, Spielmeyer W, Dolferus R (2010) Breeding for improved water productivity in temperate cereals: phenotyping, quantitative trait loci, markers and the selection environment. Funct Plant Biol 37:85–97

Tuberosa R, Salvi S (2006) Genomics-based approaches to improve drought tolerance of crops. Trends Plant Sci 11:405–412

# Chapter 2
# Stacking of Complex Traits Through Physiological Prebreeding

**Puja Srivastava and Navtej Singh Bains**

**Abstract** There are immense prebreeding opportunities for a crop like wheat. The vast number of progenitor species and equally large number of nonprogenitor, while nonavailability of wild hexploid wheat, make it a right target for prebreeding investments. Lot more has been achieved in terms of biotic stresses be it gene scouting, mapping, and transfers. However, abiotic stresses are an area that needs more genes from the wild which are gold mines for abiotic stress tolerances. This chapter includes the breeding opportunity, methods, and probable strategy for pre breeding in wheat, while elucidating each with examples from our own breeding program.

**Keywords** Wheat · Introgression · Wild species · Stacking

## 2.1 Introduction

The wheat yield demand to be fulfilled by 2050 should be such that total production over the next 50 years will need to exceed the total produced in the 10,000 years since agriculture began (Moore 2015). The quest for accelerating genetic gains is one of the most rapidly evolving science and technology areas given that genomic and marker-assisted selection, bioinformatics, and genome editing have great potential in wheat breeding. More specifically, with the advances in genomics, we are just beginning to understand the wealth of genetic diversity available within wheat relatives, may it be wild ancestors, landraces, or exotic relatives. This genetic diversity, if tapped for traits of interest, can serve as the building blocks for future cultivars to be deliberately deployed to increase the productivity and resilience of agricultural systems. Prebreeding is described as an activity under which we scout and transfer desirable and at times hidden traits from wild non-adapted relatives of a crop species to an adapted breeding line. These can then be utilized by cultivar

---

P. Srivastava (✉) · N. S. Bains
Department of Plant Breeding and Genetics, PAU, Ludhiana, Punjab, India
e-mail: pujasrivastava@pau.edu

 
H. M. Mamrutha et al. (eds.), *Translating Physiological Tools to Augment Crop Breeding*, https://doi.org/10.1007/978-981-19-7498-4_2

breeders in a more efficient manner. Prebreeding bridges the gap between germpalsm repositories and active breeding programmes.

## 2.2 Prebreeding: Historical Antecedents

The early half of the last century witnessed extensive work in the area wheat germpalsm identification and development. Where, gene pool (Harlan and de Wet 1971) provided a universal framework for the utilization of crop wild relatives in breeding. While evolutionary and cytogenetical relationships of cultivated wheats with wild species of the primary, secondary, and tertiary gene pools was also established (Kihara 1924, 1944; Dvořák 1976) and methods for gene transfer published (Feldman and Sears 1981; Jauhar 1993; Mujeeb-Kazi 1998). While the later half was much invested on exploitation of variation within cultivated gene pool. Exceptional in this regards was first the advent of dwarfing genes in wheat and their world wide extension through CIMMYT, Mexico. Followed by similar success through "Veery derivatives," with Attila and sibs taking the centre stage. The basis of green revolution was the dwarfing genes that in turn changed the entire wheat physiology and its responses to inputs, while for Veery derivatives, the basis formed the 1B.1R translocation in the background of Russian winter wheat variety KavKaz (Zeller and Hossam 1983). This stream of breeding lines had better uptake efficacy, plasticity in phenology and tolerance to biotic (rusts, powdery mildew, etc.), and abiotic stresses (heat, drought, and marginal soils) recognized by the fact that these cultivars (e.g., PBW343) spread across all environments and climatic zones in India and remained the queen cultivar for more than 15 years from its release in the year 1996 and is still cultivated in some parts of India (Sharma et al. 2021). In the background of worldwide commercialization, credited as successful example of plant breeding, there exists a prebreeding history of not less than 30 years.

## 2.3 Physiological Traits in Wheat and Breeding Conduits

Physiological traits are difficult to study owing to their dependency on environment and complex genetic control. All growth stages and process encompasses a crops physiology, however, one needs to target traits that have economic importance as well as show genetic variation.

In wheat most of the work of the standardization of wheat physiology has been done by Matthew P. Reynolds, Wheat Physiologist at CIMMYT (Reynolds et al. 2021). His work has well established the traits that can be targeted for improving the physiological efficacy of wheat, making it more tolerant to stress environments. These include drought and heat adaptive traits. Drought and heat are mostly twin phenomena's in India context. If drought is there, it is always coupled with high temperature. Wheat yield can substantially decrease in drought conditions. Better adaptation to drought can be obtained through genotypes PBW 343 and C306 having different mechanisms to adaptation (Srivastava et al. 2016, 2017). Deeper roots, mobilization of resources, and stay green habit are other traits that help in better adaptation to drought and heat. Traits associated with drought tolerance that are easily evaluated with the eye include rapid early ground cover by leaves, leaf glaucousness, leaf pubescence, and erect leaf profile of wheat is said to increase crop yield possible enhanced light use efficiency in high-radiation environments, spike photosynthesis, grain filling under stress, and water use efficiency (Reynolds 2002). While for heat individually, stay green habit is the most apparent trait. Canopy temperature measurements and membrane thermostability (MT) are effective for screening wheat cultivar or breeding line for their response to drought and heat.

## 2.4 Prerequisites of Prebreeding and Breeding for Physiological Traits

### *2.4.1 Genetic Variation*

While considering the adoption of physiological traits into a breeding program, breeders should establish the degree of genetic variability that exists for the trait (s) of interest along with its screening methodology and required infrastructure. There exists an extensive genetic diversity in the wheat gene pool stored in ex situ gene banks. In total, nearly 800,000 wheat germplasm accessions are held globally (FAO 1996). However, a very small percentage of this has been utilized so far.

A number of genes have been identified mapped, tagged, and transferred at PAU for wheat stripe and leaf rust and other biotic and abiotic stresses from these species. A multi-pronged, exploratory strategy for productivity enhancement, over and above all the mandatory components such as resistance to biotic and abiotic stresses has been followed at PAU. If we aim to bring this target well within our reach wide genetic input for introgression of untapped productivity alleles and physiological breeding, needs to be targeted. The two approaches are not mutually exclusive and can have synergistic overlaps. Neither is productivity to be viewed in isolation of other traits.

### *2.4.2 Trait Transfer Protocols from Wild Species*

**Wide hybridization approach** One particular line of research, based on wide hybridization (initiated in 1980s) has led to a saga of rust resistance gene transfers and has provided depth to the wheat breeding program. Wild wheat germplasm consisting of more than 1500 accessions is being maintained as an active collection

**Table 2.1** Wild wheat germpalsm at Punjab Agricultural University

| Species | Genome | Total accessions |
|---|---|---|
| *Aegilops tauschii* | DD | 488 |
| *Triticum boeoticum* | $A^bA^b$ | 165 |
| *Triticum monococcum* | $A^mA^m$ | 60 |
| *Triticum urartu* | $A^uA^u$ | 18 |
| *Aegilops speltoides* | SS | 170 |
| *Aegilops searsii* | SS | 20 |
| *Aegilops longissima* | SS | 10 |
| *Aegilops comosa* | MM | 10 |
| *Aegilops umbellulata* | UU | 10 |
| *Triticum dicoccoides* | AABB | 172 |
| *Triticum araraticum* | AAGG | 25 |
| *Aegilops markgrafii* | UUMM | 20 |
| *Aegilops ovata* | UUMM | 18 |
| *Aegilops triuncialis* | UUCC | 28 |
| *Aegilops cylindrica* | CCDD | 20 |
| *Aegilops ventricosa* | DDNN | 15 |
| *Aegilops kotschyi* | UUSS | 12 |
| *Aegilops peregrina* | UUSS | 42 |
| *Aegilops neglecta* | UUMM | 7 |
| *Aegilops triaristata* | UUMMNN | 12 |
| Misc. | – | 120 |
| Total | | 1439 |

and evaluated against changing pathogen specificities and climate regularly at PAU. This is the largest collection of wild wheats that any centre is maintaining in India (Table 2.1). Below is the picture where vernalization and photoperiod treatments are being given.

Efforts toward evaluation of wild wheat germplasm, in the earlier phase, had led to the identification of A, D, C, U, S, and M genome species as excellent sources of resistance. The maintenance and use in crossing if targeted for these wild species require vernalization and photoperiod (Fig. 2.1 and Table 2.2).

A series of studies employing various chromosome engineering strategies have resulted in introgression of more than 20 disease resistance genes from alien species to cultivated wheats. PAU has the distinction of designating the five rust resistance genes, *Yr40, Lr57,* and *Lr58* from alien sources and *Lr48* and *Lr49* from cultivated wheat.

These genes are being mobilized into elite wheat lines as discussed in section on MAS. *Lr57, Yr40,* and *Lr58* have also been transferred to US lines Jagger and Overley and germplasm lines as KS11WGGRC53-J and -O and KS11WGGRC54-J and -O released by Kansas State University, USA. These genes are also being mobilized into Australian lines at the University of Sydney, Australia. *Lr57* is now being characterized through expression profiling and histochemical studies under another national network project on *Puccinia triticina* genomics.

Vernalization treatment for six weeks at 6-8$^{0}$C
Extended photoperiod during pre-flowering phase

**Fig. 2.1** Wild Wheat being given vernalization (**a**) and extended photoperiod (**b**)

Germplasm screening and wide hybridization are also being carried out for traits other than disease resistance:

- QTLs for higher iron and zinc content in the grain (*Q GFe.pau-2A*, *Q GFe.pau-7A*, and *Q GZn.pau-7A*) have been identified from *T. boeoticum*. (Iron and zinc biofortification of cereal grains is considered as a viable public health strategy for Third World countries.)
- A major gene (*GpcB1*) for high protein content and enhanced micronutrient content, originally derived from *T. dicoccoides* has been transferred to a wide range of wheat genotypes using marker-assisted selection. Significant improvement in protein concentration is observed, and its impact on protein quality is underway.
- About 100 accessions of *Aegilops tauschii* have been characterized for cellular thermotolerance traits such as membrane thermostability and TTC cell viability. An efficient hybridization and trait transfer protocol has been designed for introgression of useful genes from *Ae. tauschii* directly to bread wheat and transfer for components of heat tolerance is under way.
- CMS lines based on different alien cytoplasms including *T. timopheevii* have been developed in wheat. Restorer gene pool has been developed by using diversifying genetic input from synthetic hexaploid wheats.
- Chromosome segment substitution lines (CSSL) of *Ae. speltoides* in durum wheat are being characterized as a potential source of genes for heat tolerance.
- Introgression of rye chromosome segments has been induced using a *Ph1* suppressor stock with the objective of transferring tolerance to micronutrient deficiencies and outcrossing promoting traits (for use in hybrid wheat development).

Complex traits like input use efficiency and abiotic stress tolerance are virtually intractable for breeding by conventional means. In case of input use efficiency and

**Table 2.2** Introgression strategies with respect to status of donor

| Target species | Strategy |
|---|---|
| Introgression from progenitor genomes | • AA and BB genome diploid progenitor genome species such as *T. urartu* (AA), *T. boeoticum* (AA), *Ae. speltoides* (BB or SS) are first crossed with tetraploid wheat, *T. durum* (AABB) |
| | • Requires repeated pollination at the time of first cross |
| | • *Ae. speltoides* pollen will dehisce after given a stimulus by rubbing the spike |
| | • Tetraploid wheat (serves as an initial or primary recipient) and therefore as a bridging species |
| | • Subsequent crosses with hexaploid wheat |
| Introgression from DD genome progenitor | • Introgression from DD genome progenitor, *Aegilops tauschii* cannot be received by *T. durum* into its AA and BB genomes |
| | • Synthetic hexaploid wheat (SHW) development provides the bridge |
| | • AABB X DD → ABD → AABBDD |
| | • Direct gene transfer into wheat also possible by DD X AABBDD cross (Fig. 2.2), Cambay et al. (2021) |
| | • Problems due to wild traits such as hard threshing coming from *Ae. tauschii* |
| From tetraploid progenitors | • Tetraploid X tetraploid crosses are easy to make |
| | • Tetraploid X hexaploid crosses are also regularly made |
| Introgression from nonprogenitor genomes | • Chromosomal transfers have to be induced through special techniques: Ph locus manipulations, irradiation-induced chromosomal breakages and translocations |
| | • Linkage drag issues |
| | • A longer and unpredictable time frame |
| | • Hybridization with nonprogenitor species received greater attention of researchers: novelty of obtaining new hybrid combinations, excellent material for cytogenetic stud |

abiotic stress tolerance techniques such as DH and MAS almost become a prerequisite. There arises an acute need of precise phenotyping modules. These would have to be designed to suit the specific requirements. Genotypic differences for nitrogen use efficiency (NUE) have been detected in cultivated wheat variety, PBW 621 (alias KACHU, DBW88, DBW 50, HD3059) has a higher efficiency and the leads have been followed up in the form of NUE QTL mapping and tagging. Tolerance to micronutrient deficiencies is high in triticale (conferred by rye genome). Triticale × wheat crosses involving a Ph locus suppressor for inducing wheat-rye translocations have been attempted, and these populations are available with PAU. Detection of translocated rye segments has been completed and uptake efficacy are under study. Water use efficiency (WUE) rather than drought tolerance is another desired trait in the context of the natural resource crunch. Traditional (C306, etc.) and exotic (Australian lines Excalibur and Kirchauff and CIMMYT line Babax) lines have already been utilized as donors as heat tolerance is the key to productivity in the Indian wheat growing environments (Table 2.3).

**Fig. 2.2** Stepwise methodology for Ae. tauschii X wheat cross

**Table 2.3** Exotic germplasm and its utilization

| S. no | Exotic germplasm resource | No. of accessions | Utilization |
|---|---|---|---|
| 1 | European winter wheat collection from Germany, France, and UK | 376 | Utilized for productivity, resistance, and dwarfing gene (Grover et al. 2018, 2019, 2022) |
| 2 | Washington State University world wheat collection | 630 | Used for diversification |
| 3 | Winter wheat collection from Kansas State University | 1200 | Used for diversification |
| 4 | Australian cultivars and stocks | 103 | Used for rust resistance genes, drought tolerance |
| 5 | Linked top cross populations from CIMMYT | >2000 lines | Used for resistance and productivity traits |
| 6 | Iranian and Mexican landrace | 1300 | Used for stress tolerances |

### *2.4.3 Marker-Assisted Selection (MAS)*

Markers once identified from wild species or elite donor traits, it becomes very easy to follow the trait and its transfer. MAS for rust resistance has become an integral

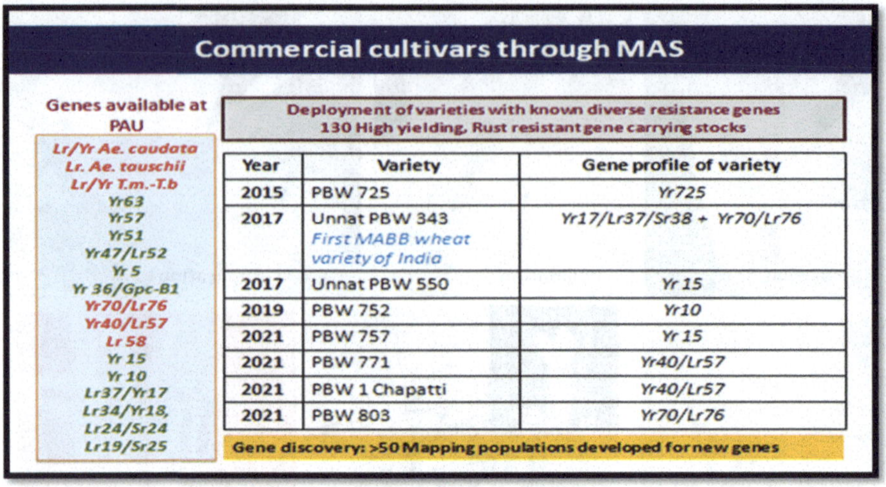


**Fig. 2.3** Available known genes from wild and cultivars released

part of the wheat breeding programme at PAU as a result of support coming from national (DBT, ICAR) and international (ACIAR) projects. Use of known rust resistance genes, mainly through markers has grown rapidly in the last 10 years and have resulted in the development of varieties at national level. Availability of multiple, individually monitorable resistance options have resulted in a qualitative change in the breeding strategy. A large set of materials have graduated to various stages of yield testing, and five cultivars have been released (Fig. 2.3). Alternatively, simultaneous transfer into an elite background from wild and mapping of the trait can be done following AB-QTL strategy. We have a rust resistance gene pool of almost 30 genes, yet emergence of the new pathotypes makes it important to scout newer sources of resistance to ensure the production remains unaffected by these biotic constraints. Since greater genotypic variation for disease resistance has been identified in progenitor species and landraces of wheat, identification of chromosomal location and marker tags of the associated novel disease resistance genes will aid in their mobilization to productive wheat genetic backgrounds and enhancing the resistance and productivity of present day wheat cultivars.

The strength of a breeding programme depends on the prebreeding germplasm. One needs to maintain an active germplasm set and keep investing in development of mapping populations.

The previously mentioned sets of germplasm resources are highly amenable for direct use in wheat breeding. These two donor germplasm sets have the advantages of allelic diversity and free recombination.

A molecular marker-oriented germplasm resource is available in the form of several mapping populations. A large set of RILs and NILs based on established donors have been generated and studied to unravel the complex genetic basis of various abiotic stress traits. Five BC-RIL populations were developed with drought

**Table 2.4** Mapping populations for physiological traits developed at PAU, Ludhiana

| S. no. | Trait | Available RIL populations | Number of RILs |
|---|---|---|---|
| 1. | Drought tolerance, heat tolerance, chapati quality | C 273/2* PBW 343; C306/2* PBW 534; C518/2* PBW 343; C 591/2* PBW 343; C 591/2* PBW 534 | 5 |
| 2. | Lodging resistance | WH 542 × PBW 343 | 1 |
| 3. | Thousand grain weight | Turaco × DL 788-2, DL 788-2 × Turaco, Rye Selection III × PBW343 + *Lr24* + *Lr28* | 3 |
| 4. | Grain micronutrient content | Synthetic Hexaploid 43 × WH 542 | 1 |
| 5. | Nitrogen use efficiency | PBW677 × PBW703 | |
| 6. | Grain traits | PBW746 × PBW761 | |

and heat-tolerant landrace-based donors. Two of these populations have been profiled with heat tolerance-related physiological and biochemical parameters. Similarly populations for drought tolerance with Indian and Australian wheat lines as donors are also available (Table 2.4).

Productivity and resistance to abiotic stresses cannot be divorced and viewed in isolation. This is particularly true of terminal heat stress tolerance and water use efficiency, being prioritized in the context of high fertility, irrigated conditions. Several populations involving pre-dwarfing era, drought-adapted parental lines, have been developed by us, but may not be appropriate as tolerance traits of the donor such as low chlorophyll content and long peduncles are not associated with high productivity. Thus creation of new populations with productive, physiologically efficient parents complementing each other for the traits related to heat tolerance or water use efficiency, etc. needs to be taken up. Similarly *chapati* quality was sacrificed at the altar of productivity and none of the modern cultivars matches *chapati* quality of land race-based obsolete cultivars. The progenitor genomes offer the greatest diversity across the entire spectrum of traits, besides relative ease of transfer and free recombination. These species hold the potential of triggering a diversity explosion in the genetically constrained cultivated wheat germplasm. A phenotypically well-characterized set of accessions from these donor species is available with PAU. Well-tested hybridization and gene transfer protocols are in place. Identified accessions of *Aegilops speltoides* and *Aegilops tauschii* are viewed as excellent sources of heat tolerance while A genome progenitors are known to confer large number of spikelets/spike. Keeping in view the hidden genes for productivity and other traits these donors may carry, it would be ideal to create introgression panels representing the entire donor genome in stable/homozygous form before evaluating for a set of relevant/indicated traits.

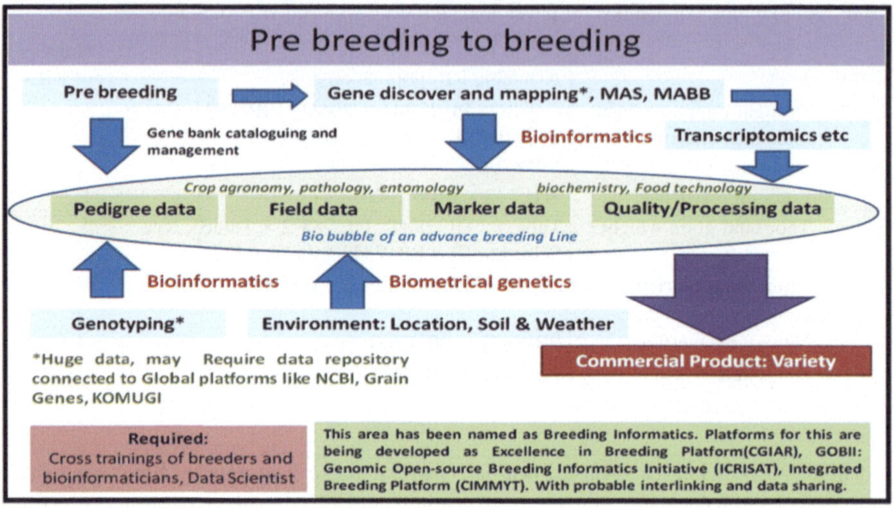


**Fig. 2.4** Schematic diagram for breeding from wild wheats

## *2.4.4 Stacking of the Physiological Traits*

The easiest method can be marker-assisted trait pyramiding. In the absence of that, convergence crosses with screening of the segregants can also be targeted. Multiparent populations (MAGIC) along with marker information can be a great help in this regard. Linked top crop population strategy was novel in terms that it maintained the balance with elite × exotic three-way crosses that were linked by the third parent. These can help in simultaneous mapping and convergence of the trait that too in an elite background. The next-generation methods like GWAS and genomic selection have also come to the rescue of prebreeders as these can also help in targeted selection of the desired alleles whereas rejecting the unfavorable ones while prebreeding. However, the promise that genome editing hold is far above all, the deleterious traits or negative alleles here can be directly knocked out before transferring. Also, removing of the linkage drag through genome editing or making desired wild allele in the elite line through base editing has already opened enormous prebreeding opportunities (Fig. 2.4).

Earlier scientist working on prebreeding might have felt left out or less acknowledged; however, new terms like Trait Discovery and Discovery Breeder and hiring of such people by private sector ring bell for great prebreeding opportunities ahead.

# References

Cambay SR, Bains NS, Sandhu SK, Puja S (2021) Reconsidering the direct cross approach for gene transfer from Aegilops tauschii to hexaploid wheat using diversified D genome donors. Cereal Res Commun 49(1):125–131

Dvořák J (1976) The relationship between the genome of Triticum urartu and the A and B genome of Triticum aestivum. Can J Genet Cytol 14:371–377

FAO (1996) The state of the World's plant genetic resources for food and agriculture. FAO, Rome

Feldman M, Sears ER (1981) The wild gene resources of wheat. Sci Am 244:102–112

Grover G, Sharma A, Gill HS, Srivastava P, Bains NS (2018) Rht8 gene as an alternate dwarfing gene in elite Indian spring wheat cultivars. PLoS One 13(6):e0199330

Grover G, Sharma A, Srivastava P, Kaur J, Bains NS (2019) Genetic analysis of stripe rust resistance in a set of European winter wheat genotypes. Euphytica 215(3):1–10

Grover G, Sharma A, Mackay I, Srivastava P, Kaur S, Kaur J, Burridge A, Allen SP, Bentley AR, Chhuneja P, Bains NS (2022) Identification of a novel stripe rust resistance gene from the European winter wheat cultivar 'Acienda': a step towards rust proofing wheat cultivation. PLoS One 17(2):e0264027

Harlan J, de Wet JMJ (1971) Towards a rational classification of cultivated plants. Taxon 20:509–517

Jauhar PP (1993) Alien gene transfer and genetic enrichment of bread wheat. In: Damania AB (ed) Biodiversity and wheat improvement. John Wiley & Sons, Chichester, pp 103–119

Kihara H (1924) Cytologische und genetische Studien bei wichtigen Getreidearten mit besonderer Rucksicht auf das Verhalten der Chromosomen und die Sterilität in den Bastarden. Mem Coll Sci Univ Kyoto Ser B1:1–200. (in German)

Kihara H (1944) Die Entdeckung der DD-Analysatoren beim Weizen. Agric Hort (Tokyo) 19:889–890. (in German)

Moore G (2015) Strategic pre-breeding for wheat improvement. Nat Plants 1:15018

Mujeeb-Kazi A (1998) Evolutionary relationships and gene transfer in the Triticeae. In: Jaradat AA (ed) Triticeae III. Science Publishers, Enfield, pp 59–65

Reynolds MP (2002) Physiological approaches to wheat breeding. In: Curtis BC, Rajaram S, Gómez Macpherson H (eds) Bread wheat: improvement and production. FAO plant production and protection series no. 30

Reynolds MP, Lewis JM, Ammar K, Basnet BR et al (2021) Harnessing translational research in wheat for climate resilience. J Exp Bot 72(14):5134–5157

Sharma A, Srivastava P, Mavi GS, Kaur S, Kaur J, Bala R, Singh TP, Sohu VS, Chhuneja P, Bains NS, Singh GP (2021) Resurrection of wheat cultivar PBW343 using marker-assisted gene pyramiding for rust resistance. Front Plant Sci 42:570408

Srivastava A, Srivastava P, Khobra R, Sharma A, Sarlach RS, Dogra A, Bains NS (2016) Association of morpho-physiological traits in recombinant inbred population of wheat under rainfed environments. Indian J Ecol 43:72–77

Srivastava A, Srivastava P, Sharma A, Sarlach RS, Bains NS (2017) Effect of stem reserve mobilization on grain filling under drought stress conditions in recombinant inbred population of wheat. J Appl Nat Sci 9(1):1–5

Zeller FJ, Hossam SLK (1983) Broadening the genetic variability of cultivated wheat by utilizing rye chromatin. In: Proceedings of the sixth international wheat genet symposium, Kyoto, Japan, pp 161–174

# Chapter 3
# Strategies to Develop Heat and Drought–Tolerant Wheat Varieties Following Physiological Breeding

**Narayana Bhat Devate, Karthik Kumar Manjunath, Rahul Ghajghate, P. Shashikumara, Uday G. Reddy, Monu Kumar, Hari Krishna, Neelu Jain, P. K. Singh, and Gyanendra Pratap Singh**

**Abstract** Worldwide, abiotic stresses including heat and drought are the major obstructions that threaten the agricultural production. Development of climate-resilient cultivars is the easy and economical way to combat drought and heat stress with limited resources. Plants do follow adaptation strategies to mitigate the impact of stress and lead to alteration in some of the morphological traits such as leaf rolling, leaf angle, cuticular wax content, stomatal conductance, deep root system, altered signalling and metabolic pathways. Targeting such traits along with the economical yield will help to identify suitable genotypes which perform better under stress environment. The basic step is to explore the available physiological trait variation among the cultivars, germplasm set and wild relatives to main stream alleles of importance to breeding material from the donor parent. Conventional and advanced breeding strategies can be implemented to develop climate-resilient cultivars with the suitable breeding and screening methods. As a key factor hybridization and selection along with the implication of advanced breeding methods like MABB, MARS, GS and transgenic approach make it easy and accurate to develop varieties in less time. Linkage, QTL and genome-wide association mapping helps to identify the genomic region of interest to target during marker-aided breeding approaches. A cocktail of breeding methods from conventional to transgenic may help in the development of high-yielding climate-resilient varieties which can help to serve

N. B. Devate · K. K. Manjunath · H. Krishna (✉) · N. Jain · P. K. Singh
ICAR-Indian Agriculture Research Institute, New Delhi, India

R. Ghajghate · P. Shashikumara
ICAR-Indian Grassland and Fodder Research Institute, Jhansi, India

U. G. Reddy
University of Agricultural Sciences, Dharwad, India

M. Kumar
ICAR-Indian Agricultural Research Institute, Jharkhand, India

G. Pratap Singh
Indian Institute of Wheat and Barley Research, Karnal, India

 
H. M. Mamrutha et al. (eds.), *Translating Physiological Tools to Augment Crop Breeding*, https://doi.org/10.1007/978-981-19-7498-4_3

farmers to escape from glitch of crop loss due to dry spell during cropping season. The recent advancement and methodologies regarding drought and heat tolerance breeding in wheat are discussed in this chapter along with the difficulties posed.

**Keywords** Drought · Heat · Wheat · Physiological breeding · Molecular markers

## 3.1 Introduction

Abiotic stresses are one of the major obstructions that threaten agricultural production and food security worldwide (Fahad et al. 2017; Hu et al. 2020). Among abiotic stress, heat stress and drought stress (HS and DS) are undoubtedly the most important stresses having huge impact on growth and productivity of the crops (Gajghate et al. 2020). Heat stress is defined as the exposure of plant to temperature for a longer period of time, above the threshold temperature of a plant to cause continual or irreversible damage to its growth and development (Hasanuzzaman et al. 2013; Gajghate et al. 2020). Global warming is the main key that kept drastic and rapid changes in the global climate, and aggravate the severity of heat and drought stress in different crops. According to IPCC 2014 reports, per decade there is rising of global air temperature by 0.18°C, which tends to enhance the temperature by 1.5–4.5 °C (IPCC 2012) above the current level in 2100 (Hansen et al. 2012). Day and night temperatures affect crop yield, especially when high-temperature stress coincides with flowering (Wheeler et al. 2000), and data shows that high temperatures can affect grain quality, including nutritional and milling properties in grain crops (Craufurd and Wheeler 2009). Increase in temperature enhances the evapotranspiration rate, thereby early consumption of soil water reserve leading to terminal drought stress. Nevertheless, global warming and climate change lead to erratic rain fall and increased dry spell leads to drought stress in the arable land.

Reduced leaf water potential and turgor pressure, stomatal closure, and decreased cell growth and cell enlargement are symptoms of drought stress in plants (Farooq et al. 2009). Drought stress affects a variety of physiological and biochemical systems in plants, including photosynthesis, chlorophyll production, nutrient metabolism, ion absorption and translocation, respiration, and carbohydrates metabolism (Farooq et al. 2009; Li et al. 2011). Low temperature can also create turgor stress at the cellular level (Thomashow 1994; Janska et al. 2009; Yadav 2010) due to poor root hydraulic conductivity and reduced root activity (Aroca et al. 2003).

Heat and drought affect the wheat crop mostly during the blooming and grain development stages, reducing yield and grain quality (lower 1000 grain weight and change in protein quality). Heat waves and drought conditions in major wheat growing belts throughout the world contributed to an annual production variation around 40% (Zampieri et al. 2017). Demand for wheat is expected to rise by 60% by 2050, while output may fall by 29% as a result of climate change-induced environmental stresses (Manickavelu et al. 2012). These findings suggest that enhancing wheat's abiotic stress tolerance is critical for global food security in the near future.

Out of all the factors for the global warming, the increment in the greenhouse gas concentration plays a major role. It indirectly influences the pattern and amount of precipitation. Over the past 250 years, a sudden rise in the concentration of the $CO_2$ (30%) and methane (150%) has been observed (Lal 2004; Friedlingstein et al. 2010), and the percentage of these greenhouse gases will keep rising in the upcoming years. Thus, the impact of heat stress can be minimized by reducing the main factor, i.e. decline in the production of the greenhouse gases (that is impossible) or in other way develop crops that are tolerant to heat and drought stress by means of selection of plant adapted to stress (morphological, physiological and biochemical mechanism), via molecular breeding, genetic engineering, etc. (Wahid et al. 2007; Hu et al. 2020).

HS and DS are the major limiting factors in the area where the crops are grown in the tropical and sub-tropical climate. Rice, wheat, maize, cotton, chickpea, peanut, sunflower and Brassica have been evaluated in response to high-temperature stress (Devasirvatham et al. 2016). The selection of heat and drought–tolerant lines is the first step to overcome the effect of abiotic stress along with the traits that are specific to heat and drought tolerance, viz. long awns, high root length, cooler canopy, waxy leaf, high WUE, etc., which helps to overcome the HS and DS conditions. Escape mechanism (earlier flowering and crop maturity) is one more major survival ability developed by crops to avoid maximum loss during water scarcity and high temperature stress (Toker et al. 2007).

In this chapter, we will discuss about the different available germplasm tolerant to heat and drought stress along with the selection of important traits. There are different screening methods to identify the heat and drought–tolerant line in different crops. Furthermore, the development of heat and drought–tolerant lines from the earlier conventional breeding methods to recently developed molecular breeding methods to identify the quantitative trait loci (QTLs) as well as to identify linked markers for marker-assisted selection (MAS) and transgenic approaches. The plant reactions to high temperatures and water deficiency are presented in this chapter, as well as the consequences for stress tolerance breeding in plants.

## 3.2 Germplasm Utilization for Improving Heat Tolerance

For the identification of donor genotypes for heat tolerance characteristics, germplasm resources were used and evaluated. A basic strategy to identify superior wheat genetic stock for improved heat tolerance is to evaluate genotypes for stable grain production under heat stress in various conditions.

Rane et al. (2007) examined advance lines and cultivars of wheat from India's several wheat breeding institutions. Genetic stocks Raj 3765 and Raj 4037 have been recognized as stable stocks for grain yield. Under field and controlled conditions, hexaploid synthetic wheat lines were screened for terminal heat tolerance, and four heat-tolerant lines, viz. ALTAR 84/Aegilops squarrosa (192); 68.111/RGB-U//WARD/3/FGO/4/RABI/5/Aegilops squarrosa (629); 68.111/RGB-U//WARD/3/

FGO/4/RABI/5/Aegilops squarrosa (878) and LCK59.6′1/Aegilops squarrosa (313) were identified (Sareen et al. 2012). Khanna-Chopra and Viswanathan (1999) screened hexaploid and tetraploid wheat cultivars and their diploid relatives under heat stress conditions for stable grain yield and its component traits. C306 and HI1136 (hexaploid wheat) *Triticum dicoccoides*, *Triticum monococcum* acc. *BSP1* and *Aegilops speltoides* ssp. *liqustica* were found to be highly heat tolerant with respect to their grain yield. Zaharieva et al. (2001) reported *Triticum dicoccum* and *Aegilops geniculata* serves as potential gene pool source for the improvement of heat and drought tolerance in wheat and could contribute to the choice of traits to be introgressed and the accessions to be used in wide hybridization programmes. Alien chromosome introgressed lines in wheat were used as useful genetic resources for the improvement of thermo-tolerance in wheat. Wheat Leymusracemosus chromosome-introgressed Chinese Spring lines were evaluated and studied for their response towards heat stress reported by Mohammed et al. (2014). Germplasm exploration has been conducted for novel traits related to drought tolerance such as *Elymus* spp. and stress tolerance traits like spike pigmentation, low seed shattering, dense pubescence and waxiness were observed (Kumar et al. 2021). Evaluation of morpho-physiological traits of *Elymus semicostatus* (Nees ex Steud.) was found to be a potential donor for drought tolerance in wheat (Kumar et al. 2022).

## 3.3 Physiological Traits to Be Targeted to Improve Yield Under Heat and Drought Tolerance

An outline of some of the main physiological traits that can be applied in current wheat breeding programmes to help boost yield and improve adaptation under heat and drought condition, respectively, is given below.

### *3.3.1 Heat Stress*

**Normalized Difference Vegetation Index (NDVI)** Ground cover is a good proxy for the Leaf Area Index (LAI) and can be estimated using NDVI measurements. Heat-induced chlorosis and "stay green" can also be detected using NDVI (Lopes and Reynolds 2012) or through other specifically developed spectral indices (Anderegg et al. 2020). Different patterns of chlorophyll loss among genotypes have been observed using sequential measurement of NDVI (Pinto et al. 2016).

**Canopy Temperature (CT)** Wheat, like most other crops, maintains a lower temperature than the ambient temperature when water is not a constraint, often by several degrees, depending on the relative humidity. Canopy temperature can be measured at high throughput using infrared (IR) thermometers or thermal (infrared) imagery (Tattaris et al. 2016) and shows a robust association with grain yield under heat stress (Amani et al. 1996). Cooler canopy temperature appears to have some common genetic basis under both heat and drought stress. CT is associated with deeper roots under drought and heat stress. Since cooler canopies are also associated with genetic variation in stomatal conductance under heat, selection for CT is also likely to improve assimilation capacity per se.

**Stem Reserve Mobilization** Grain filling from mobilized stem reserves is a constitutive trait which supports grain filling under heat stress. In wheat stems storage and remobilization of water-soluble carbohydrates have also been associated with wheat yield under heat stress, presumably compensating for reduced assimilation capacity during grain filling (Blum et al. 1994).

**Membrane Thermostability (MT)** Membrane thermostability, measured as electrolyte leakage from leaf discs over a range of temperatures, is a sensitive method to evaluate heat tolerance in plants but cannot be assayed at high throughput using electrolyte leakage (Reynolds et al. 1994).

More targeted traits to increase assimilation capacity include screening for early vigour and delayed senescence to improve LAI as well as MT, photoprotective pigments and epicuticular wax/leaf glaucousness to improve radiation use efficiency. Rubisco and its regulation and spike photosynthesis are long-term targets with potential high payoffs.

### *3.3.2 Drought Stress*

Drought stress is one of the most damaging abiotic stresses to crop yields. Cell division, leaf surface and stem growth, and root cell proliferation are all disrupted when there is not enough water for proper growth, but it also raises ROS levels and has a detrimental impact on plant growth (Gill and Tuteja 2010).

**Transpiration efficiency (TE)** Increased availability to water with a responsive root system and conserving water to guarantee that it does not run out before the crop life cycle is finished are the two main techniques for enhancing production in drought-prone settings (Condon et al. 2004). The second method is significant in situations where deep water is unavailable or when the subsoil is toxic to roots owing to heavy metals, salinity or other factors (Bennett et al. 2012). In both circumstances, efficiency in transpiration is likely to be beneficial. Despite the difficulty of measuring TE directly in the field, carbon isotope discrimination (CID) can be employed as a substitute. CID is integrative and heritable, yet it is costly to detect because it necessitates mass spectrometry. Transpiration efficiency also depends on root architecture which alters the water intake capacity and helps to combat drought and heat stress due to transpirational cooling (Danakumara et al. 2021).

**Relative water content (RWC)** In terms of the physiological consequences of cellular water shortage, relative water content (RWC) is probably the most accurate indicator of plant water status. Grain yield and RWC were found to have a positive relationship during anthesis and mid-grain fill, with the high-yielding selections having a much higher RWC than the low-yielding selections (Tahara et al. 1990).

**Shoot Dry Mass (SDM)** RWC dropped from 67.61% to 59.40%, while SDW dropped from 0.50 to 0.25 g plant$^{-1}$ (Pour-Aboughadareh et al. 2010).

**Anthocyanin** Plants transcriptionally control anthocyanin biosynthesis genes to enhance anthocyanin production during drought stress, because anthocyanins can act as osmoregulators to maintain water balance.

**Chlorophyll fluorescence** Relative chlorophyll content, initial fluorescence (Fo) and maximum fluorescence efficiency (Fv/Fm) changed significantly under drought stress. Under drought stress, the SPAD index averaged across all genotypes decreased by about 15% compared with the control. Similarly, the Fv/Fm ratios under drought decreased by 11.42% (Pour-Aboughadareh et al. 2010).

**Chlorophyll a and b Content** Drought impacts chlorophyll in distinct ways. The chlorophyll (a + b) content in wheat leaves rose slightly during the initial stage of drought (3 days). The chlorophyll content in leaves of all cultivars reduced insignificantly on the 5th day after ceasing plant watering. However, during a 7-day dry period, the chlorophyll content decreased by 13–15%. The pigment apparatus is comparably robust to dehydration in wheat cultivars contained in leaves of various ages, based on these slight variations in chlorophyll content and the stability of the chlorophyll a/b ratio were observed (Nikolaeva et al. 2010).

**Stomata Conductance** Drought stress has a big impact on stomatal conductance. Furthermore, drought-reduced stomatal conductance is greatest in *T. durum* (56.91%), *Ae. tauschii* (53.41%) and *Ae. crassa* (48.95%), as species with the AB, D and DM genomes, respectively (Pour-Aboughadareh et al. 2010).

## 3.4 Plant Adaptive Mechanisms to Heat Stress

Most of the crops for human consumption are categorized under "Thermophiles." These are the organisms (or plants) grows optimally in the temperature ranges between 30 and 65 °C or even higher (Larcher 1995; Źróbek-Sokolnik 2012). The organisms categorized under Psychrophiles (temperature ranges between 0 and 10 ° C) and Mesophyles (temperature ranges between 10 and 30 °C) barely get caught by HS, since the organisms in this category grow in temperature condition $<$30 °C. Species under the thermophilic group are further classified into heat sensitive, relative heat resistant and heat tolerant (Hasanuzzaman et al. 2013).

Survival of plants under extreme condition of HS is achieved by adapting according to the environmental conditions (Fitter and Hay 2002); these are of different types: (a) avoidance, (b) escape and (c) tolerance mechanism. The conceptual mechanism of HS adaptation has been presented in Fig. 3.1.

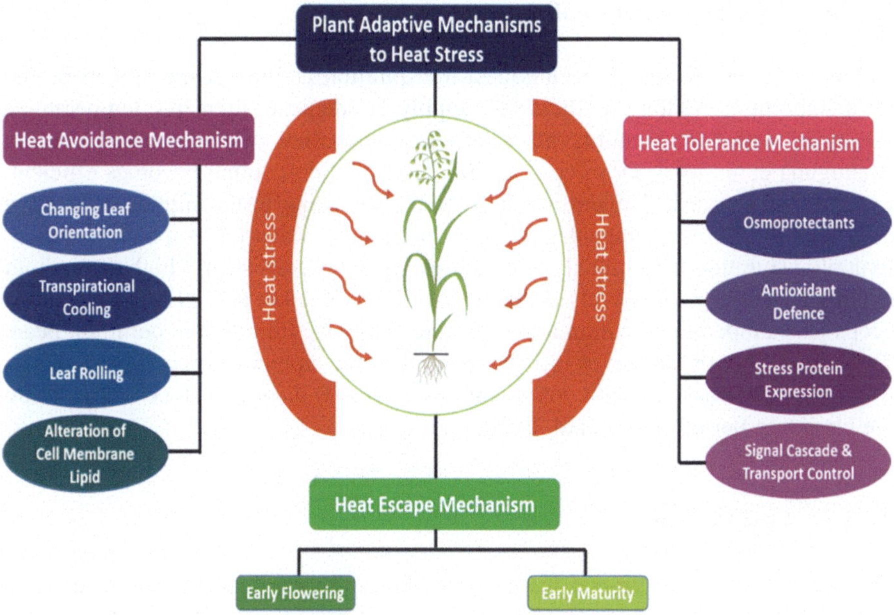


**Fig. 3.1** Schematic diagram of plant heat stress adaption

## *3.4.1 Heat Stress Avoidance*

Plants use a variety of survival mechanisms under HS conditions, including long-term evolutionary phenological and morphological adaptations as well as short-term avoidance or acclimation mechanisms in different crops, such as changing leaf orientation in sunflower (Kalyar et al. 2013), transpirational cooling in common bean (Deva et al. 2020) and alterations in membrane lipid compositions at the cellular level (Barrero-Sicilia et al. 2017). Heat-induced changes in plants include stomatal closure and decreased water loss, higher stomatal and trichomatous densities and bigger xylem vessels (Mathur et al. 2014).

### 3.4.1.1 Changing Leaf Orientation

Adjusting the leaf direction might also help in heat avoidance (Kalyar et al. 2013; Zhang et al. 2016). Many plants change their leaf angle, limiting the amount of leaf area exposed to sunshine heat (Huey et al. 2002). Vertically orientated leaves reduce midday light absorption, which reduces heating and transpirational water loss, allowing better maintenance of leaf hydration and water potential (Tozzi et al. 2013).

#### 3.4.1.2 Transpirational Cooling

Among the heat avoidance techniques, transpiration cooling is thought to be the most important. As long as the water supply is adequate, high day temperatures combined with low humidity allow efficient utilization of transpiration cooling (Bahuguna et al. 2014; Deva et al. 2020). Transpiration cooling becomes efficient, when the evaporative demand is relatively higher owing to the significant heating effect and the soil is having adequate moisture to give water for evapotranspiration. Evapotranspiration rises quickly after sunrise, remains consistently high throughout the day and then progressively declines after dusk (Khan et al. 2021). Leaf cooling keeps leaf temperatures within a photosynthetically useful range in common bean. The temperature of the leaves is 5 °C cooler than the ambient temperature. Furthermore, the differences in leaf cooling are more towards the top of the canopy during the blooming period (Deva et al. 2020).

#### 3.4.1.3 Leaf Rolling

One-way plants try to protect themselves when things get too hot by turning the edges of their leaves inward. This reduces the amount of leaf surface which is exposed in an attempt to cool off the plant. Leaf rolling causes stomata to close, reducing water loss from the leaf. Improved heat tolerance in sugarcane was associated with reduced water loss via leaf rolling and faster reversal of these effects after recovery, according to the findings. This indicated that leaf rolling is a critical adaptation for reducing water loss in the face of HS (Sadia et al. 2008).

#### 3.4.1.4 Alteration of Membrane Lipid Compositions

Plants change the activity of enzymes that impact lipid content in response to rising temperatures (Shiva et al. 2020). Biological membranes are mosaics of lipids and proteins that form highly organized structures. Elevated temperatures can modify the features of these membranes directly and effectively, including fluidity and permeability, through a holistic impact including changes in lipid content and/or interactions between lipids and particular membrane proteins. One lipid metabolic reaction that has long been known to occur in response to high temperature is a reduction in unsaturation of leaf fatty acids in membrane lipids (Higashi et al. 2015; Shiva et al. 2020). Reduced desaturation of newly synthesized fatty acids and higher lipid turnover are expected to contribute to the reduction in unsaturation, which might assist to keep membrane fluidity at high temperatures. As a result, dynamic membrane regulation not only aids cells in detecting temperature changes but also plays a role in intracellular reactions and decides a cell's fate (Niu and Xiang 2018).

### 3.4.2 *Heat Stress Escape*

In heat escape mechanism, plants complete their maturation before the impede of HS by sacrificing smaller yield loss (Adams et al. 2001; Hemantaranjan et al. 2018). Early maturity, often known as earliness, is an adaptation technique in which early heading lines complete seed setting and grain filling at favourable conditions, avoiding late heat stress. In the eastern Gangetic plains, where wheat suffers from severe temperature stress during grain filling, earliness has been recommended as a desirable breeding strategy (Joshi et al. 2007). According to Al-Karaki (2012), the length of the pre-heading phase in durum wheat influences grain production under heat stress. The early developing high yielding durum lines were shown to be suited for heat stress-affected locations in semi-arid climates.

### 3.4.3 *Heat Stress Tolerance*

Heat tolerance refers to a plant's capacity to live under extreme temperatures. In the near future, consistency and superior performance in HS conditions are critical for food security. By retaining their molecular, cellular, biochemical and physiological mechanisms at the whole plant level, heat-tolerant genotypes may function well in both humid and dry situations (Bhusal et al. 2021). To improve thermo-tolerance in rice, different strategies could be used, including (1) changing agronomic practises, such as sowing time or selecting early morning flowering cultivars; (2) inducing acclimation with growth regulators and fertilizers; (3) breeding for genetically heat resistant cultivars and (4) developing genetic modification (Khan et al. 2019a, b).

#### 3.4.3.1 Osmoprotectants

Osmoprotectants are low-molecular-weight tiny molecules that are electrically neutral, highly soluble organic chemicals that are accumulated in significant quantities to maintain homeostasis without allowing excessive amounts of salt ions to penetrate (Bougouffa et al. 2014; Ahn et al. 2011). A prevalent plant response to abiotic stressors is the accumulation of osmoprotectants (Marwein et al. 2019). Microorganisms can collect osmoprotectants in two ways: by direct acquisition from the environment if accessible, or through de novo manufacture (Galinski 1995). They comprise sugars and derivatives, amino acids and derivatives, and polyols and derivatives, among others (Paul 2013). Proline, glycine-betaine, polyamines and sugars are the most prevalent osmolytes involved in osmoregulation. The osmotic differences between the cell's surrounds and the cytoplasm are stabilized by these substances. Furthermore, they protect plant cells from oxidative stress by reducing the development of damaging reactive oxygen species (ROS) such as hydroxyl ions, superoxide ions, hydrogen peroxide and other free radicals (Sharma et al. 2019).

#### 3.4.3.2 Antioxidant Defence

Increases in temperature over a certain threshold cause oxidative stress in plants by causing the creation of reactive oxygen species (ROS), which are typically created as a result of aerobic metabolism. However, metabolic imbalances caused by changes in environmental variables encourage ROS overabsorption, which is increased when $CO_2$ supply is limited owing to stomatal closure (Suzuki et al. 2012). ROS such as superoxide ($O_2$), hydrogen peroxide ($H_2O_2$) and hydroxyl free radical (OH) are important signal transduction molecules (Sun et al. 2002; Baxter et al. 2014; Mittler 2016; Hassan et al. 2020), but their accumulation beyond a threshold level causes oxidative damage at the cellular level by inhibiting photosynthesis (Choudhury et al. 2017). ROS-scavenging enzymes like superoxide dismutase (SOD), which is the first line of defence against superoxide radicals, reduce ROS damage. Mittler et al. (2004) found that SOD-catalysed $O_2^{\bullet-}$ dismutation produces $H_2O_2$ as a reaction product, and APX (ascorbate peroxidase) lowers $H_2O_2$ (Foyer and Noctor 2005). Plants, on the other hand, contain non-enzymatic scavenging processes like proline and carotenoids that act in tandem with the enzymatic scavenging system (Hasanuzzaman et al. 2012). Understanding the expression, accumulation and developmental process of antioxidants under environmental stress conditions aids in the improvement and advancement of heat-tolerant lines (Gajghate et al. 2020).

#### 3.4.3.3 Expression of Stress Protein

Important enzymes, metabolic proteins (functional proteins) and regulatory proteins are triggered by heat stress (Lata et al. 2011; Nakashima et al. 2012). Transcription factors (TFs), which interact with cis-acting elements present in the promoter region of various target stress-responsive genes in the signal transduction processes, play a critical role in the conversion of stress signal perception to stress-responsive gene expression, thus activating a signalling cascade and a whole network of genes that act together in enhancing plant tolerance to harsh environmental conditions (Akhtar et al. 2012). In comparison to mammals and yeasts, TFs account for around 7% of the coding sequences in plant genomes, and many of them belong to major gene families, such as the heat stress transcription factors (HSFs) family (Baniwal et al. 2004; Udvardi et al. 2007). Heat shock proteins (Hsp) are classified into five groups depending on their molecular mass: (1) Hsp100, (2) Hsp90, (3) Hsp70, (4) Hsp60 and (5) small heat shock proteins (HSFs) (sHsps).

The protective mechanism of pathways leading to the expression of genes to synthesize heat shock proteins is made up of temperature sensing that is linked to signal transmission to heat shock factors, where gene expression is activated by binding to the heat shock element in DNA (Larkindale et al. 2005). Heat shock proteins serve as molecular chaperones, controlling protein folding and accumulation, as well as protein localisation and degradation in all plant and animal species (Lindquist and Craig 1988; Panaretou and Zhai 2008; Hu et al. 2009; Gupta et al.

2010). As chaperones, these proteins prevent other proteins from aggregating irreversibly and aid in protein refolding under heat stress conditions (Tripp et al. 2009; Morrow and Tanguay 2012; Bhattacharya 2019).

#### 3.4.3.4 Signalling Cascade and Transport Control

Heat-responsive signalling is also essential for protecting photosynthetic machinery from heat stress. Although the mechanisms of plant heat sensing and signalling remain unknown, many signalling molecules (e.g., $Ca^{2+}$ and ROS), hormones (e.g., ABA), protein kinases (e.g., calcium-dependent protein kinases (CDPK) and mitogen-activated protein kinases (MAPK)) and transcription factors (e.g., heat shock factor (HSF)) are known to participate in heat signal transduction.

The sensing and transmission of heat signals are critical elements in plant stress tolerance. A plasma membrane channel that triggers an inward calcium flow, a histone sensor in the nucleus and two unfolded protein sensors in the endoplasmic reticulum and the cytoplasm are postulated to activate the heat stress response (Mittler et al. 2012). $Ca^{2+}$ signalling, G protein-mediated signalling and kinase signalling are believed to constitute the downstream heat-responsive signalling pathways.

$Ca^{2+}$ signalling pathways are well known that calcium inward flux acts as one of plants' key heat sensors (Mittler et al. 2012), and that it modulates many signalling pathways in response to heat stress (Goraya et al. 2017). By binding to $Ca^{2+}$, calcium-binding proteins play a role in calcium cell signalling pathways, which are implicated in the activation of numerous kinases and transcription factors under heat stress (Hu et al. 2015).

G-protein-mediated signalling is well known for being part of several signalling pathways and constitutes one of the most essential cell signalling cascades (Kim et al. 1995; Tuteja 2009). Previous research has linked several G protein isoforms (G alpha, G beta, G gamma, and Rab7) to plant heat tolerance (Peng et al. 2011; Yadav et al. 2012; Misra et al. 2007; Kang et al. 2015), as well as G protein in *P. oleracea* (Yang et al. 2012) and Rab1C in *G. max* (Ahsan et al. 2010).

Multiple protein kinases are activated by the heat signals given by second messengers (e.g., CDPK and MAPK). Nucleoside diphosphate kinases (NDPKs) were detected in higher abundance in heat-stressed leaves of *G. max* (Ahsan et al. 2010) and *O. sativa* (Ahsan et al. 2010). Plant heat tolerance is linked to NDPK, which interacts with $H_2O_2$-mediated mitogen-activated protein kinase (MAPK) signalling (Moon et al. 2003). Plants have been shown to raise their NDPK levels in response to various environmental conditions such as drought, salt and cold (Lee et al. 2007; Dooki et al. 2006).

Heat-Responsive Transcription Factors: WRKY, HsfA1 and MYB are transcription factors (TFs) involved in plant heat signal transduction (Nakashima et al. 2014). The TFs "WRKY" and "MYB" were upregulated in *P. oleracea* leaves after being exposed to 35 °C (Yang et al. 2012). MYB proteins are involved in signal transduction and secondary metabolite production (Samad et al. 2017). *O. sativa* tolerance to heat is improved by overexpression of OsMYB55, which enhances amino acid metabolism through transcription activation (El-Kereamy et al. 2012). WRKY is a crucial component of abscisic acid signalling and is involved in a variety of stress tolerance mechanisms (Samad et al. 2017; Rushton et al. 2012). *O. sativa* overexpressing OsWRKY11 showed improved drought and heat tolerance (Wu et al. 2009). All these indicate that MYB proteins and WRKY can act as very promising targets to improve crop heat stress tolerance.

## 3.5 Heat and Drought Stress Tolerance Screening Methods

### *3.5.1 Direct Screening for Yield in Stress Environments Under Natural Conditions*

Breeders make deliberate choices in order to achieve maximum output in their target environments. Drought and heat stress occur when water supply is below and temperature is above the normal range for the impact zone during crucial growth phases of crop. The possibility of aberrant seasons varies in accordance with the part of the world, along with wheat, is often cultivated in areas where the chance of adverse growth conditions is high (Langridge and Reynolds 2021). Screening for varieties in such natural condition is considered to be true estimates of complex traits governed by multi-genes. However, the estimation may be biased due to change in the temperature and precipitation pattern in year to year.

The most reliable technique for achieving broad adaptation among varieties to perform better under poor conditions is to select them in a range of differentiating habitats that provide information on performance under both normal and adverse conditions. However, this is laborious and cost-ineffective method. To overcome this, breeding programmes should be aimed for a limited number of test locations that can provide a fair indicator of overall performance. National varietal trials can help breeders to figure out where they should test their germplasm (Eichi et al. 2020).

Researchers have also aimed to promote the selection of germplasm that is both globally adaptive and drought resilient. Screening the offspring of biparental, multi-parental or association mapping panels in a variety of contexts to find genetic areas linked to superior performance in various environments. The mapping of loci associated to drought and heat tolerance has been the subject of several recent reviews (for example, Khadka et al. 2020; Goel et al. 2020). A meta-analysis incorporating data from numerous studies identified 854 QTL, including 502 drought-related and 234 heat-related regions (Acuña-Galindo et al. 2015). Few of these studies have yielded unambiguous selection goals for breeders, and they had limited impact on actual breeding. Part of the problem may be due to the oversimplified perspective of drought and heat stress that has served as the foundation for many research (Khadka et al. 2020; Goel et al. 2020). Heat and drought are difficult traits to screen because studies have shown that alleles are helpful when deep roots effectively absorb water as it disappears down the soil profile, but the same alleles are detrimental when plants are subjected to both heat and drought stress in situations where moisture at unreachable depth (Bonneau et al. 2013). Due to these difficulties, it has become mandatory to use artificial screening methods in order to find out stress resilient genotypes.

### *3.5.2 Artificial or Controlled Environments*

To create effective experimental conditions for wheat research, it is vital to examine the crop's vast variety of development patterns, which are divided into three categories: winter, spring and facultative (Braun et al. 2010). Two key factors influence the three growth habits: first, differential sensitivity to day length (daily photoperiod is a function of latitude and sowing date), and second, whether the crop is seeded before the cold winter season, which is typical of winter wheat. This is mostly due to significant effect genes like photoperiod sensitivity (Ppd) and vernalization (Vrn), which control adaptation to a protracted cold period that occurs after planting but before reproductive development begins (Kamran et al. 2014). The worldwide scope of breeding influences in spring wheat attained during and after the Green Revolution may be attributed in part to the fact that spring wheat was developed to be photoperiod and vernalization insensitive (Reynolds and Borlaug 2006). As a result, centralized research and breeding are now possible at various locations where factors relevant to target population environments (TPEs) may be substantially reproduced (Gerard et al. 2020). No selection condition can ever truly simulate the diversity represented by a TPE, particularly for uncontrolled elements like soil type, seasonal fluctuation in weather fluxes and so on.

Temperature, irrigation and humidity status can be maintained based on the requirement in the phytotron facility, green house and shade nets with drip irrigations. Use of artificial screening methods ensures maintenance of required micro environmental conditions and homogeneous soil condition for uniformity of treatments. Artificial light and temperature make it possible to overcome actions of vernalization and photoperiod induction of flowering in any time period of year. The disadvantage is being limited space and unnatural condition will not allow screening for huge number of germplasm and hinders direct application in field condition before validation. In fact, complex trait responses in controlled environments are rarely similar to field observations (Poorter et al. 2016). For example, root growth in pots cannot be assumed to be equivalent to root growth in the field, where temperature and depth profiles are dramatically different (Füllner et al. 2012). According to Langridge and Reynolds (2015), huge expenditure in controlled environment with pot research using crop genotypes largely deceives the evaluation of genetically complex traits.

## 3.6 Breeding Strategies and Methods

### *3.6.1 Breeding for Physiological Traits for Improving Heat Tolerance in Wheat*

Heat tolerance breeding in wheat strategy is based on conventional selection methods; however, with advances in our understanding of the biochemical, physiological and molecular basis of heat tolerance, physiological traits are now being used in addition to traditional methods for genotype selection to improve heat tolerance. To increase heat tolerance, Cossani and Reynolds (2012) developed selection methods for physiological traits such as radiation use efficiency, light interception and total assimilate partitioning. Explained digital imaging technology to identify a genotype for early vigour or ground cover which indicates efficiency of genotype to produce biomass and canopy area. Leaf greenness (SPAD), PSII's maximal photochemical efficiency (Fv/Fm) and photosynthetic rate (A) in relation to grain weight become crucial heat tolerance selection criteria. Haque et al. (2019) investigated the impact of heat stress on the early stages of grain filling. Four wheat cultivars (BARI GOM-25, BARI GOM-26, BARI GOM-27 and BARI GOM-28) revealed considerable genotypic diversity, with BARI GOM-28 showing superior heat stress tolerance than the other cultivars based on physiological and biochemical traits. Correlation studies between chlorophyll content and thylakoid membrane environment revealed loss of chlorophyll under heat stress condition (Ristic et al. 2007). The use of a chlorophyll metre to measure chlorophyll concentration will be advantageous for high-throughput screening of wheat for heat tolerance. Deeper root systems provide better access to soil moisture, which aids crop canopies in meeting increased evaporative demands associated with drought and heat stress (Cossani and Reynolds 2012). Reynolds et al. (2007) found a positive correlation between canopy temperature at crown root stage and grain filling stage and grain yield and biomass under heat stress condition. Wheat yields under heat stress have also been linked to the storage and remobilization of water-soluble carbohydrates in wheat stems. Water soluble carbohydrates preserved in stem act as useful resource for grain filling and avoid natural senescence of green biomass. Grain filling under heat stress is similarly linked to the ability for stem reserve remobilization and therefore to thermal tolerance in spring wheat (Fokar et al. 1998).

### *3.6.2 Conventional Approaches for Developing Heat and Drought–Tolerant Lines*

Heat and drought tolerance is a complicated phenomenon that is difficult to evaluate since it is influenced by a large number of genes. Many morphophysiological traits have been linked to heat stress performance in wheat. Longer leaf chlorophyll retention, canopy temperature depression, photosynthetic rate and leaf senescence have been described as indicators of heat tolerance metabolism. Heat tolerance has a significant impact on production parameters such as biomass, 1000 grain weight and grain yield. High tillering ability, relatively higher grain weight, acceptable spike fertility, higher spike number, grains per spike, early ground cover and other features have been linked to yield under heat stress according to several researchers.

#### 3.6.2.1 Breeding for Heat and Drought Tolerance

For traditional breeding, an index for measuring heat stress is a must. To find heat-tolerant genotypes, yield performance under heat stress conditions must be stable. As a result, the relative performance of yield and relative traits in heat-stressed and non-stressed conditions has long been employed as a marker for identifying heat-tolerant wheat genotypes. This heat sensitivity index has been found to be a good predictor of yield stability as well as a proxy for heat tolerance. Heat stress tolerance in wheat was also measured using membrane thermostability and chlorophyll fluorescence, which had high genetic relationships with grain production. Wheat genotypes are often examined across area and time by manipulating planting dates or selecting particular field test locations for heat treatments. Alternatively, heat stress can also be simulated under plastic film-covered shelters.

Wild species populations typically have substantial intra-species diversity for tolerance traits that are superior to those seen in contemporary cultivars. Indeed, *Triticum dicoccoides* and *Triticum monococcum* have been identified as prospective sources of germplasm for improving bread wheat heat tolerance. *Aegilops speltoides*, *Aegilops longissima* and *Aegilops searsii* all showed varying degrees of heat tolerance. Due to the constraints of traditional breeding methods, only a tiny part of the observed genetic diversity in heat tolerance has been used.

### *3.6.3 Molecular Breeding Strategies*

#### 3.6.3.1 Applications of Biochemical Markers to Improve Drought Tolerance

Drought stress triggers the expression of numerous genes, which alters the metabolism of many biochemicals such as key enzymes, transcription factors, hormones, amino acids and carbohydrates (Yang et al. 2010). A few examples are proline, abscisic acid (ABA), raffinose, mannitol, late embryogenesis abundant (LEA) proteins, glycine-betaine, tryptophan, trehalose and superoxide dismutase (Hameed et al. 2011; Nio et al. 2011). These biomolecules are involved in osmotic adjustment, membrane stability, anti-oxidation, reactive oxygen species (ROS) scavenging and gene regulation, in dehydration avoidance or tolerance (Ashraf 2010; Yang et al. 2010). However, the precise drought-related functions of most of these metabolites are yet to be known. This demands further research to uncover their roles, allowing for informed modulation of the genetic diversity present in the expression of their respective genes under stress. Yield advantage from accumulation of biochemicals should be taken into account in breeding programmes because use of energy to accumulate photo-assimilates and osmotic adjustment ensures survival of the plant at the expense of grain yield.

Transgenics derived in other crop species than the *Triticum* genome have been employed to learn more about the signalling and metabolism of these drought-related biochemicals. Transgenic wheat lines expressing the barley (*Hordeum vulgare* L.) gene *HVA1*, which encodes for a number of late embryogenesis abundant proteins that function as osmoprotectants, demonstrated improved water use efficiency, biomass accumulation and root weight when exposed to water stress (Sivamani et al. 2000). In transgenic lines, however, a proline inducing gene (*P5CS*) boosted drought tolerance, most likely as a result of proline's antioxidant protection of cells against oxidative damage by oxygen free radicals (Vendruscolo et al. 2007). Wheat drought tolerance was also increased by an *E. coli* mannitol biosynthesis (mtlD) gene that served as an osmoprotectant (Abebe et al. 2003). As most studies are conducted on seedling plants under artificial conditions, the potential contribution of individual genes to drought resistance enhancement may be exaggerated. However, this may not be a true representation of the expression of the trans-genes under natural field conditions. Intensive biochemical accumulation-based screening of various wheat germplasm might improve the introgression of the genes involved in utilizing traditional breeding procedures.

#### 3.6.3.2 Molecular Marker

Recent advances in knowledge of molecular markers have improved the powerful tools for breeders to recognize and select complex traits like drought tolerance. The use of marker trait association in breeding might minimize environmental effect, which is a key flaw in traditional quantitative trait breeding (Tuberosa and Salvi 2006). Marker-assisted selection, which includes methods such as marker-assisted backcross breeding (MABB), marker-assisted recurrent selection (MARS) and genomic selection (GS), improves the efficiency of selection considerably.

##### 3.6.3.2.1 Marker-Assisted Backcross Breeding

The MABB method yields a cultivar with the main gene/QTL from the donor background but the rest of the genomic makeup from the recurrent parent (Hospital et al. 1997). MABB has mostly been utilized for biotic stresses breeding but under exploited for complex traits like drought and heat tolerance. Complexity of the traits accompanies with epistatic interactions among QTLs with small or no effect on phenotype in different genetic backgrounds. Few identified QTLs for drought and heat are given in Table 3.1. Elite Indian variety GW322 was transferred with QTLs pertaining to drought and heat tolerance traits from HI1500. Genomic regions linked with drought and heat tolerance traits, viz. NDVI, stay green, chlorophyll content/ chlorophyll fluorescence and yield, were transferred and background selection with 109 polymorphic SSR markers (Todkar et al. 2020). From the same source, another Indian cultivar HD2733 was improved with the drought-tolerant QTL through MABB (Rai et al. 2018). For high-yielding recurrent parent background selection, SSR or SNPs covering whole genome is used. Marker-assisted foreground selection for tracking QTLs linked to drought-related physiological traits and agronomic traits like yield and its related traits for drought and heat tolerance in the backcrossed and advanced population like BC1F2 and BC2F2. With previously reported SSR markers linked to important traits such as grain filling duration, heat susceptibility index (HSI), grain filling duration, single kernel weight of main spike and kernel weight under heat stress, marker-assisted selection (MAS) was successfully used to screen 25 bread wheat genotypes to heat stress (Sadat et al. 2013). Marker-trait associations (MTAs) of meta-QTLs identified for different physiological and agronomic traits located in different genomic regions were evaluated and validated under rain fed and late sown environments (Sinha et al. 2018). Six potential desiccation-tolerant wheat genotypes under drought stress were evaluated and found confirmed greater tolerance under conditions of drought, making them ideal for increasing the productivity of rainfed and arid regions (Mujtaba et al. 2018). Four wheat cultivars viz. HUW468, HUW234, DBW17 and K307 were introduced with Qyld.csdh.7AL to generate high-yielding drought tolerant genotype (Gautam et al. 2020). Seven stable drought-related QTLs were identified across environment for DH (*QDh.iari_6D*), GWPS (*QGWPS.iari_5B*), PH (*QPh.iari_4B-2, QPh.iari_4B-3*) and NDVI (*QNdvi1.iari_5D, QNdvi3.iari_5A*) using SSR and SNP markers in the backcross inbred line population developed from HD2733/2*C306 in view of utilizing them in the upcoming study (Puttamadanayaka et al. 2020).

#### 3.6.3.2.2 Marker-Assisted Recurrent Selection

Plant breeders may now choose plants based on genotypic values produced by markers since molecular markers are abundant and widely available (Bernardo 2008). Marker-assisted recurrent selection, which involves random intermating between individuals selected based on their marker genotypes, is one of them (Hospital et al. 1997). MARS includes estimating marker effects for a number of minor impact QTLs or major QTLs, followed by two or three rounds of recombination (Eathington et al. 2007). MARS was used in biotic stress tolerance such as crown rot disease to more effectively pyramid minor genes controlling the resistance (Rahman et al. 2020). For drought and heat tolerance, MARS was exercised by conducting multi-location phenotyping under drought (restricted irrigated) and irrigated conditions, progenies carrying desirable QTLs were detected. Multi-location data and genotyping using SSR markers associated to stress adaptive traits were used to identify the best progenies (Jain et al. 2014). A practical utilization of MARS for improvement of drought tolerance in wheat was conducted and identified 51 QTLs by Harikrishna (2017).

**Table 3.1** Drought (D) and heat (H)–tolerant QTLs reported in wheat

| Condition | Mapping population | Trait | QTL | Marker | Chr. location | Chr. position | PVE (%) | Reference |
|---|---|---|---|---|---|---|---|---|
| DH | SeriM82/Babax (RILs) | DHE | | *7D-acc/cat-10* | 7D | 8.7 cM | 19.6 | Tahmasebi et al. (2016) |
| DH | Harry/Wesley (RILs) | PH | DH_PH_1 | *Xsnp124* | 1A | 81.7 cM | 8 | Sallam et al. (2022) |
| | | | DH_PH_2 | *XSNP5525/XSNP5526* | 7A | 61 cM | 7 | |
| | | | DH_PH_3 | *XSNP5629* | 7A | 78.8 cM | 9 | |
| DH | SYND/Weebill1 (RILs) | YLD | | *2265648\|F\|0-60:A > G-60: A > G* | 6D | 73 cM | 26.6 | Liu et al. (2019) |
| | | NDVI | | *RAC875_c57371_238* | 5D | 110–112 cM | 10.15 | |
| | | | | *Kukri_rep_c102608_599* | | | | |
| | | | | *RAC875_rep_c109661_107* | | | | |
| DH | Ningchun4/ Ningchun27 (RILs) | TGW | Qtgw-5D.1 | *wmc357-wmc97* | 5D | | 27.53 | Shi et al. (2021) |
| | | CC | Qcc-2A.3 | *wmc644-gwm558* | 2A | | 30.97 | |
| | | CC | Qcc-2A.6 | *wmc794-gwm382* | 2A | | 32.57 | |
| D | Kukri/Excalibur (DH) | 9 traits | 98 QTLs | | 5A and 7A | | | Gahlaut et al. (2017) |
| D | Reeder/albaney (RIL) | GVW | QTW. ndsu.7B | | 7B | 29.11–40.11 | 14.86 | Rabbi et al. (2021) |
| | | YLD | QYL.ndsu.2B | | 2B | 81.31–83.431 | 13.94 | |
| D | | GY | qGYWD.3B.2 | Xgpw7774 | | 97.6 | 19.6 | Shukla et al. (2015) |
| | | TGY | 4A | Xwmc420 90.4 | | 90.4 | 20 | Kirigwi et al. (2007) |
| | | DTH | 4A-a | Xgwm397 | | 6.0 | 23 | |

| | | | | | | | | |
|---|---|---|---|---|---|---|---|---|
| | | | | | | | | Pinto et al. (2010) |
| | | DTM | Qyld. csdh.7AL | Xgwm322 | | 155.9 | 20 | Quarrie et al. (2006) |
| | | | 3B | Xbarc101 | | 86.1 | 45.2 | Golabadi et al. (2011) |
| | | | QTgw-7D-b | XC29-P13 | | 12.5 | 21.9 | Lopes et al. (2013) |
| | | | QDh-7D.b | XC29-P13 | | 12.5 | 22.7 | Lopes et al. (2013) |
| | | | QHd. idw-2A.2 | Xwmc177 | | 46.1 | 32.2 | Maccaferri et al. (2008) |
| | | | QDm-7D.b | X7D-acc/cat-10 | | 2.7 | 22.7 | Lopes et al. (2013) |
| D | | Stem reserve mobilization | QSrm.ipk-2D | Xgwm249a | | 142.0 | 42.2 | Salem et al. (2007) |
| | | Water Soluble Carbohydrate | QSrm.ipk-5D | Xfbb238b | | 19.0 | 37.5 | Salem et al. (2007) |
| | | SPAD/Chlorophyll Content | QSrm.ipk-7D | Xfbb189b | | 338.0 | 21.0 | Salem et al. (2007) |
| | | | QWsc-c.aww-3A | Xwmc0388A | | 64.9 | 19.0 | Bennett et al. (2012) |
| | | | Qchl.ksu-3B | Xbarc68 | | 67.2 | 59.1 | Kumar et al. (2012) |
| H | NW1014/ HUW464 (RIL) | TGW | 3 QTLs | | 2BL, 7BL and 7DS | | | Paliwal et al. (2012) |
| | | GFD | 1 | | 2BL | | 22.36 | |
| | | CTD | 1 | | 7BL | | 17.89 | |
| | | DM | 1 | | 7DS | | 8.12 | |
| | | YLD | 2 | | 7BL | | | |

(continued)

**Table 3.1** (continued)

| Condition | Mapping population | Trait | QTL | Marker | Chr. location | Chr. position | PVE (%) | Reference |
|---|---|---|---|---|---|---|---|---|
| H | Kauz/ MTRWA116 | SSI | | *Gmw190* | 1B | 0-14 | 44.3 | Mohammadi et al. (2008) |
| | | | | *Gmw133A* | 5B | 112-132 | 27.3 | |
| | | | | *Gmw63B* | 7B | 68-86 | 34 | |

#### 3.6.3.2.3 Genome-Wide Association Studies (GWAS)

Identification of linked marker to the trait of interest is the prime objective before marker assisted breeding. GWAS is one among the most important method of identifying MTAs using high-throughput genome-wide markers. Such study for drought and heat tolerance was done to find out putative region of the genome governing stress tolerant activity. Flowering time (FT), leaf relative water content (RWC), osmotic adjustment (OA), chlorophyll content (SPAD) and leaf rolling (LR) in durum wheat were investigated using a genome-wide association study (GWAS), which found 15 significant QTLs for OA (global $R^2 = 63.6\%$) (Condorell et al. 2022). Genetic variation in a panel of 290 wheat lines to water deficit condition was enumerated, and the QTLs governing the trait were identified via GWAS. Evaluation for phenotypic traits was conducted under well-irrigated and drought-stressed conditions, and GWAS was performed using 15,737 SNP markers. significant marker–trait associations underlying different mechanisms for adaptation to drought-prone environments were identified (Abou-Elwafa and Shehzad 2021). A total of 62 significant marker-trait associations (MTAs) for traits like days-to-heading (DTH), number of days-to-maturity (DTM), plant height (PHT), spike length (SPL), number of kernels per spike (KPS), thousand kernel weight (TKW), and grain yield (GYLD) were detected using the Diversity Arrays Technology sequencing (DArTseq) data from 93 diverse bread wheat lines evaluated under normal and drought stress condition (Mwadzingeni et al. 2017). Under seedling heat stress, a genome-wide association mapping study (GWAS) using 90k SNP chips discovered QTLs and SNP markers linked with seedling heat tolerance in a population of 200 different hard winter wheat. Marker-assisted selection (MAS) of seedling heat tolerance may be employed using the SNP markers discovered (Maulana et al. 2018). Using 251 recombinant inbred lines (RILs) produced from a hybrid HD 2808/HUW510, quantitative trait loci (QTLs) linked with heat-tolerant characteristics under terminal heat stress were found. The use of composite interval mapping resulted in the discovery of 40 QTLs, 17 of which were linked to timely planting and 23 to heat stress upon late sowing (Bhusal et al. 2018). Such identified markers and QTLs can be used in the marker assisted breeding for the transfer of genomic region related to stress tolerance.

#### 3.6.3.2.4 Genomic Selection

In marker-assisted selection, main concern is given to the major QTLs, and minor ones having significant effects are neglected (Zhao et al. 2014). To overcome this, genomic selection was proposed, a form of marker-assisted selection, which encompasses the marker data of the entire genome and uses them in the selection of superior lines. With the help of suitable statistical methodologies, effects associated with each marker loci are predicted (Meuwissen et al. 2001). Genomic estimated breeding value (GEBV) of the progeny is the basis of selection in GS. With the utilization of off season and speed breeding facilities, quick genetic gain and improvement in complex traits can be obtained (Mir et al. 2012).

Abiotic stress tolerance has been achieved via GS in important crops such as rice (Bhandari et al. 2019) and maize (Shikha et al. 2017; Xu et al. 2018; Wang et al. 2019). Experiments were also conducted to assess the efficacy of several GS models for abiotic stresses like drought and heat stress in wheat. Rutkoski et al. (2016) evaluated 557 wheat lines for canopy temperature and NDVI in optimal, early heat, late heat, drought and severe drought environments, found that secondary traits improved grain yield prediction accuracy by 70% through genomic prediction models. Similarly, Haghighattalab et al. (2017) used UAS imagery and its relationship with grain yield; they evaluated potential of UAS imagery for predicting GY at plot level and found high correlation between imagery-derived phenotypic traits and grain yield under irrigated and drought condition. Dunckel et al. (2017) used five different genomic prediction models to examine some identified lines from the DH and RIL population for grain production and other agronomic parameters under irrigated, heat and drought-stressed conditions and found that these models had moderate prediction accuracy. Sun et al. (2017) compared three statistical models namely, simple repeatability (SR), multi-trait (MT) and random regression (RR) for their predictive abilities for grain yield under five different drought and heat stress environments and found that predictive abilities were enhanced by an average of ~70%, by including secondary trait BLUPs from SR, MT or RR models in multivariate pedigree and GS models. Crain et al. (2018) evaluated 1170 advanced wheat breeding lines in drought and heat environment and concluded that advances in yield prediction models and huge data generation through genomics and phonemics will make GS feasible for plant breeders to achieve enhanced genetic gain in every possible environment. A Durum panel was evaluated in well-watered, heat-stress and drought stress environment, and reaction norm model for genomic prediction was applied and found that in all cross-validation schemes, prediction accuracy was improved by addition of G × E interaction terms to the model by Sukumaran et al. (2018).

#### 3.6.3.2.5 Genetic Engineering and Gene Editing

Drought and heat stress–tolerant germplasm might be developed quickly via genetic engineering. Although phenotyping has mostly been done under greenhouse conditions, there is a substantial literature on candidate genes and the examination of transgenic line performance under drought stress (Shinwari et al. 2020). Much of the research has focused on the transcriptional regulation of drought responses, and the findings have contributed to a better understanding of drought tolerance, while emphasizing the complexity of the plant response. Several recent studies have described advances in the creation of transgenic drought tolerance in wheat (Khan et al. 2019a, b). Khan et al. (2019a, b) listed 32 genes in transgenic wheat that exhibit enhanced drought tolerance–related traits. With the advent of gene editing, a possible approach to engineer novel stress tolerance in wheat became available (Kumar et al. 2019). Along with the exploration of altered or novel gene expression of drought and heat tolerance, we may add proteomics and metabolomics studies of drought responses in wheat (Langridge and Reynolds 2015; Goel et al. 2020). These studies have offered precise information on the nature of drought responses and, in many cases, have identified potentially valuable research targets. However, large-scale evaluations of transgenic material across many habitats and over several years are required to acquire acceptability in breeding programmes. Unfortunately, the current regulations about transgenics in many countries make it difficult to achieve the required extensive field evaluation. In addition, transgenic wheat is not grown currently anywhere in the world. Acquired knowledge from transgenic study may assist the conventional and marker-assisted breeding and may have sole contribution in near future.

### *3.6.4 Difficulties in Breeding for Drought and Heat Tolerance*

Heat and drought stress occurs irregularly in the cultivated land due to altered natural calamities. Hence, identifying drought-prone areas in which the variety should be developed is a difficult task because it differs from place to place and time to time. Maintaining the plant population in controlled conditions like moisture stress and temperature in both field and green house conditions needs sophisticated facility and care. The plant breeder should use combined selection devices with other useful resources to estimate combined heat and drought stress tolerance because isolated occurrence of heat and drought is much more lesser as compared to combined heat and drought stress occurrence. Estimating combined heat and drought–resistant characters is complex, laborious and time consuming. Another major disadvantage is the inverse relationship between heat and drought–resistant characters and yield potentiality of the crop so breeder has to make additional breeding efforts to increase the yielding ability of the plant (Ranjith and Rao 2021). So, combined breeding scheme is to be developed by plant breeder to enhance yield and abiotic resistance characters. Usage of primitive or wild cultivars as a source of heat and drought resistance needs extensive evaluation of germplasm line, and transfer of traits to the elite cultivar leads to huge amount of linkage drag. Unlike qualitative traits, heat and drought resistance is governed by quantitative traits influenced by the environment and may leads to deceived results in the end.

## 3.7 Conclusion

A major staple food crop like wheat, grown worldwide in varied environmental condition, is more prone to the abiotic stress due to scattered rainfall and global warming. Extended cultivation of wheat in semi-arid tropics necessitates the drought and heat stress cultivars. Physiological factors from root to spike influencing drought and heat stress are having immense importance in breeding climate resilient varieties. Selection for cultivars that are able to yield well under combined heat and drought stress occurred due to below average rainfall in semi-arid tropics has been a high priority for breeding programmes and has been a strong focus for many research projects. Improving target oriented physiological traits related to drought and heat tolerance can enhance the tolerance to abiotic stress among the cultivars. Approaches for identification and utilization of germplasm sources having importance in stress breeding through conventional and marker-assisted breeding enhanced the knowledge and some advance in climate resilience. However, the extensive genetic mapping work has revealed the complexity of the heat and drought responses, but utilization of them in practical breeding programme is meagre. A cocktail of breeding methods from conventional to transgenic may help in the development of high-yielding climate-resilient varieties which can help to serve farmers to escape from glitch of crop loss due to dry spell during cropping season.

## References

Abebe T, Guenzi AC, Martin B, Cushman JC (2003) Tolerance of mannitol-accumulating transgenic wheat to water stress and salinity. Plant Physiol 131:1748–1755

Abou-Elwafa SF, Shehzad T (2021) Genetic diversity, GWAS and prediction for drought and terminal heat stress tolerance in bread wheat (*Triticum aestivum L.*). Genet Resour Crop Evol 68(2):711–728

Acuña-Galindo MA, Mason RE, Subramanian NK, Hays DB (2015) Meta-analysis of wheat QTL regions associated with adaptation to drought and heat stress. Crop Sci 55:477–492. https://doi.org/10.2135/cropsci2013.11.0793

Adams SR, Cockshull KE, Cave CRJ (2001) Effect of temperature on the growth and development of tomato fruits. Ann Bot 88(5):869–877

Ahn C, Park U, Park PB (2011) Increased salt and drought tolerance by D-ononitol production in transgenic Arabidopsis thaliana. Biochem Biophys Res Commun 415(4):669–674

Ahsan N, Donnart T, Nouri MZ, Komatsu S (2010) Tissue-specific defense and thermo-adaptive mechanisms of soybean seedlings under heat stress revealed by proteomic approach. J Proteome Res 9(8):4189–4204

Akhtar M, Jaiswal A, Taj G, Jaiswal JP, Qureshi MI, Singh NK (2012) DREB1/CBF transcription factors: their structure, function and role in abiotic stress tolerance in plants. J Genet 91(3): 385–395

Al-Karaki GN (2012) Phenological development-yield relationships in durum wheat cultivars under late-season high-temperature stress in a semiarid environment. International Scholarly Research Notices 2012

Amani I, Fischer RA, Reynolds MP (1996) Canopy temperature depression association with yield of irrigated spring wheat cultivars in a hot climate. J Agron Crop Sci 176(2):119–129

Anderegg J, Yu K, Aasen H, Walter A, Liebisch F, Hund A (2020) Spectral vegetation indices to track senescence dynamics in diverse wheat germplasm. Front Plant Sci 10:1749

Aroca R, Vernieri P, Irigoyen JJ, Sancher-Diaz M, Tognoni F, Pardossi A (2003) Involvement of abscisic acid in leaf and root of maize (*Zea mays L.*) in avoiding chilling-induced water stress. Plant Sci 165:671–679. https://doi.org/10.1016/S0168-9452(03)00257-7

Ashraf M (2010) Inducing drought tolerance in plants: recent advances. Biotechnol Adv 28:169–183

Bahuguna RN, Jagadish KSV, Coast O, Wassmann R (2014) Plant abiotic stress: temperature extremes. In: Encyclopedia of agriculture and food systems. Elsevier, Amsterdam, pp 330–334

Baniwal SK, Bharti K, Chan KY, Fauth M, Ganguli A, Kotak S, Mishra SK, Nover L, Port M, Scharf KD, Tripp J, Weber C, Zielinski D, von Koskull-Döring P (2004) Heat stress response in plants: a complex game with chaperones and more than twenty heat stress transcription factors. J Biosci 29(4):471–487

Barrero-Sicilia C, Silvestre S, Haslam RP, Michaelson LV (2017) Lipid remodelling: unravelling the response to cold stress in Arabidopsis and its extremophile relative Eutrema salsugineum. Plant Sci 263:194–200

Baxter A, Mittler R, Suzuki N (2014) ROS as key players in plant stress signalling. J Exp Bot 65: 1229–1240. https://doi.org/10.1093/jxb/ert375

Bennett D, Izanloo A, Reynolds M, Kuchel H, Langridge P, Schnurbusch T (2012) Genetic dissection of grain yield and physical grain quality in bread wheat (Triticum aestivum L.) under water-limited environments. Theor Appl Genet 125:255–271

Bernardo R (2008) Molecular markers and selection for complex traits in plants: learning from the last 20 years. Crop Sci 48:1649–1664

Bhandari A, Bartholomé J, Cao-Hamadoun TV, Kumari N, Frouin J, Kumar A, Ahmadi N (2019) Selection of trait-specific markers and multi-environment models improve genomic predictive ability in rice. PLoS One 14(5):1–21. https://doi.org/10.1371/journal.pone.0208871

Bhattacharya A (2019) Effect of high temperature on crop productivity and metabolism of macro molecules. Academic, New York

Bhusal N, Sharma P, Sareen S, Sarial AK (2018) Mapping QTLs for chlorophyll content and chlorophyll fluorescence in wheat under heat stress. Biol Plant 62(4):721–731

Bhusal N, Poudel RR, Panthi S, Khanal N (2021) Chapter 8—Prospection of heat tolerance in the context of global warming in wheat for food security. In: Sareen S, Sharma P, Singh C, Jasrotia P, Pratap Singh G et al (eds) Woodhead publishing series in food science, technology and nutrition. Woodhead Publishing, Cambridge, pp 123–143

Blum A, Sinmena B, Mayer J, Golan G, Shpiler L (1994) Stem reserve mobilisation supports wheat-grain filling under heat stress. Funct Plant Biol 21(6):771–781

Bonneau J, Taylor J, Parent B, Bennett D, Reynolds M, Feuillet C, Langridge P, Mather DE (2013) Multi-environment analysis and improved mapping of a yield-related QTL on chromosome 3B of wheat. Theor Appl Genet 126:747–761. https://doi.org/10.1007/s00122-012-2015-3

Bougouffa S, Radovanovic A, Essack M, Bajic V B (2014) DEOP: a database on osmoprotectants and associated pathways. Database bau100. https://doi.org/10.1093/database/bau100

Braun HJ, Atlin G, Payne T (2010) Multi-location testing as a tool to identify plant response to global climate change. In: Reynolds MP (ed) Climate change & crop production. CABI, Wallingford, p 115. https://doi.org/10.1079/9781845936334.0115

Choudhury FK, Rivero RM, Blumwald E, Mittler R (2017) Reactive oxygen species, abiotic stress and stress combination. Plant J 90(5):856–867

Condon AG, Richards RA, Rebetzke GJ, Farquhar GD (2004) Breeding for high water-use efficiency. J Exp Bot 55(407):2447–2460

Condorell GE, Newcomb M, Groli EL, Maccaferri M, Forestan C, Babaeian E, Tuberosa R (2022) Genome wide association study uncovers the QTLome for osmotic adjustment and related drought adaptive traits in durum wheat. Genes 13(2):293

Cossani CM, Reynolds MP (2012) Physiological traits for improving heat tolerance in wheat. Plant Physiol 160:1710–1718

Crain J, Mondal S, Rutkoski J, Singh RP, Poland J (2018) Combining high-throughput phenotyping and genomic information to increase prediction and selection accuracy in wheat breeding. Plant Genome 11(1):1–14. https://doi.org/10.3835/plantgenome2017.05.0043

Craufurd PQ, Wheeler TR (2009) Climate change and the flowering time of annual crops. J Exp Bot 60:2529–2539

Danakumara T, Kumari J, Singh AK, Sinha SK, Pradhan AK, Sharma S, Prasad PV (2021) Genetic dissection of seedling root system architectural traits in a diverse panel of hexaploid wheat through multi-locus genome-wide association mapping for improving drought tolerance. Int J Mol Sci 22(13):7188

Deva CR, Urban MO, Challinor AJ, Falloon P, Svitákova L (2020) Enhanced leaf cooling is a pathway to heat tolerance in common bean. Front Plant Sci 11:19

Devasirvatham V, Tan DK, Trethowan RM (2016) Breeding strategies for enhanced plant tolerance to heat stress. In: Advances in plant breeding strategies: agronomic, abiotic and biotic stress traits. Springer, Cham, pp 447–469

Dooki AD, Mayer-Posner FJ, Askari H, Zaiee AA, Salekdeh GH (2006) Proteomic responses of rice young panicles to salinity. Proteomics 6(24):6498–6507. https://doi.org/10.1002/pmic.200600367

Dunckel S, Crossa J, Wu S, Bonnett D, Poland J (2017) Genomic selection for increased yield in synthetic-derived wheat. Crop Sci 57(2):713–725

Eathington SR, Crosbie TM, Edwards MD, Reiter RS, Bull JK (2007) Molecular markersin a commercial breeding program. Crop Sci 47:S154–S163

Eichi VR, Okamoto M, Garnett T, Eckermann P, Darrier B, Riboni M, Langridge P (2020) Strengths and weaknesses of national variety trial data for multi-environment analysis: a case study on grain yield and protein content. Agronomy 10:753. https://doi.org/10.3390/agronomy10050753

El-Kereamy A, Bi YM, Ranathunge K, Beatty PH, Good AG, Rothstein SJ (2012) The rice R2R3-MYB transcription factor OsMYB55 is involved in the tolerance to high temperature and modulates amino acid metabolism. PLoS One 7(12):e52030. https://doi.org/10.1371/journal.pone.0052030

Fahad S, Bajwa AA, Nazir U, Anjum SA, Farooq A, Zohaib A, Sadia S, Nasim W, Adkins S, Saud S, Ihsan MZ, Alharby H, Wu C, Wang D, Huang J (2017) Crop production under drought and heat stress: plant responses and management options. Front Plant Sci 8:1147

Farooq M, Wahid A, Kobayashi N, Fujita D, Basra SMA (2009) Plant drought stress: effects, mechanisms and management. Agron Sustain Dev 29:185–212. https://doi.org/10.1051/agro:2008021

Fitter AH, Hay RKM (2002) Environmental physiology of plants, 3rd edn. Academic, London

Fokar M, Blum A, Nguyen HT (1998) Heat tolerance in spring wheat. II: grain filling. Euphytica 104:9–15

Foyer CH, Noctor G (2005) Redox homeostasis and antioxidant signaling: a metabolic interface between stress perception and physiological responses. Plant Cell 17:1866–1875. https://doi.org/10.1105/tpc.105.033589

Friedlingstein P, Houghton RA, Marland G, Hackler J, Boden TA, Conway TJ, Canadell JG, Raupach MR, Ciais P, Le Quéré C (2010) Update on $CO_2$ emissions. Nat Geosci 3(12):811–812

Füllner K, Temperton VM, Rascher U, Jahnke S, Rist R, Schurr U, Kuhn AJ (2012) Vertical gradient in soil temperature stimulates development and increases biomass accumulation in barley. Plant Cell Environ 35(5):884–892. https://doi.org/10.1111/j.1365-3040.2011.02460.x

Gahlaut V, Jaiswal V, Tyagi BS, Singh G, Sareen S, Balyan HS, Gupta PK (2017) QTL mapping for nine drought-responsive agronomic traits in bread wheat under irrigated and rain-fed environments. PLoS One 12(8):e0182857

Gajghate R, Chourasiya D, Harikrishna SRK (2020) Plant morphological, physiological traits associated with adaptation against heat stress in wheat and maize. In: Giri B, Sharma MP (eds) Plant stress biology. Springer, Singapore. https://doi.org/10.1007/978-981-15-9380-2_3

Galinski EA (1995) In: Poole RK (ed) Advances in microbial physiology. Elsevier, Amsterdam, pp 273–328

Gautam T, Dhillon GS, Saripalli G, Rakhi VPS, Prasad P, Kaur S, Chhuneja P, Sharma PK, Balyan HS, Gupta PK (2020) Marker-assisted pyramiding of genes/QTL for grain quality and rust resistance in wheat (Triticum aestivum L.). Mol Breed 40:49

Gerard GS, Crespo-Herrera LA, José Cross J, Mondal S, Velu G, Juliana P, Huerta-Espino J, Vargas M, Rhandawa MS, Bhavani S, Braun H, Singh RP (2020) Grain yield genetic gains and changes in physiological related traits for CIMMYT's high rainfall wheat screening nursery tested across international environments. Field Crop Res 249:107742. https://doi.org/10.1016/j.fcr.2020.107742

Gill SS, Tuteja N (2010) Reactive oxygen species and antioxidant machinery in abiotic stress tolerance in crop plants. Plant Physiol Biochem 48(12):909–930

Goel S, Singh K, Grewal S, Nath M (2020) Impact of "omics" in improving drought tolerance in wheat. Crit Rev Plant Sci 39:222–235. https://doi.org/10.1080/07352689.2020.1778924

Golabadi M, Arzani A, Mirmohammadi Maibody SAM, Tabatabaei BES, Mohammadi SA (2011) Identification of microsatellite markers linked with yield components under drought stress at terminal growth stages in durum wheat. Euphytica 177:207–221

Goraya GK, Kaur B, Asthir B, Bala S, Kaur G, Farooq M (2017) Rapid injuries of high temperature in plants. J Plant Biol 60(4):298–305

Gupta SC, Sharma A, Mishra M, Mishra R, Chowdhuri DK (2010) Heat shock proteins in toxicology: how close and how far? Life Sci 86(11–12):377–384

Haghighattalab A, Crain J, Mondal S, Rutkoski J, Singh RP, Poland J (2017) Application of geographically weighted regression to improve grain yield prediction from unmanned aerial system imagery. Crop Sci 57(5):2478–2489

Hameed A, Bibi N, Akhter J, Iqbal N (2011) Differential changes in antioxidants, proteases, and lipid peroxidation in flag leaves of wheat genotypes under different levels of water deficit conditions. Plant Physiol Biochem 49:178–185

Hansen J, Sato M, Ruedy R (2012) Perception of climate change. Proc Natl Acad Sci 109(37): E2415–E2423

Haque MS, Tanjima U, Sushmoy DR, Hossain AKZM (2019) Heat stress alters chlorophyll fluorescence, photosynthesis and antioxidative enzyme activities in wheat cultivars. Fundam Appl Agric 4(2):858–866

Harikrishna (2017) Marker assisted recurrent selection for drought tolerance in wheat. Thesis dissertation. Indian Agricultural Research Institute, New Delhi

Hasanuzzaman M, Hossain MA, da Silva JAT, Fujita M (2012) Plant response and tolerance to abiotic oxidative stress: antioxidant defense is a key factor. In: Crop stress and its management: perspectives and strategies. Springer, Dordrecht, pp 261–315

Hasanuzzaman M, Nahar K, Alam MM, Roychowdhury R, Fujita M (2013) Physiological, biochemical, and molecular mechanisms of heat stress tolerance in plants. Int J Mol Sci 14(5): 9643–9684. https://doi.org/10.3390/ijms14059643

Hassan AH, Hozzein WN, Mousa AS, Rabie W, Alkhalifah DHM, Selim S, AbdElgawad H (2020) Heat stress as an innovative approach to enhance the antioxidant production in *Pseudooceanicola* and *Bacillus* isolates. Sci Rep 10(1):1–13

Hemantaranjan A, Malik CP, Bhanu AN (2018) Physiology of heat stress and tolerance mechanism—an overview. J Plant Sci Res 33(1):55–68

Higashi Y, Okazaki Y, Myouga F, Shinozaki K, Saito K (2015) Landscape of the lipidome and transcriptome under heat stress in *Arabidopsis thaliana*. Sci Rep 5(1):1–11

Hospital F, Moreau L, Lacoudre F, Charcosset A, Gallais A (1997) More on the efficiency of marker-assisted selection. Theor Appl Genet 95:1181–1189

Hu W, Hu G, Han B (2009) Genome-wide survey and expression profiling of heat shock proteins and heat shock factors revealed overlapped and stress specific response under abiotic stresses in rice. Plant Sci 176(4):583–590

Hu X, Wu L, Zhao F, Zhang D, Li N, Zhu G, Li C, Wang W (2015) Phosphoproteomic analysis of the response of maize leaves to drought, heat and their combination stress. Front Plant Sci 6:298

Hu S, Ding Y, Zhu C (2020) Sensitivity and responses of chloroplasts to heat stress in plants. Front Plant Sci 11:375

Huey RB, Carlson M, Crozier L, Frazier M, Hamilton H, Harley C, Hoang A, Kingsolver JG (2002) Plants versus animals: do they deal with stress in different ways? Integr Comp Biol 42(3): 415–423

IPCC (2014) Summary for policymakers. In: Stocker TF, Qin D, Plattner GK, Tignor M, Allen SK, Boschung J, Nauels A, Xia Y, Bex V, Midgley PM (eds) Climate change 2013: the physical science basis. Contribution of working group I to the fifth assessment report of the intergovernmental panel on climate change. Cambridge University Press, Cambridge

IPCC (Intergovernmental Panel on Climate Change) (2012) Managing the risks of extreme events and disasters to advance climate change adaptation. In: Field CB, Barros V, Stocker TF, Qin D, Dokken DJ, Ebi KL, Mastrandrea MD, Mach KJ, Plattner GK, Allen SK, Tignor M, Midgley PM (eds) A special report of working groups I and II of the intergovernmental panel on climate change. Cambridge University Press, Cambridge, p 582

Jain N, Singh GP, Singh PK, Ramya P, Krishna H, Ramya KT, Todkar L, Amasiddha B, Kumar KP, Vijay P, Jadon V (2014) Molecular approaches for wheat improvement under drought and heat stress. Indian J Genet Plant Breed 74(4s):578

Janska A, Mars P, Zelenkova S, Ovesna J (2009) Cold stress and acclimation—what is important for metabolic adjustment? Plant Biol 12:395–405. https://doi.org/10.1111/j.1438-8677.2009.00299.x

Joshi AK, Mishra B, Chatrath R, Ferrara GO, Singh RP (2007) Wheat improvement in India: present status, emerging challenges and future prospects. Euphytica 157(3):431–446

Kalyar T, Rauf S, da Silva JAT, Iqbal Z (2013) Variation in leaf orientation and its related traits in sunflower (*Helianthus annuus L.*) breeding populations under high temperature. Field Crops Res 150:91–98

Kamran A, Iqubal M, Spaner D (2014) Flowering time in wheat (*Triticum aestivum L.*): a key factor for global adaptability. Euphytica 197:1–26. https://doi.org/10.1007/s10681-014-1075-7

Kang CH, Lee SY, Park JH, Lee Y, Jung HS, Chi YH, Jung YJ, Chae HB, Shin MR, Kim WY, Yun DJ, Lee SY (2015) Stress-driven structural and functional switching of Ypt1p from a GTPase to a molecular chaperone mediates thermo tolerance in *Saccharomyces cerevisiae*. FASEB J 29(11):4424–4434. https://doi.org/10.1096/fj.15-270140

Khadka K, Raizada MN, Navabi A (2020) Recent progress in germplasm evaluation and gene mapping to enable breeding of drought-tolerant wheat. Front Plant Sci 11:1149. https://doi.org/10.3389/fpls.2020.01149

Khan S, Anwar S, Ashraf MY, Khaliq B, Sun M, Hussain S, Gao ZQ, Noor H, Alam S (2019a) Mechanisms and adaptation strategies to improve heat tolerance in rice. A review. Plants 8(11): 508

Khan S, Anwar S, Yu S, Sun M, Yang Z, Gao Z-Q (2019b) Development of drought-tolerant transgenic wheat: achievement and limitations. Int J Mol Sci 20:3350. https://doi.org/10.3390/ijms20133350

Khan A, Chatterjee S, Weng Y (2021) Future research for tropical UHI. In: Khan A, Chatterjee S, Weng Y (eds) Urban heat island modeling for tropical climates. Elsevier, Amsterdam, pp 215–230

Khanna-Chopra R, Viswanathan C (1999) Evaluation of heat stress tolerance in irrigated environment of T. aestivum and related species. I. Stability in yield and yield components. Euphytica 106:169–180

Kim WY, Cheong NE, Lee DC, Je DY, Bahk JD, Cho MJ, Lee SY (1995) Cloning and sequencing analysis of a full-length cDNA encoding a G protein alpha subunit, SGA1, from soybean. Plant Physiol 108(3):1315–1316. https://doi.org/10.1104/pp.108.3.1315

Kirigwi FM, Van Ginkel M, Brown-Guedira G, Gill BS, Paulsen GM, Fritz AK (2007) Markers associated with a QTL for grain yield in wheat under drought. Mol Breed 20:401–413

Kumar S, Sehgal SK, Kumar U, Prasad PVV, Joshi AK, Gill BS (2012) Genomic characterization of drought tolerance-related traits in spring wheat. Euphytica 186:265–276

Kumar R, Kaur A, Pandey A, Mamrutha HM, Singh GP (2019) CRISPR-based genome editing in wheat: a comprehensive review and future prospects. Mol Biol Rep 46:3557–3569. https://doi.org/10.1007/s11033-019-04761-3

Kumar A, Sharma A, Sharma R, Srivastva P, Choudhary A (2021) Exploration of wheat wild relative diversity from Lahaul valley: a cold arid desert of Indian Himalayas. Cereal Res Commun 50:305–320

Kumar A, Sharma A, Sharma R, Choudhary A, Srivastava P, Kaur H, Padhy AK (2022) Morpho-physiological evaluation of Elymus semicostatus (Nees ex Steud.) Melderis as potential donor for drought tolerance in Wheat (*Triticum aestivum L.*). Genet Resour Crop Evol 69(1):411–430

Lal R (2004) Soil carbon sequestration to mitigate climate change. Geoderma 123:1–22. https://doi.org/10.1016/j.geoderma.2004.01.032

Langridge P, Reynolds MP (2015) Genomic tools to assist breeding for drought tolerance. Curr Opin Biotechnol 32:130–135. https://doi.org/10.1016/j.copbio.2014.11.027

Langridge P, Reynolds M (2021) Breeding for drought and heat tolerance in wheat. Theor Appl Genet 134(6):1753–1769

Larcher W (1995) Plant physiological ecology. Springer, Berlin

Larkindale J, Hall JD, Knight MR, Vierling E (2005) Heat stress phenotypes of Arabidopsis mutants implicate multiple signaling pathways in the acquisition of thermotolerance. Plant Physiol 138(2):882–897

Lata C, Bhutty S, Bahadur RP, Majee M, Prasad M (2011) Association of an SNP in a novel DREB2-like gene SiDREB2 with stress tolerance in foxtail millet [*Setaria italica* (L.)]. J Exp Bot 62(10):3387–3401

Lee D, Ahşan N, Lee S, Kang KY, Bahk JD, Lee I, Lee BA (2007) Proteomic approach in analyzing heat-responsive proteins in rice leaves. Proteomics 7:3369–3383. https://doi.org/10.1002/pmic.200700266

Li C, Jiang D, Wollenweber B, Li Y, Dai T, Cao W (2011) Waterlogging pretreatment during vegetative growth improves tolerance to waterlogging after anthesis in wheat. Plant Sci 180: 672–678. https://doi.org/10.1016/j.plantsci.2011.01.009

Lindquist S, Craig EA (1988) The heat-shock proteins. Annu Rev Genet 22(1):631–677

Liu C, Sukumaran S, Claverie E, Sansaloni C, Dreisigacker S, Reynolds M (2019) Genetic dissection of heat and drought stress QTLs in phenology-controlled synthetic-derived recombinant inbred lines in spring wheat. Mol Breed 39(3):1–18

Lopes MS, Reynolds MP (2012) Stay-green in spring wheat can be determined by spectral reflectance measurements (normalized difference vegetation index) independently from phenology. J Exp Bot 63(10):3789–3798

Lopes MS, Reynolds MP, McIntyre CL, Mathews KL, Jalal Kamali MR, Mossad M, Feltaous Y, Tahir IS, Chatrath R, Ogbonnaya F, Baum M (2013) QTL for yield and associated traits in the Seri/Babax population grown across several environments in Mexico, in the West Asia, North Africa, and South Asia regions. Theor Appl Genet 126(4):971–984

Maccaferri M, Sanguineti MC, Corneti S, Ortega JL, Salem MB, Bort J, DeAmbrogio E, del Moral LF, Demontis A, El-Ahmed A, Maalouf F (2008) Quantitative trait loci for grain yield and adaptation of durum wheat (*Triticum durum Desf.*) across a wide range of water availability. Genetics 178(1):489–511

Manickavelu A, Kawaura K, Oishi K, Shin-I T, Kohara Y, Yahiaoui N, Keller B, Abe R, Suzuki A, Nagayama T, Yano K (2012) Comprehensive functional analyses of expressed sequence tags in common wheat (*Triticum aestivum*). DNA Res 19(2):165–177. https://doi.org/10.1093/dnares/dss001

Marwein R, Debbarma J, Sarki YN, Baruah I, Saikia B, Boruah HPD, Velmurugan N, Chikkaputtaiah C (2019) Genetic engineering/genome editing approaches to modulate signaling processes in abiotic stress tolerance. In: Plant signaling molecules. Woodhead Publishing, Cambridge, pp 63–82

Mathur S, Agrawal D, Jajoo A (2014) Photosynthesis: response to high temperature stress. J Photochem Photobiol B Biol 137:116–126

Maulana F, Ayalew H, Anderson JD, Kumssa TT, Huang W, Ma XF (2018) Genome-wide association mapping of seedling heat tolerance in winter wheat. Front Plant Sci 9:1272

Meuwissen TH, Hayes BJ, Goddard M (2001) Prediction of total genetic value using genome-wide dense marker maps. Genetics 157(4):1819–1829

Mir RR, Zaman-Allah M, Sreenivasulu N, Trethowan R, Varshney RK (2012) Integrated genomics, physiology and breeding approaches for improving drought tolerance in crops. Theor Appl Genet 125:625–645

Misra S, Wu Y, Venkataraman G, Sopory SK, Tuteja N (2007) Heterotrimeric G-protein complex and G-protein-coupled receptor from a legume (*Pisum sativum*): role in salinity and heat stress and cross-talk with phospholipase C. Plant J 51(4):656–669. https://doi.org/10.1111/j.1365-313X.2007.03169.x

Mittler R (2016) ROS are good. Trends Plant Sci 22:11–19. https://doi.org/10.1016/j.tplants.2016.08.002

Mittler R, Vanderauwera S, Gollery M, Van Breusegem F (2004) Reactive oxygen gene network of plants. Trends Plant Sci 9:490–498. https://doi.org/10.1016/j.tplants.2004.08.009

Mittler R, Finka A, Goloubinoff P (2012) How do plants feel the heat? Trends Biochem Sci 37(3): 118–125. https://doi.org/10.1016/j.tibs.2011.11.007

Mohammadi VA, Zali AA, Bihamta MR (2008) Mapping QTLs for heat tolerance in wheat. J Agric Sci Technol 10(3):261–267

Mohammed YSA, Tahir ISA, Kamal NM, Eltayeb AE, Ali AM, Tsujimoto H (2014) Impact of wheat-Leymus racemosus added chromosomes on wheat adaptation and tolerance to heat stress. Breed Sci 63:450–460

Moon H, Lee B, Choi G, Shin D, Prasad DT, Lee O, Kwak SS, Kim DH, Nam J, Bahk J, Hong JC (2003) NDP kinase 2 interacts with two oxidative stress-activated MAPKs to regulate cellular redox state and enhances multiple stress tolerance in transgenic plants. Proc Natl Acad Sci 100(1):358–363. https://doi.org/10.1073/pnas.252641899

Morrow G, Tanguay RM (2012) Small heat shock protein expression and functions during development. Int J Biochem Cell Biol 44(10):1613–1621

Mujtaba SM, Faisal S, Khan MA, Shirazi MU, Khan MA (2018) Evaluation of drought tolerant wheat genotypes using morpho-physiological indices as screening tools. Pak J Bot 50(1):51–58

Mwadzingeni L, Shimelis H, Rees DJG, Tsilo TJ (2017) Genome-wide association analysis of agronomic traits in wheat under drought-stressed and non-stressed conditions. PLoS One 12(2): e0171692

Nakashima K, Takasaki H, Mizoi J, Shinozaki K, Yamaguchi-Shinozaki K (2012) NAC transcription factors in plant abiotic stress responses. Biochim Biophys Acta 1819(2):97–103

Nakashima K, Yamaguchi-Shinozaki K, Shinozaki K (2014) The transcriptional regulatory network in the drought response and its crosstalk in abiotic stress responses including drought, cold, and heat. Front Plant Sci 5:170

Nikolaeva MK, Maevskaya SN, Shugaev AG, Bukhov NG (2010) Effect of drought on chlorophyll content and antioxidant enzyme activities in leaves of three wheat cultivars varying in productivity. Russ J Plant Physiol 57(1):87–95

Nio S, Cawthray G, Wade L, Colmer T (2011) Pattern of solutes accumulated during leaf osmotic adjustment as related to duration of water deficit for wheat at the reproductive stage. Plant Physiol Biochem 49:1126–1137

Niu Y, Xiang Y (2018) An overview of biomembrane functions in plant responses to high-temperature stress. Front Plant Sci 9:915

Paliwal R, Röder MS, Kumar U, Srivastava JP, Joshi AK (2012) QTL mapping of terminal heat tolerance in hexaploid wheat (*T. aestivum L.*). Theor Appl Genet 125(3):561–575

Panaretou B, Zhai C (2008) The heat shock proteins: their roles as multi-component machines for protein folding. Fungal Biol Rev 22(3–4):110–119

Paul D (2013) Osmotic stress adaptations in rhizobacteria. J Basic Microbiol 53(2):101–110. https://doi.org/10.1002/jobm.201100288

Peng X, Zeng X, Ding X, Li S, Yu C, Zhu Y (2011) Ectopic expression of a vesicle trafficking gene, OsRab7, from Oryza sativa, confers tolerance to several abiotic stresses in Escherichia coli. Afr J Biotechnol 10(36):6940–6946

Pinto RS, Reynolds MP, Mathews KL, McIntyre CL, Olivares-Villegas JJ, Chapman SC (2010) Heat and drought adaptive QTL in a wheat population designed to minimize confounding agronomic effects. Theor Appl Genet 121:1001–1021

Pinto RS, Lopes MS, Collins NC, Reynolds MP (2016) Modelling and genetic dissection of staygreen under heat stress. Theor Appl Genet 129(11):2055–2074

Poorter H, Fiorani F, Pieruschka R, Wojciechowski T, van der Putten WH, Kleyer M, Schurr U, Postma J (2016) Pampered inside, pestered outside? Differences and similarities between plants growing in controlled conditions and in the feld. New Phytol 212(4):838–855. https://doi.org/10.1111/nph.14243

Pour-Aboughadareh A, Ahmadi J, Balouchi HR (2010) Screening wheat parents of mapping population for heat and drought tolerance, detection of wheat genetic variation. Int J Nucl Quantum Eng 4(1):63–73

Puttamadanayaka S, Balaramaiah M, Biradar S, Parmeshwarappa SV, Sinha N, Prasad SV, Prabhu KV (2020) Mapping genomic regions of moisture deficit stress tolerance using backcross inbred lines in wheat (*Triticum aestivum L.*). Sci Rep 10(1):1–17

Quarrie SA, Pekic Quarrie S, Radosevic R, Rancic D, Kaminska A, Barnes JD, Leverington M, Ceoloni C, Dodig D (2006) Dissecting a wheat QTL for yield present in a range of environments: from the QTL to candidate genes. J Exp Bot 57(11):2627–2637

Rabbi SM, Kumar A, Mohajeri Naraghi S, Sapkota S, Alamri MS, Elias EM, Mergoum M (2021) Identification of main-effect and environmental interaction QTL and their candidate genes for drought tolerance in a wheat RIL population between two elite spring cultivars. Front Genet 12: 941

Rahman M, Davies P, Bansal U, Pasam R, Hayden M, Trethowan R (2020) Marker-assisted recurrent selection improves the crown rot resistance of bread wheat. Mol Breed 40(3):1–14

Rai N, Bellundagi A, Kumar PK, Kalasapura Thimmappa R, Rani S, Sinha N, Prabhu KV (2018) Marker-assisted backcross breeding for improvement of drought tolerance in bread wheat (*Triticum aestivum L. emThell).* Plant Breed 137(4):514–526

Rane J, Pannu RK, Sohu VS, Saini RS, Mishra B, Shoran J, Crossa J, Vargas M, Joshi AK (2007) Performance of yield and stability of advanced wheat genotypes under heat stress environments of the Indo-Gangetic plains. Crop Sci 47:1561–1573

Ranjith P, Rao MS (2021) Breeding for drought resistance. In: Plant breeding-current and future views. IntechOpen, London

Reynolds MP, Borlaug NE (2006) Impacts of breeding on international collaborative wheat improvement. J Agric Sci 144(1). https://doi.org/10.1017/S0021859606005867

Reynolds MP, Balota M, Delgado MIB, Amani I, Fischer RA (1994) Physiological and morphological traits associated with spring wheat yield under hot, irrigated conditions. Funct Plant Biol 21(6):717–730

Reynolds MP, Pierre CS, Saad ASI, Vargas M, Condon AG (2007) Evaluating potential genetic gains in wheat associated with stress-adaptive trait expression in elite genetic resources under drought and heat stress. Crop Sci 47(S3):S172–S189

Ristic Z, Bukovnik U, Vara Prasad PV (2007) Correlation between heat stability of thylakoid membranes and loss of chlorophyll in winter wheat under heat stress. Crop Sci 47:2067–2073

Rushton DL, Tripathi P, Rabara RC, Lin J, Ringler P, Boken AK, Langum TJ, Smidt L, Boomsma DD, Emme NJ, Chen X, Finer JJ, Shen QJ, Rushton PJ (2012) WRKY transcription factors: key components in abscisic acid signalling. Plant Biotechnol J 10(1):2–11. https://doi.org/10.1111/j.1467-7652.2011.00634.x

Rutkoski J, Poland J, Mondal S, Autrique E, Pérez LG, Crossa J, Reynolds M, Singh R (2016) Canopy temperature and vegetation indices from high-throughput phenotyping improve accuracy of pedigree and genomic selection for grain yield in wheat. G3 (Bethesda) 6:2799–2808

Sadat S, Saeid KA, Bihamta MR, Torabi S, Salekdeh SGH, Ayeneh GAL (2013) Marker assisted selection for heat tolerance in bread wheat. World Appl Sci J 21(8):1181–1189

Sadia G, Abdul W, Muhammad A, Muhammad A (2008) Changes in growth and leaf water status of sugarcane (*Saccharum officinarum*) during heat stress and recovery. Int J Agric Biol 10(2): 191–195

Salem KFM, Roder MS, Borner (2007) A Identification and mapping quantitative trait loci for stem reserve mobilisation in wheat (*Triticum aestivum L.*). Cereal Res Commun 35:1367–1374

Sallam A, Eltaher S, Alqudah AM, Belamkar V, Baenziger PS (2022) Combined GWAS and QTL mapping revealed candidate genes and SNP network controlling recovery and tolerance traits associated with drought tolerance in seedling winter wheat. Genomics 114(3):110358

Samad A, Sajad M, Nazaruddin N, Fauzi IA, Murad A, Zainal Z, Ismail I (2017) MicroRNA and transcription factor: key players in plant regulatory network. Front Plant Sci 8:565. https://doi.org/10.3389/fpls.2017.00565

Sareen S, Tyagi BS, Tiwari V, Sharma I (2012) Response estimation of wheat synthetic lines to terminal heat stress using stress indices. J Agric Sci 4:97–104

Sharma A, Shahzad B, Kumar V, Kohli SK, Sidhu GPS, Bali AS, Handa N, Kapoor D, Bhardwaj R, Zheng B (2019) Phytohormones regulate accumulation of osmolytes under abiotic stress. Biomolecules 9(7):285

Shi L, Bai H, Zhu Y, Lv X, Li S, Hui J, Dong J (2021) Mapping and genetic overlapping of QTLs related to drought and heat tolerances in spring wheat

Shikha M, Kanika A, Rao AR, Mallikarjuna MG, Gupta HS, Nepolean T (2017) Genomic selection for drought tolerance using genome-wide SNPs in maize. Front Plant Sci 8:550–561. https://doi.org/10.3389/fpls.2017.00550

Shinwari ZK, Jan SA, Nakashima K, Yamaguchi-Shinozaki K (2020) Genetic engineering approaches to understanding drought tolerance in plants. Plant Biotechnol Rep 14:151–162. https://doi.org/10.1007/s11816-020-00598-6

Shiva S, Samarakoon T, Lowe KA, Roach C, Vu HS, Colter M, Porras H, Hwang C, Roth MR, Tamura P, Li M (2020) Leaf lipid alterations in response to heat stress of Arabidopsis thaliana. Plants 9(7):845

Shukla S, Singh K, Patil RV, Kadam S, Bharti S, Prasad P, Singh NK, Khanna-Chopra R (2015) Genomic regions associated with grain yield under drought stress in wheat (*Triticum aestivum L.*). Euphytica 203(2):449–467

Sinha N, Priyanka V, Ramya KT, Leena T, Bhat JA, Harikrishna PKV (2018) Assessment of marker-trait associations for drought and heat tolerance in bread wheat. Cereal Res Commun 46(4):639–649

Sivamani E, Bahieldin A, Wraith JM, Al-Niemi T, Dyer WE, Ho THD, Qu R (2000) Improved biomass productivity and water use efficiency under water deficit conditions in transgenic wheat constitutively expressing the barley *HVA1* gene. Plant Sci 155:1–9

Sukumaran S, Jarquin D, Crossa J, Reynolds M (2018) Genomic-enabled prediction accuracies increased by modeling genotype × environment interaction in durum wheat. Plant Genome 11(2):1–11. https://doi.org/10.3835/plantgenome2017.12.0112

Sun W, Van Montagu M, Verbruggen N (2002) Small heat shock proteins and stress tolerance in plants. Biochim Biophys Acta 1577(1):1–9

Sun J, Rutkoski JE, Poland JA, Crossa J, Jannink JL, Sorrells ME (2017) Multitrait, random regression, or simple repeatability model in high-throughput phenotyping data improve genomic prediction for wheat grain yield. Plant Genome 10(2):1–12. https://doi.org/10.3835/plantgenome2016.11.0111

Suzuki N, Koussevitzky SHAI, Mittler RON, Miller GAD (2012) ROS and redox signalling in the response of plants to abiotic stress. Plant Cell Environ 35(2):259–270

Tahara M, Carver BF, Johnson RC, Smith EL (1990) Relationship between relative water content during reproductive development and winter wheat grain yield. Euphytica 49(3):255–262

Tahmasebi S, Heidari B, Pakniyat H, McIntyre CL (2016) Mapping QTLs associated with agronomic and physiological traits under terminal drought and heat stress conditions in wheat (Triticum aestivum L.). Genome 60(1):26–45

Tattaris M, Reynolds MP, Chapman SC (2016) A direct comparison of remote sensing approaches for high-throughput phenotyping in plant breeding. Front Plant Sci 7:1131

Thomashow MF (1994) *Arabidopsis thaliana* as a model for studying mechanisms of plant cold tolerance. In: Meyerowitz EM, Somerville CR (eds) Arabidopsis. Cold Spring Harbor Laboratory Press, Cold Spring Harbor, pp 807–834

Todkar L, Harikrishna GP, Jain N, Singh PK, Prabhu KV (2020) Introgression of drought tolerance QTLs through marker assisted backcross breeding in wheat (Triticum aestivum L.). Indian J Genet 80(2):209–212

Toker C, Canci H, Yildirim T (2007) Evaluation of perennial wild Cicer species for drought resistance. Genet Resour Crop Evol 54(8):1781–1786

Tozzi ES, Easlon HM, Richards JH (2013) Interactive effects of water, light and heat stress on photosynthesis in Fremont cottonwood. Plant Cell Environ 36(8):1423–1434

Tripp J, Mishra SK, Scharf KD (2009) Functional dissection of the cytosolic chaperone network in tomato mesophyll protoplasts. Plant Cell Environ 32(2):123–133

Tuberosa R, Salvi S (2006) Genomics approaches to improve drought tolerance in crops. Trends Plant Sci 11:405–412

Tuteja N (2009) Signaling through G protein coupled receptors. Plant Signal Behav 4(10):942–947. https://doi.org/10.4161/psb.4.10.9530

Udvardi MK, Kakar K, Wandrey M, Montanari O, Murray J, Andriankaja A, Zhang JY, Benedito V, Hofer JM, Chueng F, Town CD (2007) Legume transcription factors: global regulators of plant development and response to the environment. Plant Physiol 144(2):538–549

Vendruscolo ECG, Schuster I, Pileggi M, Scapim CA, Molinari HBC, Marur CJ, Vieira LGE (2007) Stress-induced synthesis of proline confers tolerance to water deficit in transgenic wheat. J Plant Physiol 164:1367–1376

Wahid A, Gelani S, Ashraf M, Foolad MR (2007) Heat tolerance in plants: an overview. Environ Exp Bot 61(3):199–223

Wang N, Liu B, Liang X, Zhou Y, Song J, Yang J, Yong H, Weng J, Zhang D, Li M, Nair S (2019) Genome-wide association study and genomic prediction analyses of drought stress tolerance in China in a collection of off-PVP maize inbred lines. Mol Breed 39(8):113–132

Wheeler TR, Craufurd PQ, Ellis RH, Porter JR, Prasad PV (2000) Temperature variability and the yield of annual crops. Agric Ecosyst Environ 82(1-3):159–167

Wu X, Shiroto Y, Kishitani S, Ito Y, Toriyama K (2009) Enhanced heat and drought tolerance in transgenic rice seedlings overexpressing OsWRKY11 under the control of HSP101 promoter. Plant Cell Rep 28(1):21–30. https://doi.org/10.1007/s00299-008-0614-x

Xu C, Zhang H, Sun J, Guo Z, Zou C, Li WX, Xie C, Huang C, Xu R, Liao H, Wang J (2018) Genome-wide association study dissects yield components associated with low-phosphorus stress tolerance in maize. Theor Appl Genet 131(8):1699–1714

Yadav SK (2010) Cold stress tolerance mechanisms in plants. A review. Agron Sustain Dev 30: 515–527. https://doi.org/10.1051/agro/2009050

Yadav DK, Islam SM, Tuteja N (2012) Rice heterotrimeric G-protein gamma subunits (RGG1 and RGG2) are differentially regulated under abiotic stress. Plant Signal Behav 7(7):733–740. https://doi.org/10.4161/psb.20356

Yang S, Vanderbeld B, Wan J, Huang Y (2010) Narrowing down the targets: towards successful genetic engineering of drought-tolerant crops. Mol Plant 3:469–490

Yang Y, Chen J, Liu Q, Ben C, Todd CD, Shi J, Yang Y, Hu X (2012) Comparative proteomic analysis of the thermotolerant plant Portulaca oleracea acclimation to combined high temperature and humidity stress. J Proteome Res 11(7):3605–3623

Zaharieva M, Gaulin E, Havaux M, Acevedo E, Monneveux P (2001) Drought and heat responses in the wild wheat relative Aegilops geniculata Roth: potential interest for wheat improvement. Crop Sci 41:1321–1329

Zampieri M, Ceglar A, Dentener F, Toreti A (2017) Wheat yield loss attributable to heat waves, drought and water excess at the global, national and subnational scales. Environ Res Lett 12: 064008. https://doi.org/10.1088/1748-9326/aa723b

Zhang N, Carlucci P, Nguyen J, Hayes-Jackson J W, Tonsor S (2016) Contrasting avoidance-tolerance in heat stress response from thermally contrasting climates in Arabidopsis thaliana. *bioRxiv* 044461

Zhao Y, Mette MF, Gowda M et al (2014) Bridging the gap between marker-assisted and genomic selection of heading time and plant height in hybrid wheat. Heredity 112:638–645. https://doi.org/10.1038/hdy.2014.1

Źróbek-Sokolnik A (2012) Temperature stress and responses of plants. In: Ahmad P, Prasad MNV (eds) Environmental adaptations and stress tolerance of plants in the era of climate change. Springer, New York, pp 113–134

# Chapter 4
# Developing Crop Varieties by Physiological Breeding for Improving Plant Nutrition

**Sukumar Taria, Shashi Meena, Shivani Nagar, Sudhir Kumar, and Ajay Arora**

**Abstract** Living organism acquires and utilizes the nutrients from soil through a process called as nutrition for growth and development. These nutrients may be acquired from various sources. Plants and cyanobacteria harness the sunlight energy, which is converted to chemical energy by a process called as photosynthesis. Plants are the primary producers and are the main source of nutrients for most living organisms. Animals, including humans, require many nutrients for their vital functions and proper growth and developmental process that are only synthesized by plants. Therefore, to fulfill the global demand, there will be a need of adapting and improving crop varieties according to the changing climatic conditions; hence, it is an "endless talk." For millions of years, humans have selected desirable traits of crop plants, and presently the major cereal crops are domesticated from wild relatives which are the main source for human and animal nutrition. Thus, several tactics such as conventional and molecular breeding, transgenic approaches, and various agronomical practices provide a new area for the development of nutrient supplemented crops, hence ensuring food security. This chapter mainly focuses on methods currently used for improving nutritional quality of plants and how they are being subjugated to increase the level of particular nutrients in desirable plant species to improve human health.

**Keywords** Plant nutrition · Molecular breeding · Physiological breeding · Genome editing · Nutrient use efficiency · Biofortification

---

S. Taria · S. Meena · S. Nagar · S. Kumar · A. Arora (✉)
Division of Plant Physiology, ICAR-Indian Agricultural Research Institute (ICAR-IARI), New Delhi, India
e-mail: ajayarora@iari.res.in



H. M. Mamrutha et al. (eds.), *Translating Physiological Tools to Augment Crop Breeding*, https://doi.org/10.1007/978-981-19-7498-4_4

## 4.1 Introduction

The human population has currently increasing within this last hundreds of year. Thus, major priority of developing country is to produce nutrient-rich food for healthy nutrition. Thus, need of hour is to design the technologies for desired characteristics of plants and breed diversified species, hence producing nutrient efficient crop varieties having good yield potential under this global climate change scenario. Plant breeders and agriculturists around the world will have to do regular research to enhance the dimension of breeding programs to increase crop yield potentials (Tester and Langridge 2010). In addition to this, "hidden hunger" or "micronutrient deficiency" in developing countries is another challenge to agricultural research. Though there was significant poverty alleviation and reduction in hunger through the Green Revolution, some issues related to nutrition are still unsolved (Pingali 2012). Nearly 815 million people in the world are facing problem of malnutrition and become malnourished due to deficiency of essential mineral nutrients required for growth, development, and proper function of the body (UNICEF and World Health Organization 2017). Therefore, biofortification approaches are evolving as a potential plans to confront micronutrient deficiencies (Garg et al. 2018). Biofortification is a long-term, cost effective strategy that can overcome nutrient deficiency more sustainably, thus helping in combat hidden hunger. However, this approach has no additional costs for further processing (Garg et al. 2018). This approach is mainly focusing on elevating vitamins and minerals levels in crops through various technological interventions. We are able to make nutrient-rich food through intervening this approach and also reduce the cost of external modifications required to make nutrient-rich crops. Various approaches such as traditional/conventional plant breeding, molecular breeding, transgenic techniques, genomic, agronomical practices, and physiological breeding provide an emerging way for the development of nutrient-rich crops. However, the crop species diversity exploited to produce nutritionally rich crops has been restricted. Biofortification is mainly focused on important cereal crops where various strategies have been used in current years to upsurge the nutrient content for better nutritional security. To ensure food security, there is an urgent demand for another Green Revolution and that should be focused on plant breeding program. "Orphan" or underutilized crops such as finger millet, teff, yams, and various food legumes should be the part of breeding programs to ensure nutritional security for global population approaching 9 billion in 2050 because, they are rich in nutritional values and abiotic stress tolerance and have the potential to improve human health in a positive way (Myers et al. 2015). The development of the technologies like traditional/conventional plant breeding, molecular breeding, physiological breeding, transgenic techniques, genome editing, agronomical practices and genomic approaches offers the possibility of improving the desired characteristics of crop plants.

### 4.1.1 *Traditional/Conventional Plant Breeding*

To augment nutritional value of plants, this approach has been the main strategy. This strategy is based on the knowledge obtained through the study of the inheritance patterns of any character and existing genetic variability. Appropriate breeding strategies select individuals or families based on phenotype. In this method, crossing is carried out between individuals of the same species or closely related species, and subsequently hybrids are selected to finally obtain a generation carrying the desirable trait. Traditional breeding has been extremely successful in generating new varieties adapted to different biotic and abiotic stresses. Despite their success, this method does have certain drawbacks. Moreover, desirable traits are mainly present in landraces or other closely related species, therefore to introduce the trait of interest, need more selection to remove undesirable traits or to reduce the linkage drag in the noncommercial parent. Additional limitation is the effect of the changing climate on the trait of interest. Therefore, breeders must differentiate between the real genetic potential of the plant and the impact of the environment on the plant. Hence, it is somewhat difficult to cross inter-species, since they produce few viable seeds from which selection is to be made for next generation. Thus, these drawbacks are nowadays being overcome by the increasing use of molecular markers associated with traits of interest. Selection can be made based on molecular markers which eliminate the problem of environmental effects and allow selection on a genome-wide basis. Therefore, selection for the elimination of undesirable traits is more effective through molecular markers approaches.

#### 4.1.1.1 Molecular Breeding

This involves the use of molecular markers for the selection of complex traits where multiple genes are interacted to give particular traits of interest, thus eliminating the problem of environmental effects. The crop improvement programs have been completely revolutionized by next-generation sequencing technologies, which empowered us to understand the genetic background behind complex traits. Earlier there was less genomic information available for almost all other legumes except soybean and were termed as "orphans" in the genomic era (Varshney et al. 2009). However, a constant and joined research efforts from different international research organizations within last decade, now we have genomic resources available for several grain legumes (Varshney et al. 2018). In addition to the genome sequences, other omics approaches such as metabolomics, proteomics, and ionomics will help us in the functional characterization of certain traits and accelerate the program of crop improvement. Breeders used the available genomic information and already mapped traits like drought tolerance and disease resistance in chickpea, rust resistance, and oil quality in groundnut, and screened pigeon pea for hybrid purity (Varshney 2016). The availability of genomic resources facilitates MAS, and also contributed to advancement of new-generation technologies such as genome-wide association studies (GWAS), genotype by sequencing (GBS), genomic selection (GS), and genome editing by CRISPR-cas9 in plants and helps plant breeding to reach to the next generation for selection (Langridge and Fleury 2011).

Though marker-based approaches/MAS selection have incredible prospective in accelerating plant breeding, environmental effects on genotypes and interactions (G × E) can make them unsuccessful as crops undergo significant changes in gene expression patterns during different morpho-physiological stages and stress-related changes (Xu and Crouch 2008). It has been investigated that the physiological traits are influenced by climatic changes and gene expression pattern, which makes their uses in breeding program a little complex. Marker-assisted selection (MAS) is mostly successful only when the transferability of the concerned trait to the next generation is high. Among the large numbers of QTLs identified so far, only few QTLs have been cloned and analyzed for their functions. Photosynthetic traits (such as photosynthetic rate, transpiration rate, chlorophyll content, stomatal resistance, and so on) can also be used to identify QTLs (Graham et al. 2008). Due to the availability of genomic resources for food legumes, they can be treated as model crops. They can also enable molecular breeding of these crops and can play a central role in producing climate-resilient and nutritionally superior crops (Cannon et al. 2009). Finally, molecular breeding approaches not only just complement conventional breeding but also provide important genes, metabolites, markers, and screening tools to drive improvement of crops (Varshney et al. 2018).

#### 4.1.1.2 Physiological Breeding

In the past, plant breeding involves multiple crosses for generating variations in the progenies and screening them for higher yield potentials. From a breeder's view, a physiological parameter correlates with major economic traits that are wanted to be combined into their breeding programs. Besides this, they should be a heritable, stable performance across multiple environments. However, high variability exits in most of the physiological traits and needs high-throughput screening techniques to select them to identify suitable traits. To exploit the benefits of physiology in breeding, phenomics tool for real-time phenotyping can play an important and valuable role in achieving this.

One of the major limitations of integrating plant physiological studies with breeding programs is a laborious and time-sensitive measurement of certain physiological traits such as photosynthetic traits. Therefore, a limited number of genotypes are often used in physiological studies for better output. Therefore, it is utmost necessary to improve various experimental designs and correct sampling for the successful mapping of particular traits for an association mapping panel (Shunmugam et al. 2018). The conventional breeding programs proved to be a successful and profitable solution to improve micronutrient content in the long term, but its application is limited to genetic variability that exists for the micronutrients in the plant gene pool and the time needed to produce cultivars with the improved traits. However, in certain cases, this can be done by crossing distantly related species and thus introgressing desired traits into cultivars will be useful for crop improvement. Most often, it would be difficult to cross for specific trait using traditional methods due to its various limitations such as sexual incompatibility, time requirement and efforts may be quite unrealistic such as improving wheat grains with Se content (Lyons et al. 2005) and soybean seeds with different fatty acid contents like oleic, linoleic, and linolenic acid (Oliva et al. 2006). Hence, the transgenic-based approach provides a better platform for oil quality improvement in soybean due to limited variability, heritability, and linkage drag.

### *4.1.2 Transgenic Approach*

Transgenic approaches refer to advanced techniques capable of transferring genes from unrelated organisms or species directly to recipient organisms to improve the trait of interest. It can also be used for the concurrent incorporation of genes from any other organism into plants, thus resulting in the enhancement of nutrient concentration, their bioavailability, and finally reduction in the concentration of anti-nutrients which limit the bioavailability of nutrients in plants. This method can also be used to redistribute micronutrients between tissues. In addition, genetic modifications can be targeted for redistribution of micronutrients between tissues, to enhance the concentration of micronutrients, to increase the efficiency of metabolic pathways in edible tissues, or even reconstruction of selected pathways (Shewmaker et al. 1999; Agrawal et al. 2005). The development of biofortified crops through a transgenic method can be a valid alternative when there is a minimum or no genetic variation in nutrient content among plant varieties (Zhu et al. 2007; Brinch-Pedersen et al. 2007). This approach mainly relies on genetic pool for the transfer of traits and desirable gene expression from one plant species to another which is independent of their evolutionary and taxonomic status. Additionally, if the specific micronutrient does not naturally present in crops, the only possible choice to fortify these crops with the particular nutrient is transgenic approaches (Pérez-Massot et al. 2013). The key for the development of transgenic crops is to identify and characterize the function of genes and then utilize these genes to engineer plant metabolism for traits of interest (Christou and Twyman 2004). Likewise, with the help of metabolic engineering, it is now possible to introduce the alternative pathways that existed in microorganisms like bacteria into crops (Newell-McGloughlin 2008). This process leads to the development of transgenically biofortified crops which require a sufficient amount of time, effort, and investment during the research and development stage, but in a real sense, it is a cost-effective and viable approach (Hefferon 2016). Furthermore, the transgenic approach has no taxonomic constraints, where synthetic genes can be constructed and used. Transgenically biofortified crops hold a great potential to reduce malnutrition among its consumers, especially poor people in growing countries (Tucker 2003). Various crops have been genetically modified to augment their micronutrient contents, vitamins, minerals, etc. by introgressing various genes from various sources to enhance the nutritional value of the food crops. It has been found that genes encode for minerals, albumin for essential amino acids, and essential fatty acids have been widely reported as targets for biofortification. Several reports are available for nutrient-rich crops produced by this method such as high lysine content maize, high unsaturated fatty acid soybean, high provitamin A and iron-rich cassava, and high provitamin A Golden rice (Garg et al. 2018).

#### 4.1.2.1 Genome Editing

There is an urgent need of quality germplasm with traits variability to develop elite cultivars to resist biotic and abiotic stresses, which is a major limitation in conventional plant breeding. The genome editing technologies facilitate the targeted modifications at the genomic loci in an efficient and precise manner (Gaikwad et al. 2020).

TALENs Christian et al. (2010), the first-generation genome editing technology, have been used for modifying and enhancing the nutritional value of major food crops. The research has been carried out and found that soybean oil contains high levels of polyunsaturated fatty acids, i.e., linoleic and linolenic acid, which results in oxidative instability. Hence, this method leads to increases in the levels of *trans*-fatty acids, which is linked to cardiovascular disease. Changing the composition of soybean oil by declining the levels of linoleic and linolenic acids as well as by increasing the level of oleic acid, help to reduce the need for hydrogenation. Soybean mutant lines with high oleic acid and low linoleic acid contents were produced by commencing mutation by delivering TALENs in the fatty acid desaturase 2 genes (*FAD2-1A* and *FAD2-1B*) (Haun et al. 2014) and fatty acid desaturase 3A (*FAD3A*) gene. The insertion of TALENs directly into *fad2-1a fad2-1b* soybean plants and *FAD2*-2 gene using CRISPR/Cas9 genome editing technology (Al Amin et al. 2019), thus improving the quality of the soybean oil in terms of shelf-life and heat stability. The same research has been carried out in peanuts in which gene *ahFAD2* was mutated using CRISPR/Cas9 (Yuan et al. 2019).

Amylose and resistant starch play a substantial role in the improvement of human health and thus reduce the risk of diseases including high blood pressure, diabetes, and colon cancer (Chen et al. 2012). It has been shown that the CRISPR/Cas9 is used to develop rice rich in high amylose content by targeting starch branching enzymes, *SBEI* and *SBEIIb* (Sun et al. 2017). This technology was also used to develop rice lines with high oleic acid and low linoleic acid rice bran oil (RBO) by disrupting the *OsFAD2-1* gene as this gene encodes for the enzyme fatty acid desaturase 2 (FAD2) that converts the oleic acid into linoleic acid in plants (Abe et al. 2018). The important target of biofortification is to enhance the level of β-carotene in major food crops since it is a precursor of pro-vitamin A. The β-carotene was successfully accumulated in rice callus as observed by Endo et al. (2019). This can be due to the identification and modification of the putative ortholog of the cauliflower gene, *Osor*, via CRISPR/Cas9 genome editing technology. Similarly, red grain coloration of wild rice is controlled by two complementary genes, *Rc* and *Rd* (Furukawa et al. 2007). Worldwide, aromatic rice is prevalent for its characteristic grain fragrance. In excess of 100 volatile compounds were responsible for the flavor in cooked grains. The main compound responsible for fragrance in aromatic rice is 2-acetyl-1-pyrroline. This fragrance compound is produced by defective *badh2* allele which encodes BADH-2 results in the synthesis of 2-acetyl-1-pyrroline. Shan et al. (2015) reported that aromatic rice can be produced through the targeted knockout of *OsBADH2* in non-fragrant variety using the TALEN method.

In 2016, DuPont Pioneer developed the first corn waxy hybrid using CRISPR/Cas9, silencing the maize waxy gene *Wx1*, thus eliminating amylose and making more than 97% amylopectin from the kernel (Van Steenhuyse 2017). Such varieties with higher amylopectin starch content have greater digestibility and have many usages in industries. The commercial hybrids for release with having all these traits are planned for 2020 (Van Steenhuyse 2017). Similarly, scientists from Swedish Agricultural University researched on developing high-amylopectin potatoes by knocking out the gene encodes for GBSS via CRISPR/Cas9 (Andersson et al. 2017). Cold storage induces the increase of reducing sugars which have a greater effect on the quality of the potato tubers. During processing, reducing sugars react with free amino acids, resulting in the accumulation of acrylamide, a potent carcinogen which ultimately results in brown and bitter-tasting products, which is undesirable. Hence, a TALENs technique is utilized to reduce the deposition of reducing sugars by silencing the vacuolar invertase (VInv) gene (Clasen et al. 2016). Conventional breeding and mutagenesis approaches failed to develop wheat varieties with low-immunogenic properties because of the complexity of the Gli-2 locus and the high copy number of alpha-gliadin gene. CRISPR/Cas9 technology has been used to develop low gluten, transgene-free wheat lines for patients with celiac disease (Sánchez-León et al. 2018). The CRISPR/Cas9 genome editing system is used to produce mustard lines with high oleic acid (more than 80%) (Okuzaki et al. 2018; Huang et al. 2020) by modifying *FAD2* gene which encodes for fatty acid desaturase-2.

Genome editing technologies become popular genomic tools for producing biofortified crops due to its high efficiency, specificity, and amenability to multiplexing. This technology resulted in the production of mutant lines which will provide variability in the traits important for breeding programs in an unprecedented way. The major drawback of this technology is its lower acceptance and the more time requirement for the development of stable transgenic for better performance in field (Carciofi et al. 2012). Another hindrance in its wide acceptance is the different processes for the approval and commercialization of these transgenic crops adopted by different countries, and moreover, these processes are very costly and time-consuming (Watanabe et al. 2005). For example, after 8 years of research work, the scientific details of the Golden rice were first published in 2000 in *Science* (Ye et al. 2000). However, it is still not ready for commercialization for farmers' use due to issues with its yield. Its commercialization is also being detained due to stringent regulatory processes, hence does not get approval from governments.

### 4.1.3 Agronomic Approach

The nutritionally rich crops can be produced by agronomic methods that require the temporary application of nutrients to improve the nutritional status of crops (Cakmak and Kutman 2018). It has been reported that the organic forms of minerals are more available for a man as compared to the inorganic forms due to its easy absorption in body and less excreted (Daniels 1996). This method simply involves the application of mineral fertilizers and consequently to enhance in their solubilization for its movement from the soil to the edible plant parts. It has been investigated that macronutrients for instance nitrogen, phosphorus, and potassium (NPK) play a major role in achieving higher yields (Jan et al. 2008). In the late 1960s, the application of fertilizers containing NPK increases agricultural productivity in several countries of the world which resulted in Green Revolution and saved the world from starvation. Micronutrients such as Zn, Fe, I, Se, Cu, Mn, Co, Mo, and Ni are present in varying concentrations in the different parts of certain plants and are mainly absorbed from the soil. Foliar application of fertilizers can improve micronutrient status in soil which ultimately reduces the human micronutrient deficiency (Cakmak 2008). It has been noted that application of soluble inorganic fertilizers to roots, leaves, and other plant parts can make the particular micronutrient available to the edible tissues. Improving the nutritional quality (biofortification) through agronomic approaches is an easy and low-cost process, but it needs special attention in terms of source of nutrients, method of application, and its effect on environment. For proper effect, these should be applied frequently in every growing season. Some of the developed countries such as Finland (Aro et al. 1995), turkey (Cakmak et al. 1999) and China (Xin-Min et al. 1997) have practiced the use of mineral fertilizers, i.e., Se, Zinc, and Iodine, respectively, which represent that it is a feasible and less cost input process. Besides the fertilizers, plant growth promoting soil microorganisms like different species of genera *Bacillus*, *Pseudomonas*, *Rhizobium*, *Azotobacter*, etc. can be used to improve their nutritional status as they enhance the mobilization of nutrients from the soil to the edible plant parts. $N_2$-fixing bacteria play major role in enhancing the productivity of the crop under nitrogen-limited conditions (Hardarson and Broughton 2003). Mycorrhizal fungi are associated with many crops that play beneficial role in the degradation of organic compounds, hence increasing the mineral concentration in edible parts by secreting organic acids, siderophores, and enzymes in the rhizosphere (Rengel et al. 1999; Cavagnaro 2008). Agronomic biofortification has been done for many crops, thus improving the human nutritional status. The iron and zinc deficiency in rice grain can be reduced by agronomic biofortification, thus acting as an alternative strategy for the improvement of the nutritional value of the crop. Foliar application of iron was an operative way to increase rice grain Fe concentration (Fang et al. 2008; Yuan et al. 2013). Similarly, ferrous sulfate can be used to fortify the germinating brown rice plantlets with high iron content (Yuan et al. 2013). It has been reported that foliar application of zinc testified as anominal agronomic intervention to enhance Zn content of grain and its bioavailability (Ram et al. 2016). Besides foliar application of Zn, soil application fertilizer proves to be an important strategy to enhance zinc content in rice grains under low zinc levels in soil (Guo et al. 2016). Selenium, also known as a trace element and verified to be a strong antioxidant, has also been enhanced by foliar application or as fertilizer in rice (Fang et al. 2008). This method of fortification has been very effectively and economically utilized in the improvement of wheat grain quality. It has been investigated that in turkey (Cakmak 2008), consumption of zinc-enriched wheat grains improves human nutritional status. Zinc contents in grains can be enhanced by the use of exogenous applications of zinc-containing fertilizers. Similarly, biofortification through agronomical approaches of wheat grains with Se have been done in Finland (Aro et al. 1995). Apart from fertilizers (chemical and organic), the role of biofertilizers in promoting the grain yield in wheat has been investigated by several researchers. For example, iron-biofortified wheat grains have been developed through integration of organic and chemical fertilizer. Microorganisms like *Bacillus aryabhattai* (Ramesh et al. 2014; Ramzani et al. 2016) have been used for increasing the zinc content in grains. Similar results have been observed in the maize crop (Fahad et al. 2015). Furthermore, zinc and iron concentration in barley grains have been improved by the application of biofertilizers along with inorganic fertilizers and vermicompost (Maleki et al. 2011). Researchers are intended to enhance the nutritional quality of the sorghum through agronomic biofortification method which includes the use of a combination of plant growth-promoting bacteria and arbuscular mycorrhizal fungi (AMF) (Dhawi et al. 2016). Moreover, $N_2$ fixing bacteria alone and in combination with phosphate-solubilizing bacteria (PSB) improved grain yield and protein content in sorghum by improving the mineral nutrition status of the soil (Patidar and Mali 2004). Selenium-enriched soybean, chickpea and potato (Poggi et al. 2000), and mustard (Yasin et al. 2015) have been produced by the foliar application of a particular mineral. Biofortification of chickpea for iron and zinc has been concentrated by using AMF (Pellegrino and Bedini 2014). The same results have been observed in the field peas (Poblaciones and Rengel 2016) and common beans (Ram et al. 2016) where foliar applications of zinc alone or in combination with soil lead to the high zinc content. Furthermore, it has been researched out that in common bean, the use of both organic and chemical fertilizers stimulated the uptake process of some of the mineral nutrients, i.e., N, P, K, Cu, Mn, and Zn (Westermann et al. 2011). Agronomic biofortification in canola resulted in increased protein, oleic acid, and linoleic acid content in the seed, thus proving highly effective in improving the yield and nutritional value of canola oil (Nosheen et al. 2011). Potato fortified with zinc by using the foliar spray of zinc fertilizer has been proved an effective way of enhancing tuber zinc concentration. But by practicing various fertilizers, zinc nitrate proved to be more effective as compared to zinc oxide and zinc sulfate as foliar fertilizers for increasing zinc concentrations (White et al. 2017). Foliar spray of selenium along with humic acid was an effective way to increase the selenium content of potatoes (Poggi et al. 2000). Biofortification of sweet potato with β-carotene has been demonstrated with irrigation and chemical fertilizer applications (Laurie et al. 2012). There is high variability in the success rate of agronomic biofortification due to the mobility's difference of minerals, accumulation among plant species, soil composition in the particular geographic location of each crop. This method is cost-ineffective and labor-intensive as it requires continuous inputs, in the form of application of micronutrients to the soil or plant regularly. Another limitation of this method is accumulation of desired specific nutrients in the leaves and non-edible portions of plants; therefore, this technique is applicable to certain minerals and specific plant species. Furthermore, the bioavailability of minerals is deterred by the presence of antinutrient compounds like phytic acid in the soil, which is another major challenge (Frossard et al. 2000). Besides, regular addition of fertilizers in soil and water poses adverse effects on the environment (Waters and Sankaran 2011).

### 4.1.4 Genomic Approach

For the development of high-nutrient cultivars, there is a need for appropriate breeding strategy, genetic diversity availability, and modern genomics information. Over the years, sufficient genetic variability for nutritional quality traits has already been explored in cultivars, direct progenitors, and wild relatives. Moreover, a vast genetic diversity across the kingdoms can be explored and utilized to produce nutritionally rich crops/cultivars through genetic engineering and genome editing. To understand the genetic basis of nutritional quality traits, an important requirement is knowledge about the genetic pool and their interaction with the environment; hence, it is of the utmost importance for an efficient breeding program. The combination of genomics with crop physiology together with precise high-throughput phenotyping and advanced breeding methodologies has effectively revealed the genes and the metabolic pathways of quality traits. The identification and the amalgamation of nutritional quality traits in the cultivars are being enabled by several approaches, each having its pros and cons. One of the approaches is bi-parental QTL mapping that involves the identification of trait of interest loci which can be facilitated with QTL mapping. In the bi-parental QTL mapping approach, two parents which differ for the target trait are required to cross and to produce a mapping population of the mortal (F2, F3, etc.) or immortal (RILs, NILs, etc.) in nature. QTLs can also be identified with natural populations by using association mapping (AM), viz. genome-wide AM, candidate-gene-based AM, and bulk segregant analysis (BSA). The accuracy of the identified QTLs depends on the nearness or closeness of the marker with the linked locus/QTLs. The QTLs identified with functional markers/genic markers are more consistence because these markers are synthesized from the transcribed region and/or part of the identified locus. Once genomic regions/QTLs associated with nutritional traits are identified, can utilize for the development of nutritionally rich cultivars. MAS is suggested for those traits which are controlled by a few loci. However, if the traits are transferred from landraces or distant relatives to the next generation, marker-assisted backcross breeding (MABB) would be suitable for the trait transfer. Furthermore, if the nutritional trait is controlled by a large number of genes with lesser genetic variation, then genomic selection (GS) would be the best way to transfer the target of interest

(Prasanna et al. 2020). Some of the elite maize hybrids (e.g., PusaVivek QPM-9 Improved, PusaVivek Hybrid-27, Pusa HQPM5 Improved, and "Pusa HQPM7" Improved) rich in nutritional traits such as QPM and pro-vitamin A have been developed by MAS and released for commercial cultivation from ICAR-IARI, Delhi, India namely hybrid "PusaVivek QPM-9 Improved" (Guzman et al. 2016; Prasanna et al. 2020).

Similarly, MAS has also been used for the introgression of major gene *Gpc-B1* (high protein content) (Kumar et al. 2011; Tyagi et al. 2014; Vishwakarma et al. 2016) in durum wheat and developed a cultivar named "Desert King-High Protein" which had been developed at the University of California wheat breeding program (Gaikwad et al. 2020). However, the same method has also been used in the development of nutritionally rich cultivars in other food crops with little success rate. The reason for less success rate is that most of these traits are controlled by multiple genes having little effect individually in the trait expression. Therefore, genomic selection (GS) is preferable where complex quantitative traits are governed numerous minor QTLs (Lau and Latif 2019). In addition to the genes or major QTLs responsible for nutritional qualities, large numbers of minor QTLs are there that have effects on the trait expression. Recently, the work has been carried out in rice that the betaine aldehyde dehydrogenase (BADH 2) gene solely is not responsible for the variation of aroma in rice because background minor QTLs contribute to the overall fragrance in rice (Lau and Latif 2019). Therefore, amalgamated effects of minor QTLs on the nutritional traits can be effectively realized by employing the recurrent selection under breeding cycles under genomic selection.

## 4.2 Mineral Nutrition Under Climate Change Conditions

Cultivated land for farmer was not changed considerably for last 25 years globally (O'Mara 2012), but crop yield had increased substantially with time. This is due to the modern agricultural practices and improved breeding approaches for trait of interest to improve productivity. Arias et al. (2021) reported that global surface temperature was 1.09 °C more in 2011–2020 than in 1850–1859, while the $CO_2$ concentration of 410 ppm in 2019 has been reported. This combined action of increasing atmospheric $CO_2$ and temperature has multiple effects on plant growth and productivity and so for mineral nutrient uptake capacity of plants. Again, crop yield and nutritional quality are dependent upon climatic conditions and therefore food production for human consumption is under threat in this climate change scenario.

Increasing atmospheric $CO_2$ concentration and temperature can affect (a) availability of nutrients to plant root, (b) uptake of the mineral nutrients by plant, (c) translocation of the mineral nutrients from plant root to the grains, (d) accumulation of carbohydrate in the amyloplast and in vacuole (Ebi et al. 2021). Rising level of atmospheric $CO_2$ enhances the photosynthetic rate (through carbon fixation), leading to greater release of the carbon to the soil affecting soil microbial properties and pH. This action affects the partitioning or subdividing of the nutrients between the soil particle and soil water. Higher $CO_2$ fixation enhances WUE, thus affecting the nutrient availability and causes alternation in the C/N balance. Higher $CO_2$ decreased the transpiration though reduced stomatal conductance, while increase in temperature enhanced the transpiration, affecting nutrient mass flow to the root surface (Pilbeam 2015). Moreover, reaction kinetic upsurge with temperature and therefore, rate of the reaction that mobilize the nutrient from soil water will increase (Farhat et al. 2021).

Most of the experiment measured the grain quality with respect to elevated $CO_2$ level (Ebi et al. 2021). Study showed that increase in temperature can worsen the grain quality, while other studies claimed that it can amend the negative effect of enhanced $CO_2$ on grain quality (Köhler et al. 2019; Wang et al. 2019a, 2020). There may be multiple effect of combined rise of atmospheric $CO_2$ with rising temperature, affecting the crop yield and productivity including the nutritional qualities.

### *4.2.1 Rise in $CO_2$ Level and Mineral Nutrition*

It is well known that increase in atmospheric $CO_2$ levels enhances the plant growth through carbon assimilation, a key process in light-driven photosynthesis. This in turn is associated with decrease in grain nutritional quality (Dong et al. 2018) by decreasing the protein concentration. This proposed that elevated $CO_2$ changes the equilibrium between the carbon assimilation and nutrient uptake (Nakandalage and Seneweera 2018). Under elevated $CO_2$, decreased stomatal conductance has been reported (Long et al. 2004), resulted in decline in transpiration rate as well. It is also reported that increased $CO_2$ concentration inside the leaves causes reduced stomatal aperture (Zhang et al. 2018b). This will ultimately affect availability of the nutrients, which acquire through the mass flow movement of water such as available soluble nitrogen.

Increase in $CO_2$ levels also caused decline in other mineral nutrients such as Ca concentration in edible parts of food crops, while additionally decrease in Mg, P, and K contents (Taub and Wang 2008) have been reported. Decrease in Co, Fe, Mn, Ni, and Zn were also being testified under this condition (McGrath and Lobell 2013). Predicted future $CO_2$ enhancement and temperature predicted that there could be possibly 9–14% drop in proteins as well as decrease in Zn and Fe levels in grain by 2050 for Indo-Gangetic plains of India (Patra 2010). Despite decline in Fe and Zn concentration in grain under elevated levels of $CO_2$, genotypic differences within the species is observed with respect to Zn and Fe. So it is possible to breed against this risk (Myers et al. 2014). Again it is reported that, elevated $CO_2$ also imposed inhibition of nitrate photo assimilation in wheat, thus influencing mineral nutrient levels in plant (Bloom et al. 2002). These effects seem to be due to the more use of the reducing equivalent (NADPH and reduced ferredoxin) for carbon fixation, and thus unavailability of the same imposes limitation for nitrogen assimilation in chloroplast (Baysdorfer and Robinson 1985; Peirson and Elliott 1988), decreasing the grain quality.

### 4.2.2 Effect of Elevated $CO_2$ on Arbuscular Mycorrhiza (AM)

It is well known that arbuscular mycorrhiza is involved in the uptake of mineral nutrients. The association of these fungi in the roots helps in the uptake of the phosphorous from low P soil as well as help in the uptake of $Zn^{2+}$ and $NH_4^+$. It is reported that formation of AM associations is inhibited by the presence of high phosphate concentrations in soil is induced by the warmer temperature (Graham et al. 1982). Increased $CO_2$ concentration enhanced the growth of the AM fungi (Drigo et al. 2010). It is reported that mycorrhizal tomato plant had higher biomass as compared to the non-mycorrhizal plants (Cavagnaro et al. 2007). Enhanced $CO_2$ concentration stimulates the plant growth, which in turn increases the demand of available nitrogen. This promotes the activity of AM fungi, which transfer ammonium to the plants (Cheng et al. 2012). However, results showed that though there is a significant growth under elevated $CO_2$ level in leguminous plant, growth is still limited by N availability through nodule N fixation. This is due to the acclimation of plants to enhanced $CO_2$ level called as $CO_2$ acclimation with time. Thus, supply of the nitrogen as $NH_4NO_3$ overcame this acclimation response (Sanz-Sáez et al. 2010).

### 4.2.3 High Temperature and Mineral Nutrition

Decline in nutrient uptake under elevated temperature might be due the reduction in rhizosphere or nutrient acquisition per unit root (Bassirirad 2000; Giri et al. 2017). This may be due to the decrease in carbon labile pool. This decrease in overall carbohydrates in the roots might occur due to lesser partitioning of photosynthates from the shoot toward the root. Moreover, this reduction in nutrient uptake may be due to increase root respiration causing decline in available carbohydrate for nutrient uptake. It is reported that direct root damage may be a factor contribute to the decline rate of nutrient absorption elevated high temperature (Huang et al. 2012). This may affect the nutrient acquisition proteins involved in uptake of mineral nutrients.

Warmer condition also reduces the precipitation and evapotranspiration, which ultimately affect the nutrient availability by affecting the mass flow. However, warmer condition enhanced diffusion of nutrients compared to cooler conditions, causing movement of the ion quicker to the roots which will compensate for the loss of bulk flow movement of water (Brouder and Volenec 2008).

### 4.2.4 Interactive Effect of Atmospheric $CO_2$ Rise and Temperature on Plant Growth

It is reported that present atmospheric $CO_2$ concentration is approaching to Km value of Rubisco. Thus, enhanced $CO_2$ level should increase the carboxylation rate in plants (Long et al. 2004). Reports stated that if nitrogen (N) concentration is not limiting, enhanced $CO_2$ level increased the root growth as compared to shoot growth (De Graaff et al. 2006). In contrast, warming temperature alone (without $CO_2$ rise) enhanced the shoot growth in comparison to the root growth by enhancing the soil microbial activities which make the nutrients available to plants (Dieleman et al. 2012). However, under combined action of higher $CO_2$ level and warmer temperature, there was still enhanced root growth, but not to the same magnitude to the level by enhanced $CO_2$ alone (Dieleman et al. 2012). Enhanced $CO_2$ level also likely to decline nitrogen concentration in leaves by decreasing the movement of soluble nitrate ($NO_3^-$) by mass flow of the water across the root (Taub and Wang 2008; McGrath and Lobell 2013) or by changing in the root architecture affecting the N uptakes (Taub and Wang 2008).

## 4.3 Breeding Nutrient Efficient Crops

Nutrient use efficiency is typically divided into two components such as nutrient acquisition efficiency (amount of nutrient uptake by the plants in relation to initial supply) and nutrient utilization efficiency (biomass produced per unit of nutrient incorporated by the plants). The goal of breeding programs is based on the making productive genotypes across a wide range of environment. However, genotype and environment interaction (G × E) hinders this developmental program (Aspinwall et al. 2015). Thus, there is a need of selection of genotypes that are adaptive to current climate change scenarios. Some of the plant traits are constant across the environment such as herbicide resistance, whereas other shows plasticity behaviors. Experiments have to be carried out to identify the traits which are plastic in nature and can be breed into the crops to give optimum performance under this global climate change. In case of small grain crops like cereals, tiller number is more plastic than seed size across wide range of environment. Therefore, for enhancement of the nutritional qualities in crops under climate change can be carried out by some of the following approaches.

### *4.3.1 Exploiting Genetic Diversity for Nutritional Traits*

Exploitation of genetic variation within the gene pool is the easiest strategy to identify the donor for trait of interest for concerned traits (Dwivedi et al. 2012). As Fe and Zn content is likely to decrease under elevated under $CO_2$, mineral nutrients were analyzed for a collection of common bean and found that Fe content was ranged from 34 to 89 ppm (mg/kg grain) and Zn content was found that 21–54 ppm (Beebe et al. 2000). Islam et al. (2004) also investigated the variation of mineral contents for common bean. These studies can be performed under different simulated environmental condition to find out the effective donor for the nutritional qualities and current breeding tools can be employed to transfer the gene of interest for concern nutritional traits.

But, when sufficient genetic variation is not available for the nutritional traits of interest, insertion of foreign gene of concern trait can be targeted to the crop through genetic engineering.

### 4.3.2 Approaches for Enhancing Nutritional Quality

Selection of the traits associated with the plant nutrition is utmost important in breeding nutrient efficient crops. There are two methods to identify the traits associated with plant nutrition such as Quantitative Trait Loci (QTL) and Genome-Wide Association Studies (GWAS) (Mitchell-Olds 2010). In biparental QTL approach, two parents differ with respect to trait of interest are crossed and $F_2$ generation are analyzed with the polymorphic markers that differentiate the parents. Traits linked with markers will segregate in a statistically significant manner (Miles and Wayne 2008). If any QTLs are found for the trait of interest, then a database search for similar DNA sequences that have been identified previously can be performed to find out what genes are involved in that particular trait of interest. Then, these genes can be manipulated in breeding programs using MAS for improving nutrition under climate change condition.

GWAS has more potential than the QTL analysis, but it is limited to model species where genomes have been mapped, and it is also not possible in polyploid species (Koprivova et al. 2014). Moreover, mRNA sequences are now used in the identification of polymorphic markers in polyploid species (Koprivova et al. 2014). Again, genotype factor of the G $\times$ E interaction is relatively constant, and each genotype differs with respect to nutrient acquisition and utilization efficiency. Thus, experiment can be carried out to identify the QTLs that with high or low nutrient supply. The QTLs that are found in both conditions (high and low nutrient condition) are related to efficiency, which can be breed to enhance nutritional quality (Pilbeam 2015). Studies reported that 15 QTLs were identified in different potassium treatment studies, which can be used for MAS of K efficiency (Kong et al. 2013).

QTLs can also be identified through association mapping (genome-wide association mapping and candidate gene-based association mapping) and bulk segregant analysis (BSA). In genome-wide AM, whole-genome polymorphism is measured, while in candidate gene-based AM, specific genomic region is used for mapping of locus (Gaikwad et al. 2020). In bulk segregant analysis (BSA), target genomic regions linked with trait of interest are identified. Then, this trait is concurrently transferred from wild source into recipient parents using backcross program. Generally, QTLs identified through the functional markers/genic markers are more precise because these markers are from transcribed region or parts of locus (Gaikwad et al. 2020). Therefore, single-nucleotide polymorphism (SNP) genotyping with proper phenotyping with modern statistical tools will be helpful for fine mapping of the QTLs of target traits. MAS is used when nutritional traits are controlled by the few loci in elite genetic background, marker-assisted backcross breeding (MABB) is used where donor parents are land races or distant relative. Genomic selection (GS) based on the genomic estimated breeding values (GEBVs) is gaining importance now-a-days due to low cost of genotyping.

## *4.3.3 Engineering Pathways and Nutrient Use Efficiency (NUE)*

It is well known that crop yield can be improved by enhancing the carbon fixation, which is the primary process of photosynthesis. Engineering of cyanobacterial Rubisco can increase the photosynthesis as compared to $C_3$ plant as the cyanobacterial Rubisco has higher carbon fixation capacity under higher atmospheric $CO_2$ level. However, enhanced crop yield will impose demands of nutrients. It is also reported that higher $CO_2$ level is closely related to bigger demands for nutrients and water which is the result of enhanced plant growth (Briat et al. 2015). Thus, nutrient use efficiency has to be maintained as a minimum requirement for enhancing crop yield. Most importantly, higher nutrient use efficiency (NUE) will help in reducing the emission of greenhouse gases, which otherwise causes global warming.

Several efforts have been made through transfer of gene of interest which contributes to the traits of interest through transgenic approach, where sufficient genetic variation for gene pool is not available. Phytoene synthase 1 (*psy1*) gene has been transferred from the maize to groundnut to enhance the level of β-carotene (Bhatnagar et al. 2011). Significant research has been carried out for carotene biofortification to enhance vitamin A content in plants.

## *4.3.4 Targeting Traits for Breeding Nutrient Efficient Crop*

### 4.3.4.1 Root Characteristics and NUE

Nitrogen use efficiency (NUE) can be enhanced by the roots that grow deeper in the soil, whereas phosphate use efficiency (PUE) can be enhanced by breeding the crops with root that proliferates in the upper surface of the soil. Similarly, potassium use efficiency can be upgraded by the root with intermediate between the PUE and NUE (White et al. 2013). Decrease in transpiration upon increased $CO_2$ uptake may be advantageous to deep-rooted species (Nelson et al. 2004). If increase in temperature favors the diffusion of nutrients in soil over the bulk flow, root systems with higher root length density, higher root hair density, and longer root hairs have been reported for enhancing NUE (Zhang et al. 2007).

Increase in the crop growth may occur under higher $CO_2$ concentration, but this may lead to lower N concentration in the crop yield. This could decrease the contents of the proteins and other essential nutrients in the grain, thus imposing harmful effects to human nutrition. However, application of higher level of fertilizer can improve the nutrient concentration in grain, but it will also lead toward the emission of greenhouse gases. Therefore, improvement of nutrient utilization efficiency will be needed. Nutrient utilization efficiency can be enhanced by the breeding for suitable root system. Therefore, target of better root system, better remobilization of nutrients to grain, and plasticity of these traits will increase the grain yield under this climate changing situation (Pilbeam 2015).

#### 4.3.4.2 Staygreen and NUE

Staygreen character is the trait which helps in maintaining the photosynthetic leaf area (Borrell et al. 2001) for longer duration of time, i.e., by maintaining higher leaf area duration (LAD). As staygreen cultivars have higher sink strength by forming more grain, this would also have the potential to acquire more mineral nutrients from soil. Thus by integrating this trait, it can lead to enhanced nutrition under climate change situation.

#### 4.3.4.3 Nutrient Harvest Index (NHI) and NUE

Enhanced $CO_2$ concentration can lead to increase in carbon assimilation per unit of leaf nitrogen in $C_3$ crop, thus enhancing the nitrogen utilization efficiency, a component of nitrogen use efficiency. However, this will also lead toward the decrease in nutrient concentration as compared to lower atmospheric $CO_2$ concentration. This will impose serious limitation to the human nutrition. Thus, cultivars can be selected based on the nutrient remobilization efficiency to grain and one of the traits is nutrient harvest index (NHI) (Ciampitti and Vyn 2012; Hawkesford 2012). As grain yield per N content in the shoot and grain nitrogen concentration is in negative correlation, traits such as specific uptake rate can be selected for enhanced grain nutrition. Thus, redistribution of essential mineral nutrients is also required for enhancing yield under higher $CO_2$ concentration. This redistribution is also dependent on metallothionein genes (Schiller et al. 2014) which may be major goal for plant breeding.

## 4.4 Biofortification: Overview

There are numerous approaches to enhance nutritional quality, but biofortification of iron and zinc is the most challenging. Modern breeding technologies help in the accumulation of essential elements and bioavailability while reducing the gathering of antinutrients such as phytate and hazardous heavy metals, which are frequently transported alongside iron and zinc. Sperotto et al. (2012) emphasized biofortification work in common beans (*Phaseolus vulgaris*) and showed that model species lessons could aid disclose pathways and responsible genes for Fe, Zn, and antinutrient modification in beans. The prospect of targeted tissue-specific biofortification of daily bean seed coat and cotyledons and decreased phytate, polyphenol antinutrients in the same tissues has been examined. Other group explores the function of root vacuoles for symplastic nutrient concentration and potentially harmful trace elements that impact shoot accumulation. Loss-of-function mutants and natural variability in vacuolar transporters reveal that the higher the number of root vacuoles, the less it gets loaded into the xylem and further translocation into the shoots and seed. As a result, biofortification treatments should involve root storage capacity adjustment. Some workers highlight the significance of raising plants' nicotianamine (NA) levels by overexpressing nicotianamine synthase (NAS) genes. NA is a metal chelator that helps in metal transfer. Plants that collect more NA have higher concentration of Fe and Zn in their edible sections and are more resilient to Fe deficiency. Several practical techniques may be utilized to boost plant nutrition. Increased seed storage protein (SSP) content is one of the critical breeding techniques for boosting rice nutritional quality owing to rice SSP for human and animal nutrition. Chen et al. (2018) analyze albumin, globulin, prolamin, glutelin, and total SSP levels in milled rice from 527 rice accessions farmed in two separate settings. Scientists uncovered additional SNPs and putative genes linked with rice seed protein concentration and composition by combining these nutritional properties with genome sequencing data. This could be relevant rice molecular breeding experiments in future aimed at boosting quality. Eating fruits and vegetables every day is well acknowledged to lower the risk of a variety of chronic diseases. The high concentration of numerous health-promoting chemicals, such as fiber, polyphenols, and vitamins, is commonly connected to the health advantages of vegetable and fruit crops. Plant breeding activities in developing nations, where cereal grain is the main source of sustenance for most of the population, have typically enhanced crop output and improved quality throughout the last 40–50 years to fill people's calorific needs. In about half of Middle Eastern and Central Asian countries, wheat accounts for about half of daily calorie consumption, and it accounts for roughly 70% in rural areas (Cakmak 2008). Despite acceptable essential food consumption, more than half of the world's population is deficient in zinc, iron, and vitamin A, despite a linear expansion in the food supply throughout the years. Malnutrition driven by a deficiency of micronutrients (mainly Zn and Fe) has impacted a considerable percentage of the world population. The phenomenon is known as "hidden hunger," which is especially widespread among teens (Harding et al. 2018). In 2017, over 70% of Asian children and 27% of African children under the age of 5 years were malnourished or unwell. More than two billion people globally were revealed to be zinc, iron, and zinc oxide deficient. Se, Cu, and Mn deficits were also identified in the diet. According to World Health Organization (WHO) investigation, it has adverse effects on the immune system and individual mental faculties. Today's high-yielding wheat cultivars are dangerous for human consumption due to micronutrient deficits, mainly in Zn and Fe. Zn and Fe in commercial wheat cultivars range from 20 to 35 mg/kg (Rengel et al. 1999). Wheat includes antinutritional chemicals such as phenolic and phytic acid components that decrease the availability of Zn and Fe in the human digestive system (Welch and Graham 2004). For human nutritional diets, these chemicals are insufficient. Long-term consumption of wheat-based diets may lead to malnutrition; approximately 20% of children under the age of 5 years die due to malnutrition (Prentice 2008). Anemia is a disorder caused by the deficiency of iron that affects roughly 25% of the world's population. Blindness, sickness, early mortality, and poor mental health are caused by the shortage of micronutrients. Almost 17.3 % of the world's population is in danger of not obtaining adequate zinc (Wessells and Brown 2012). Human Zn insufficiency affects roughly 2 billion people worldwide, with infants under 5 years and pregnant women suffering the lion's share of the burden (Bhutta and Salam 2012). According to needed plasma Zn levels, roughly 40% of Pakistani children and women are Zn deficient. Inadequate Zn availability leads to a broad spectrum of illnesses in the world's population (Black 2014). In order to grow and heal wounds, Zn is essential for bodily tissue mending and normal embryonic development. Calcium (Ca), one of the humans' most necessary dietary components, is also accessible in lower amounts than the RDA of 800–1300 mg per capita (Kranz et al. 2007). Calcium intake in adolescence is crucial for minimizing bone loss, rickets, and osteoporosis. Simultaneously, lower consumption exacerbates health conditions such as hypocalcemia, hypertension, colon cancer, bone deterioration, and fractures connected with aging (Centeno et al. 2009). Wheat being the world's major grain crop may have a micronutrient deficiency. It is the principal source of proteins, micronutrients, and calories for most of the world's population, especially in underdeveloped areas (Shewry 2009). Influential individuals in emerging countries do not acquire enough energy from their meals to fulfill their essential demands (Kennedy et al. 2002).

### *4.4.1 Micronutrient Absorption Mechanism*

Understanding the absorption pathways is critical for boosting cereal grain micronutrient content (Bouis and Welch 2010). Micronutrient accumulation in grains is affected by the absorption of micronutrients from the rhizosphere into the roots. Bioavailable nutrients deposited in grains are translocated and remobilized from the source. Zn and Fe are acquired from the soil in two ways: directly by ZRT and ZIPS transporters or indirectly by phytosiderophores that YSL transporters absorb chelate Zn and $Fe^{+}$ (Sperotto et al. 2012). To absorb minerals, monocots such as wheat require chelating reactions. Metal chelators, like nicotinamide transport Zn and Fe into vacuoles through the roots. They were taken to the xylem, where Fe chelates with citrate and Zn travels as a cation or chelates with citrate. Transport from xylem to phloem happens in the basal section of the shoot and during grain filling stages when translocation from leaves occurs. Several transporters were transferred from the maternal tissue to the endosperm and then into embryo's aleurone layer. The aleurone layer in wheat grains transfers the majority of the Fe and Zn lost during milling. Balmer et al. (2006) opined that Ferritin is a more bioavailable version of ferritin present in the endosperm. Iron is coupled to phytate within the storage vacuoles, making it less accessible to humans (Borg et al. 2012). The bioavailability of Fe and Zn is influenced by tissue location as well as the total Fe and Zn content of the grain.

Zn and Fe may be chelated by phytosiderophores, allowing cereal crops to absorb them more effectively (Marschner and Römheld 1994). A lack of Zn and Fe enhances the synthesis and release of phytosiderophores from wheat plant roots (Mori et al. 1991), they examined the influence of deoxymugineic acid on Zn transport from roots to rice crop shoots. Suzuki et al. (2008) evaluated the influence of deoxymugineic acid on Zn transport from roots to rice crop shoots. Application of Fe-deoxymugineic acid enhances Fe uptake and transport in plants. Another element related to plant Fe and Zn metabolism is sulfur. The amount of S contained in cereal crops impacts plants' capacity to absorb and store Fe. Celletti et al. (2016) discovered that phytosiderophores, which assist the roots, absorb Fe, positively associate S levels, boosting the efficiency with which Fe is used.

### *4.4.2 Fertilizer's Impact on Nutrition Composition*

Increasing food output demands a considerable increase in nitrogen (N) fertilizer. Wheat crops necessitate approximately 21% of the fertilizers used in Indian agriculture. Nitrogenous fertilizers account for approximately 73% of wheat fertilizer use (FAO 2005). Fertilizers account for increase in post-Green Revolution food crop output. Nitrogenous fertilizers contribute to more than one-third of this growth. Nitrogen is an important component of chlorophyll and plays a function in the metabolic process. Protein is necessary for all metabolic functions in plants. Nitrogen supply is vital for improving yield per unit area in a field crop like wheat. N influences the number of tillers per $m^2$, the number of spikelets per spike, the number of grains per spike, the spike length, and the 1000-grain weight. Sufficient N delivery to agricultural plants is vital for growing leaf and floret primordia during their early growth stage. Nitrogen levels also boost grain filling rates (Eichenauer et al. 1986). As per Comfort et al. (1988), high rates of N treatment may prevent more depth root development, resulting in less optimal utilization of deeper nitrate in soil and water sources. The availability of nitrogen affected the leaf area index (LAI), which enhanced the head and biomass output (Whitfield and Smith 1989) showed that increasing the N supply rate from 2.37 t/ha with 37 kg of N to 2.80 t/ha with 120 kg of N per ha boosted grain yields. Lloveras et al. (2004) studied the impact of additional N fertilizer on the quality through top dressing and production of bread-making wheat under irrigated Mediterranean conditions. They noticed that applying top-dressed N boosted yields by 100 kg N per acre more rapidly. A 200 kg N per ha top-dressed N treatment would be acceptable for creating high-quality bread-making wheat while reducing $NO_3$ leaching losses. Nitrogen feeding promotes grain quality and vigor through enhancing grain protein levels. Protein feeding boosts the proportion of final germination while shortening T50 and mean germination time. A nitrogen rate of 140 kg/ha enhances seed germination and seed weight. Changing nitrogen rates largely influenced grain yield by increasing spikelets, grains per spike, and 1000-grain weight. According to Singh et al. (2010), Haryana farmers use an average of 165.7 kg of nitrogen per ha and 12.5 kg of phosphorus per ha of wheat, with only 9.2% applying K-fertilizer. The average wheat yield in Haryana is 4.2 t/ha; this translates to around 80 kg of N and 12.6 kg of P eliminated in grain. Chen et al. (2017) did a greenhouse experiment with selenium rates of Se1 (0.74 mg/kg) or Se2 (2.60 mg/kg), with each Se condition obtaining either N1 (100 mg/kg) or N2 (200 mg/kg) nitrogen rates. In the Se1 and Se2 treatments, boosting nitrogen rates enhanced grain yield by 13.2% and 24.0%, respectively. Increased nitrogen levels may boost root selenium absorption, resulting in increased selenium levels in wheat grains and leaves. Abd El-kader and El-Basioni (2013) looked at biofertilizers and micronutrient combinations with less mineral nitrogen. It saves roughly 50% of the N fertilizer while enhancing soil N, P, grain, and straw yield. Plants may affect multiple Zn and Fe mobility and absorption stages in plants (Marschner 1994). They controlled root development and generated organic material exudates in the roots (Marschner 1994). Plant nitrogen levels may impact the expression of Zn and Fe transporter proteins localized on the root cell membrane, such as the ZIP transporter protein family (Grotz and Guerinot 2006). The transfer of Zn and Fe from vegetative tissues to grains via the phloem may impact nitrogen status (Haydon and Cobbett 2007). Zinc and Fe transporter proteins have been discovered on the plasma membrane of phloem cells, indicating that they are necessary for Zn and Fe trafficking into grains via the phloem. Haydon and Cobbett (2007) examined root absorption and root-to-shoot translocation of 65Zn in durum wheat (*Triticum durum*) seedlings precultured with various N treatments. The findings demonstrate that the N-nutritional status of wheat affects crucial phases in the journey of Zn from the growth medium to the grain, such as absorption, xylem transport, and phloem translocation. Thus, nitrogen is essential for Zn absorption and accumulation in plants, which is especially relevant in Zn biofortification of food crops. Increased nitrogen tissue content enhances the amount of Zn absorption transporters in the root, including ZIPs such as IRT1 and other unidentified proteins (Ishimaru et al. 2005). The induction of root-to-shoot Zn transport by N may be connected to enhanced xylem transport. Furthermore, Zn xylem loading increases the synthesis of nitrogenous compounds such as nicotinamide and deoxymugineic acid, which speed up Zn transport in plants due to N triggered transporter proteins involved in xylem loading (Suzuki et al. 2008). Kutman et al. (2010) investigated how various Zn and N treatments influenced total absorption, remobilization, and Fe, Zn, and N partitioning in durum wheat. Plants were raised in greenhouses with low or high N and Zn inputs, and samples were obtained at eight various periods of growth. Heavy N treatment boosted Fe and Zn absorption per wheat plant by four times, but growth affects were much reduced. When N and Zn supplies were ample, postanthesis shoot absorption of Zn and Fe for grain was almost 50% and 80%, respectively, demonstrating that Zn contributes more remobilization to grain accumulation than Fe. Approximately 60% of Zn and 40% of Fe initially reserved in vegetative cover were retranslocated to grains. Approximately 80% of the entire shoot Zn and 60% of the total shoot Fe were collected with grains. As a result, all of these statistics were substantially lower with the low N treatment. Consequently, they observed that N nutrition is critical for Zn and Fe acquisition and grain distribution in wheat harvests. Ammonical form of nitrogen reduces soil pH speeds up the solubilization of Mn, Cu, Zn, and Fe in alkaline soils and promotes plant absorption, while $NO_3$ limits absorption. Optimal N application promotes Cu, Fe, Zn, and Mn transportability from root to shoot, resulting in biofortification of micronutrients in food crops and reduced malnutrition (hidden hunger) in humans. Zhang et al. (2012) found that the source–sink relationship of photosynthates and nitrogen in grains could alter micronutrient and protein levels. Adequate N treatment may enhance grain yield while simultaneously improving Fe, Zn, Mn, Cu, and protein content. Barunawati et al. (2013) studied the effects of nitrate- or ammonium-related nitrogeneous fertilizers on Fe, Zn, and Cu accumulation in wheat flag leaves and grains, as well as metal chelators. Nitrogen addition boosted the quantity of nitrogen and the metal chelator nicotianamine (NA) in green leaves while preserving the amount of 2′--deoxymugineic acid (DMA). According to the data, increasing the N nutritional quality of flag leaves promotes the accumulation of Fe, Zn, and Cu in flag leaves and the pool of NA. Sheoran et al. (2015) undertook a pot culture experiment to examine the influence of N on wheat micronutrient absorption. With each sequential nitrogen treatment, wheat grain and straw's Zn, Cu, Fe, and Mn absorption climbed rapidly, reaching up to 200 mg/kg of soil.

### *4.4.3 Nutritional Content and Quality Improvement: Future Challenges*

Many investigators studied a wide variety of biofortification in crops, including agronomic, conventional, transgenics and biotechnology, crop breeding, and fertilization. The authors explore drawbacks of each technique and the consumer acceptance challenges. They analyze the present state of knowledge as well as the future potential for enhancing the productivity and nutritional quality of fodder crops, which have gotten less attention than cereals, fruits, and vegetables. The authors also underline how data from model plants and grain crops can be utilized by genetics, and bioinformatics to enhance fodder crops for food security. Hameed et al. (2018) offer a summary of nutritional enhancement operations in potatoes (*Solanum tuberosum*) employing conventional breeding and genetic engineering approaches, such as boosting nutrient concentration and lowering antinutrients. Exogenous glycine application may promote the accumulation of health-promoting chemicals and improve antioxidant activity in hydroponically produced lettuce. The wheat grain's major carbohydrate component is starch. In addition to the major components, amylose and amylopectin may interact with smaller components such as lipids, proteins, and phosphorus. The availability of phosphorus may impact wheat grain production and quality. Zhang (2017) employed three doses of phosphorus fertilizer in wheat fields. They showed that phosphorus greatly affects starch biosynthesis gene expression, starch biosynthesis, breakdown, and micro-structure in wheat grains. The findings stressed on employing the proper amount of phosphorus fertilizer to boost wheat yield and starch quality. *OsNRT2.1*, a high-affinity nitrate transporter, is involved in nitrogen absorption and translocation in rice. Overexpression of *OsNRT2.1* promotes nitrate absorption and Mn build up in rice grains. The considerable rise in Mn induced by *OsNRT2.1* overexpression is most likely the consequence of increased expression of Mn transporter genes such as *OsNRAMP*3, 5, and 6. This study offers a different approach for enhancing Mn absorption in plants, which may impact grain quality.

## 4.5 Improvement of Plant Nutrient and Nutritional Quality Through Genome Editing

CRISPR/Cas9 has evolved as cutting-edge technology in the field of biotechnology. It allows scientists to manipulate/bring precise changes in the genome of an organism. The technology of the CRISPR-Cas9 based editing system is under continuous advancement, and its usages are rapidly growing in plant genome editing for functional genomics studies and crop improvement. In CRISPR-Cas9 systems, customized small single guide RNA (sgRNA) directs Cas9 nuclease at a precise genomic location. Cas9 makes a double-stranded break in the target DNA near the protospacer adjacent motif (PAM) sequence. These breaks are subsequently repaired by the homologous recombination or non-homologous end-joining (NHEJ) repair mechanism in cell (Chen et al., 2019). The NHEJ repair is the error-prone pathway that creates random INDELS and results in frameshift mutations and targeted gene knock-outs (Jinek et al., 2012; Feng et al., 2013). SpCas9 (Streptococcus pyogenes) the most widely used Cas9 is usually limited to performing genome editing at sites with PAM sequence "NGG". To extend the applicability of genome editing in any region of a genome, numerous Cas9 orthologs and engineered variants of Cas9 have been developed with different PAM such as Staphylococcus aureus Cas9 (SaCas9), Neisseria meningitides (NmCas9), and Cas9-VQR (D1135V/R1335Q/T1337R) (Kleinstiver et al., 2015; Hu et al., 2018). Further, engineered variants of Cas9 enzymes expanded the scope of targeted genome editing as they can identify different PAMs i.e. EQR-Cas9 (NGAG PAM), VRERCas9 (NGCG PAM), VQR-Cas9 (NGA PAM), xCas9 (NG, GAA, and GTA PAM), SpCas9-NG (NG PAM), and SaKKH-Cas9 (NNNRRT PAM) (Kleinstiver et al. 2015; Nishimasu et al. 2018). Developing a multiplex editing system allows targeted editing of the same locus at multiple sites or targeting multiple genes at one time. Multiplex genome editing systems have been used successfully by many researchers for the rapid progress of bacterial blight resistance, yield improvement (Xu et al., 2016; Oliva et al., 2019). Further to develop and expand multiplex genome editing efficiency, the tRNA processing system has been used to generate multiple mRNAs from a single polycistronic gene by using an endogenous tRNA-processing system, which specifically cleaves both ends of the tRNA precursor. This system could improve the targeting and multiplex editing competency of the CRISPR/Cas9 system (Xie et al., 2015).The CRISPR-Cas12a/Cpf1 has been identified, leading to the improvement of a new advanced genome editing system. In this system, base editors have evolved as more efficient and accurate genome editing with low off-target activity. These tools might quicken the pace of crop improvement (Tang et al., 2017b). Targeting mutations in untranslated region of coding genes (Mao et al., 2018), promoter regions (Seth, 2016), microRNAs (Chang et al., 2016), non-coding RNAs (ncRNAs) (Li et al., 2018) may give different phenotype.In the last few years, scientist have published numbers of research articles demonstrating successful targeted mutagenesis by CRISPR/Cas9 technology for a wide range of traits in major crops, such as rice, rapeseed/mustard, corn, soybean etc. CRISPR/Cas9 mediated targeted mutation in the promoter region of the OsSWEET14 and OsSWEET11 caused in bacterial leaf blight resistance (Jiang et al., 2013). Blast resistance of variety of rice named Kuiku131 was enhanced by mutating ERF transcription factor gene OsERF922 was mutated by CRISPR/ Cas9 (Wang et al., 2016). Knock-out of the thermo-sensitive genic male-sterile (TGMS) gene tms5 of 11 fertile elite cultivars produced TGMS lines with good agronomic characteristics (Zhou et al., 2016). Knock-out of Os8N3 gene through CRISPR/Cas9-mediated displayed decreased sucrose concentration in the embryo sacs and defective grain filling, signifying that Os8N3 plays an major role in sucrose transport during the early stage of rice grain filling but, it also conferred resistance to Xanthomonas oryzae pv. Oryzae (Kim et al., 2019). Triple mutant generated through simultaneous knock-out of grain development genes GW2, GW5, and TGW6 in rice brought about 29.8% increase in thousand-grain weight (Xu et al., 2016). Zhang et al. (2019) reported upgraded rice salinity tolerance by targeting the OsRR22 gene in rice. CRISPR/Cas9 is widely used for producing mutation in coding genes/RNA of crop plants.

### *4.5.1 Nutrient*

Under the present climate change scenario, it is challenging to keep the agriculture systems sustainable. Improvement of nutrient use efficiency is essential for climate-resilient crop production. Thus, maintaining crop yield stability with nutrient use efficiency is desirable in all crops. CRISPR/Cas9 genome editing of critical regulators of nutrient uptake and signaling pathway gene can improve NUE. PHOSPHATE 1 (PHO1) gene plays a vital role in the uptake and transfer of a phosphate. Zhao et al. (2019) developed deletion mutants of *SlPHO1;1* through CRISPR, which has shown general phenotypes of Pi starvation which is decreased shoot fresh weight and increased fresh root weight, leading to more root/shoot ratio. In mutants, a higher accumulation of anthocyanin and soluble Pi content was observed in the root compared to shoot.

To improve the nitrogen use efficiency, CRISPR/Cas9-based editing system has been used to make targeted substitution in *NRT1.1B, a nitrogen transporter gene*. In rice, through base editing, 1.4–11.5% C/T substitution was achieved in the *NRT1.1B* gene of rice, which improved NUE (Hu et al. 2015). Natural mutant of the abnormal cytokinin response1 repressor1 (are1) gene exhibits increased NUE, delayed senescence, and increased grain yield in rice. In wheat, CRISPR/Cas9-mediated targeted mutagenesis of the *TaARE1* gene led to improved tolerance to nitrogen starvation and grain yield in field conditions (Zhang et al. 2021). Present study by Gao et al. (2020) revealed that natural variation between India and japonica subspecies of Asian rice at the MYB61 locus leads to differential nitrogen use and cellulose biogenesis. MYB61, a transcriptional factor regulator of cellulose synthesis, is directly regulated by GROWTH-REGULATING FACTOR4 (GRF4), which synchronizes cellulosic biomass production and N utilization. The indica allele of MYB61 resulted in higher NUE and increased grain yield at reduced nitrogen supply in India genotypes. Genome editing can be utilized to improve NUE in japonica rice by a targeted mutation in MYB61.

### *4.5.2 Improvement of Nutritional Quality*

Quality traits like nutrient content, starch content, low antinutritional factor, and storability are improved through genome editing in many crops. Cereal foods that are higher in amylose and resistant starch are nutritious and offer greater potential to improve human health and lower the risk of severe noninfectious diseases. The mutation of starch branching enzyme gene *SBEIIb* produces high-amylose and resistant starch rice, which improves the starch's delicate structure and nutritional properties having health benefits for the chronically diseased person (Sun et al. 2016, 2017). In other study, *SBEI* and *SBEIIb* genes were knocked-out (starch branching enzyme). Knockout lines of *SBEIIb* had a high amylose (25%) and resistant starch (9.8%) content (Sun et al. 2017). The potato knockout lines of all four alleles of *the StGBSS* gene had high-amylopectin potato (Andersson et al. 2017). Similarly, CRISPR/Cas9 genome editing mediated targeted mutagenesis of *TaSBEIIa* in a modern winter wheat *cv*Zhengmai 7698 (ZM) and a spring wheat *cv* Bobwhite was carried out (Li et al. 2021). The triple-null lines of *TaSBEIIa* had significantly high amylose, resistant starch, protein, and soluble pentosan contents, which improve the human health. In contrast, knocking out the GBSS gene led to low-amylose and high amylopectin in rice. DuPont Pioneer, now known as CortevaAgriScience, targeted maize waxy gene *Wx1* to increase grain amylopectin, which has improved digestibility and commercial use potential. Silencing of *Waxy* gene by CRISPR/Cas9 led to a decrease in amylose content and thus improved cooking quality (Zhang et al. 2018a).

Wheat storage protein, gluten, can trigger celiac diseases, an autoimmune disorder in some individuals. To reduce gluten contents in wheat protein, CRISPR/Cas9 mutant lines of gliadin genes were generated. The transgene-free mutant lines had reduced gliadin content in wheat gluten, which led to reduced immunoreactivity by 85% (Sánchez-León et al. 2018). Specific targeted editing of *ZmMADS47* gene encoding a MADS-box protein using RNA interference (RNAi) and CRISPR/Cas9 led to decrease in zein content by 16.8% and 12.5%, respectively. High oleic/low linoleic content is essential for oil quality. Oleic acid is catalyzed to linoleic acid by enzyme fatty acid desaturase 2 (FAD2) in plants. In the oilseed plants like *Brassica napus*, *Camelina sativa* silencing FAD2 improved oleic acid content while decreasing polyunsaturated fatty acids (Morineau et al. 2017; Okuzaki et al. 2018). Targeted mutagenesis of *OsFAD2-1* via CRISPR-Cas9 editing technology led to the creation of *OsFAD2-1* knockout rice plants with a twofold increase in oleic acid contents with no detectable linoleic acid, thereby improving the fatty acid composition of rice bran oil (Abe et al. 2018).

### *4.5.3 Antinutritional Factor*

Antinutritional factors compounds in plant foods diminish the nutrient availability of plants by limiting the absorption of nutrients and micronutrients. Some antinutritional factors like glucosinolates, phytic acid, saponins, lectins, tannins, amylase inhibitors, and protease inhibitors have been shown to decrease the availability of nutrients and resulted in growth inhibition. CRISPR/Cas9 genome editing can be utilized for bringing reduction in antinutritional factors in plants.

Phytic acid is the chief storage form of phosphate in cereal grains, but, at the same time, it is known as an antinutrient by reducing availability of other nutrients by inhibiting the absorption of phosphorus, other minerals, and protein. The development of crop plants with reduced IP6 (inositol hexakisphosphate) content could augment phosphate and mineral bioavailability. Inositol trisphosphate 5/6 kinases (ITPK) enzymes participate in phosphorylation of inositol phosphate to inositol hexakisphosphate, storage form of phosphate in cereal grains. When the *IPK1* gene encoding for inositol-phosphate 2-kinase was blocked using ZFNs in maize, it significantly reduced phytate content.

Similarly, knockout lines of *Zea mays* for genes *IPK1A*, *IPK*, and *MRP4* (multidrug resistance-associated protein 4) had reduced phytate biosynthetic pathway and phytic acid content (Liang et al. 2014). In Barley, homozygous mutants for the *HvITPK1* gene was generated by CRISPR/Cas9 (Vlčko and Ohnoutkova 2020). Mutant lines contained altered phosphate levels in the mature grains, ranges from 65% to 174% of the wild-type content. Among 11 mutants, maximum increase in phosphate content (74%) was detected in the homozygous deletion mutant itpk1-14. Interestingly, mutant insertion lines have shown that more tolerance to salinity stress than deletion mutants, along with a decrease in grain phosphorus content.

Greater accumulation of heavy metals is highly toxic to crop plants at high tissue concentrations, and it leads to the toxicity of heavy metals. Metal toxicity severely affects plant growth, development, and yield. It causes oxidative stress and disturbs cellular ionic homeostasis, leading to cellular damage. CRISPR/Cas9 system may hold potential for the improvement of reduction in metal toxicity in plants. The plant genes regulating heavy metal uptake and its transport are potential candidate genes for CRISPR/Cas9 genome editing. Cadmium (Cd) is highly toxic and hazardous for human health (Clemens et al. 2013). Rice is a primary source of calorie intake, and its grains frequently contain excessive amounts of Cd. Therefore, Cd accumulation should be controlled in rice. In rice root cells, *OsNramp5* is a crucial gene involved in controlling the uptake of Cd, Mn, and other metal ions. Knocking out the metal transporter gene *OsNramp5* by CRISPR/Cas9 reduced Cd accumulation in rice without any significant effect on yield. The mutants contained less than 0.05 mg/kg Cd compared to 0.33–2.90 mg/kg in the control plants under field conditions (Ishikawa et al. 2012; Tang et al. 2017a). *lcd-kmt2,* an *Osnramp5* mutant has been identified and registered as a Japanese variety, KoshihikariKan No. 1 (Ishikawa et al. 2012) and has been permeated for practical usage in Japan (Ishikawa et al. 2016). However, in other studies, knockout mutants of *OsNRAMP5* showed lesser growth and yield (Sasaki et al. 2012; Wang et al. 2019b). In another study by Songmei et al. (2019), mutant lines were generated for one *OsLCT1* and two *OsNramp5* using CRISPR/Cas9-mediated mutagenesis. In *OsLCT1* mutants, grain Cd accumulation was reduced by 40% (0.17 mg/kg) of the Cd concentration of the wild-type and two *OsNramp5* mutants also showed lesser grain Cd accumulation (<0.06 mg/kg). However, only one mutant *nramp5* × *7* showed normal growth and yield. Thus, CRISPR-aided genome engineering holds the potential to develop plants with a high level of nutrients and a reduced amount of toxic metal and antinutritional factors for better health.

Under storage conditions, potato tubers accumulate steroidal glycoalkaloids (SGAs) a-solanine and a-chaconine. Accumulation of these compounds confers bitter taste and exhibits toxicity. Thus reduction in the content of these antinutritional compounds would benefit health. The CRISPR/Cas9 and TALEN knockout lines of the *SSR2* gene (sterol side chain reductase 2) and *St16DOX* gene (encoding for the steroid 16a-hydroxylase) have reduced the accumulation of solanine and chaconine compounds (Nakayasu et al. 2017; Yasumoto et al. 2019).

### 4.5.4 *Biofortification*

Approximately three billion people suffer from deficiency of micronutrients and vitamins, which affects their growth, development, and immunity and increases the risk for infectious illness. Micronutrient deficiency is more noticeable in developing countries than in developed countries. Biofortification is the way to enhance the level of bioavailable micronutrients in edible parts of staple food crops such as rice, wheat, maize, potato and is considered an effective strategy to provide balanced diets with enriched levels of vitamins and minerals for better human health. The deficiency of micronutrients is attributed due to its lower uptake of nutrients by plants, accumulation in edible plant parts, and presence of high level of inhibitors affecting their absorption. Development in plant science made it possible to produce biofortified crops such as rice, wheat, and maize to solve the problem of nutrition deficiencies. CRISPR/Cas9 genome editing can be a powerful tool in improving the micronutrient and vitamin content by manipulating genes of biosynthesis pathway enzymes. Successful attempts of increasing carotenoid content by CRISPR/Cas9 have been reported in rice and banana. Kaur et al. (2018) successfully demonstrate that genome editing through CRISPR/Cas9 can be applied as an effective tool for banana genome modification by creating mutation in phytoene desaturase of banana *cv* Rasthali. Further, in 2020 they were able to develop the β-carotene-enriched Cavendish banana cultivar (*cv*) Grand Naine (AAA genome). Metabolic profiling of the fruit pulp of selected, edited lines showed increased accumulation of β-carotene content (sixfold ~24μg/g) compared with the unedited plants without any significant effect on the agro-morphological parameters in edited lines. The *Osor* gene, an ortholog of the Orange (Or) gene in cauliflower, was targeted in rice using CRISPR/Cas9. Alternatively, it is responsible for β-carotene accumulation in cauliflower curd. The directed modification of *Osorvia* CRISPR/Cas9-based genome editing resulted in enhanced β-carotene accumulation in rice callus (Endo et al. 2019). Dong et al. (2020) achieved the targeted insertion of a 5.2 kb carotenoid biosynthesis cassette at two genomic safe harbors in rice using CRISPR-Cas9 genome editing. They obtained marker-free rice plants with high β-carotenoid content ($7.90 \pm 0.19$ μg $g^{-1}$ dry weights) in the seeds with no detectable changes in morphology or yield penalty and no off-target mutations by Cas9 in the engineered plants. These results showed that CRISPR-Cas9 genome editing is an auspicious technology for the genetic improvement of cultivated crops. The targeted gene expression strategy can be utilized to insert genes involved in the uptake and consequent translocation of micronutrients to improve their content in plants.

## 4.6 Conclusion

Improvement in NUE may be realized in low-intensity agriculture by increasing the efficiency of uptake of nutrients by breeding for suitable root systems. However, in more intensive agriculture, increased nitrogen utilization efficiency from remobilization of N and micronutrients within the plants will be required. Prolonged root uptake and better remobilization of nutrients are targets for breeding, and there is sufficient plasticity of these characteristics in crop species for increases in efficiency to be achieved. Exploitation of genetic variation within the gene pool is the easiest strategy to identify the donor for trait of interest for concerned traits as Fe and Zn contents are likely to decrease under elevated $CO_2$. The studies can be performed under different simulated environmental condition to find out the effective donor for the nutritional qualities and current breeding tools can be employed to transfer the gene of interest for concerned nutritional traits. But, when sufficient genetic variation is not available for the nutritional traits of interest, insertion of foreign gene of concern trait can be targeted to the crop through genetic engineering.

## References

Abd El-kader SM, El-Basioni BMM (2013) Precision farming solution in Egypt using the wireless sensor network technology. Egypt Inf J 14(3):221–233

Abe K, Araki E, Suzuki Y, Toki S, Saika H (2018) Production of high oleic/low linoleic rice by genome editing. Plant Physiol Biochem 131:58–62

Agrawal PK, Kohli A, Twyman RM, Christou P (2005) Transformation of plants with multiple cassettes generates simple transgene integration patterns and high expression levels. Mol Breed 16(3):247–260

Al Amin N, Ahmad N, Wu N, Pu X, Ma T, Du Y, Bo X, Wang N, Sharif R, Wang P (2019) CRISPR-Cas9 mediated targeted disruption of FAD2–2 microsomal omega-6 desaturase in soybean (Glycine max. L). BMC Biotechnol 19(1):1–10

Andersson M, Turesson H, Nicolia A, Falt AS, Samuelsson M, Hofvander P (2017) Efficient targeted multiallelic mutagenesis in tetraploid potato (*Solanum tuberosum*) by transient CRISPR-Cas9 expression in protoplasts. Plant Cell Rep 36:117–128

Arias P, Bellouin N, Coppola E, Jones R, Krinner G, Marotzke J, Naik V, Palmer M, Plattner GK, Rogelj J, Rojas M (2021) Climate change 2021: the physical science basis. contribution of working group14 I to the sixth assessment report of the intergovernmental panel on climate change; technical summary

Aro A, Alfthan G, Varo P (1995) Effects of supplementation of fertilizers on human selenium status in Finland. Analyst 120(3):841–843

Aspinwall MJ, Loik ME, Resco de Dios V, Tjoelker MG, Payton PR, Tissue DT (2015) Utilizing intraspecific variation in phenotypic plasticity to bolster agricultural and forest productivity under climate change. Plant Cell Environ 38(9):1752–1764

Balmer Y, Vensel WH, Dupont FM, Buchanan BB, Hurkman WJ (2006) Proteome of amyloplasts isolated from developing wheat endosperm presents evidence of broad metabolic capability. J Exp Bot 57:1591–1602

Barunawati N, HettwerGiehl RF, Bauer B, Wirén NV (2013) The influence of inorganic nitrogen fertilizer forms on micronutrient retranslocation and accumulation in grains of winter wheat. Front Plant Sci 4:320

Bassirirad H (2000) Kinetics of nutrient uptake by roots: responses to global change. New Phytol 147(1):155–169

Baysdorfer C, Robinson JM (1985) Metabolic interactions between spinach leaf nitrite reductase and ferredoxin-NADP reductase: competition for reduced ferredoxin. Plant Physiol 77(2): 318–320

Beebe S, Gonzalez AV, Rengifo J (2000) Research on trace minerals in the common bean. Food Nutr Bull 21(4):387–391

Bhatnagar M, Bhatnagar-Mathur P, Reddy DS, Anjaiah V, Sharma KK (2011) Crop biofortification through genetic engineering: present status and future directions

Bhutta ZA, Salam RA (2012) Global nutrition epidemiology and trends. Ann Nutr Metab 61(Suppl 1):19–27

Black RE (2014) Global distribution and disease burden related to micronutrient deficiencies. In: International nutrition: achieving millennium goals and beyond, vol 78. Karger, Basel, pp 21–28

Bloom AJ, Smart DR, Nguyen DT, Searles PS (2002) Nitrogen assimilation and growth of wheat under elevated carbon dioxide. Proc Natl Acad Sci 99(3):1730–1735

Borg S, Brinch-Pedersen H, Tauris B, Madsen LH, Darbani B, Noeparvar S, Holm PB (2012) Wheat ferritins: improving the iron content of the wheat grain. J Cereal Sci 56:204–213

Borrell A, Hammer G, Van Oosterom ERIK (2001) Stay-green: a consequence of the balance between supply and demand for nitrogen during grain filling? Ann Appl Biol 138(1):91–95

Bouis HE, Welch RM (2010) Biofortification—a sustainable agricultural strategy for reducing micronutrient malnutrition in the global south. Crop Sci 50:S-20–S-32

Briat JF, Dubos C, Gaymard F (2015) Iron nutrition, biomass production, and plant product quality. Trends Plant Sci 20(1):33–40

Brinch-Pedersen H, Borg S, Tauris B, Holm PB (2007) Molecular genetic approaches to increasing mineral availability and vitamin content of cereals. J Cereal Sci 46(3):308–326

Brouder SM, Volenec JJ (2008) Impact of climate change on crop nutrient and water use efficiencies. Physiol Plant 133(4):705–724

Cakmak I (2008) Enrichment of cereal grains with zinc: agronomic or genetic biofortification? Plant Soil 302(1):1–17

Cakmak I, Kutman UÁ (2018) Agronomic biofortification of cereals with zinc: a review. Eur J Soil Sci 69(1):172–180

Cakmak I, Kalaycı M, Ekiz H, Braun HJ, Kılınç Y, Yılmaz A (1999) Zinc deficiency as a practical problem in plant and human nutrition in Turkey: a NATO-science for stability project. Field Crops Res 60(1–2):175–188

Cannon SB, May GD, Jackson SA (2009) Three sequenced legume genomes and many crop species: rich opportunities for translational genomics. Plant Physiol 151(3):970–977

Carciofi M, Blennow A, Jensen SL, Shaik SS, Henriksen A, Buleon A, Holm PB, Hebelstrup KM (2012) Concerted suppression of all starch branching enzyme genes in barley produces amylose-only starch granules. BMC Plant Biol 12(1):1–16

Cavagnaro TR (2008) The role of arbuscular mycorrhizas in improving plant zinc nutrition under low soil zinc concentrations: a review. Plant Soil 304(1):315–325

Cavagnaro TR, Sokolow SK, Jackson LE (2007) Mycorrhizal effects on growth and nutrition of tomato under elevated atmospheric carbon dioxide. Funct Plant Biol 34(8):730–736

Celletti S, Pii Y, Mimmo T, Cesco S, Astolfi S (2016) The characterization of the adaptive responses of durum wheat to different Fe availability highlights an optimum Fe requirement threshold. Plant Physiol Biochem 109:300–307

Centeno V, de Barboza GD, Marchionatti A, Rodriguez V, de Talamoni NT (2009) Molecular mechanisms triggered by low-calcium diets. Nutr Res Rev 22(2):163–174

Chang H, Yi B, Ma R, Zhang X, Zhao H, Xi Y (2016) CRISPR/cas9, a novel genomic tool to knock down microRNA in vitro and in vivo. Sci Rep 6(1):1–12

Chen L, Magliano DJ, Zimmet PZ (2012) The worldwide epidemiology of type 2 diabetes mellitus—present and future perspectives. Nat Rev Endocrinol 8(4):228–236

Chen YP, Peng Q, Liang DL, Song WW, Lei LM, Yu DS (2017) Effects of nitrogen application on selenium uptake, translocation and distribution in winter wheat. Huanjingkexue = Huanjingkexue 38(2):825–831

Chen P, Shen Z, Ming L, Li Y, Dan W, Lou G et al (2018) Genetic basis of variation in rice seed storage protein (albumin, globulin, prolamin, and glutelin) content revealed by genome-wide association analysis. Front Plant Sci 9:612

Chen K, Wang Y, Zhang R, Zhang H, Gao C (2019) CRISPR/Cas genome editing and precision plant breeding in agriculture. Annu Rev Plant Biol 70:667–697

Cheng L, Booker FL, Tu C, Burkey KO, Zhou L, Shew HD, Rufty TW, Hu S (2012) Arbuscular mycorrhizal fungi increase organic carbon decomposition under elevated $CO_2$. Science 337(6098):1084–1087

Christian M, Cermak T, Doyle EL, Schmidt C, Zhang F, Hummel A, Bogdanove AJ, Voytas DF (2010) Targeting DNA double-strand breaks with TAL effector nucleases. Genetics 186(2): 757–761

Christou P, Twyman RM (2004) The potential of genetically enhanced plants to address food insecurity. Nutr Res Rev 17(1):23–42

Ciampitti IA, Vyn TJ (2012) Physiological perspectives of changes over time in maize yield dependency on nitrogen uptake and associated nitrogen efficiencies: a review. Field Crops Res 133:48–67

Clasen BM, Stoddard TJ, Luo S, Demorest ZL, Li J, Cedrone F, Tibebu R, Davison S, Ray EE, Daulhac A, Coffman A, Yabandith A, Retterath A, Haun W, Baltes NJ, Mathis L, Voytas DF, Zhanget F (2016) Improving cold storage and processing traits in potato through targeted gene knockout. Plant Biotechnol J 14(1):169–176

Clemens S, Aarts MG, Thomine S, Verbruggen N (2013) Plant science: the key to preventing slow cadmium poisoning. Trends Plant Sci 18(2):92–99

Comfort SD, Malzer GL, Busch R (1988) Nitrogen fertilization of spring wheat genotypes: influence on root growth and soil water depletion. Agron J 80:114–120

Daniels LA (1996) Selenium metabolism and bioavailability. Biol Trace Elem Res 54(3):185–199

De Graaff MA, Van Groenigen KJ, Six J, Hungate B, Van Kessel C (2006) Interactions between plant growth and soil nutrient cycling under elevated $CO_2$: a meta-analysis. Global Change Biol 12(11):2077–2091

Dhawi F, Datta R, Ramakrishna W (2016) Mycorrhiza and heavy metal resistant bacteria enhance growth, nutrient uptake and alter metabolic profile of sorghum grown in marginal soil. Chemosphere 157:33–41

Dieleman WI, Vicca S, Dijkstra FA, Hagedorn F, Hovenden MJ, Larsen KS, Morgan JA, Volder A, Beier C, Dukes JS, King J, Leuzinger S, Linder S, Luo Y, Oren R, Angelis PD, Tingey D, Hoosbeek MR (2012) Simple additive effects are rare: a quantitative review of plant biomass and soil process responses to combined manipulations of $CO_2$ and temperature. Global Change Biol 18(9):2681–2693

Dong J, Gruda N, Lam SK, Li X, Duan Z (2018) Effects of elevated $CO_2$ on nutritional quality of vegetables: a review. Front Plant Sci 9:924

Dong OX, Yu S, Jain R, Zhang N, Duong PQ, Butler C, Li Y, Lipzen A, Martin JA, Kerrie Barry W, Schmutz J, Tian L, Ronald PC (2020) Marker-free carotenoid-enriched rice generated through targeted gene insertion using CRISPR-Cas9. Nat Commun 11(1):1–10

Drigo B, Pijl AS, Duyts H, Kielak AM, Gamper HA, Houtekamer MJ, Boschker HTS, Bodelier PLE, Whiteley AS, van Veen JA, Kowalchuk GA (2010) Shifting carbon flow from roots into associated microbial communities in response to elevated atmospheric $CO_2$. Proc Natl Acad Sci 107(24):10938–10942

Dwivedi SL, Sahrawat KL, Rai KN, Blair MW, Andersson MS, Pfeiffer W (2012) Nutritionally enhanced staple food crops. In: Plant breeding reviews. Wiley, Chichester, pp 169–292

Ebi KL, Anderson CL, Hess JJ, Kim SH, Loladze I, Neumann RB, Singh D, Ziska L, Wood R (2021) Nutritional quality of crops in a high CO2 world: an agenda for research and technology development. Environ Res Lett 16(6):064045

Eichenauer J, Natt C, Hoefner W (1986) Variability of the yield structures of spring wheat ears caused by nitrogen supply, thermoperiod and growth regulators. Z Pflanzen 149:147–156

Endo A, Saika H, Takemura M, Misawa N, Toki S (2019) A novel approach to carotenoid accumulation in rice callus by mimicking the cauliflower Orange mutation via genome editing. Rice 12(1):1–5

Fahad S, Hussain S, Saud S, Hassan S, Shan D, Chen Y, Deng N, Khan F, Wu C, Wu W, Shah F, Ullah B, Yousaf M, Ali S, Huang J (2015) Grain cadmium and zinc concentrations in maize influenced by genotypic variations and zinc fertilization. CLEAN 43(10):1433–1440

Fang Y, Wang L, Xin Z, Zhao L, An X, Hu Q (2008) Effect of foliar application of zinc, selenium, and iron fertilizers on nutrients concentration and yield of rice grain in China. J Agric Food Chem 56(6):2079–2084

FAO (2005) FAOSTAT: FAO statistical databases. http://faostat3.fao.org/home/index.html

Farhat YA, Kim SH, Seyfferth AL, Zhang L, Neumann RB (2021) Altered arsenic availability, uptake, and allocation in rice under elevated temperature. Sci Total Environ 763:143049

Feng Z, Zhang B, Ding W, Liu X, Yang DL, Wei P, Cao F, Zhu S, Zhang F, Mao Y, Zhu JK (2013) Efficient genome editing in plants using a CRISPR/Cas system. Cell Res 23(10):1229–1232

Frossard E, Bucher M, Mächler F, Mozafar A, Hurrell R (2000) Potential for increasing the content and bioavailability of Fe, Zn and Ca in plants for human nutrition. J Sci Food Agric 80(7): 861–879

Furukawa T, Maekawa M, Oki T, Suda I, Iida S, Shimada H, Takamure I, Kadowaki K (2007) The Rc and Rd genes are involved in proanthocyanidin synthesis in rice pericarp. Plant J 49(1): 91–102

Gaikwad KB, Rani S, Kumar M, Gupta V, Babu PH, Bainsla NK, Yadav R (2020) Enhancing the nutritional quality of major food crops through conventional and genomics-assisted breeding. Front Nutr 7:533453

Gao Y, Xu Z, Zhang L, Li S, Wang S, Yang H, Liu X, Zeng D, Liu Q, Qian Q, Zhang B (2020) MYB61 is regulated by GRF4 and promotes nitrogen utilization and biomass production in rice. Nat Commun 11(1):1–12

Garg M, Sharma N, Sharma S, Kapoor P, Kumar A, Chunduri V, Arora P (2018) Biofortified crops generated by breeding, agronomy, and transgenic approaches are improving lives of millions of people around the world. Front Nutr 5:12

Giri A, Heckathorn S, Mishra S, Krause C (2017) Heat stress decreases levels of nutrient-uptake and-assimilation proteins in tomato roots. Plants 6(1):6

Graham JH, Leonard RT, Menge JA (1982) Interaction of light intensity and soil temperature with phosphorus inhibition of vesicular-arbuscular mycorrhiza formation. New Phytol 91(4): 683–690

Graham J, Ratnaparkhe MB, Powell W (2008) Molecular mapping and breeding of physiological traits. In: Kole C, Abbott AG (eds) Principles and practices of plant genomics. Science Publishers, Enfield, pp 217–241

Grotz N, Guerinot ML (2006) Molecular aspects of Cu, Fe and Zn homeostasis in plants. Biochim Biophys Acta 1763(7):595–608

Guo JX, Feng XM, Hu XY, Tian GL, Ling N, Wang JH, Shen QR, Guo SW (2016) Effects of soil zinc availability, nitrogen fertilizer rate and zinc fertilizer application method on zinc biofortification of rice. J Agric Sci 154(4):584–597

Guzman C, Peña RJ, Singh R, Autrique E, Dreisigacker S, Crossa J, Rutkoski J, Poland J, Battenfield S (2016) Wheat quality improvement at CIMMYT and the use of genomic selection on it. Appl Transl Genomics 11:3–8. https://doi.org/10.1016/j.atg.2016.10.004

Hameed A, Zaidi SSEA, Shakir S, Mansoor S (2018) Applications of new breeding technologies for potato improvement. Front Plant Sci 9:925

Hardarson GG, Broughton WJ (eds) (2003) Maximising the use of biological nitrogen fixation in agriculture, vol 99. Springer Science & Business Media, Berlin

Harding KL, Aguayo VM, Webb P (2018) Hidden hunger in South Asia: a review of recent trends and persistent challenges. Public Health Nutr 21(4):785–795

Haun W, Coffman A, Clasen BM, Demorest ZL, Lowy A, Ray E, Retterath A, Stoddard T, Juillerat A, Cedrone F, Mathis L, Voytas DF, Zhang F (2014) Improved soybean oil quality by targeted mutagenesis of the fatty acid desaturase 2 gene family. Plant Biotechnol J 12(7): 934–940

Hawkesford MJ (2012) Improving nutrient use efficiency in crops. eLS. John Wiley & Sons, Chichester

Haydon MJ, Cobbett CS (2007) Transporters of ligands for essential metal ions in plants. New Phytol 174:499–506

Hefferon KL (2016) Can biofortified crops help attain food security? Curr Mol Biol Rep 2(4): 180–185

Hu B, Wang W, Ou S, Tang J, Li H, Che R, Zhang Z, Chai X, Wang H, Wang Y, Liang C (2015) Variation in NRT1.1B contributes to nitrate-use divergence between rice subspecies. Nat Genet 47(7):834–838

Hu JH, Miller SM, Geurts MH, Tang W, Chen L, Sun N, Zeina CM, Gao X, Rees HA, Lin Z, Liu DR (2018) Evolved Cas9 variants with broad PAM compatibility and high DNA specificity. Nature 556(7699):57–63

Huang B, Rachmilevitch S, Xu J (2012) Root carbon and protein metabolism associated with heat tolerance. J Exp Bot 63(9):3455–3465

Huang H, Cui T, Zhang L, Yang Q, Yang Y, Xie K, Fan C, Zhou Y (2020) Modifications of fatty acid profile through targeted mutation at BnaFAD2 gene with CRISPR/Cas9-mediated gene editing in Brassica napus. Theor Appl Genet 133:2401–2411

Ishikawa S, Ishimaru Y, Igura M, Kuramata M, Abe T, Senoura T, Hase Y, Arao T, Nishizawa NK, Nakanishi H (2012) Ion-beam irradiation, gene identification, and marker-assisted breeding in the development of low-cadmium rice. Proc Natl Acad Sci U S A 109:19166–19171

Ishikawa S, Makino T, Ito M, Harada K, Nakada H, Nishida I, Nishimura M, Tokunaga T, Shirao K, Yoshizawa C, Matsuyama M, Abe T, Arao T (2016) Low-cadmium rice (*Oryza sativa* L.) cultivar can simultaneously reduce arsenic and cadmium concentrations in rice grains. Soil Sci Plant Nutr 62:327–339

Ishimaru Y, Suzuki M, Kobayashi T, Takahashi M, Nakanishi H, Mori S, Nishizawa NK (2005) OsZIP4, a novel zinc-regulated zinc transporter in rice. J Exp Bot 56(422):3207–3214

Islam FA, Beebe S, Muñoz M, Tohme J, Redden RJ, Basford KE (2004) Using molecular markers to assess the effect of introgression on quantitative attributes of common bean in the Andean gene pool. Theor Appl Genet 108(2):243–252

Jan WE, James G, Mark AS, Zbigniew K, Winiwarter W (2008) How a century of ammonia synthesis inferences the world. Nat Geosci 1:636–639

Jiang W, Zhou H, Bi H, Fromm M, Yang B, Weeks DP (2013) Demonstration of CRISPR/Cas9/sgRNA-mediated targeted gene modification in Arabidopsis, tobacco, sorghum and rice. Nucleic Acids Res 41(20):e188

Jinek M, Chylinski K, Fonfara I, Hauer M, Doudna JA, Charpentier E (2012) A programmable dual-RNA–guided DNA endonuclease in adaptive bacterial immunity. Science 337(6096): 816–821

Kaur N, Alok A, Kaur N, Pandey P, Awasthi P, Tiwari S (2018) CRISPR/Cas9-mediated efficient editing in phytoene desaturase (PDS) demonstrates precise manipulation in banana cv. Rasthali genome. Funct Integr Genomics 18(1):89–99

Kennedy T, Naeem S, Howe KM, Knops JMH, Timan D, Reich P (2002) Biodiversity as a barrier to ecological invasion. Nature 417:636–638

Kim YA, Moon H, Park CJ (2019) CRISPR/Cas9-targeted mutagenesis of Os8N3 in rice to confer resistance to Xanthomonas oryzae pv. oryzae. Rice 12(1):1–13

Kleinstiver BP, Prew MS, Tsai SQ, Topkar VV, Nguyen NT, Zheng Z, Gonzales AP, Li Z, Peterson RT, Yeh JRJ, Aryee MJ (2015) Engineered CRISPR-Cas9 nucleases with altered PAM specificities. Nature 523(7561):481–485

Köhler IH, Huber SC, Bernacchi CJ, Baxter IR (2019) Increased temperatures may safeguard the nutritional quality of crops under future elevated CO2 concentrations. Plant J 97(5):872–886

Kong FM, Guo Y, Liang X, Wu CH, Wang YY, Zhao Y, Li SS (2013) Potassium (K) effects and QTL mapping for K efficiency traits at seedling and adult stages in wheat. Plant Soil 373(1): 877–892

Koprivova A, Harper AL, Trick M, Bancroft I, Kopriva S (2014) Dissection of the control of anion homeostasis by associative transcriptomics in Brassica napus. Plant Physiol 166(1):442–450

Kranz S, Lin PJ, Wagstaff DA (2007) Children's dairy intake in the United States: too little, too fat? J Pediatr 151(6):642–646

Kumar J, Jaiswal V, Kumar A, Kumar N, Mir RR, Kumar S, Dhariwal R, Tyagi S, Khandelwal M, Prabhu KV, Prasad R, Balyana HS, Gupta PK (2011) Introgression of a major gene for high grain protein content in some Indian bread wheat cultivars. Field Crops Res 123(3):226–233

Kutman UB, Yildiz B, Ozturk L, Cakmak I (2010) Biofortification of durum wheat with zinc through soil and foliar applications of nitrogen. Cereal Chem 87:1–9

Langridge P, Fleury D (2011) Making the most of 'omics' for crop breeding. Trends Biotechnol 29(1):33–40

Lau WCP, Latif MA (2019) Current breeding approaches for developing rice with improved grain and nutritional qualities. In: Quality breeding in field crops. Springer, Cham, pp 199–216

Laurie SM, Faber M, Van Jaarsveld PJ, Laurie RN, Du Plooy CP, Modisane PC (2012) β-Carotene yield and productivity of orange-fleshed sweet potato (Ipomoea batatas L. Lam.) as influenced by irrigation and fertilizer application treatments. Sci Hortic 142:180–184

Li X, Wang Y, Liu Y, Yang B, Wang X, Wei J, Lu Z, Zhang Y, Wu J, Huang X, Yang L, Chen J (2018) Base editing with a Cpf1-cytidine deaminase fusion. Nat Biotechnol 36:324–327. https://doi.org/10.1038/nbt.4102

Li J, Jiao G, Sun Y, Chen J, Zhong Y, Yan L, Jiang D, Ma Y, Xia L (2021) Modification of starch composition, structure and properties through editing of TaSBEIIa in both winter and spring wheat varieties by CRISPR/Cas9. Plant Biotechnol J 19:937–951

Liang Z, Zhang K, Chen K, Gao C (2014) Targeted mutagenesis in Zea mays using TALENs and the CRISPR/Cas system. J Genet Genomics 41:63–68. https://doi.org/10.1016/j.jgg.2013.12.001

Lloveras J, Manent J, Viudas J, López A, Santiveri P (2004) Seeding rate influence on yield and yield components of irrigated winter wheat in a Mediterranean climate. Agron J 96:1258–1265

Long SP, Ainsworth EA, Rogers A, Ort DR (2004) Rising atmospheric carbon dioxide: plants FACE the future. Annu Rev Plant Biol 55:591–628

Lyons G, Ortiz-Monasterio I, Stangoulis J, Graham R (2005) Selenium concentration in wheat grain: is there sufficient genotypic variation to use in breeding? Plant Soil 269(1):369–380

Maleki, FS, Chaichi MR, Mazaheri D, Tavakkol, AR, Savaghebi G (2011) Barley grain mineral analysis as affected by different fertilizing systems and by drought stress

Mao Y, Yang X, Zhou Y, Zhang Z, Botella JR, Zhu JK (2018) Manipulating plant RNA-silencing pathways to improve the gene editing efficiency of CRISPR/Cas9 systems. Genome Biol 19(1): 1–15

Marschner H, Römheld V (1994) Strategies of plants for acquisition of iron. Plant Soil 165(2): 261–274

McGrath JM, Lobell DB (2013) Reduction of transpiration and altered nutrient allocation contribute to nutrient decline of crops grown in elevated CO2 concentrations. Plant Cell Environ 36(3): 697–705

Miles C, Wayne M (2008) Quantitative trait locus (QTL) analysis. Nat Educ 1(1):208

Mitchell-Olds T (2010) Complex-trait analysis in plants. Genome Biol 11(4):1–3

Mori S, Nishizawa N, Hayashi H, Chino M, Yoshimura E, Ishihara J (1991) Why are young rice plants highly susceptible to iron deficiency? In: Iron nutrition and interactions in plants. Springer, Dordrecht, pp 175–188

Morineau C, Bellec Y, Tellier F, Gissot L, Kelemen Z, Nogué F, Faure JD (2017) Selective gene dosage by CRISPR-Cas9 genome editing in hexaploid Camelina sativa. Plant Biotechnol J 15(6):729–739

Myers SS, Zanobetti A, Kloog I, Huybers P, Leakey AD, Bloom AJ, Carlisle E, Dietterich LH, Fitzgerald G, Hasegawa T, Holbrook NM, Nelson RL, Ottman MJ, Raboy V, Sakai H, Sartor KA, Schwartz J, Seneweera S, Tausz M, Usui Y (2014) Increasing $CO_2$ threatens human nutrition. Nature 510(7503):139–142

Myers SS, Wessells KR, Kloog I, Zanobetti A, Schwartz J (2015) Rising atmospheric $CO_2$ increases global threat of zinc deficiency. Lancet Glob Health 3(10):e639

Nakandalage N, Seneweera S (2018) Micronutrients use efficiency of crop-plants under changing climate. In: Plant micronutrient use efficiency. Academic, New York, pp 209–224

Nakayasu M, Umemoto N, Ohyama K, Fujimoto Y, Lee HJ, Watanabe B, Muranaka T, Saito K, Sugimoto Y, Mizutani M (2017) A dioxygenase catalyzes steroid 16α-hydroxylation in steroidal glycoalkaloid biosynthesis. Plant Physiol 175(1):120–133

Nelson JA, Morgan JA, LeCain DR, Mosier AR, Milchunas DG, Parton BA (2004) Elevated $CO_2$ increases soil moisture and enhances plant water relations in a long-term field study in semi-arid shortgrass steppe of Colorado. Plant Soil 259(1):169–179

Newell-McGloughlin M (2008) Nutritionally improved agricultural crops. Plant Physiol 147(3): 939–953

Nishimasu H, Shi X, Ishiguro S, Gao L, Hirano S, Okazaki S, Noda T, Abudayyeh OO, Gootenberg JS, Mori H, Oura S (2018) Engineered CRISPR-Cas9 nuclease with expanded targeting space. Science 361(6408):1259–1262

Nosheen A, Bano A, Ullah F (2011) Nutritive value of canola (Brassica napus L.) as affected by plant growth promoting rhizobacteria. Eur J Lipid Sci Technol 113(11):1342–1346
O'Mara FP (2012) The role of grasslands in food security and climate change. Ann Bot 110(6): 1263–1270
Okuzaki A, Ogawa T, Koizuka C, Kaneko K, Inaba M, Imamura J, Koizuka N (2018) CRISPR/Cas9-mediated genome editing of the fatty acid desaturase 2 gene in Brassica napus. Plant Physiol Biochem 131:63–69
Oliva ML, Shannon JG, Sleper DA, Ellersieck MR, Cardinal AJ, Paris RL, Lee JD (2006) Stability of fatty acid profile in soybean genotypes with modified seed oil composition. Crop Sci 46(5): 2069–2075
Oliva R, Ji C, Atienza-Grande G, Huguet-Tapia JC, Perez-Quintero A, Li T, Eom JS, Li C, Nguyen H, Liu B, Auguy F (2019) Broad-spectrum resistance to bacterial blight in rice using genome editing. Nat Biotechnol 37(11):1344–1350
Patidar M, Mali AL (2004) Effect of farmyard manure, fertility levels and bio-fertilizers on growth, yield and quality of sorghum (Sorghum bicolor). Indian J Agron 49(2):117–120
Patra AK (2010) Nutritional security: a missing link in climate change debates. Curr Sci 99(7): 887–889
Peirson DR, Elliott JR (1988) Effect of nitrite and bicarbonate on nitrite utilization in leaf tissue of bush bean (*Phaseolus vulgaris*). J Plant Physiol 133(4):425–429
Pellegrino E, Bedini S (2014) Enhancing ecosystem services in sustainable agriculture: biofertilization and biofortification of chickpea (Cicer arietinum L.) by arbuscular mycorrhizal fungi. Soil Biol Biochem 68:429–439
Pérez-Massot E, Banakar R, Gómez-Galera S, Zorrilla-López U, Sanahuja G, Arjó G et al (2013) The contribution of transgenic plants to better health through improved nutrition: opportunities and constraints. Genes Nutr 8(1):29–41
Pilbeam DJ (2015) Breeding crops for improved mineral nutrition under climate change conditions. J Exp Bot 66(12):3511–3521
Pingali PL (2012) Green revolution: impacts, limits, and the path ahead. Proc Natl Acad Sci 109(31):12302–12308
Poblaciones MJ, Rengel Z (2016) Soil and foliar zinc biofortification in field pea (Pisum sativum L.): grain accumulation and bioavailability in raw and cooked grains. Food Chem 212:427–433
Poggi V, Arcioni A, Filippini P, Pifferi PG (2000) Foliar application of selenite and selenate to potato (*Solanum tuberosum*): effect of a ligand agent on selenium content of tubers. J Agric Food Chem 48(10):4749–4751
Prasanna BM, Palacios-Rojas N, Hossain F, Muthusamy V, Menkir A, Dhliwayo T et al (2020) Molecular breeding for nutritionally enriched maize: status and prospects. Front Genet 10:1392
Prentice A (2008) Vitamin D deficiency: a global perspective. Nutr Rev 66:S153–S164
Ram H, Rashid A, Zhang W, Duarte AP, Phattarakul N, Simunji S, Kalayci M, Freitas R, Rerkasem B, Bal RS, Mahmood K, Savasli E, Lungu O, Wang ZH, de Barros VLNP, Malik SS, Arisoy RZ, Guo JX, Sohu VS, Zou CQ, Cakmak I (2016) Biofortification of wheat, rice and common bean by applying foliar zinc fertilizer along with pesticides in seven countries. Plant Soil 403(1):389–401
Ramesh A, Sharma SK, Sharma MP, Yadav N, Joshi OP (2014) Inoculation of zinc solubilizing Bacillus aryabhattai strains for improved growth, mobilization and biofortification of zinc in soybean and wheat cultivated in Vertisols of central India. Appl Soil Ecol 73:87–96
Ramzani PMA, Khalid M, Naveed M, Ahmad R, Shahid M (2016) Iron biofortification of wheat grains through integrated use of organic and chemical fertilizers in pH affected calcareous soil. Plant Physiol Biochem 104:284–293
Rengel Z, Batten GD, Crowley DD (1999) Agronomic approaches for improving the micronutrient density in edible portions of field crops. Field Crops Res 60(1-2):27–40
Sánchez-León S, Gil-Humanes J, Ozuna CV, Giménez MJ, Sousa C, Voytas DF, Barro F (2018) Low-gluten, nontransgenic wheat engineered with CRISPR/Cas9. Plant Biotechnol J 16(4): 902–910

Sanz-Sáez Á, Erice G, Aranjuelo I, Nogués S, Irigoyen JJ, Sánchez-Díaz M (2010) Photosynthetic down-regulation under elevated $CO_2$ exposure can be prevented by nitrogen supply in nodulated alfalfa. J Plant Physiol 167(18):1558–1565

Sasaki A, Yamaji N, Yokosho K, Ma JF (2012) Nramp5 is a major transporter responsible for manganese and cadmium uptake in rice. Plant Cell 24:2155–2167

Schiller M, Hegelund JN, Pedas P, Kichey T, Laursen KH, Husted S, Schjoerring JK (2014) Barley metallothioneins differ in ontogenetic pattern and response to metals. Plant Cell Environ 37(2): 353–367

Seth K (2016) Current status of potential applications of repurposed Cas9 for structural and functional genomics of plants. Biochem Biophys Res Commun 480(4):499–507

Shan Q, Zhang Y, Chen K, Zhang K, Gao C (2015) Creation of fragrant rice by targeted knockout of the Os BADH 2 gene using TALEN technology. Plant Biotechnol J 13(6):791–800

Sheoran HS, Duhan BS, Grewal KS, Sheoran S (2015) Grain yield and NPK uptake of wheat (*Triticum aestivum* L.) as influenced by nitrogen, vermicompost and herbicide (*Clodinafop propargyl*). Afr J Agric Res 10(42):3952–3961

Shewmaker CK, Sheehy JA, Daley M, Colburn S, Ke DY (1999) Seed-specific overexpression of phytoene synthase: increase in carotenoids and other metabolic effects. Plant J 20(4):401–412

Shewry PR (2009) Wheat. J Exp Bot 60:1537–1553

Shunmugam AS, Kannan U, Jiang Y, Daba KA, Gorim LY (2018) Physiology based approaches for breeding of next-generation food legumes. Plants 7(3):72

Singh S, Malik R, Dhankar J, Yadav A, Garg R, Kamboj LO (2010) Nutrient use pattern in the irrigated rice-wheat cropping system in the Indo-Gangetic plains of Haryana, India. Exp Agric 46(2):191–209

Songmei L, Jie J, Yang L, Jun M, Shouling X, Yuanyuan T, Youfa L, Quigyao S, Jiazhong H (2019) Characterization and evaluation of OsLCT1 and OsNramp5 mutants generated through CRISPR/Cas9-mediated mutagenesis for breeding low Cd rice. Rice Sci 26(2):88–97

Sperotto RA, Vasconcelos MW, Grusak MA, Fett JP (2012) Effects of different Fe supplies on mineral partitioning and remobilization during the reproductive development of rice (*Oryza sativa L.*). Rice 5:27

Sun Y, Zhang X, Wu C, He Y, Ma Y, Hou H, Guo X, Du W, Zhao Y, Xia L (2016) Engineering herbicide-resistant rice plants through CRISPR/Cas9-mediated homologous recombination of acetolactate synthase. Mol Plant 9:628–631

Sun Y, Jiao G, Liu Z, Zhang X, Li J, Guo X, Du W, Du J, Francis F, Zhao Y, Xia L (2017) Generation of high-amylose rice through CRISPR/Cas9-mediated targeted mutagenesis of starch branching enzymes. Front Plant Sci 8:1298

Suzuki M, Tsukamoto T, Inoue H, Watanabe S, Matsuhashi S, Takahashi M, Nakanishi H, Mori S, Nishizawa NK (2008) Deoxymugineic acid increases Zn translocation in Zn-deficient rice plants. Plant Mol Biol 66:609–617

Tang L, Mao B, Li Y, Lv Q, Zhang L, Chen C, He H, Wang W, Zeng X, Shao Y, Pan Y (2017a) Knockout of OsNramp5 using the CRISPR/Cas9 system produces low Cd-accumulating indica rice without compromising yield. Sci Rep 7(1):1–12

Tang X, Lowder LG, Zhang T, Malzahn AA, Zheng X, Voytas DF, Zhong Z, Chen Y, Ren Q, Li Q, Kirkland ER (2017b) A CRISPR–Cpf1 system for efficient genome editing and transcriptional repression in plants. Nat Plants 3(3):1–5

Taub DR, Wang X (2008) Why are nitrogen concentrations in plant tissues lower under elevated $CO_2$? A critical examination of the hypotheses. J Integr Plant Biol 50(11):1365–1374

Tester M, Langridge P (2010) Breeding technologies to increase crop production in a changing world. Science 327(5967):818–822

Tucker G (2003) Nutritional enhancement of plants. Curr Opin Biotechnol 14(2):221–225

Tyagi S, Mir RR, Kaur H, Chhuneja P, Ramesh B, Balyan HS, Gupta PK (2014) Marker-assisted pyramiding of eight QTLs/genes for seven different traits in common wheat (Triticum aestivum L.). Mol Breed 34(1):167–175

UNICEF & World Health Organization (2017) The state of food security and nutrition in the world 2017: building resilience for peace and food security

Van Steenhuyse L (2017) DuPont Pioneer's next generation of waxy corn shows the green side of CRISPR-Cas9. BioVox

Varshney RK (2016) Exciting journey of 10 years from genomes to fields and markets: some success stories of genomics-assisted breeding in chickpea, pigeonpea and groundnut. Plant Sci 242:98–107

Varshney RK, Close TJ, Singh NK, Hoisington DA, Cook DR (2009) Orphan legume crops enter the genomics era! Curr Opin Plant Biol 12(2):202–210

Varshney RK, Singh VK, Kumar A, Powell W, Sorrells ME (2018) Can genomics deliver climate-change ready crops? Curr Opin Plant Biol 45:205–211

Vishwakarma MK, Arun B, Mishra VK, Yadav PS, Kumar H, Joshi AK (2016) Marker-assisted improvement of grain protein content and grain weight in Indian bread wheat. Euphytica 208(2): 313–321

Vlčko T, Ohnoutkova L (2020) Allelic Variants of CRISPR/Cas9 induced mutation in an inositol trisphosphate 5/6 kinase gene manifest different phenotypes in barley. Plants 9(2):195

Wang F, Wang C, Liu P, Lei C, Hao W, Gao Y, Liu Y, Zhao K (2016) Enhanced rice blast resistance by CRISPR/Cas9-targeted mutagenesis of the ERF transcription factor gene OsERF922. PLoS One 11(4):e0154027

Wang J, Hasegawa T, Li L, Lam SK, Zhang X, Liu X, Pan G (2019a) Changes in grain protein and amino acids composition of wheat and rice under short-term increased [CO2] and temperature of canopy air in a paddy from East China. New Phytol 222(2):726–734

Wang T, Li Y, Fu Y, Xie H, Song S, Qiu M, Wen J, Chen M, Ge C, Tian Y, Li C, Yuan D, Wang J, Li L (2019b) Mutation at different sites of metal transporter gene OsNramp5 affects Cd accumulation and related agronomic traits in rice (*Oryza sativa L.*). Front Plant Sci 10:1081

Wang J, Li L, Lam SK, Liu X, Pan G (2020) Responses of wheat and rice grain mineral quality to elevated carbon dioxide and canopy warming. Field Crops Res 249:107753

Watanabe KN, Sassa Y, Suda E, Chen CH, Inaba M, Kikuchi A (2005) Global political, economic, social and technological issues on trasngenic crops. Plant Biotechnol 22(5):515–522

Waters BM, Sankaran RP (2011) Moving micronutrients from the soil to the seeds: genes and physiological processes from a biofortification perspective. Plant Sci 180(4):562–574

Welch RM, Graham RD (2004) Breeding for micronutrients in staple food crops from a human nutrition perspective. J Exp Bot 55(396):353–364

Wessells KR, Brown KH (2012) Estimating the global prevalence of zinc deficiency: results based on zinc availability in national food supplies and the prevalence of stunting. PLoS One 7(11): e50565

Westermann D, Terán H, Muñoz-Perea C, Singh S (2011) Plant and seed nutrient uptake in common bean in seven organic and conventional production systems. Can J Plant Sci 91(6): 1089–1099

White PJ, George TS, Gregory PJ, Bengough AG, Hallett PD, McKenzie BM (2013) Matching roots to their environment. Ann Bot 112(2):207–222

White PJ, Thompson JA, Wright G, Rasmussen SK (2017) Biofortifying Scottish potatoes with zinc. Plant Soil 411(1–2):151–165

Whitfield DM, Smith CJ (1989) Effects of irrigation and nitrogen on growth, light interception and efficiency of light conversion in wheat. Field Crops Res 20:279–295

Xie K, Minkenberg B, Yang Y (2015) Boosting CRISPR/Cas9 multiplex editing capability with the endogenous tRNA-processing system. Proc Natl Acad Sci 112(11):3570–3575

Xin-Min J, Xue-Yi C, Ji-Yong J, Tai M, James DW, Rakeman MA, Dou ZH, Mamette M, Amette K, Zhang ML, Delong GR (1997) Dynamics of environmental supplementation of iodine: four years' experience of iodination of irrigation water in Hotien, Xinjiang, China. Arch Environ Health 52(6):399–408

Xu Y, Crouch JH (2008) Marker-assisted selection in plant breeding: from publications to practice. Crop science 48(2):391–407

Xu R, Yang Y, Qin R, Li H, Qiu C, Li L, Wei P, Yang J (2016) Rapid improvement of grain weight via highly efficient CRISPR/Cas9-mediated multiplex genome editing in rice. J Genet Genom 43:529–532

Yasin M, El Mehdawi AF, Jahn CE, Anwar A, Turner MF, Faisal M, Pilon-Smits EA (2015) Seleniferous soils as a source for production of selenium-enriched foods and potential of bacteria to enhance plant selenium uptake. Plant Soil 386(1):385–394

Yasumoto S, Umemoto N, Lee HJ, Nakayasu M, Sawai S, Sakuma T, Yamamoto T, Mizutani M, Saito K, Muranaka T (2019) Efficient genome engineering using Platinum TALEN in potato. Plant Biotechnol 2019:19–0805

Ye X, Al-Babili S, Klöti A, Zhang J, Lucca P, Beyer P, Potrykus I (2000) Engineering the provitamin A (β-carotene) biosynthetic pathway into (carotenoid-free) rice endosperm. Science 287(5451):303–305

Yuan L, Wu L, Yang C, Lv Q (2013) Effects of iron and zinc foliar applications on rice plants and their grain accumulation and grain nutritional quality. J Sci Food Agric 93(2):254–261

Yuan M, Zhu J, Gong L, He L, Lee C, Han S et al (2019) Mutagenesis of FAD2 genes in peanut with CRISPR/Cas9 based gene editing. BMC Biotechnol 19(1):1–7

Zhang MY (2017) Investigations of the effects of Zn biofortification in different wheat genotypes. Yangzhou University, Yangzhou

Zhang H, Rong H, Pilbeam D (2007) Signalling mechanisms underlying the morphological responses of the root system to nitrogen in Arabidopsis thaliana. J Exp Bot 58(9):2329–2338

Zhang Y, Zhang Y, Liu N, Su D, Xue Q, Stewart BA, Wang Z (2012) Effect of source–sink manipulation on accumulation of micronutrients and protein in wheat grains. J Plant Nutr Soil Sci 175(4):622–629

Zhang J, Zhang H, Botella JR, Zhu JK (2018a) Generation of new glutinous rice by CRISPR/Cas9-targeted mutagenesis of the waxy gene in elite rice varieties. J Integr Plant Biol 60(5):369–375

Zhang J, De-oliveira-Ceciliato P, Takahashi Y, Schulze S, Dubeaux G, Hauser F, Azoulay-Shemer-T, Tõldsepp K, Kollist H, Rappel W, Julian I, Schroeder JI (2018b) Insights into the molecular mechanisms of $CO_2$-mediated regulation of stomatal movements. Curr Biol 28(23):R1356–R1363

Zhang A, Liu Y, Wang F, Li T, Chen Z, Kong D, Bi J, Zhang F, Luo X, Wang J, Tang J, Yu X, Liu G, Luo L (2019) Enhanced rice salinity tolerance via CRISPR/Cas9-targeted mutagenesis of the OsRR22 gene. Mol Breed 39(3):1–10

Zhang J, Zhang H, Li S, Li J, Yan L, Xia L (2021) Increasing yield potential through manipulating of an ARE1 ortholog related to nitrogen use efficiency in wheat by CRISPR/Cas9. J Integr Plant Biol 63(9):1649–1663

Zhao P, You Q, Lei M (2019) A CRISPR/Cas9 deletion into the phosphate transporter SlPHO1; 1 reveals its role in phosphate nutrition of tomato seedlings. Physiol Plant 167(4):556–563

Zhou H, He M, Li J, Chen L, Huang Z, Zheng S, Zhu L, Ni E, Jiang D, Zhao B, Zhuang C (2016) Development of commercial thermo-sensitive genic male sterile rice accelerates hybrid rice breeding using the CRISPR/Cas9-mediated TMS5 editing system. Sci Rep 6(1):1–12

Zhu C, Naqvi S, Gomez-Galera S, Pelacho AM, Capell T, Christou P (2007) Transgenic strategies for the nutritional enhancement of plants. Trends Plant Sci 12(12):548–555

# Chapter 5
# Role of Transpiration in Regulating Leaf Temperature and its Application in Physiological Breeding

**S. R. W. M. C. J. K. Ranawana, Helen Bramley, Jairo A. Palta, and Kadambot H. M. Siddique**

**Abstract** Global warming and consequent changes in climate will put intense pressure on crop production, undermining global food security. Adaptation strategies are needed to minimise the adverse impact of high temperature on crops, and physiological trait-based breeding is considered as a promising strategy in this regard. Identifying and assessing physiological traits associated with improved crop performance under warmer climatic conditions will assist physiological breeding programmes. In this context, it is vital to assess the role of transpiration in ameliorating leaf temperature as a potential heat avoidance strategy as the literature is limited and controversial. The complex relationship between transpiration, the physical environment, and plant anatomical and morpho-physiological attributes could explain the contrasting views on the role of transpiration in ameliorating leaf temperature. A comprehensive examination of transpiration and leaf cooling in relation to its controlling factors will assist in unravelling this complex relationship. This chapter discusses the different heat-dissipating mechanisms, the contribution of transpirational cooling as a heat avoidance strategy evidenced by previous research findings on different crops, the complex relationship of transpiration and leaf

S. R. W. M. C. J. K. Ranawana
The UWA Institute of Agriculture, The University of Western Australia, Perth, WA, Australia

Department of Export Agriculture, Faculty of Animal Science and Export Agriculture, Uva Wellassa University, Badulla, Sri Lanka

H. Bramley
Plant Breeding Institute, School of Life and Environmental Sciences, The University of Sydney, Narrabri, NSW, Australia

J. A. Palta
The UWA Institute of Agriculture, The University of Western Australia, Perth, WA, Australia

CSIRO Agriculture, Wembley, WA, Australia

K. H. M. Siddique (✉)
The UWA Institute of Agriculture, The University of Western Australia, Perth, WA, Australia
e-mail: kadambot.siddique@uwa.edu.au



H. M. Mamrutha et al. (eds.), *Translating Physiological Tools to Augment Crop Breeding*, https://doi.org/10.1007/978-981-19-7498-4_5

temperature with morpho-physiological attributes and environmental parameters, and the application of transpirational cooling in physiological breeding.

**Keywords** Heat avoidance · Transpiration · Canopy temperature · Morpho-physiological attributes · Physiological breeding

## 5.1 Introduction

Climate change models consistently predict a dramatic increase in air temperatures and atmospheric vapour pressure deficit (VPD) (IPCC 2007), which could put global crop production at risk due to morpho-anatomical, physiological, biochemical, and molecular changes (Xu et al. 1995; Tewari and Tripathy 1998; Zhao et al. 2007; Farooq et al. 2011; Asseng et al. 2015). Heat-stress sensitive crops such as wheat (Wollenweber et al. 2003) would be highly affected by elevated temperatures under future climate scenarios. For example, a 2 °C increase in average temperatures in the wheat-growing regions of Australia could reduce yields by up to 50% (Asseng et al. 2011). Temperatures above optimum will accelerate plant development in crops such as wheat, limiting their ability to accumulate biomass to support grain growth (Asseng et al. 2011). Baker et al. (1992) reported a 7–8% grain yield reduction in rice per 1 °C increment in air temperature. We predict that global warming will put immense pressure on future food security, highlighting the current need to identify potential strategies to mitigate the adverse effect of increased temperatures on crops.

Different mechanisms can minimise heat-stress damage in crops, including heat escape, heat avoidance, and heat tolerance (Ishimaru et al. 2010; Jagadish et al. 2010; Julia and Dingkuhn 2012, 2013; Hirabayashi et al. 2014). In general, heat tolerance is emphasised when considering the heat-stress responses of crops, with few studies investigating heat avoidance or escape phenomena, which are also highly effective under field conditions (Prasad et al. 2017). Transpirational cooling is a heat avoidance strategy that helps plants to regulate their temperatures. Transpiration assists crops in dissipating excessive heat through evaporative cooling, allowing their canopies to remain a few degrees cooler than air temperature. Although water loss through transpiration has been considered a negative consequence of stomatal opening for $CO_2$ acquisition, transpiration plays an important role in ameliorating leaf temperature (Deva et al. 2020) and facilitating nutrient uptake through roots. Therefore, enhanced transpirational cooling may be a useful physiological trait in physiological breeding. However, transpiration and its cooling effect are complex as they are controlled by several plant anatomical and morpho-physiological attributes and the physical environment. Therefore, a better understanding of the relative role of transpiration in regulating leaf temperature in relation to its controlling factors is needed to use this trait successfully in physiological breeding.

This chapter discusses the potential role of transpiration as a heat avoidance strategy in regulating leaf temperature, its complex relationships with morpho-physiological characteristics and other environmental conditions, and the application of transpirational cooling in physiological breeding.

## 5.2 Leaf Energy Balance

There is a flow of energy between the plant and its surrounding environment (Raschke 1960; Gates 1968). Plants receive energy from solar radiation, which is used for plant physiological and metabolic processes or converted to heat, increasing leaf temperature (Gates 1968). The available energy is divided into different plant and environmental processes by a concept called *energy balance* (De Costa 2004; Fig. 5.1).

Solar radiation incident on leaves is reflected, absorbed, or transmitted. The absorbed energy can be dissipated by the emission of long-wave radiation (re-radiation), sensible heat loss (conductance and convection), and latent heat loss through transpiration (Jones 2014; Fig. 5.1). The relative contribution of each heat-dissipating mechanism differs between habitats and plant types. For example, in arid habitats, leaf physical adjustments play a prominent role in dissipating heat, while the contribution of transpirational cooling is often negligible (Ansari and Loomis 1959; Lin et al. 2017). In contrast, transpiration contributes more to leaf cooling than physical attributes when there is no water limitation (Lin et al. 2017). Moreover, leaf morpho-anatomical traits and physiological processes, mainly transpiration, can influence the amount of radiation absorbed by leaves and its dissipation, ultimately regulating leaf temperature (Gates 2003). Vaseline-treated (restricted transpiration) and control (unrestricted transpiration) leaves are often used by researchers to distinguish the thermal effects of transpiration and physical attributes (Curtis 1936; Lin et al. 2017).

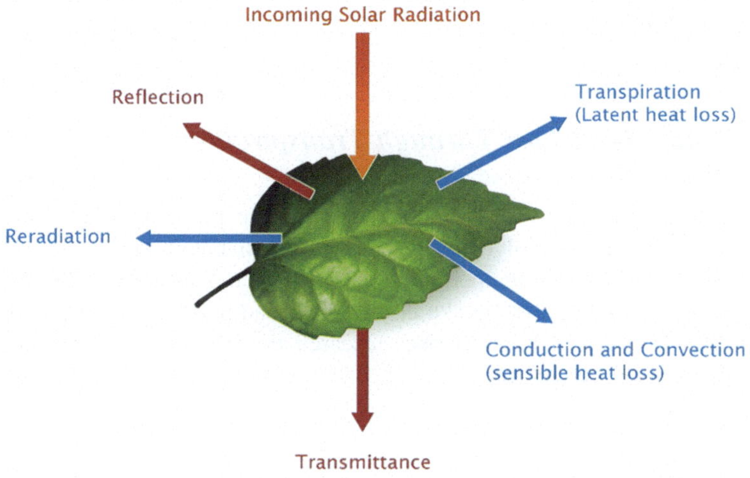


**Fig. 5.1** Leaf energy balance shows the different processes involved in dissipating the absorbed energy received through solar radiation

### 5.2.1 Re-radiation

Re-radiation is the most effective energy dissipation mode, functioning under any set of environmental conditions, while transpiration is less effective under humid and cool environmental conditions (Idso and Baker 1967).

### 5.2.2 Sensible Heat Loss

Sensible heat loss is the dissipation of heat through conduction and convection. Heat loss through convection increases with high wind speed and large differences between leaf and air temperatures (Idso and Baker 1967). It also depends on leaf size, where convective heat loss is more effective in smaller leaves due to low resistance from the boundary layer (Givnish 1987; Ball et al. 1988), the layer of still air adjacent to the leaf surface. Larger leaves have a thicker boundary layer of air adhering to the surface. In general, the boundary layer of a small leaf is ~1 mm thick, and a large leaf is 1–2 mm thick. In addition, leaves in the top and lateral periphery of the canopy are exposed to intense radiation and their heat dissipation through convection is greater than that in interior and lower peripheral leaves (Idso and Baker 1967). Under circumstances where transpiration is low (low latent heat flux—see below), the ratio between sensible heat flux to latent heat flux (*Bowen ratio*) increases (Taiz and Zeiger 2010) and heat dissipation through sensible heat loss becomes prominent. In some cases, there can be sensible heat gain rather than sensible heat loss when a leaf cools below air temperature by transpiration (Taiz and Zeiger 2010).

### 5.2.3 Latent Heat Loss Through Transpiration

Considerable heat is removed from plants during transpiration as latent heat loss. Therefore, transpiration involves heat avoidance by maintaining leaf temperatures several degrees below air temperature, minimising the damage associated with above-optimum temperatures (Drake et al. 1970; Radin 1992; Farooq et al. 2011; Xu et al. 2011). Table 5.1 summarises research findings on transpiration and its cooling effect in selected crops.

**Table 5.1** Effect of transpirational cooling on plant temperature regulation in selected crops

| Crop | Key observations/conclusion | Reference |
|---|---|---|
| Rice (*Oryza sativa*) | Panicle temperature ranged from 9.5 °C below to 2 °C above air temperature at 2 m height<br>Avoided heat stress via heat escape and transpirational cooling, even at 40 °C in all tested rice genotypes<br>Greater heat stress injury in warm-humid conditions than hot-arid conditions due to humidity effects on transpiration cooling | Julia and Dingkuhn (2013) |
| | Rice panicles maintained 6.8 °C lower temperatures than air due to low humidity and greater transpirational cooling | Matsui et al. (2007) |
| | Panicle temperature increased by 4 °C above ambient under humid conditions due to limited transpirational cooling | Tian et al. (2010) |
| Wheat (*Triticum aestivum*) | Genotypic variability in leaf temperature response to air temperature<br>Genotypes whose transpiration rate increased more in response to VPD maintained lower leaf temperatures as air temperature increased and vice versa. However, the leaf temperature response to air temperature for some genotypes did not correlate with the transpiration response to VPD. That is, the leaf temperature response to air temperature is mediated by, but not solely dependent on, the transpiration response to VPD | Ranawana (2015) |
| | Leaf temperature is closely related to transpirational cooling | Inagaki and Nachit (2008) |
| Sorghum (*Sorghum bicolor*) | Plant temperatures increased in $CO_2$-enriched treatments due to decreased transpirational cooling | Triggs et al. (2004) |
| Barley (*Hordeum vulgare*) | Transpirational cooling decreased (warmer leaves) due to the decreasing transpiration from elevated $CO_2$ | Wall et al. (2011) |
| Potato (*Solanum tuberosum*) | Inverse relationship between canopy temperature and transpiration rate | Mahmud et al. (2016) |
| | Leaf temperature increased with decreasing transpiration rates under low soil water availability | Yuan et al. (2003) |
| Cotton (*Gossypium hirsutum*) | Leaf temperature decreased with increasing transpiration rate | Pallas Jr et al. (1967) |
| Pima cotton (*Gossypium barbadense*) | Cotton leaves maintained substantially lower temperatures than air temperature | Wise et al. (2004) |

(continued)

**Table 5.1** (continued)

| Crop | Key observations/conclusion | Reference |
|---|---|---|
| | due to the large leaf size and very high $g_s$ allowing high rates of water loss for evaporative cooling | |
| Maize (*Zea mays*) | Decrease in transpiration under restricted water supply increased leaf temperature Transpirational cooling is an effective form of heat avoidance when water supply is not limited | Yadav et al. (2016) |

## 5.3 Transpiration as a Heat Avoidance Strategy

Transpiration is the evaporation of water from the leaf to the surrounding atmosphere through minute apertures located in the leaf epidermis known as stomata (Idso and Baker 1967). There are two basic requirements of transpiration (De Costa 2004):

1. Energy supply: Plant receives energy needed to convert liquid water to water vapour by solar radiation. Therefore, the rate of transpiration depends on how much radiation energy is absorbed by a plant canopy (De Costa 2004).
2. Removal of evaporated water from the leaf to the atmosphere (Fig. 5.2): Water vapour evaporated from mesophyll cells (site of evaporation within leaves)

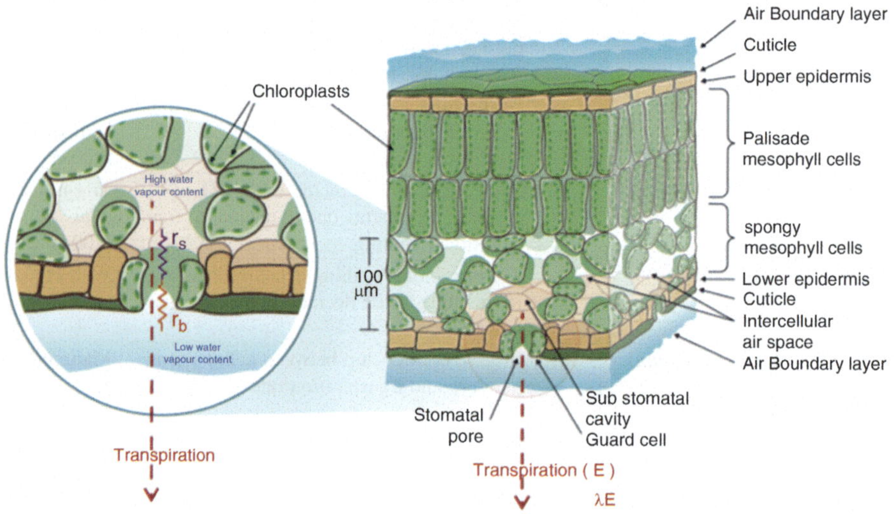


**Fig. 5.2** Schematic cross-section of a typical leaf showing the transpiration process. $r_s$ and $r_b$ are stomatal resistance and boundary layer resistance, respectively; $\lambda E$ is latent heat flux, the loss of energy from the leaf surface due to transpiration; $\lambda$ is latent heat of vaporisation; $E$ is transpiration rate. [Modified from Nobel (2005)]

(Holmgren et al. 1965) accumulates in the sub-stomatal cavity and diffuses from the sub-stomatal cavity to the leaf surface through stomata. This process is controlled by stomatal conductance ($g_s$) and the water vapour concentration gradient or leaf-to-air VPD, the driving force for transpiration. Water vapour then moves from the leaf surface to the air beyond the boundary layer, influenced by boundary layer resistance, wind speed, and relative humidity or VPD (De Costa 2004).

To convert liquid water to water vapour requires energy. Every gram of water that escapes from the leaf through transpiration carries 2.45 kJ of energy (at 20 °C) and is called the latent heat of vaporisation (Jones 2014). Therefore, transpiration is an important mode of temperature regulation in plants. However, the relative role of transpiration in ameliorating leaf temperature is controversial. Some studies have suggested that the contribution of transpiration (latent heat loss) in regulating leaf temperature is less than other heat dissipation mechanisms (Clum 1926; Curtis 1936). Conversely, Shull (1930) and Clements (1934) contend that transpiration plays a major role in dissipating heat energy and lowering leaf temperature. Clements (1934) also pointed out that comparing leaf temperature with air temperature may underestimate the role of transpiration in leaf cooling as leaf and air absorption coefficients differ, where plants show relatively more capacity to absorb light than air. These contrasting views on the role of transpiration in ameliorating leaf temperature necessitate a closer examination of this phenomenon.

### *5.3.1 Canopy Temperature Depression (CTD)*

A crop's canopy temperature can be several degrees cooler than the air temperature due to evaporative cooling during transpiration or other heat-dissipating mechanisms (Kumar and Tripathi 1991; Farooq et al. 2011). The degree of cooling depends on the rate of transpiration (Amani et al. 1996) and the magnitude of heat flux from other heat-dissipating mechanisms. The difference between air and canopy temperature is expressed as canopy temperature depression (CTD) and is considered a selection trait in screening many crops, including wheat, for heat tolerance in hot environments with plenty of available water (Reynolds et al. 1994; Amani et al. 1996). CTD has been positively correlated with some physiological parameters such as $g_s$ (Amani et al. 1996), plant water status (Blum et al. 1982), and yield (Reynolds et al. 1994; Amani et al. 1996) and hence used as an indicator in plant water relations studies (Feng et al. 2009). Leaf temperature and $g_s$ strongly correlated for wheat under a cyclic drought treatment, but not under well-watered conditions (Izanloo et al. 2008). Recent studies confirmed that wheat genotypes with low canopy temperature are more superior than genotypes with high canopy temperature for physiological and metabolic aspects such as chlorophyll content, leaf functional duration, transpiration rate, photosynthetic rate, protein content and superoxide dismutase, catalase and peroxidase activities (Feng et al. 2009).

Genotypic variation in canopy temperature has been widely observed in many crops (Blum et al. 1989; Amani et al. 1996; Fischer et al. 1998; Ayeneh et al. 2002; Takai et al. 2010; Ranawana 2015), which likely exist due to anatomical and morpho-physiological factors contributing to transpirational cooling and the other heat-dissipating mechanisms.

### 5.3.2 *Complex Relationship of Transpiration and Leaf Temperature with Plant and Physical Factors*

The relationship between leaf/canopy temperature, air temperature, and transpiration is complex (Fig. 5.3) and arises from the relationship between transpiration and the physical environment and plant anatomical and morpho-physiological attributes.

Therefore, assessing the role of transpiration in mediating leaf temperature is challenging due to the confounding effects of other variables. A comprehensive examination of transpiration and leaf cooling under varying levels of controlling factors such as plant anatomical and morpho-physiological attributes and atmospheric and soil conditions will assist in unravelling this complex relationship.

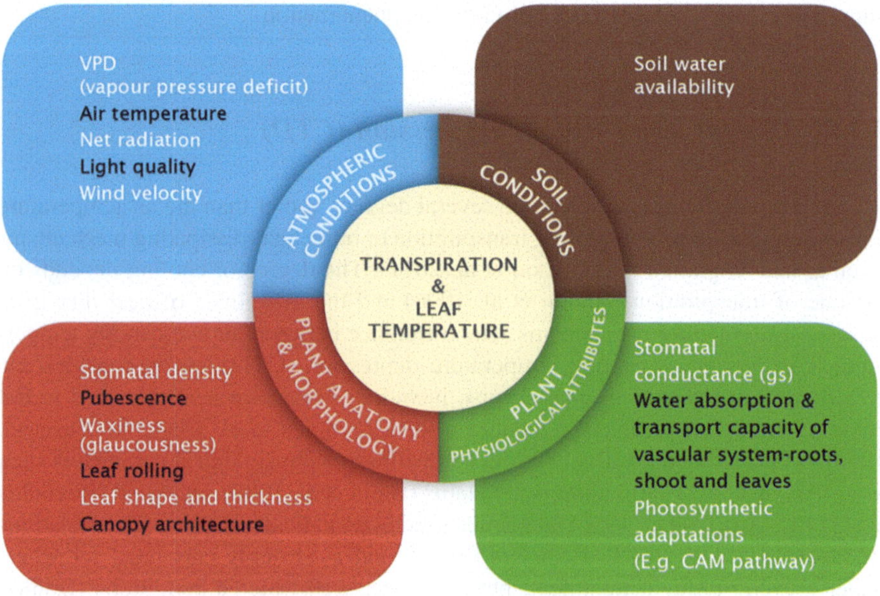


**Fig. 5.3** Relationship between biophysical factors and transpiration and leaf temperature

#### 5.3.2.1 Transpiration and Leaf Temperature Affected by Plant Physiological Attributes

The resistance or conductance for water vapour diffusion and water absorption and transport capacity of vascular roots, shoots, and leaves affect transpiration rate.

##### 5.3.2.1.1 Stomatal Conductance ($g_s$)/Stomatal Resistance ($r_s$)

Stomata are the predominant transpiration sites (De Costa 2004), and the turgor pressure of two flanking guard cells controls their opening (Fig. 5.2). The number of stomata per unit leaf area (stomatal density) and the degree of stomatal aperture determine stomatal conductance ($g_s$), which measures the diffusion rate of water through stomata. The reciprocal of $g_s$ is stomatal resistance ($r_s$).

Stomatal functioning involves biophysical and physiological processes that are not yet fully resolved. As reviewed by Dodd (2003), environmental (light, $CO_2$ concentration, temperature and evaporative demand, i.e., VPD) and hormonal factors (ABA, auxins, cytokinins, ethylene, and gibberellins) and their interactions influence stomatal functioning. Amani et al. (1996) suggested a direct influence of $g_s$ on transpirational cooling, where canopy temperature negatively correlates with $g_s$. Some scientists suggest that plants that can keep their stomata open at elevated temperatures can better regulate their temperatures (Porch and Hall 2013; Prasad et al. 2017). Plants that close stomata to control water used through transpiration under water-stressed and high VPD conditions may also increase leaf temperature. However, the relationship between $g_s$ and canopy temperature depends on VPD, where a strong relationship exists in high VPD environments compared with low VPD environments (Fig. 5.4).

Besides stomatal control, other biophysical factors influence the transpiration rate. Jarvis and McNaughton (1986) introduced the Omega factor or **decoupling**

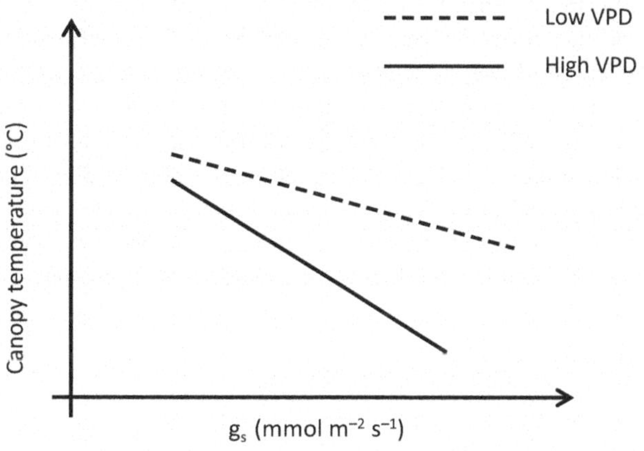


**Fig. 5.4** Schematic relationship between canopy temperature and stomatal conductance ($g_s$) in relation to VPD. [Adapted from Cossani et al. (2012)]

coefficient ($\Omega$) as an index ranging from 0 to 1 to determine the degree of stomatal control of transpiration. $\Omega = 0$ represents 100% stomatal control of transpiration (Martin et al. 1999), and $\Omega = 1$ represents situations with no stomatal control over transpiration. Variations in $g_s$ have less effect on transpiration under low boundary layer conductance ($g_b$) than under high boundary conductance (Pallardy et al. 1994).

#### 5.3.2.1.2 Leaf Boundary Layer Conductance ($g_b$)/Resistance ($r_b$)

The leaf boundary layer affects transpiration and leaf temperature, reducing the conductance to water vapour diffusion and sensible heat loss. Leaf size and leaf physical attributes such as pubescence and wind speed regulate the $g_b$ of individual leaves (De Costa 2004). Therefore, genotypes contrasting in these physical attributes may have different boundary layers, leading to differences in transpiration and leaf cooling. At the plant canopy level, in addition to wind speed, $g_b$ is influenced by the 'surface roughness' of the canopy. Grass cover or cereal crops have smooth canopy surfaces with a thick boundary layer (low $g_b$), while a forest has a rough canopy surface with a thin boundary layer (high $g_b$) (De Costa 2004). Condon et al. (1990) suggested that $g_b$ has a greater effect on canopy water used through transpiration than stomatal control under well-watered conditions. In contrast, stomatal control may be more prominent under water-stressed conditions (Condon et al. 1990).

#### 5.3.2.1.3 Leaf Mesophyll Conductance ($g_m$)/Resistance ($r_m$)

Non-stomatal regulation of transpiration by mesophyll cell walls exerting resistance to water vapour diffusion (mesophyll resistance) has been debated strongly (Livingston and Brown 1912; Jarvis and Slatyer 1970; Farquhar and Raschke 1978). Livingston and Brown (1912) found that transpiration reduced with declining leaf water content, independently of stomatal aperture, possibly due to non-stomatal control of transpiration by mesophyll cell walls. Usually, boundary layer resistance and mesophyll resistance are small compared with stomatal resistance. However, a recent study on *Agapanthus praecox* Wild. revealed that mesophyll resistance may be a significant factor regulating transpiration in addition to stomatal resistance (Canny 2012). Such control over transpiration from mesophyll cells may be significant under excessive evaporative demand and limited water supply (Jarvis and Slatyer 1970) or when stomata fail to regulate water used through transpiration (Canny 2012), as observed with leaf aging in wheat (Frank et al. 1973). Under such conditions, mesophyll control over transpiration may affect leaf temperature.

#### 5.3.2.1.4 Leaf Cuticular Resistance ($r_c$)

The cuticle on the leaf surface serves as a barrier for water escape, exerting resistance to water vapour transfer to the atmosphere. However, it is not entirely impermeable to the passage of water as there are hydrophilic components embedded within the cuticle. When stomata are open, the proportion of water escaping through the cuticle is very low relative to the total transpiration flux (Jarvis and Slatyer 1970). However, cuticular transpiration substantially contributes to total water escape under water stress conditions and during minimum stomatal aperture at night (Bengtson et al. 1978; Rawson and Clarke 1988; Caird et al. 2007; Knipfer and Fricke 2011). Cuticular transpiration may vary depending on the amount, composition and pattern of wax deposition on leaves (Bengtson et al. 1978). For instance, Bengtson et al. (1978) found a 14% decrease in cuticular transpiration with a 24% increase in the total amount of epicuticular wax in primary leaves of two oat varieties. Genotypic variation exists for wax deposition on wheat leaves (Johnson et al. 1983; Uddin and Marshall 1988); in some genotypes, the leaves become waxier under drought or heat stress (such as Espada) (Australian Grain Technologies 2010a), but little is known about the influence of these variations on transpiration due to the difficulty in separating water loss from cuticular and stomatal evaporation.

#### 5.3.2.1.5 Water Absorption and Transport Capacity of the Vascular System: Roots, Stems, and Leaves

The vascular system of a plant comprises roots, stems, and leaves and plays a crucial role in absorbing and delivering water to the site of transpiration.

##### *5.3.2.1.5.1 Root Traits Associated with Improved Transpirational Cooling through Better Water Absorption*

Although hidden underground, a plant's root system plays a vital role in transpirational cooling by exploring soil water stored in deep layers, especially when water is scarce. Therefore, vigorous and deep root systems are linked to better water productivity (Palta and Watt 2009; Richards et al. 2010) and maintenance of cooler canopies (Palta et al. 2011; Cossani et al. 2012). Strong geotropism in wheat promotes deeper penetration of roots (Araki and Iijima 2001) with variation in seminal and nodal axile root geotropism (Araki and Iijima 2001; Manschadi et al. 2008). Other root traits, such as root weight density (RWD: amount of root weight per unit soil volume) and root length density (RLD: length of root per unit soil volume) (Elazab et al. 2012), which determine the spatial distribution of the root system in the soil profile also influence the amount of water absorption from soil profiles (more absorption with higher RWD and RLD) (Passioura 1983; Elazab et al. 2012). Song et al. (2010) observed an increase in specific root length (SRL) or reduced root thickness in spring wheat under water-stressed conditions. In contrast, Elazab et al. (2012) observed thicker roots (low SRL) in water-stressed wheat plants than well-watered plants. Thicker roots with greater strength can penetrate through the soil profile when mechanical barriers exist in drier soils, thus enhancing water conductance under dry soil conditions due to their ability to exploit water at deeper soil layers (Davies and Bacon 2003; Elazab et al. 2012). However, the development of an extensive and deep root system may require greater respiratory cost (Cossani et al. 2012). Hence, improving the efficiency of root systems to uptake and deliver water would be advantageous.

#### 5.3.2.1.5.2 *Efficiency of Root Water Absorption and Transport through the Vascular System*

The capability or efficiency of water absorption and transport is 'hydraulic conductance ($K$)', quantified as the ratio of liquid water flow rate through the system to the gradient in water potential driving the flow (Sack and Holbrook 2006). Plants require an efficient water absorption and delivery system to facilitate transpiration and leaf cooling. Fig. 5.5 shows the different components of $K$ in the different segments of the soil–plant–air continuum (SPAC) and their association with environmental variables and other physiological parameters.

The root system exerts the greatest limitation in the liquid phase of SPAC when the soil is well-watered, but the soil is the most limiting when the soil is dry (Steudle

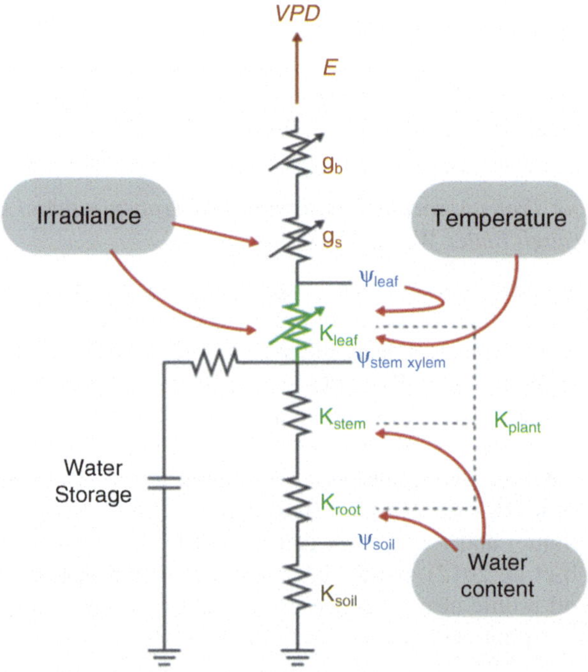


**Fig. 5.5** A schematic representation of different components of plant hydraulic conductance ($K$) in a simplified electronic circuit analogue of the whole plant system and their association with other physiological parameters and environmental variables in relation to the regulation of transpiration ($E$). $K_{soil}$, $K_{root}$, $K_{leaf}$, and $K_{plant}$ are the hydraulic conductance of soil, root, leaf, and plant, respectively. $\Psi_{soil}$, $\Psi_{stem\ xylem}$, and $\Psi_{leaf}$ are the water potentials of soil, stem xylem, and leaf, respectively, and $g_s$, $g_b$, and VPD are stomatal and boundary layer conductance and leaf-to-air vapour pressure deficit, respectively. The influence of environmental variables, irradiance, and temperature is shown in red arrows. [Adapted from Sack and Holbrook (2006) and Jones (2014)]

2000). The conductances in radial and axial water flow (root xylem) pathways in the root affect $K_{root}$ (Blum 2011). Living tissues are involved in the radial flow of water in roots; hence, $K$ in this pathway can vary depending on whether water flows through the cell-to-cell pathway or apoplastically (cell walls and intercellular spaces) (Bramley et al. 2007). In the cell-to-cell pathway, water flow can cross membranes with the involvement of aquaporins (AQPs) (Knipfer and Fricke 2011), water channel proteins located in plasma and intracellular membranes that facilitate transcellular water transport (Fricke and Chaumont 2007), which can rapidly adjust $K$ through changes in their abundance and activity (gating/open or closed state) (Tyerman et al. 2002). In wheat roots, radial water flow occurs predominantly through the cell-to-cell pathway by crossing membranes with AQPs playing a regulatory role in $K_{root}$ (Bramley et al. 2009). Axial flow of water depends on the number and diameter of xylem vessels (Bramley et al. 2007; Blum 2011), and hydraulic failures can occur when stem xylem is under high tension (more negative water potentials), leading to cavitation, the loss of hydraulic conductance due to the entry of air into xylem vessels (embolism) (Tyree and Sperry 1989). Radial hydraulic conductance is much lower than axial hydraulic conductance in roots (Steudle and Peterson 1998); hence, it may be the rate-limiting factor in root water transport (Bramley et al. 2007).

$K_{leaf}$ is the efficiency of liquid water transport from stem xylem to the evaporation site, determined by two components: resistance in the xylem ($R_x$) and resistance outside the xylem ($R_{ox}$) (Sack and Holbrook 2006; Blum 2011). $R_{ox}$, the resistance when water moves from the vascular bundle to the evaporation site (stomata) across the mesophyll contributes most to leaf hydraulic resistance (Yang and Tyree 1994; Salleo et al. 2003). $K_{leaf}$ between and within species depends on the leaf architecture/venation pattern (affects $R_x$) and AQP abundance and activity (affects $R_{ox}$) (Sack and Holbrook 2006; Blum 2011; Shatil-Cohen et al. 2011; Chaumont and Tyerman 2014; Sade et al. 2014).

$K_{root}$ and $K_{leaf}$ are highly responsive to changes in temperature, irradiance, and water supply (Tsuda and Tyree 2000; Martre et al. 2001; Sack et al. 2002; Lee et al. 2004; Sack and Holbrook 2006), and by changing $K_{root}$, plants tend to maintain internal water balance (Steudle 2000). A greater $K$ requires a smaller water potential gradient to drive water flow, minimising the drop in water potential (Tsuda and Tyree 2000) and catastrophic xylem cavitation. As these changes in $K$ in response to such environmental variables are rapid and reversible, they are mainly brought about by changes in expression (abundance) and activity of AQPs rather than anatomy (Lee et al. 2004; Ionenko et al. 2010). Roots are also known to vary their $K$ in response to salinity (Maurel and Chrispeels 2001), oxygen deficiency (Maurel and Chrispeels 2001; Bramley and Tyerman 2010), ethylene (Kamaluddin and Zwiazek 2002), nutrient deficiency (Carvajal et al. 1996; Maurel and Chrispeels 2001), and diurnally (Clarkson et al. 2000; Maurel and Chrispeels 2001), which have been attributed to changes in AQP (Aroca et al. 2012). AQP abundance/activity differs between genotypes (Bramley et al. 2007; Sadok and Sinclair 2010; Schoppach et al. 2014), and changes in response to water and salt stress (Jang et al. 2004), pH (Tournaire-Roux et al. 2003), heavy metals (Niemietz and Tyerman 2002), nutrient deficiencies (Clarkson et al. 2000), hypoxia (Bramley and Tyerman 2010), temperature (Jang et al. 2004; Ionenko et al. 2010), light (Sato-Nara et al. 2004), and abscisic acid (Hose et al. 2000; Jang et al. 2004; Aroca et al. 2006; Parent et al. 2009).

#### 5.3.2.2 Transpiration and Leaf Temperature as Affected by Environmental Parameters

As mentioned above, VPD or atmospheric demand (also known as atmospheric drought) is increasing with global warming; hence, crop production systems in different regions will face warm and dry conditions in the future, becoming more prone to edaphic drought due to less soil water availability. Therefore, it is important to understand how crops sense and respond to atmospheric and edaphic drought in terms of their physiological processes, such as transpiration and leaf temperature regulation. Changes in transpiration due to climate may influence plant temperature regulation which is likely to vary between genotypes.

##### 5.3.2.2.1 Air Temperature and Atmospheric VPD

When air becomes warm and dry, it increases the atmospheric VPD (evaporative demand or drying power of the air). Under high VPD conditions, air has more capacity to hold moisture favouring greater evapotranspiration. The leaf-to-air VPD, which is generally considered the driving force for transpiration (De Costa 2004), also increases with temperature and atmospheric VPD, increasing transpiration (Sinclair et al. 1984).

Genotypic variation in the transpiration response to VPD exists in peanut (Devi et al. 2010), pearl millet (Kholová et al. 2010), soybean (Fletcher et al. 2007), sorghum (Gholipoor et al. 2010), wheat (Schoppach and Sadok 2012; Ranawana et al. 2021), and chickpea (Zaman-Allah et al. 2011). Rawson and Clarke (1988) observed genotypic variation in the night-time transpiration response to VPD in wheat. In the above species, transpiration response to VPD falls into one of two categories: (1) transpiration increases in a simple linear fashion with VPD or (2) transpiration responds to VPD in two phases separated by a break point (BP) (called segmented or broken stick regression), where transpiration responds little to VPD after the BP.

Genotypic variation in the transpiration response to VPD in wheat could be due to differential hydraulic restrictions present in leaves and/or roots (Schoppach and Sadok 2012), supported by the finding of a narrower root metaxylem and lower flow rate of water through roots in the wheat genotype RAC 875, which has a lower transpiration response to VPD with no BP (Schoppach et al. 2014). Such hydraulic restrictions that limit flow rate can affect transpirational cooling. The lower root water transport was attributed to less AQP activity through inhibition effects on transpiration with known AQP blockers. However, caution must be taken in interpreting these results because the compounds are non-specific, and most of the concentrations used are much greater than the potency needed for AQP inhibition

(Niemietz and Tyerman 2002) and considered toxic (Zhang and Tyerman 1999). In soybean, variation in the transpiration response to VPD was linked to leaf level hydraulic restrictions (Sinclair et al. 2008; Sadok and Sinclair 2010). Gholipoor et al. (2010) suggested root- and leaf-based restrictions for water flow in sorghum genotypes with low BP. In pearl millet (Kholová et al. 2010), differential accumulation of leaf abscisic acid may explain the genotypic variability in transpiration response to VPD. However, stomata are expected to play a role in mediating transpiration response to VPD as it is the major pathway for water vapour escape via transpiration. Ranawana et al. (2021) suggested that stomata coordinate with plant hydraulics to regulate transpiration response to VPD in wheat using their research findings and a theoretical model. They concluded that a simple linear increase in transpiration in response to VPD occurs due to a responsive $g_s$ to VPD and a segmented linear transpiration response to VPD is a function of $g_s$. The BP of the segmented linear response depends on the plant's capacity to meet the transpirational demand set by the shoots (Ranawana et al. 2021). The possibility of using contrasting transpiration responses in physiological breeding in wheat is discussed in Sect. 5.4. Moreover, these differences in transpiration response to VPD may affect leaf temperature regulation. In a controlled environment study, leaf temperature increased more slowly in some wheat genotypes as air temperature increased, with greater transpiration response to VPD under well-watered conditions, with some exceptions (Ranawana 2015). For example, Gladius maintained lower leaf temperatures as air temperature increased under well-watered conditions despite having a low transpiration to VPD response. Gladius is also known to exhibit heat tolerance (Fleury et al. 2010) and has waxy leaves and stems (Australian Grain Technologies 2010b), which help to reflect incident radiation, keeping the leaf cooler. In contrast, Drysdale could not maintain proportionally lower leaf temperatures as air temperature increased under well-watered conditions despite having the highest transpiration to VPD response (Ranawana 2015).

There is some controversy about how stomata respond to changing atmospheric VPD. For instance, Rawson et al. (1977) found that $g_s$ remained unchanged in response to VPD. Mott and Parkhurst (1991) studied stomatal response to VPD in some grain legumes using air and a helium:oxygen mixture (79:21 v/v, with $CO_2$ and water vapour added). They concluded that stomata are unresponsive to changes in humidity, and stomatal closure occurs due to leaf transpiration rather than humidity. Conversely, Lange et al. (1971) found a direct response, with $g_s$ decreasing with decreasing humidity. In support, Turner et al. (1984) observed decreasing $g_s$ with increasing VPD in several herbaceous and woody species. Ranawana et al. (2021) found three patterns of $g_s$ response to VPD in wheat: (1) no significant linear $g_s$ response to VPD (e.g., Gladius and Mace); (2) $g_s$ declined linearly at similar rates in response to VPD (e.g., Excalibur, Long Reach-Envoy, Drysdale, and Espada); (3) $g_s$ declined in two linear phases separated by a break point at 2.4 kPa VPD (e.g., Glennson 81 and Sonora 64).

Supporting the $g_s$ response to VPD, Oren et al. (1999) undertook an empirical analysis of 40 plant species, including grasses and deciduous and evergreen trees, to show that the magnitude of the reduction in $g_s$ to increasing in VPD or 'stomatal sensitivity' is proportional to the magnitude of $g_s$ at low VPD (≤1 kPa), which was supported by Ranawana et al.'s (2021) findings in wheat.

The decreasing $g_s$ with increasing VPD may be due to one of two mechanisms:

1. A direct response of stomatal guard cells to changing VPD regardless of bulk leaf water potential, called 'feed forward response'(Farquhar 1978). Lange et al. (1971) described the guard cells as 'humidity sensors' and suggested that the decrease in $g_s$ in response to humidity is due to increased 'peristomatal transpiration' (water escape from guard cells and subsidiary cells).
2. An indirect response of stomata to changing VPD through changes in leaf water potential caused by changes in transpiration is known as a 'feedback mechanism' (Mott and Parkhurst 1991; Monteith 1995; Matzner and Comstock 2001).

Stomatal closure under high VPD is important for maintaining plant water balance, although it can restrict the $CO_2$ gain affecting photosynthesis (Sperry 2000). Such stomatal control would prevent transpiration exceeding the critical level that leads to greater xylem tension and xylem cavitation (Breshears et al. 2013).

Stomatal closure under high VPD can increase leaf temperature by several degrees, increasing leaf-to-air VPD. Therefore, leaf-to-air VPD cannot be considered an independent driving variable. Tardieu and Simonneau (1998) proposed that leaf-to-air VPD is 'a cause and a consequence' of transpiration or water flow through the plant. Conversely, some scientists argue that partial stomatal closure under high VPD does not significantly affect leaf temperature. For instance, Allen et al. (1998) found that soybean leaves under elevated $CO_2$ were only 0.5 °C warmer than leaves under ambient $CO_2$, even with a 50% reduction in $g_s$. Deva et al. (2020) observed a stronger cooling response to VPD in heat-tolerant common bean genotypes and suggested that it may be due to higher stomatal conductance and enhanced transpiration. Heat-tolerant genotypes may be able to keep their stomata open under high VPD and maintain cooler canopies through transpiration.

#### 5.3.2.2.2 Soil Water Availability

As soil available water declines, $g_s$ decreases in some crops (Henson et al. 1989; Siddique et al. 1990; Premachandra et al. 1992; Elazab et al. 2012), affecting transpiration rate and leaf cooling. As stomata are the common pathway for $CO_2$ intake and water escape, a trade-off is involved, especially under water-stressed conditions (Cornic 2000; Lawlor and Cornic 2002). However, under well-watered conditions, stomatal regulation maintains an optimum level of internal $CO_2$ (Cossani et al. 2012).

Frequent soil water limitations can inhibit leaf cooling via latent heat loss (Monteith and Unsworth 2013). Maricle et al. (2007) revealed that latent heat loss from leaves decreased with decreasing soil water potential (indicator of soil water availability) with up to 4 °C temperature increases in leaves of 13 estuarine C4 grass species. They concluded that sensible and radiative heat losses are the main pathway for heat dissipation under reduced soil water potential rather than latent heat loss.

Moreover, transpiration and allied traits degrade in some plant species (e.g., pineapple) with unique photosynthetic adaptations (Crassulacean acid metabolism) to conserve water (Lin et al. 2017).

#### 5.3.2.3 Transpiration and Leaf Temperature as Affected by Plant Anatomy and Morphology

Various anatomical and morphological attributes of plants affect transpirational cooling, but it is challenging to identify the relative role of individual attributes due to the simultaneous expression of some attributes, such as leaf colour and pubescence.

##### 5.3.2.3.1 Transpiration, Leaf Temperature, and Glaucousness

Glaucousness is the deposition of epicuticular wax on the leaves, stem, and ears of cereal crops, giving a bluish-white colour (Richards et al. 1986). Epicuticular wax refers to the smooth film of lipids deposited on the cuticle (thin waxy covering on the outer surface of the leaf), and the amorphous mixture of lipids embedded in the cuticle is known as intracuticular wax. Epicuticular wax also refers to wax crystals on the surface of the cuticle (Barthlott et al. 1998).

The maximum appearance of glaucousness in wheat is on the flag leaf sheath and abaxial surface of the flag leaf lamina in some genotypes (Richards et al. 1986, 2010). Richards et al. (1986) revealed that the temperature of glaucous wheat lines was 0.7 °C and 0.3 °C cooler than non-glaucous lines under water-stressed and well-watered conditions, respectively. However, glaucous lines had lower day and night transpiration rates and photosynthetic rates of ears than non-glaucous lines in water-stressed and well-watered treatments. It was, therefore, suggested that the temperature reduction in glaucous lines was mainly due to spectral reflectance (Johnson et al. 1983). That is, the presence of wax would be advantageous to maintain cooler canopies, especially when transpirational cooling is reduced under water-stressed conditions. However, Figueiredo et al. (2012) did not observe a significant reduction in leaf temperature with respect to epicuticular wax in *Jatropha mollissima* and *J. curcas*.

Crop genotypes differ in waxiness and heterogeneous composition of wax, and the pattern of wax deposition may create differential leaf cooling among genotypes due to its variation in spectral reflectance properties and water permeability. Wax deposition on stomatal guard cells and other cells increases under low humidity conditions and is sometimes absent in high humidity environments (Fuchigami et al. 1981; Grantz 1990). Glaucousness also helps conserve water, improving water use efficiency (Richards et al. 1986). Therefore, wax content is considered a trait related to increased drought tolerance in some plants (Yang et al. 2011). Despite many studies on the heterogeneous composition of the epicuticular and intra cuticular wax layers, reliable information on their relationship with leaf transpirational cooling is rare, particularly for crops.

#### 5.3.2.3.2 Transpiration, Leaf Temperature, and Pubescence

The presence of leaf hairs in *Encelia farinosa*, a desert species, decreased light absorption and increased reflectance (Ehleringer et al. 1976). The presence of leaf hairs creates a thick boundary layer (Schuepp 1993), reducing the transpiration rate (Wuenscher 1970) and possibly keeping the leaf humid. However, the density of leaf hairs and wind speed will influence whether airflow over the leaf surface is smooth or turbulent, affecting the size of the boundary layer. In some plant species, such as Proteaceae, pubescence positively and linearly correlated with stomatal density, with no effect on transpiration (Skelton et al. 2012). A similar phenomenon may occur among other crop species. Pubescent wheat varieties with leaf hairs may reflect radiation, but there will be a trade-off between leaf cooling via improved reflection and transpirational cooling from a thicker boundary layer. Leaf pubescence and its impact on transpiration and leaf temperature have not been widely examined in crops such as wheat (Loss and Siddique 1994).

#### 5.3.2.3.3 Transpiration Leaf Temperature and Leaf Rolling

Leaf rolling is a plant's response to abiotic stress factors such as drought, high temperature and radiation (Kadioglu and Terzi 2007). Leaf rolling could reduce transpiration by reducing the effective leaf area for transpiration and radiation interception (Clarke 1986; Loss and Siddique 1994). The effective transpiring leaf area is halved with complete leaf rolling, enclosing the adaxial surface of the leaf with more stomatal density and higher $g_s$ in cereal crops (Sojka 1985). Reduced transpiration is also related to the variation in stomatal distribution and the degree and pattern of stomatal opening in rolled leaves (Heckathorn and DeLucia 1991). Despite the reduction in transpirational cooling, leaf rolling may avoid the excessive radiation load incident on the leaf. Furthermore, leaf rolling exposes the abaxial side of the leaf with greater waxiness (Richards et al. 2010), helping to reflect excess radiation. However, there is much inconsistency in the literature on the role of leaf rolling in regulating leaf temperature. For example, Heckathorn and DeLucia (1991) observed reduced leaf temperature with leaf rolling in two C4 prairie grasses, *Andropogon gerardiz* and *Spartina pectinate*, and suggested that it was mainly due to increased sensible heat loss at reduced incident solar irradiance. However, Turner et al. (1986) observed a 1.6 °C increase in rice per unit of leaf rolling scored on a visual scale and suggested that this was mainly due to reduced transpiration with leaf rolling.

#### 5.3.2.3.4 Transpiration, Leaf Temperature, and Canopy Architecture

Canopy architecture affects the radiation distribution within the canopy, air and soil temperature, and VPD, thus influencing the leaf's microclimate and affecting gas exchange, water relations (Neufeld et al. 1988) and leaf temperature (Comstock and Mahall 1985). Leaf orientation or angle can vary depending on genotype, stage of development, and leaf position on the plant. For example, some wheat genotypes (e.g., Yecora 70) have planophile (horizontal) leaf orientation, while others (e.g., Ciano 67) have erectophile orientation (vertical) (Pinter Jr et al. 1985). Increased radiation heat load or interception in planophile leaf orientation (Ball et al. 1988) may increase leaf temperature compared with erectophile orientation. Transpiration rates are higher in planophile leaves than erectophile leaves (Stockle 1992), which may contribute to differences in transpirational heat loss of the two leaf forms. Therefore, the relative effect of both factors (radiation heat load and transpirational cooling) must be considered when selecting a particular leaf form in crop breeding targeting effective leaf temperature regulation.

#### 5.3.2.3.5 Transpiration, Leaf Temperature, and Leaf Thickness

There are different views on the effect of leaf thickness on transpiration and leaf temperature. Lin et al. (2017) speculate that the abaxial epidermis functions as an insulation layer for the leaf, extending the vapour path from stomata to air, such that leaf temperature positively correlates with the thickness of the abaxial epidermis. However, some scientists suggest that thick leaves may withstand heat damage better than thin leaves due to their heat-storing capacity (Groom et al. 2004; Leigh et al. 2012).

#### 5.3.2.3.6 Transpiration, Leaf Temperature, and Leaf Size

Theoretically, small leaves are coupled with air temperature due to lower resistance imposed by a thin boundary layer (Lin et al. 2017). In contrast, larger leaves are decoupled from the surrounding air due to a thick boundary layer, such that leaf temperature is several degrees warmer than air. The thin boundary layer of small leaves facilitates transpiration in winter and sensible heat loss in summer. Such a strategy would benefit crops using the available water while efficiently regulating leaf temperature. Maintaining higher transpiration rates in winter when water is sufficiently available and evaporative demand is low would support nutrient acquisition in winter, whereas sensible heat loss reduces the need for transpirational cooling in summer when water is scarce (Yates et al. 2010).

## 5.4 Transpirational Cooling Under Future Climate: Applications in Physiological Breeding

There has been increased attention on finding strategies to minimise the negative impact of heat stress on global crop production, including the potential benefits of a physiological-trait-based breeding approach over breeding for yield per se (Reynolds and Trethowan 2007; Mohammadi et al. 2012). Heat avoidance through transpirational cooling is a powerful selection tool in breeding for improving heat and drought adaptation (Deva et al. 2020). For example, plant breeders used the transpirational cooling trait to screen spring wheat cultivars for heat tolerance (Porch and Hall 2013). Enhanced transpirational cooling is a useful trait for identifying bean genotypes with greater thermal plasticity to adapt to climate change (McClean et al. 2011). The role of transpirational cooling in temperature regulation in rice crops has also been investigated (Weerakoon et al. 2008; Fukuoka et al. 2012). There is some evidence that transpirational cooling is an important physiological mechanism for avoiding heat stress in potato, maize, and various legumes (Kumar et al. 2017; Deva et al. 2020). For transpirational cooling to effectively regulate leaf temperature, genotypes need other supportive traits, such as enhanced access to water through a deep, extensive root system, higher hydraulic conductivity, and greater osmotic adjustments, which would lead to more stable cell turgor.

Physiologists and breeders should pay due attention to transpirational cooling in heat-susceptible organs, such as panicles, positioned above the canopy. Rice panicles transpire through permanently open epidermal pores (Julia and Dingkuhn 2013) and are not regulated like stomata. Transpirational cooling in rice panicles reduces high temperature-induced spikelet sterility (Weerakoon et al. 2008; Fukuoka et al. 2012). Therefore, screening varieties based on transpiration in reproductive structures would be a key trait for developing rice varieties that can avoid heat damage in a warming climate.

The role of transpiration in regulating leaf temperature may be affected by future climatic conditions, which are predicted to be warmer and drier with limited soil water (Wetherald and Manabe 2002; IPCC 2007). Heat avoidance via transpirational cooling may not be effective under water-limited conditions as it relies on water availability. The predicted increased occurrence of drought spells with higher intensity (Wetherald and Manabe 2002) will result in a trade-off between water conservation and transpirational cooling. Current or anticipated water scarcity has diverted researchers attention to water-saving strategies which will reduce transpiration. However, they may warm the plants unless other physical attributes adequately regulate leaf temperature. In such situations, maximising other heat-dissipating mechanisms such as sensible heat loss and reflectance and improving heat tolerance will be more beneficial by screening for traits such as waxiness (e.g., Gladius) and heat tolerance. However, the benefit of greater reflection and associated cooling may depend on the level of wax expression and/or type of wax. Wax expression is constitutive in some genotypes but can be induced or increased with heat and drought (e.g., Espada) (Australian Grain Technologies 2010a). Traits like leaf size can also impact transpirational cooling and sensible heat loss. When water is not limited, transpirational cooling is effective for large leaves due to greater surface area for water escape. However, when water is scarce, imposing limits on transpirational cooling, large leaves may be at a disadvantage due to their thick boundary layer, preventing sensible heat loss (Jones 2014).

Furthermore, elevated $CO_2$ under future climate scenarios will decrease stomatal conductance and, therefore, transpiration (Robredo et al. 2007; Wall et al. 2011), which might affect the relative contribution of transpiration in regulating leaf temperature (Wall et al. 2011). Latent heat flux through transpiration decreased by 11.8–13.8% in sorghum under $CO_2$-enriched conditions, resulting in warmer plant temperatures (Triggs et al. 2004). Yoshimoto et al. (2005) reported 1–2 °C leaf temperature increments in rice due to decreased stomatal conductance under elevated $CO_2$. Therefore, it is better to consider the role of transpiration in regulating leaf temperatures in relation to changes in climatic conditions when using this trait in physiological breeding.

Contrasting transpiration to VPD response patterns can be used to develop new cultivars for specific environmental conditions. For example, lower transpiration to VPD responses (greater hydraulic restrictions) may be suited to rainfed agricultural systems, where soil water conservation is needed for later use during the grain filling stage. In contrast, the segmented linear transpiration to VPD may be suited to environments with intermittent in-season rainfall as rapid increases in transpiration under low to medium VPD levels (after rainfall events) would maximise gas exchange and restrict transpiration under high VPD levels (beyond VPD breakpoint) during seasonal droughts (Ranawana et al. 2021). The relative contribution of transpirational cooling may differ with these different transpiration response patterns, which needs to be considered in physiological breeding.

## 5.5 Conclusion

Enhanced transpiration cooling may induce heat tolerance under hot and irrigated conditions but also sensitivity under hot and dry conditions due to early soil water depletion. Therefore, breeding for heat tolerance via enhanced cooling requires particular attention to the target population of environments (TPEs). Against this backdrop, models are best used to understand the trade-offs of enhanced transpirational cooling under changing TPEs and assess its costs and benefits in physiological breeding.

## References

Allen LH, Valle RR, Jones JW, Jones PH (1998) Soybean leaf water potential responses to carbon dioxide and drought. Agron J 90:375–383

Amani I, Fischer RA, Reynolds MP (1996) Canopy temperature depression association with yield of irrigated spring wheat cultivars in a hot climate. J Agron Crop Sci 176:119–129
Ansari AQ, Loomis WE (1959) Leaf temperatures. Am J Bot 46(10):713–717
Araki H, Iijima M (2001) Deep rooting in winter wheat: rooting nodes of deep roots in two cultivars with deep and shallow root systems. Plant Prod Sci 4:215–219
Aroca R, Ferrante A, Vernieri P, Chrispeels MJ (2006) Drought, abscisic acid and transpiration rate effects on the regulation of PIP aquaporin gene expression and abundance in *Phaseolus vulgaris* plants. Ann Bot 98:1301–1310
Aroca R, Porcel R, Ruiz-Lozano JM (2012) Regulation of root water uptake under abiotic stress conditions. J Exp Bot 63:43–57
Asseng S, Foster I, Turner NC (2011) The impact of temperature variability on wheat yields. Glob Change Biol 17:997–1012
Asseng S, Ewert F, Martre P, Rotter RP, Lobell DB, Cammarano D, Kimball BA, Ottman MJ, Wall GW, White JW, Reynolds MP, Alderman PD, Prasad PVV, Aggarwal PK, Anothai J, Basso B, Biernath C, Challinor AJ, De Sanctis G, Doltra J, Fereres E, Garcia-Vila M, Gayler S, Hoogenboom G, Hunt LA, Izaurralde RC, Jabloun M, Jones CD, Kersebaum KC, Koehler AK, Muller C, Naresh Kumar S, Nendel C, O'Leary G, Olesen JE, Palosuo T, Priesack E, Eyshi Rezaei E, Ruane AC, Semenov MA, Shcherbak I, Stockle C, Stratonovitch P, Streck T, Supit I, Tao F, Thorburn PJ, Waha K, Wang E, Wallach D, Wolf J, Zhao Z, Zhu Y (2015) Rising temperatures reduce global wheat production. Nat Clim Chang 5:143–147
Australian Grain Technologies (2010a) Espada-fact sheet. AGT, Roseworthy, SA
Australian Grain Technologies (2010b) Gladius: wheat variety fact sheet for South Australia. AGT, Roseworthy, SA
Ayeneh A, van Ginkel M, Reynolds MP, Ammar K (2002) Comparison of leaf, spike, peduncle and canopy temperature depression in wheat under heat stress. Field Crops Res 79:173–184
Baker JT, Allen LH Jr, Boote KJ (1992) Response of rice to carbon dioxide and temperature. Agric For Meteorol 60(3–4):153–166
Ball MC, Cowan IR, Farquhar GD (1988) Maintenance of leaf temperature and the optimisation of carbon gain in relation to water loss in a tropical mangrove forest. Funct Plant Biol 15:263–276
Barthlott W, Neinhuis C, Cutler D, Ditsch F, Meusel I, Theisen I, Wilhelmi H (1998) Classification and terminology of plant epicuticular waxes. Bot J Linn Soc 126:237–260
Bengtson C, Larsson S, Liljenberg C (1978) Effects of water stress on cuticular transpiration rate and amount and composition of epicuticular wax in seedlings of six oat varieties. Physiol Plant 44:319–324
Blum A (2011) Plant breeding for water-limited environments. Springer, New York
Blum A, Mayer J, Gozlan G (1982) Infrared thermal sensing of plant canopies as a screening technique for dehydration avoidance in wheat. Field Crops Res 5:137–146
Blum A, Shpiler L, Golan G, Mayer J (1989) Yield stability and canopy temperature of wheat genotypes under drought-stress. Field Crops Res 22:289–296
Bramley H, Tyerman S (2010) Root water transport under waterlogged conditions and the roles of aquaporins. In: Mancuso S, Shabala S (eds) Waterlogging signalling and tolerance in plants. Springer, Berlin, Heidelberg, pp 151–180
Bramley H, Turner DW, Tyerman SD, Turner NC (2007) Water flow in the roots of crop species: the influence of root structure, aquaporin activity, and waterlogging. Adv Agron 96:133–196
Bramley H, Turner NC, Turner DW, Tyerman SD (2009) Roles of morphology, anatomy, and aquaporins in determining contrasting hydraulic behavior of roots. Plant Physiol 150:348–364
Breshears DD, Adams HD, Eamus D, McDowell NG, Law DJ, Will RE, Williams AP, Zou CB (2013) The critical amplifying role of increasing atmospheric moisture demand on tree mortality and associated regional die-off. Front Plant Sci 4:1–4
Caird MA, Richards JH, Donovan LA (2007) Nighttime stomatal conductance and transpiration in $C_3$ and $C_4$ plants. Plant Physiol 143:4–10
Canny M (2012) Water loss from leaf mesophyll stripped of the epidermis. Funct Plant Biol 39: 421–434

Carvajal M, Cooke DT, Clarkson DT (1996) Responses of wheat plants to nutrient deprivation may involve the regulation of water-channel function. Planta 199:372–381

Chaumont F, Tyerman SD (2014) Aquaporins: highly regulated channels controlling plant water relations. Plant Physiol 164:1600–1618

Clarke JM (1986) Effect of leaf rolling on leaf water loss in *Triticum* spp. Can J Plant Sci 66:885–891

Clarkson DT, Carvajal M, Henzler T, Waterhouse RN, Smyth AJ, Cooke DT, Steudle E (2000) Root hydraulic conductance: diurnal aquaporin expression and the effects of nutrient stress. J Exp Bot 51:61–70

Clements HF (1934) Significance of transpiration. Plant Physiol 9:165–172

Clum HH (1926) The effect of transpiration and environmental factors on leaf temperatures I. transpiration. Am J Bot 13:194–216

Comstock JP, Mahall BE (1985) Drought and changes in leaf orientation for two California chaparral shrubs: *Ceanothus megacarpus* and *Ceanothus crassifolius*. Oecologia 65:531–535

Condon AG, Farquhar GD, Richards RA (1990) Genotypic variation in carbon isotope discrimination and transpiration efficiency in wheat. Leaf gas exchange and whole plant studies. Funct Plant Biol 17:9–22

Cornic G (2000) Drought stress inhibits photosynthesis by decreasing stomatal aperture – not by affecting ATP synthesis. Trends Plant Sci 5:187–188

Cossani CM, Pietragalla J, Reynolds MP (2012) Canopy temperature and plant water relations traits. In: Reynolds MP, Pask AJD, Mullan DM (eds) Physiological breeding I: interdisciplinary approaches to improve crop adaptation. CIMMYT, Mexico, D.F, pp 58–66

Curtis OF (1936) Transpiration and the cooling of leaves. Am J Bot 23:7–10

Davies WJ, Bacon MA (2003) Adaptation of roots to drought. In: De Kroon H, Visser EJW (eds) Root ecology. Springer, Berlin, Heidelberg, pp 173–192

De Costa WAJM (2004) Plant water relations: principles and applications. University of Peradeniya, Peradeniya, Sri Lanka, p 615

Deva CR, Urban MO, Challinor AJ, Falloon P, Svitákova L (2020) Enhanced leaf cooling is a pathway to heat tolerance in common bean. Front Plant Sci 11:19. https://doi.org/10.3389/fpls.2020.00019

Devi MJ, Sinclair TR, Vadez V (2010) Genotypic variation in peanut for transpiration response to vapor pressure deficit. Crop Sci 50:191–196

Dodd IC (2003) Hormonal interactions and stomatal responses. J Plant Growth Regul 22:32–46

Drake BG, Raschke K, Salisbury FB (1970) Temperature and transpiration resistances of *Xanthium* leaves as affected by air temperature, humidity, and wind speed. Plant Physiol 46:324–330

Ehleringer J, Bjorkman O, Mooney HA (1976) Leaf pubescence: effects on absorptance and photosynthesis in a desert shrub. Science 192:376–377

Elazab A, Molero G, Serret MD, Araus JL (2012) Root traits and $\delta^{13}$C and $\delta^{18}$O of durum wheat under different water regimes. Funct Plant Biol 39:379–393

Farooq M, Bramley H, Palta JA, Siddique KHM (2011) Heat stress in wheat during reproductive and grain-filling phases. Crit Rev Plant Sci 30:491–507

Farquhar GD (1978) Feedforward responses of stomata to humidity. Funct Plant Biol 5:787–800

Farquhar GD, Raschke K (1978) On the resistance totranspiration of the sites of evaporation within the leaf. Plant Physiol 61:1000–1005

Feng B, Yu H, Hu Y, Gao X, Gao J, Gao D, Zhang S (2009) The physiological characteristics of the low canopy temperature wheat (*Triticum aestivum* L.) genotypes under simulated drought condition. Acta Physiol Plant 31:1229–1235

Figueiredo KV, Oliveira MT, Oliveira AFM, Silva GC, Santos MG (2012) Epicuticular-wax removal influences gas exchange and water relations in the leaves of an exotic and native species from a Brazilian semiarid region under induced drought stress. Aust J Bot 60:685–692

Fischer RA, Rees D, Sayre KD, Lu Z-M, Condon AG, Saavedra AL (1998) Wheat yield progress associated with higher stomatal conductance and photosynthetic rate, and cooler canopies. Crop Sci 38:1467–1475

Fletcher AL, Sinclair TR, Allen LH Jr (2007) Transpiration responses to vapor pressure deficit in well watered 'slow-wilting' and commercial soybean. Environ Exp Bot 61:145–151

Fleury D, Jefferies S, Kuchel H, Langridge P (2010) Genetic and genomic tools to improve drought tolerance in wheat. J Exp Bot 61:3211–3222

Frank AB, Power JF, Willis WO (1973) Effect of temperature and plant water stress on photosynthesis, diffusion resistance, and leaf water potential in spring wheat. Agron J 65:777–780

Fricke W, Chaumont F (2007) Solute and water relations of growing plant cells. In: Verbelen JP, Vissenberg K (eds) The expanding cell. Springer, Berlin, Heidelberg, pp 7–31

Fuchigami LH, Cheng TY, Soeldner A (1981) Abaxial transpiration and water loss in aseptically cultured plum. J Am Soc Hortic Sci 106:519–522

Fukuoka M, Yoshimoto M, Hasegawa T (2012) Varietal range in transpiration conductance of flowering rice panicle and its impact on panicle temperature. Plant Prod Sci 15(4):258–264

Gates DM (1968) Transpiration and leaf temperature. Annu Rev Plant Physiol 19:211–238

Gates DM (2003) Biophysical ecology. Dover Publications, Mineola, New York

Gholipoor M, Prasad PVV, Mutava RN, Sinclair TR (2010) Genetic variability of transpiration response to vapor pressure deficit among sorghum genotypes. Field Crops Res 119:85–90

Givnish TJ (1987) Comparative studies of leaf form: assessing the relative roles of selective pressures and phylogenetic constraints. New Phytol 106:131–160

Grantz DA (1990) Plant response to atmospheric humidity. Plant Cell Environ 13:667–679

Groom PK, Lamont BB, Leighton S, Leighton P, Burrows C (2004) Heatdamage in sclerophylls is influenced by their leaf properties and plantenvironment. Ecoscience 11:94–101

Heckathorn SA, DeLucia EH (1991) Effect of leaf rolling on gas exchange and leaf temperature of *Andropogon gerardii* and *Spartina pectinata*. Bot Gaz 152(3):263–268

Henson IE, Jensen CR, Turner NC (1989) Leaf gas exchange and water relations of lupins and wheat. I. Shoot responses to soil water deficits. Funct Plant Biol 16:401–413

Hirabayashi H, Sasaki K, Kambe T, Gannaban RB, Miras MA, Mendioro MS, Simon EV, Lumanglas PD, Fujita D, Takemoto-Kuno Y, Takeuchi Y, Kaji R, Kondo M, Kobayashi N, Ogawa T, Ando I, Jagadish KSV, Ishimaru T (2014) *qEMF3*, a novel QTL for early-morning flowering trait from wild rice, *Oryza officinalis*, to mitigate heat stress damage at flowering in rice (*Oryza sativa* L.). J Exp Bot 66:1227–1236

Holmgren P, Jarvis PG, Jarvis MS (1965) Resistances to carbon dioxide and water vapour transfer in leaves of different plant species. Physiol Plant 18:557–573

Hose E, Steudle E, Hartung W (2000) Abscisic acid and hydraulic conductivity of maize roots: a study using cell- and root-pressure probes. Planta 211:874–882

Idso SB, Baker DG (1967) Relative importance of reradiation, convection, and transpiration in heat transfer from plants. Plant Physiol 42:631–640

Inagaki MN, Nachit MM (2008) Visual monitoring of water deficit stress using infrared thermography in wheat. In: Appels R, Eastwood R, Lagudah E, Langridge P, Mackay M, McIntyre L, Sharp P (eds) The $11^{th}$ international wheat genetics symposium proceedings. Sydney University Press, Sydney, NSW, pp 1–3

Ionenko IF, Anisimov AV, Dautova NR (2010) Effect of temperature on water transport through aquaporins. Biol Plant 54:488–494

IPCC (2007) Climate change: the physical science basis. Contribution of working group I to the fourth assessment report of the intergovernmental panel on climate change. Cambridge University Press, Cambridge

Ishimaru T, Hirabayashi H, Ida M, Takai T, San-Oh YA, Yoshinaga S, Ando I, Ogawa T, Motohiko K (2010) A genetic resource for early-morning flowering trait of wild rice *Oryza officinalis* to mitigate high temperature-induced spikelet sterility at anthesis. Ann Bot 106:515–520

Izanloo A, Condon AG, Langridge P, Tester M, Schnurbusch T (2008) Different mechanisms of adaptation to cyclic water stress in two South Australian bread wheat cultivars. J Exp Bot 59: 3327–3346

Jagadish SVK, Muthurajan R, Oane R, Wheeler TR, Heuer S, Bennett J, Craufurd PQ (2010) Physiological and proteomic approaches to dissect reproductive stage heat tolerance in rice (*Oryza sativa* L.). J Exp Bot 61:143–156

Jang JY, Kim DG, Kim YO, Kim JS, Kang H (2004) An expression analysis of a gene family encoding plasma membrane aquaporins in response to abiotic stresses in *Arabidopsis Thaliana*. Plant Mol Biol 54:713–725

Jarvis PG, McNaughton KG (1986) Stomatal control of transpiration: scaling up from leaf to region. In: Macfadyen A, Ford E (eds) Advances in ecological research. Academic Press Inc, London, pp 1–49

Jarvis PG, Slatyer RO (1970) The role of the mesophyll cell wall in leaf transpiration. Planta 90: 303–322

Johnson DA, Richards RA, Turner NC (1983) Yield, water relations, gas exchange, and surface reflectances of near-isogenic wheat lines differing in glaucousness. Crop Sci 23:318–325

Jones HG (2014) Plants and microclimate: a quantitative approach to environmental plant physiology. Cambridge University Press, Cambridge

Julia C, Dingkuhn M (2012) Variation in time of day of anthesis in rice in different climatic environments. Euro J Agron 43:166–174

Julia C, Dingkuhn M (2013) Predicting temperature induced sterility of rice spikelets requires simulation of crop-generated microclimate. Eur J Agron 49:50–60

Kadioglu A, Terzi R (2007) A dehydration avoidance mechanism: leaf rolling. Bot Rev 73:290–302

Kamaluddin M, Zwiazek JJ (2002) Ethylene enhances water transport in hypoxic aspen. Plant Physiol 128:962–969

Kholová J, Hash CT, Kumar PL, Yadav RS, Kočová M, Vadez V (2010) Terminal drought-tolerant pearl millet [*Pennisetum glaucum* (L.) R. Br.] have high leaf ABA and limit transpiration at high vapour pressure deficit. J Exp Bot 61:1431–1440

Knipfer T, Fricke W (2011) Water uptake by seminal and adventitious roots in relation to whole-plant water flow in barley (*Hordeum vulgare* L.). J Exp Bot 62:717–733

Kumar A, Tripathi RP (1991) Relationships between leaf water potential, canopy temperature and transpiration in irrigated and nonirrigated wheat. J Agron Crop Sci 166:19–23

Kumar M, Govindasamy V, Rane J, Singh A, Choudhary RL, Raina SK, George P, Aher LK, Singh NP (2017) Canopy temperature depression (CTD) and canopy greennessassociated with variation in seed yield of soybean genotypes grown in semi-aridenvironment. South Afr J Bot 113: 230–238

Lange OL, Lösch R, Schulze ED, Kappen L (1971) Responses of stomata to changes in humidity. Planta 100:76–86

Lawlor DW, Cornic G (2002) Photosynthetic carbon assimilation and associated metabolism in relation to water deficits in higher plants. Plant Cell Environ 25:275–294

Lee SH, Singh AP, Chung GC, Ahn SJ, Noh EK, Steudle E (2004) Exposure of roots of cucumber (*Cucumis sativus*) to low temperature severely reduces root pressure, hydraulic conductivity and active transport of nutrients. Physiol Plant 120:413–420

Leigh A, Sevanto S, Ball MC, Close JD, Ellsworth DS, Knight CA, Nicotra AB, Vogel S (2012) Do thick leaves avoid thermal damage in critically low wind speeds? New Phytol 194(2):477–487

Lin H, Chen Y, Zhang H, Fu P, Fan Z (2017) Stronger cooling effects of transpiration and leaf physical traits of plants from a hot dry habitat than from a hot wet habitat. Funct Ecol 31(12): 2202–2211

Livingston BE, Brown WH (1912) Relation of the daily march of transpiration to variations in the water content of foliage leaves. Bot Gaz 53:309–330

Loss SP, Siddique KHM (1994) Morphological and physiological traits associated with wheat yield increases in Mediterranean environments. Adv Agron 52:229–276

Mahmud AA, Hossain MM, Karim MA, Mian MK, Zakaria M, Kadian MS (2016) Plant water relations and canopy temperature depression for assessing water stress tolerance of potato. Indian J Plant Physiol 21(1):56–63

Manschadi AM, Hammer GL, Christopher JT, de Voil P (2008) Genotypic variation in seedling root architectural traits and implications for drought adaptation in wheat (*Triticum aestivum* L.). Plant Soil 303:115–129

Maricle BR, Cobos DR, Campbell CS (2007) Biophysical and morphological leaf adaptations to drought and salinity in salt marsh grasses. Environ Exp Bot 60:458–467

Martin TA, Hinckley TM, Meinzer FC, Sprugel DG (1999) Boundary layer conductance, leaf temperature and transpiration of *Abies amabilis* branches. Tree Physiol 19:435–443

Martre P, North GB, Nobel PS (2001) Hydraulic conductance and mercury-sensitive water transport for roots of *Opuntia acanthocarpa* in relation to soil drying and rewetting. Plant Physiol 126: 352–362

Matsui T, Kobayasi K, Yoshimoto M, Hasegawa T (2007) Stability of rice pollination in the field under hot and dry conditions in the Riverina region of New South Wales, Australia. Plant Prod Sci 10(1):57–63

Matzner S, Comstock J (2001) The temperature dependence of shoot hydraulic resistance: implications for stomatal behaviour and hydraulic limitation. Plant Cell Environ 24:1299–1307

Maurel C, Chrispeels MJ (2001) Aquaporins. A molecular entry into plant water relations. Plant Physiol 125:135–138

McClean PE, Burridge J, Beebe S, Rao IM, Porch TG (2011) Cropimprovement in the era of climate change: an integrated, multi-disciplinaryapproach for common bean (*Phaseolus vulgaris)*. Funct Plant Biol 38:927–933

Mohammadi M, Karimizadeh R, Sabaghnia N, Shefazadeh MK (2012) Effective application of canopy temperature for wheat genotypes screening under different water availability in warm environments. Bulg J Agric Sci 18(6):934–941

Monteith JL (1995) A reinterpretation of stomatal responses to humidity. Plant Cell Environ 18: 357–364

Monteith JL, Unsworth MH (2013) Principles of environmental physics: plants, animals, and the atmosphere, 4th edn. Academic Press, Cambridge

Mott KA, Parkhurst DF (1991) Stomatal responses to humidity in air and helox. Plant Cell Environ 14:509–515

Neufeld HS, Meinzer FC, Wisdom CS, Sharifi MR, Rundel PW, Neufeld MS, Goldring Y, Cunningham GL (1988) Conopy architecture of *Larrea tridentata* (DC.) Cov., a desert shrub: foliage orientation and direct beam radiation interception. Oecologia 75:54–60

Niemietz CM, Tyerman SD (2002) New potent inhibitors of aquaporins: silver and gold compounds inhibit aquaporins of plant and human origin. FEBS Lett 531:443–447

Nobel PS (2005) Physicochemical and environmental plant physiology. Academic Press, Cambridge

Oren R, Sperry JS, Katul GG, Pataki DE, Ewers BE, Phillips N, Schäfer KVR (1999) Survey and synthesis of intra- and interspecific variation in stomatal sensitivity to vapour pressure deficit. Plant Cell Environ 22:1515–1526

Pallardy SG, Cermak J, Ewers FW, Kaufmann MR, Parker WC, Sperry JS (1994) Water transport dynamics in trees and stands. In: Smith WK, Hinckley TM (eds) Resource physiology of conifers: acquisition, allocation and utilization. Academic Press, San Diego, pp 301–389

Pallas JE Jr, Michel BE, Harris DG (1967) Photosynthesis, transpiration, leaf temperature, and stomatal activity of cotton plants under varying water potentials. Plant Physiol 42(1):76–88

Palta JA, Watt M (2009) Vigorous crop root systems: form and function for improving the capture of water and nutrients. In: Sadras V, Calderini D (eds) Crop physiology: applications for genetic improvement and agronomy. Academic Press, London, pp 309–325

Palta JA, Chen X, Milroy SP, Rebetzke GJ, Dreccer MF, Watt M (2011) Large root systems: are they useful in adapting wheat to dry environments? Funct Plant Biol 38:347–354

Parent B, Hachez C, Redondo E, Simonneau T, Chaumont F, Tardieu F (2009) Drought and abscisic acid effects on aquaporin content translate into changes in hydraulic conductivity and leaf growth rate: a trans-scale approach. Plant Physiol 149:2000–2012

Passioura JB (1983) Roots and drought resistance. Agric Water Manag 7:265–280

Pinter PJ Jr, Jackson RD, Ezra CE, Gausman HW (1985) Sun-angle and canopy-architecture effects on the spectral reflectance of six wheat cultivars. Int J Remote Sens 6:1813–1825
Porch TG, Hall AE (2013) Heat tolerance. In: Kole C (ed) Genomics and breeding for climate-resilient crops. Springer, Berlin, Heidelberg
Prasad PV, Bheemanahalli R, Jagadish SK (2017) Field crops and the fear of heat stress—opportunities, challenges and future directions. Field Crops Res 200:14–121
Premachandra GS, Saneoka H, Fujita K, Ogata S (1992) Osmotic adjustment and stomatal response to water deficits in maize. J Exp Bot 43(11):1451–1456
Radin JW (1992) Reconciling water-use efficiencies of cotton in field and laboratory. Crop Sci 32: 13–18
Ranawana SRWMCJK (2015) Assessing the role of transpiration in ameliorating leaf temperatures in Wheat (*Triticum aestivum* L.) in relation to changing environmental conditions. Doctoral dissertation, University of Western Australia
Ranawana SRWMCJK, Siddique KHM, Palta JA, Stefanova K, Bramley H (2021) Stomata coordinate with plant hydraulics to regulate transpiration response to vapour pressure deficit in wheat. Funct Plant Biol 48:839–850
Raschke K (1960) Heat transfer between the plant and the environment. Ann Rev Plant Physiol 11: 111–126
Rawson HM, Clarke JM (1988) Nocturnal transpiration in wheat. Funct Plant Biol 15:397–406
Rawson HM, Begg JE, Woodward RG (1977) The effect of atmospheric humidity on photosynthesis, transpiration and water use efficiency of leaves of several plant species. Planta 134:5–10
Reynolds MP, Trethowan RM (2007) Physiological interventions in breeding for adaptation to abiotic stress. In: Spiertz JHJ, Struik PC, van Laar HH (eds) Scale and complexity in plant systems research, gene–plant–crop relations. Springer, pp 129–146
Reynolds MP, Balota M, Delgado MIB, Amani I, Fischer RA (1994) Physiological and morphological traits associated with spring wheat yield under hot, irrigated conditions. Funct Plant Biol 21:717–730
Richards RA, Rawson HM, Johnson DA (1986) Glaucousness in wheat: its development and effect on water-use efficiency, gas exchange and photosynthetic tissue temperatures. Funct Plant Biol 13:465–473
Richards RA, Rebetzke GJ, Watt M, Condon AG, Spielmeyer W, Dolferus R (2010) Breeding for improved water productivity in temperate cereals: phenotyping, quantitative trait loci, markers and the selection environment. Funct Plant Biol 37:85–97
Robredo A, Pérez-López U, de la Maza HS, González-Moro B, Lacuesta M, Mena-Petite A, Muñoz-Rueda A (2007) Elevated $CO_2$ alleviates the impact of drought on barley improving water status by lowering stomatal conductance and delaying its effects on photosynthesis. Environ Exp Bot 59(3):252–263
Sack L, Holbrook NM (2006) Leaf hydraulics. Annu Rev Plant Biol 57:361–381
Sack L, Melcher PJ, Zwieniecki MA, Holbrook NM (2002) The hydraulic conductance of the angiosperm leaf lamina: a comparison of three measurement methods. J Exp Bot 53:2177–2184
Sade N, Shatil-Cohen A, Attia Z, Maurel C, Boursiac Y, Kelly G, Granot D, Yaaran A, Lerner S, Moshelion M (2014) The role of plasma membrane aquaporins in regulating the bundle sheath-mesophyll continuum and leaf hydraulics. Plant Physiol 166:1609–1620
Sadok W, Sinclair TR (2010) Genetic variability of transpiration response of soybean [*Glycine max* (L.) Merr.] shoots to leaf hydraulic conductance inhibitor $AgNO_3$. Crop Sci 50:1423–1430
Salleo S, Raimondo F, Trifilo P, Nardini A (2003) Axial-to-radial water permeability of leaf major veins: a possible determinant of the impact of vein embolism on leaf hydraulics? Plant Cell Environ 26:1749–1758
Sato-Nara K, Nagasaka A, Yamashita H, Ishida J, Enju A, Seki M, Shinozaki K, Suzuki H (2004) Identification of genes regulated by dark adaptation and far-red light illumination in roots of *Arabidopsis thaliana*. Plant Cell Environ 27:1387–1394

Schoppach R, Sadok (2012) Differential sensitivities of transpiration to evaporative demand and soil water deficit among wheat elite cultivars indicate different strategies for drought tolerance. Environ Exp Bot 84:1–10

Schoppach R, Wauthelet D, Jeanguenin L, Sadok W (2014) Conservative water use under high evaporative demand associated with smaller root metaxylem and limited trans-membrane water transport in wheat. Funct Plant Biol 41:257–269

Schuepp PH (1993) Tansley review no. 59. Leaf boundary layers. New Phytol 125:477–507

Shatil-Cohen A, Attia Z, Moshelion M (2011) Bundle-sheath cell regulation of xylem-mesophyll water transport via aquaporins under drought stress: a target of xylem-borne ABA? Plant J 67: 72–80

Shull CA (1930) The mass factor in the energy relations of leaves. Plant Physiol 5:279–282

Siddique KHM, Tennant D, Perry MW, Belford RK (1990) Water use and water use efficiency of old and modern wheat cultivars in a Mediterranean-type environment. Aust J Agric Res 41:431–447

Sinclair TR, Tanner CB, Bennett JM (1984) Water-use efficiency in crop production. Bio Sci 34: 36–40

Sinclair TR, Zwieniecki MA, Holbrook NM (2008) Low leaf hydraulic conductance associated with drought tolerance in soybean. Physiol Plant 132:446–451

Skelton RP, Midgley JJ, Nyaga JM, Johnson SD, Cramer MD (2012) Is leaf pubescence of Cape Proteaceae a xeromorphic or radiation-protective trait? Aust J Bot 60:104–113

Sojka RE (1985) Field evaluation of drought response in small-grain cereals. In: Russel GE (ed) Progress in plant breeding-1. Butterworth and Co, London, pp 165–191

Song L, Zhang DW, Li FM, Fan XW, Ma Q, Turner NC (2010) Soil water availability alters the inter- and intra-cultivar competition of three spring wheat cultivars bred in different eras. J Agron Crop Sci 196:323–335

Sperry JS (2000) Hydraulic constraints on plant gas exchange. Agric For Meteorol 104:13–23

Steudle E (2000) Water uptake by roots: effects of water deficit. J Exp Bot 51:1531–1542

Steudle E, Peterson CA (1998) How does water get through roots? J Exp Bot 49:775–788

Stockle CO (1992) Canopy photosynthesis and transpiration estimates using radiation interception models with different levels of detail. Ecol Model 60(1):31–44

Taiz L, Zeiger E (2010) Plant physiology. Sinauer Associates, Sunderland, MA

Takai T, Yano M, Yamamoto T (2010) Canopy temperature on clear and cloudy days can be used to estimate varietal differences in stomatal conductance in rice. Field Crops Res 115(2):165–170

Tardieu F, Simonneau T (1998) Variability among species of stomatal control under fluctuating soil water status and evaporative demand: modelling isohydric and anisohydric behaviours. J Exp Bot 49:419–432

Tewari AK, Tripathy BC (1998) Temperature-stress-induced impairment of chlorophyll biosynthetic reactions in cucumber and wheat. Plant Physiol 117:851–858

Tian X, Matsui T, Li S, Yoshimoto M, Kobayasi K, Hasegawa T (2010) Heat-induced floret sterility of hybrid rice (*Oryza sativa* L.) cultivars under humid and low wind conditionsin the field of Jianghan Basin, China. Plant Prod Sci 13(3):243–251

Tournaire-Roux C, Sutka M, Javot H, Gout E, Gerbeau P, Luu D-T, Bligny R, Maurel C (2003) Cytosolic pH regulates root water transport during anoxic stress through gating of aquaporins. Nature 425:393–397

Triggs JM, Kimball BA, Pinter PJ Jr, Wall GW, Conley MM, Brooks TJ, LaMorte RL, Adam NR, Ottman MJ, Matthias AD, Leavitt SW (2004) Free-air $CO_2$ enrichment effects on the energy balance and evapotranspiration of sorghum. Agric For Meteorol 124(1–2):63–79

Tsuda M, Tyree MT (2000) Plant hydraulic conductance measured by the high pressure flow meter in crop plants. J Exp Bot 51:823–828

Turner NC, Schulze ED, Gollan T (1984) The responses of stomata and leaf gas exchange to vapour pressure deficits and soil water content. I. Species comparisons at high soil water contents. Oecologia 63:338–342

Turner NC, O'Toole JC, Cruz RT, Namuco OS, Ahmad S (1986) Responses of seven diverse rice cultivars to water deficits I. stress development, canopy temperature, leaf rolling and growth. Field Crops Res 13:257–271

Tyerman SD, Niemietz CM, Bramley H (2002) Plant aquaporins: multifunctional water and solute channels with expanding roles. Plant Cell Environ 25:173–194

Tyree MT, Sperry JS (1989) Vulnerability of xylem to cavitation and embolism. Annu Rev Plant Biol 40:19–36

Uddin MN, Marshall DR (1988) Variation in epicuticular wax content in wheat. Euphytica 38:3–9

Wall GW, Garcia RL, Wechsung F, Kimball BA (2011) Elevated atmospheric $CO_2$ and drought effects on leaf gas exchange properties of barley. Agric Ecosyst Environ 144(1):390–404

Weerakoon WMW, Maruyama A, Ohba K (2008) Impact of humidity on temperature-induced grain sterility in rice (*Oryza sativa* L). J Agron Crop Sci 194:135–140

Wetherald RT, Manabe S (2002) Simulation of hydrologic changes associated with global warming. J Geophys Res 107:4379

Wise RR, Olson AJ, Schrader SM, Sharkey TD (2004) Electron transport is the functional limitation of photosynthesis in field-grown Pima cotton plants at high temperature. Plant Cell Environ 27(6):717–724

Wollenweber B, Porter JR, Schellberg J (2003) Lack of interaction between extreme high-temperature events at vegetative and reproductive growth stages in wheat. J Agron Crop Sci 189:142–150

Wuenscher JE (1970) The effect of leaf hairs of *Verbascum thapsus* on leaf energy exchange. New Phytol 69:65–73

Xu Q, Paulsen AQ, Guikema JA, Paulsen GM (1995) Functional and ultrastructural injury to photosynthesis in wheat by high temperature during maturation. Environ Exp Bot 35:43–54

Xu Y, Zhan C, Huang B (2011) Heat shock proteins in association with heat tolerance in grasses. Intl J Proteomics 2011:529648. https://doi.org/10.1155/2011/529648

Yadav SK, Tiwari YK, Kumar DP, Shanker AK, Lakshmi NJ, Vanaja M, Maheswari M (2016) Genotypic variation in physiological traits under high temperature stress in maize. Agric Res 5(2):119–126

Yang S, Tyree MT (1994) Hydraulic architecture of *Acer saccharum* and *A. rubrum*: comparison of branches to whole trees and the contribution of leaves to hydraulic resistance. J Exp Bot 45:179–186

Yang J, Ordiz MI, Jaworski JG, Beachy RN (2011) Induced accumulation of cuticular waxes enhances drought tolerance in *Arabidopsis* by changes in development of stomata. Plant Physiol Biochem 49:1448–1455

Yates MJ, Anthony Verboom G, Rebelo AG, Cramer MD (2010) Ecophysiological significance of leaf size variation in Proteaceae from the Cape Floristic region. Funct Ecol 24(3):485–492

Yoshimoto M, Oue H, Takahashi H, Kobayashi K (2005) The effect of FACE (free-air $CO_2$ enrichment) on temperature andtranspiration of rice panicles at flowering stage. J Agric Meteorol 60:597–600

Yuan BZ, Nishiyama S, Kang Y (2003) Effects of different irrigation regimes on the growth and yield of drip-irrigated potato. Agric Water Manag 63(3):153–167

Zaman-Allah M, Jenkinson DM, Vadez V (2011) Chickpea genotypes contrasting for seed yield under terminal drought stress in the field differ for traits related to the control of water use. Funct Plant Biol 38:270–281

Zhang WH, Tyerman SD (1999) Inhibition of water channels by $HgCl_2$ in intact wheat root cells. Plant Physiol 120:849–858

Zhao H, Dai T, Jing Q, Jiang D, Cao W (2007) Leaf senescence and grain filling affected by post-anthesis high temperatures in two different wheat cultivars. Plant Growth Regul 51:149–158

# Chapter 6
# Photosynthesis as a Trait for Improving Yield Potential in Crops

**Preety Rani, Kapil Deswal, Zeenat Wadhwa, Rinki Khobra, Harohalli Masthigowda Mamrutha, Gopalareddy Krishnappa, and Gyanendra Pratap Singh**

**Abstract** Photosynthesis is one of the most significant physiological processes on earth for agricultural sustainability. In field crops the correlation among photosynthesis, biomass, and yield is very significant, and there are supporting evidences that improved yields can be obtained by increasing photosynthetic efficiency. However, new approaches and technology will be necessary to meet the set targets for feeding the world's rising population. For improving photosynthetic efficiency, manipulation can be done at biochemical, anatomical, and molecular levels. Currently, new technologies such as genome editing, synthetic biology, and system biology can open new avenues to hasten the research. This chapter highlights the current status of photosynthetic research and methodologies for improving the photosynthetic efficiency with multidisciplinary approaches.

**Keywords** Photosynthesis · Photosynthetic efficiency · Yield improvement · $C_4$ enzymes · Leaf anatomy

---

P. Rani · K. Deswal
ICAR-Indian Institute of Wheat and Barley Research, Karnal, Haryana, India

Chaudhary Charan Singh Agricultural University, Hisar, Haryana, India

Z. Wadhwa · R. Khobra (✉) · H. M. Mamrutha
ICAR-Indian Institute of Wheat and Barley Research, Karnal, Haryana, India
e-mail: Rinki@icar.gov.in

G. Krishnappa
ICAR-Indian Institute of Wheat and Barley Research, Karnal, Haryana, India

ICAR-Sugarcane Breeding Institute, Coimbatore, Tamil Nadu, India

G. P. Singh
ICAR-Indian Institute of Wheat and Barley Research, Karnal, Haryana, India

ICAR-National Bureau of Plant Genetic Resources, New Delhi, India

 
H. M. Mamrutha et al. (eds.), *Translating Physiological Tools to Augment Crop Breeding*, https://doi.org/10.1007/978-981-19-7498-4_6

## 6.1 Introduction

Photosynthesis is a complex biochemical phenomenon operating in the plant system which ultimately determines the biomass, harvest index and yield potential of the crop (Simkin et al. 2019). The capacity of plant canopy to capture solar radiations and convert it to biomass determines its photosynthetic efficiency. This efficiency is pivotal for global food security and agricultural sustainability. But research evidences highlighted that photosynthesis is contributing to yield at its lower potential than expected (Gu et al. 2017). Yield improvements done by the adoption of Green Revolution dwarfing genes to wheat and rice are becoming exhausted and improvements in biomass and radiation use efficiency are now sought in these crops (Evans 2013). But with the increasing changes in global climate, the crop productivity is decreasing significantly. Improvement in photosynthetic efficiency now appears to offer an exciting opportunity to tackle the challenge of global climate change. The previous literature evidences proves that there is a lot of scope for improving the photosynthesis both in $C_3$ and $C_4$ plants (McDonald et al. 2011; Evans 2013).

It has been reported that in C3 plants the photosynthetic efficiency is less than 4.6%, whereas in C4 plants this percentage can reach up to 6% (Cui 2021). So there is a vast scope for increasing the yield potential of plant through enhanced photosynthesis. But for this to be pragmatic in real sense photosynthesis needs to be an advance system than the normal photosynthesis. To develop an advance photosynthetic system, understanding of photosynthesis-related components such as photosystems (I and II), chlorophyll fluorescence, photochemical efficiency (Fv/Fm), non-photochemical quenching (NPQ), Rubisco carboxylation efficiency and velocity, photosynthetically active radiation (PAR), radiation use efficiency (RUE), and other chlorophyll fluorescence parameters is very much necessary (Robles-Zazueta et al. 2021). However, instruments like chlorophyll fluorometer (CFL), infrared gas analyzer (IRGA) along with hyper spectral technology are also available for measuring the photosynthetic efficiency.

A number of research strategies for improving photosynthesis efficiency have been followed, such as reducing the antenna size of light-harvesting complex, engineering photorespiratory bypasses, introducing machinery of algal $CO_2$-concentrating mechanisms, and exploring the mechanisms of accelerating recovery from non-photochemical quenching (Orr et al. 2017; Batista-Silva et al. 2020). Molecular approaches to improve photosynthesis have been proved to be successful in developing transgenics with enhanced photosynthetic efficiency with increased yield potential and plant traits. Altering the regulatory process governing photosynthesis may also provide an added route to increase photosynthetic efficiency (Simkin et al. 2019). The key enzymes of Calvin–Benson cycle, i.e., Rubisco and Sedheptulose 1,7-bisphoshatase, can be upregulated to improve the process of photosynthesis (Hammel et al. 2020). The improved agronomic practices such as use of inorganic fertilizers and slow releasing of $CO_2$ fertilizer also helped in increasing photosynthetic efficiency (Richards 2000). Despite the selection of high yielders, genetic gains are still far below the actual potential. So, the physiological breeding approaches should be explored to increase the biomass and yield and to identify the most suitable physiological trait for future endeavors. Hence, attention should be given to the enhancement of photosynthesis efficacy as a strategy for the optimization of crop productivity.

## 6.2 Phenotyping for Photosynthesis/Photosynthetic Efficiency (PSE) under Controlled and Field Conditions

Photosynthetic reactions can be measured at cellular, leaf, and plant level with low- to medium-throughput phenotyping techniques (Murchie et al. 2018), and at ecosystem scale using sensors mounted on micro-meteorological stations (Baldocchi 2003), and biome photosynthesis using chlorophyll fluorescence information collected from satellite sensors, and productivity can be estimated (Parazoo et al. 2014; Duveiller and Cescatti 2016; Zhang et al. 2016). Some of the phenotyping tools used to measure photosynthesis and photosynthetic efficiency have been described below.

### *6.2.1 Precise Studying of Photosynthetic Efficiency (PSE)*

#### 6.2.1.1 Portable Photosynthesis System (PP)

Photosynthesis estimate is expensive, time demanding, and highly dependent on meteorological conditions. As a result, it is preferred for very precise phenotyping of small populations rather than huge populations. The methodology of PP system analysis is nondispersive infrared, designed as an absolute absorptiometer with microprocessor-controlled linearization. The gas analysis system is the most important component of any leaf gas exchange system. The system's gas mixing technique and $CO_2$ regulator provides a precise, steady, and consistent supply of $CO_2$. Each heteroatomic gas molecule, including $CO_2$, has distinct absorption spectra in the infrared area. $CO_2$ absorbs most strongly at 4.25 m, with secondary peaks at 2.66, 2.77, and 14.99 m. The amount of $CO_2$ pumped into and moved out of the closed leaf chamber is monitored in the photosynthesis system, and the difference is utilized to calculate the amount of $CO_2$ fixed by photosynthesis using an infrared gas analyzer (IRGA) (Nataraja and James 1999). Each gas analyzer incorporates an infrared source, highly polished gold-plated sample cells, and detectors designed for $CO_2$ (4.26 m) and $H_2O$ (2.60 m). The higher rate of photosynthesis represents more drought tolerance of the genotype under stress conditions.

Several photosynthesis systems are available, such as CIRAS-3 PP System (Fig. 6.1), TARGAS PP System 1, CIRAS-4 PP System, Licor, and few others manufactured and designed PP Systems by International, Inc., USA. To get maximal photosynthesis and avoid inaccuracy, photosynthesis should be measured at noon in

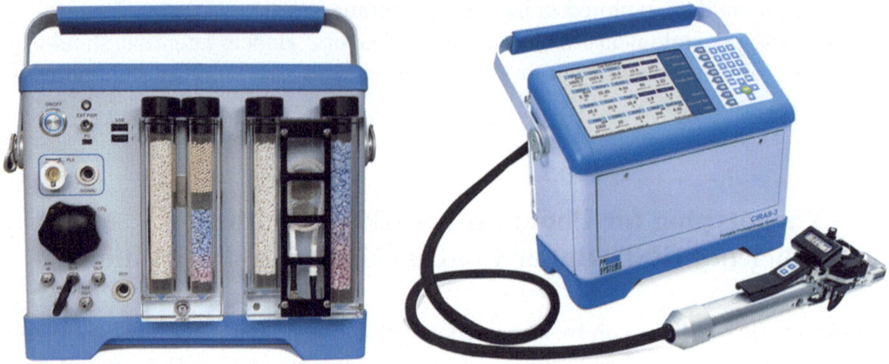

**Fig. 6.1** CIRAS portable photosynthesis system

the open field conditions. However, under controlled conditions, the data recording can be done at any time.

## *6.2.2 Rapid Screening for Photosynthetic Efficiency*

### 6.2.2.1 Chlorophyll Fluorescence (CFL)

Chlorophyll fluorescence is one of the most commonly used noninvasive parameter employed in field phenotyping. It is used to indirectly measure genotype's photosynthetic efficiency, primarily in terms of Photosystem II (PSII) function. The greatest quantity of photons needed for photochemistry is calculated as the Fv/Fm ratio, where Fv stands for variable fluorescence and Fm stands for maximal fluorescence. CFL evaluates the Fv/Fm ratio when the leaf is exposed to light immediately after dark adaption.

When photons strike the leaf surface, they are mostly dissipated through two processes: (a) photochemical quenching in the form of photosynthesis and (b) nonphotochemical quenching in the form of heat and fluorescence. The Fv/Fm ratio obtained under nonstress conditions with maximal photon consumption for photochemistry is 0.79–0.84. Before CFL measurement, dark adaptation (15–20 min) of the leaf is required primarily to transmit all of PSII's electrons to PSI for $CO_2$ fixation so that PSII is capable of capturing the most photons. The timing of the observation should be closely adhered to as that of the photosynthetic measurement. When the plant is stressed, the PSII efficiency will be reduced and hence will get less value of the Fv/Fm ratio compared to tolerant genotypes (Kate and Giles 2000). Photosynthesis measurement using IRGA under field condition is a laborious and time-consuming process. Hence, CFL (Fig. 6.2) measurement is used as a common process for assessing photosynthetic variability in breeding programs where the population size is large.

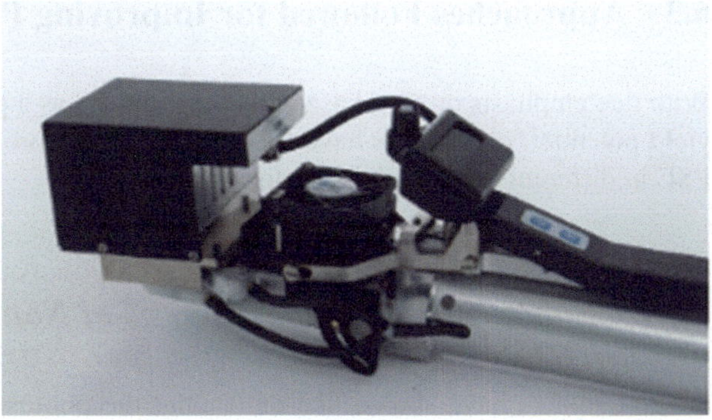

**Fig. 6.2** Chlorophyll fluorescence meter

### 6.2.3 *Imaging Techniques for Rapid Screening of PSE*

Phenomics facility includes the use of imaging methods such as fluorescence imaging, luminescence imaging, hyperspectral imaging, near-infrared/infrared (NIR/IR) imaging, which can show biochemical changes in the surface proximity that are not observable with visible light imaging. Image and data processing, particularly artificial intelligence approaches, connects the observed signals to the biological features of the samples.

LemnaTec is the global leader in automated solutions that not only save time and labor but also give a considerably larger base of data than visual scoring. The PhenoAIxpert range of systems (LemnaTec) works for small to big samples, low to high throughput, in the lab, growth chamber, and greenhouse. The system comprised of a sensor bay and sensor exchange system. The system bay carries one measuring head where all cameras/sensors are combined and the exchange system connect all the sensors at single station and the support system attaches all of them one after the next for data measurement. The system have different types of camera modules, such as Visible-light (VIS) Camera Module (size, count, color, morphology, texture, movement), NIR Camera Module (reflectance in the water band at 1450 nm), IR Camera Module (thermal radiation in 7500–13,000 nm of range), CFL Kinetics Module (PAM imaging, chlorophyll status and activity), Hyperspectral Imaging Module (spectrally resolved reflectance), Multispectral Imaging Module (reflectance at a series of distinct wavelengths), and 3D Laser Scanning Module (3D point cloud, height, and angle). For photosynthetic measurements, chlorophyll fluorescence parameters (such as quantum yield) and the images from the chlorophyll fluorescence cameras are further processed by the software (LemnaTec) which extracts significant variation in PSE among the study population.

## 6.3 Approaches Followed for Improving PSE in Crops

With due emphasis on trait-based breeding and PS as a potential trait for improving yield potential, researchers have followed several approaches (Fig. 6.3) to improve PSE in different crops which are discussed below.

### *6.3.1 Enhancing the Photosynthesis of Nonlaminar Organ*

Several studies have shown that enhancing the photosynthesis in the organ other than leaves contributes significantly for improving photosynthesis. The flag leaf synthesizes the most part of assimilates in order to fill the grain (Evans et al. 1972); however, pre-anthesis reserves (Gebbing and Schnyder 1999) and ear photosynthesis (Tambussi et al. 2007) also contribute to the grain yield in wheat. Earlier studies have proposed that flag leaf is the key source of photosynthates at dough-developing stage, but at grain filling stage awns play fundamental role as their chloroplasts are more active during later stages, and also the moving distance of assimilates between awns and grains is also very less (Li et al. 2006). Chaudhary et al. (2021) estimated the role of awns in grain filling. Thousand grain weight was measured at maturity stage under three treatments, by removing flag leaf, awns, and by shading the spike of three wheat cvs. PBW343, HD2329, and K7903. Thousand grain weight analysis showed that flag leaf, awn, and spike contribute to grain filling in different proportions depending upon the wheat cultivar and climatic conditions. In cv. PBW343, shading of spike caused maximum reduction in grain weight followed by flag leaf and awns, whereas this pattern deviates for the other cultivars. The data suggests that the awn vividly contributes to the grain yield. According to Grundbacher (1963), wheat awns acts as compensation source of assimilate supply under stress conditions.

### *6.3.2 Introduction of $C_4$-like Anatomy in $C_3$ Plants*

Engineering the $C_4$ pathway into $C_3$ crops is proposed to be an important way to improve photosynthetic efficiency and consequently crop yield. The $C_4$ plants contain kranz anatomy in which the mesophyll cell is also surrounded by the bundle sheath cell which helps in providing the maximum $CO_2$ to the PEP enzyme. The complete transfer of this kranz anatomy to $C_3$ plants is quite difficult, but there are some available methods by which we can achieve $C_4$ anatomy in $C_3$ plants up to some extent. Jiang et al. (2021) created a rice plant which contains proto-kranz anatomy, which is a unique approach for engineering $C_4$ rice that achieved the transition from $C_3$ to proto-kranz and enhanced the productivity of transgenic rice resulting from both photosynthetic $CO_2$-concentrating effect and improved energy

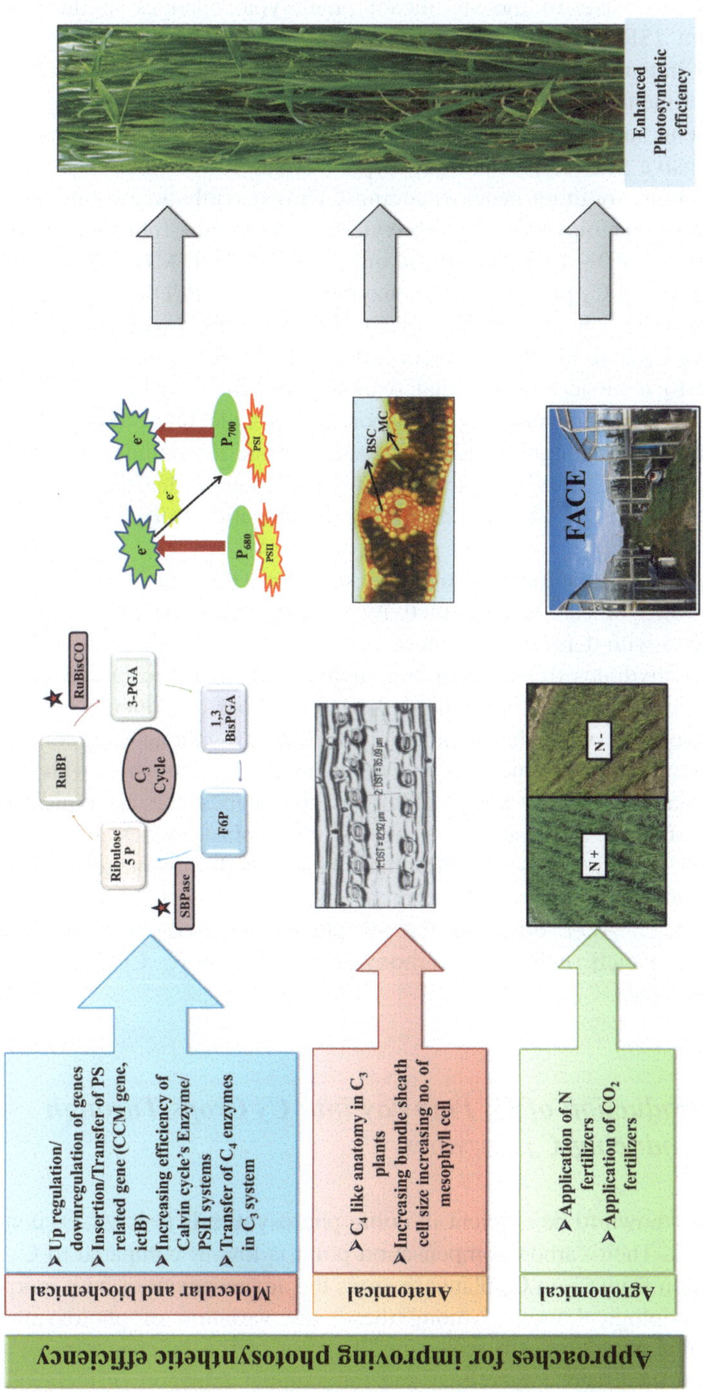


**Fig. 6.3** Different possible approaches for improving photosynthetic efficiency in crops

balance. On the basis of the significant phenotypic changes in the transgenic rice sorghum (SR) hypothesized that the introduction of exogenous sorghum DNA could integrate some photosynthetic traits of sorghum into rice. Previous studies experimentally confirmed that spike-stalk injection could integrate donor DNA fragments in the genomes of variants. The spike-stalk injection method also caused extensive phenotypic and genotypic variations for rice by genome-wide comparison. Four sorghum genes encoding $C_4$ photosynthetic enzymes, including PEPC pPhosphoenolpyruvate carboxylase), PEPCK (phosphoenolpyruvate carboxykinase), PPDK (pyruvate Pi dikinase), and NADP-ME, injected were not able to integrate all $C_4$ photosynthetic enzymes into rice. PPDK is the key enzyme for $C_4$ photosynthesis in mesophyll cells. Conversely, in BCs, a high level of PEPCK is required for $C_4$ acid. PPDK acts in concert with PEPCK to catalyze the conversion of pyruvate to oxaloacetate and that oxaloacetate can then be used to produce aspartate prior to the biosynthesis of transport amino acids. Interestingly, the number of bundle sheaths was significantly increased, but the interbundle sheath distance was not decreased. Collectively, these results demonstrate that the introduction of exogenous sorghum DNA to rice can induce the transition from $C_3$ leaf anatomy to proto-kranz.

Notably, the introduction of exogenous sorghum DNA into rice resulted in the improved photosynthesis and productivity of rice. Enhanced photosynthetic efficiency together with delayed leaf senescence led to significantly increased accumulation of carbohydrates (both starch and sugars) and improved biomass and grain yields when SR was grown in the field. Additional lines of evidence indicated that these improvements were due to not only the photosynthetic $CO_2$-concentrating effect but also an improvement in energy balance. (1) Compared with wild-type plants, SR plants displayed many additional phenotypes that are typical of $CO_2$ enrichment, such as increase in the size of chloroplasts and starch grains, Chl contents, and carbohydrates. (2) SR plants had growth advantages under higher light conditions, supported by the photosynthetic rate and light-response curves, and higher $A_{max}$. AQY observed for SR plants increased the Rubisco content and activity, which enhanced carboxylation efficiency, $CO_2$ fixation, and photosynthetic rate.

### *6.3.3 Introduction of $C_4$ Pathway into $C_3$ Crops Through Introducing $C_4$ Enzymes*

$C_4$ plants are known to be efficient in doing photosynthesis in low $CO_2$ conditions than $C_3$ plants. Their carbon compensation point is low as compared to $C_3$ plants. The conversion from $C_3$ to $C_4$ plants requires the advancement of both morphological and physiological traits. Among these, the variation of photosynthetically active vascular bundle sheath cells, modification in the biochemistry of several enzymes, and increased intercellular and intracellular transport of metabolites are of fundamental significance. A better understanding of the initial events that occurred during the $C_3$ evolution to $C_3$–$C_4$ intermediates and then to $C_4$ plants can contribute to increasing photosynthetic efficiency in $C_3$ plants. Two strategies were evolved to deal with the increasing oxygen–$CO_2$ ratio. First, Rubisco kinetic properties were changed to improve its ability to distinguish between $CO_2$ and oxygen. Second, $CO_2$-concentrating mechanisms were evolved to allow Rubisco to operate in a $CO_2$-rich space. The $CO_2$-concentrating mechanism has evolved multiple times among terrestrial plant species but always involves phosphoenolpyruvate (PEP) carboxylase fixing bicarbonate into a four-carbon acid. This gives rise to the descriptive term $C_4$ plants. $C_4$ photosynthesis requires significant additional ATP, the plants benefit from boosted biomass production and improvements in nitrogen and water use efficiencies. Using complementary approaches, including genome and transcriptomics analysis, the international $C_4$ rice consortium is working toward introducing the $C_4$ mechanism into rice. This involves five countries and seven institutions. This research initiative has already produced exciting results, including the identification of metabolite transporters and transcription factors. Ermakova et al. (2021) installed enzymes for $C_4$ pathway in (rice) *Oryza sativa* spp. *japonica* cultivar, Kitaake plant, with a single construct containing the coding regions of carbonic anhydrase, phosphoenolpyruvate (PEP) carboxylase, NADP-malate dehydrogenase, pyruvate orthophosphate dikinase, and NADP-malic enzyme from *Zea mays*, driven by cell-preferential promoters. Gene expression, protein accumulation, and enzyme activity were confirmed for all five transgenes, and intercellular localization of proteins was analyzed. $^{13}CO_2$ labeling demonstrated a tenfold increase in flux through PEP carboxylase, exceeding the increase in measured in vitro enzyme activity, and estimated to be about 2% of the maize photosynthetic flux. Flux from malate via pyruvate to PEP remained low and proportionate with the low NADP-malic enzyme activity observed in the transgenic lines. Physiological perturbations were minor and RNA sequencing revealed no substantive effects of transgene expression on other endogenous rice transcripts associated with photosynthesis. These results provide promise that, with enhanced levels of the $C_4$ proteins introduced, a functional $C_4$ pathway is achievable in rice.

### *6.3.4 Introduction of Carbon Concentrating Mechanism*

$CO_2$-concentrating mechanism (CCM) in plants suggests a significant potential to improve the photosynthetic efficiency of $C_3$ plants and could translate into substantial increases in crop yield. In organisms operating a biophysical CCM, this mechanism efficiently ambiances a high turnover rate of Rubisco with higher $CO_2$ concentrations to maximize carboxylation rates. A critical feature of both native biophysical CCMs and one engineered into a $C_3$ plant chloroplast is functional bicarbonate ($HCO_3^-$) transporters and vectorial $CO_2$ to $HCO_3^-$ converters. Engineering strategies aim to locate these transporters and conversion systems to the $C_3$ chloroplast, enabling elevation of $HCO_3^-$ concentrations within the chloroplast stroma. Several CCM components have been identified in proteobacteria, cyanobacteria, and microalgae as likely candidates for this approach, yet their successful functional expression in $C_3$ plant chloroplasts remains elusive.

### 6.3.5 *Modifying the Rubisco*

Rubisco, a key regulator enzyme of the Calvin–Benson cycle (CBC), is accountable for the assimilation of $CO_2$ and is acknowledged as the most plentiful protein in nature (Ellis 1979; Raven 2013; Erb and Zarzycki 2018). Rubisco shows slower catalytic rate than most enzymes involved in the central metabolism of plants, and it has long been considered as a limiting step for photosynthesis and hence for primary productivity (Pottier et al. 2018). Therefore, a number of metabolic engineering and synthetic biology strategies have been proposed to improve Rubisco's ability to fix $CO_2$ (Weigmann 2019). However, there are some technical challenges, suggesting that Rubisco is already operating at or near to its physiological optimum in plants (Long et al. 2015). The main challenge in modifying Rubisco is that the enzyme is a strong hexadecameric complex (550 kDa) which has multiple numbers of large and small subunits. It is having eight copies of a large subunit (RbcL), encoded by the chloroplast genome, and further eight copies of a small subunit (RbcS), encoded by the nuclear genome (Pottier et al. 2018). Moreover, Rubisco exhibits extensive natural diversity regarding subunit stoichiometry and carboxylation kinetics (Whitney and Sharwood 2008; Sharwood 2017). Further development has been made by manipulating the small subunit of Rubisco, an approach that has provided higher Rubisco catalytic turnover rates in rice (Ishikawa et al. 2011; Ogawa et al. 2012) and Arabidopsis (Izumi et al. 2012; Atkinson et al. 2017). In addition, the introduction of a foreign Rubisco from *Synechococcus elongatus* PCC7942 into tobacco allowed complete assemblage of Rubisco with functional activity and photosynthetic aptitude, supporting autotrophic growth (Lin et al. 2014; Occhialini et al. 2016). It is also already known that the folding and assembly of Rubisco large and small subunits in L8S8 holoenzymes within the chloroplast stroma involves many auxiliary factors, including chaperones such as bundle sheath defective protein (BSD2) (Aigner et al. 2017; Bracher et al. 2017; Conlan and Whitney 2021; Conlan et al. 2019, Salvucci and Crafts-Brandner 2004; Kurek et al. 2007; Fukayama et al. 2012; Yamori et al. 2012). These findings, combined with those of previous studies, indicate that engineering Rubisco activity increases photosynthesis by means of synthetic biology.

### 6.3.6 *Optimization of Calvin Cycle's Enzymes*

Calvin cycle's enzymes are more important enzymes from the view of photosynthesis. In most of the important cereal crops, the effects of these enzymes have not explored much. It has been shown through many studies under controlled condition that overexpressing SBPase enhances the rate of photosynthesis and total biomass in transgenic tobacco and tomato plants (Lefebvre et al. 2005; Ding et al. 2016). Moreover, under the conditions of abiotic stresses, the overexpression of SBPase in rice plants boosted photosynthesis rate under both high-temperature and salt stress (Feng et al. 2007) by providing higher restoration of RuBP in the stroma of the chloroplast. In transgenic wheat also they reported enhanced leaf photosynthesis by overexpressing the SBPase activity, which also exhibits increased total biomass and dry seed yield (Driever et al. 2017). In addition, increase in SBPase activity in tomato plants resulted in higher photosynthetic efficiency by higher RuBP regeneration capacity.

The effects of increased SBPase in wheat were studied through the generation of transgenic wheat plants expressing a *Brachypodium distachyon* gene for SBPase, to elevate the total SBPase levels and activity. They showed that for an important crop such as wheat, increase in SBPase activity consistently resulted in an increase in leaf $CO_2$ assimilation rate, predominantly when measured under high $CO_2$. Notably, the greatest increases in SBPase activity resulted not only in an increase in total biomass but also in an increase in total seed weight (30–40% higher) than wild-type. Total seed weight was found to be improved in both the experiments with the highest SBPase activity. This can be accomplished either through a higher number of seeds being formed per ear (fewer tillers, at high plant density), or a larger number of ears being produced per plant (more tillers, at lower plant density). These results indicate that the positive effect of increased SBPase activity can be achieved at different plant densities. It also supports the contention that increasing SBPase activity in wheat has a positive effect on leaf photosynthetic capacity and leads to an increased yield potential (Driever et al. 2017).

To avoid bottlenecks in different parts of the CBC, additional efforts have been pursued to implement engineering principles to other CBC targets (Simkin et al. 2015, 2017). These include the enzymes fructose-1,6-biphosphate aldolase (FBPA) and photorespiratory glycine decarboxylase-H (GDH-H), and the last was shown to increase photosynthesis and biomass when overexpressed in transgenic tobacco plants (López-Calcagno et al. 2019). In this respect, intensive effort has been placed into multi-gene manipulation of photosynthetic carbon assimilation to improve crop yield (Simkin et al. 2015). Co-overexpression of SBPase and FBPA has enhanced the photosynthesis and yield in transgenic tobacco (Simkin et al. 2015, 2017). In addition, coexpression of GDC-H with SBPase and FBPA resulted in a positive impact on leaf area and biomass in Arabidopsis (Simkin et al. 2017). However, despite promising results obtained so far, multi-gene manipulations are still in development and still need to be tested under field conditions. Salesse-Smith et al. (2018) overexpressed the Rubisco large (LS) and small (SS) subunits with the Rubisco assembly chaperone RAF1. While overexpression of LS and/or SS had no discernable impact on Rubisco content, addition of RAF1 overexpression resulted in >30% increase in Rubisco content.

### 6.3.7 Redesigning Photorespiration and $CO_2$ Fixation Pathways

In plants, during the process of photosynthesis some of the carbon is lost as a result of photorespiration. But now there are some strategies by which this carbon loss can be reduced, by creating photorespiratory bypass routes which cause no loss of carbon. Three photorespiratory bypasses have been designed and established in plants over the past few years. These bypasses wanted to metabolize the glycolate produced by ribulose-1,5-bisphosphate (RuBP) oxygenation, while minimizing the loss of carbon, nitrogen, energy, and by avoiding the accumulation of photorespiratory intermediates. Bypass engineered plants showed improved photosynthesis biomass and yield and reduced photorespiration (Kebeish et al. 2007; Carvalho et al. 2011; Maier et al. 2012; Dalal et al. 2015). The first bypass distracted photorespiratory glycolate into glycerate within the chloroplast, which leads to the shifting released $CO_2$ from mitochondria to chloroplasts, and reducing ammonia release. Seventy-five percent of glycolate fed into this bypass was returned to the Calvin cycle (Kebeish et al. 2007). In the second bypass, two *Escherichia coli* enzymes were used in the peroxisome to catalyze the conversion of glyoxylate into hydroxy pyruvate and $CO_2$ in a two-step process (Carvalho et al. 2011). This shortcut results in moving the position of $CO_2$ release from mitochondria to peroxisomes, reduced ammonia release and recycled 75% of the glycolate. In the third bypass, glycolate was completely oxidized into $CO_2$ inside chloroplasts by both newly introduced and native enzymes (Maier et al. 2012). Characterization of these transformed plants revealed that the first and third bypasses increased photosynthesis and biomass yield (Kebeish et al. 2007; Maier et al. 2012; Dalal et al. 2015), while the second bypass was not so effective (Carvalho et al. 2011). Shen et al. (2019) designed a different photorespiratory bypass (known as GOC bypass), which is regarded as no reducing equivalents being produced during a complete oxidation of glycolate into $CO_2$. This pathway is catalyzed by three rice-self-originating enzymes, i.e., glycolate oxidase, oxalate oxidase, and catalase which were successfully established in rice chloroplasts using a multi-gene vector. The plants with this GOC pathway showed noteworthy increases in photosynthesis efficiency, biomass yield, and nitrogen content, in both field conditions and green house chamber. Shen et al. (2019) enhanced the efficiency of photosynthesis in rice plants by creating a new photorespiratory bypass by altering the chloroplast content of the glycolic acid, glyoxylic acid, oxalic acid, carbon dioxide, and reduced respiration. Antisense P-protein potato plants containing 30–40% of wild-type (WT) levels accumulated >100-fold higher levels of glycine and displayed a significant reduction in the rate of glycine oxidation (Heineke et al. 2001; Bykova et al. 2005) and deletion of the GCS P-protein in Arabidopsis was shown to be lethal under non-photorespiratory conditions (Engel et al. 2007). Furthermore, in rice under ambient $CO_2$, knockdown of the H-protein also resulted in chlorophyll loss, protein degradation, lipid peroxidation, and an accumulation of reactive oxygen species (ROS), leading to ROS-induced senescence (Zhou et al. 2013).

### *6.3.8 Controlling the PSII Efficiency*

Photosynthetic apparatus balances the state transition between photosystem II (PSII) and PSI which regulate the excitation energy and improves the efficiency of light harvesting complex. It is a self-regulating mechanism in which the excitation energy is reversible between the light harvesting complexes. Zhou et al. (2020) demonstrated the modification of light-harvesting polymer PBF (polyboron-dipyrromethene-co-fluorene) with green light absorption and far-red emission to improve PSI and PSII activities of algae *Chlorella pyrenoidosa* for augmenting the photosynthesis. PBF could induce synergistical improvement of PSI and PSII activity through regulating their state transition. The photosynthetic efficiency is found to be increased due to enhanced oxygen evolution and the ATP and NADPH contents. The Rubisco activity was also found to be increased by 57% and the expression levels of rbcL (encoding Rubisco) and prk (encoding phosphoribulokinase) were upregulated to 3.6- and 2.5-fold, leading to obvious increase of lipid and protein production.

### *6.3.9 Insertion of Inorganic Carbon Transporter B (*ictb*) Gene*

*ictb* gene is a membrane protein that is proved to be responsible for improving photosynthesis, and this gene is found in cyanobacteria. It has been demonstrated in maize to improve the photosynthetic efficiency of the crop. Hay et al. (2017) made the transgenic plant with inserted *ictb* gene which was proved to be responsible for increased photosynthetic efficiency of soyabean plant. Under greenhouse conditions, transgenic plants showed significantly higher net assimilation rate compared to wild-type plants but low gaseous exchange. Other experiment done by Simkin et al. (2015) and Lieman-Hurwitz et al. (2003) showed that expression of the *ictB* gene in transgenic tobacco and Arabidopsis plants showed higher rates of photosynthesis without affecting the stomatal conductance compared to the WT. Koester et al. (2021) inserted *ictB* gene from *Synechococcus* sp. strain PCC 7942 into maize using *Agrobacterium*-mediated transformation. They showed extramembrane domains of ictB interacted in vitro with proteins involved in photosynthesis and carbohydrate metabolism which increased the ΦPSII and ictB localized primarily to the microsome fraction of leaf bundle-sheath cells.

### *6.3.10 Downregulation of Genes to Improve PSE*

The products of some genes in the process of photosynthesis are negative regulators, and downregulation of these enzymes can increase the photosynthetic process.

Advances in genetic engineering and genome editing technology make it conceivable to manipulate genes associated with photosynthesis with the target of improving crop yield. Several genes of interest for the improvement of photosynthesis and crop yield have been identified and manipulated to modify stomatal conductance, generate better ideotypes, reduce losses from photorespiration and respiration, enhance Rubisco efficiency, and increase sink strength and photo assimilate partitioning into the sink. Currently, CRISPR/Cas genome-editing technology has been applied for the crop improvement where excellent genes have been edited accurately. Genome editing could be used to increase the efficiency of the key $CO_2$ fixing enzyme Rubisco and thus improve photosynthesis efficiency and yield. An ambitious goal is to increase photosynthesis efficiency in $C_3$ plants such as rice by making their photosynthesis pathways more similar to those of $C_4$ plants such as maize. Zheng et al. (2021) edited hexokinase gene *OsHXK1* by knocking out via the CRISPR/Cas9 gene-editing method in the *indica* rice varieties. HXKs constitute a class of multifunctional proteins that play an important role in promoting germination, inhibiting seedling formation under high glucose concentrations, promoting vegetative growth and flowering, restoring fertility, and transducing senescence signals. HXK is a ubiquitous protein in all organisms. It plays an important role in metabolism, glucose signal transduction, and phosphorylation of glucose and fructose. The edited plants showed increased light saturation points, stomatal conductance, light tolerance, photosynthetic products, and rice yields. Moreover, transcriptome analysis showed that the expression of photosynthesis-related genes significantly increased and lead to the cultivation of high-photosynthetic efficiency and high-yielding rice varieties.

## 6.3.11 Agronomical Approaches

### 6.3.11.1 Increased Photosynthesis Through Application of Nitrogen (N)

Bassi et al. (2018) demonstrated the increase in photosynthesis through the application of N in the sugarcane plant. Two contrasting genotypes of sugarcane were supplied with low and high N conditions. N supply upregulates the working of photosynthetic apparatus and also the chlorophyll content of the sugarcane leaf. The N treatment also enhances the other processes of the plant which contributes to the photosynthesis like, the amount and activity of carboxylation enzymes, total protein, sugar content, total N, and photosynthesis-related metabolites.

### 6.3.11.2 Use of $CO_2$ Fertilizers

The research and certain experiments have proven that increased $CO_2$ can increases the rate of photosynthesis. Wang et al. (2019) improved the photosynthesis rate of *Brassica chinensis* through slow releasing of $CO_2$ fertilizer, along with photosynthetic rate and other processes of plants like chlorophyll fluorescence, chlorophyll content, leaf area, leaf mass per area, and dry matter. In this work, a slow release of $CO_2$ gas fertilizer inspired by polyphenol chemistry was prepared to provide sustainable $CO_2$ that could improve plant photosynthetic capacity and get a higher crop yield. The core–shell structure was designed to confer gas fertilizers slow-release property. Micron-sized calcium carbonate particles as a core with uniform particle size as carbon sources for plant photosynthesis and tannic acid was coated on it as a shell via oxidative oligomerization and cross-linked by polyetherimide.

### 6.3.12 Molecular Markers for Improving PSE

QTLs for photosynthesis-related traits will be beneficial for MAS in breeding programs in order to attain genotypes with increased $CO_2$ assimilation. Several investigations of QTL mapping of characteristics linked with $CO_2$ assimilation in important crops have been conducted (Morales et al. 2018). As photosynthetic traits are highly heritable and show significant variations, and at the same time, it is not easy to study as it is also highly polygenic as well as phenotypically highly responsive to environmental conditions (Zargar et al. 2017; Kaiser et al. 2018; Vico et al. 2019). Salter et al. (2020) identify QTL for Rubisco activation on chromosome 7H on barley, and it was first report on the identification of a QTL for Rubisco activation rate in planta, and this discovery opens the door to marker-assisted breeding to improve whole-canopy photosynthesis of crop plants and thereby improving photosynthetic efficiency in dynamic environments and also been recently highlighted as a key target to increase whole-canopy carbon assimilation (Murchie et al. 2018). It has been estimated that it could increase daily carbon gain by as much as 21% in wheat if Rubisco activation was instantaneous (Taylor and Long 2017). Variation in Rubisco activation kinetics has now been observed in crop species including soybean (Soleh et al. 2017), rice (Acevedo-Siaca et al. 2020), and wheat (Salter et al. 2019), and work with other species has indicated specific molecular targets and pathways that could accelerate Rubisco activation speed, with a particular focus on Rubisco's catalytic chaperone Rubiscoactivase (Rca) (in *Arabidopsis thaliana*, Mott et al. 1997; and in *Oryzasativa*, Yamori et al. 2012). Huang et al. (2018) used molecular markers associated with photosynthetic traits with link to canopy leaves, and 1366 single-nucleotide polymorphism (SNP) markers covering the whole genome of durum wheat were used to genotype 150 cultivars. A total of 120 SNP marker associations were detected on 13 of the 14 chromosomes. Among these markers, 83 were associated with the canopy leaf traits, 10 with 1000-grain weight, and 29 with kernel number per spike, and their study contributes to a better understanding of the potential and genetic basis of functional leaves, as well as to the pyramiding of favorable alleles in durum wheat breeding using marker-assisted selection for ideal plant-type and high photosynthesis efficiency. Recently, a genome-wide association study (GWAS) had countless advantage in the dissection of genetic basis of complex traits in crops. Dhanapal et al. (2016) identified the genomic loci (genes) governing photosynthetic carbon metabolism (PCM) for genetic improvement of legumes including soybean (Table 6.1).

## 6.4 Effect of Climate Change on Photosynthesis

Climate change affects plant growth, development, and yield that cause a serious threat to agriculture productivity. Numerous environmental variables like temperature, $CO_2$ concentration, and light significantly affect phenomena and functionality of photosynthesis. Recently, climate change has becoming a major concern for the agricultural scientists. According to IPCC reports long-term global warming trends showed that temperature has been increased by 0.1–0.3 °C per decade across the world, and it will increase the mean global temperature by 1.5 °C till 2052 (IPCC 2014, 2018). Industrial revolution raised the atmospheric $CO_2$ concentrations from 280 to 414 μmol mol$^{-1}$ (NOAA Mauna Loa Atmospheric Baseline Observatory 2019). It has been predicted that by the end of the twenty-first century the atmospheric $CO_2$ concentrations will increase up to 700 μmol mol$^{-1}$ (Salazar-Parra et al. 2018). However, evidences reveal that elevated $CO_2$ stimulate carbon fixation and nitrogen-use efficiency that improves net primary production (Afzal et al. 2018; Jiménez et al. 2020).

However, the response of photosynthesis with increase in the climate change varies differentially in various countries across the globe, and it is also influenced by several pollutants and temperature of the location. Due to climate change, large variability has been observed in different crop species $C_3$, $C_4$, CAM (Crassulacean Acid Metabolism), and it affects the physiological process such as stomatal conductance, leaf water content, water use efficiency, canopy temperature, chlorophyll content, leaf water potential, and transpiration which contribute to photosynthesis.

Stomatal conductance is defined as the transfer of $CO_2$ from the environment into intercellular air spaces via stomata (Flexas et al. 2012; Sun et al. 2014). As a result, improving $CO_2$ diffusion rates and enhancing gaseous exchange might be another avenue for increasing photosynthetic efficiency. According to Sekhar et al. (2021), decreased photosynthetic efficiency in plants grown under elevated $CO_2$ was found to be correlated with increased accumulation of nonstructural carbohydrate (NSC), particularly starch, which in turn increased the suppression of genes related to Rubisco as well as other proteins involved in the carbon sequestration process (Kirschbaum 2011; Takatani et al. 2014). Crafts-Brandner and Salvucci (2002) reported that heat stress may affect $CO_2$ fixation by phosphoenolpyruvate (PEP) carboxylase, transport (C4 acids) between mesophyll and bundle sheath cells, due to the differential localization of Calvin cycle and PSII in C4 plants.

Increasing temperature also has an effect on photosynthesis up to a certain optimal temperature, after which there is water deficit in plants as a result of the high temperature, and plant development is retarded. As the temperature rises, the

**Table 6.1** Different genes used in improving PSE in various crops

| Sr. no. | Gene expressed | crop | Target trait | Percentage increase | Reference |
|---|---|---|---|---|---|
| 1 | SBPase | Arabidopsis<br>Tobacco<br>Wheat | Tissue-specific expression. 37–85% increase in SBPase, activity, 37% increase in $CO_2$ assimilation.<br>Constitutive expression. 90–110% average increase in SBPase activities, increase in photosynthetic rates, increases in sucrose and starch<br>Constitutive expression | 42% increase in dry weight and a 53% increase in seed yield<br>30–34% increase in dry weight<br>Up to 90% increase in SBPase activities in some lines, increase in $CO_2$ assimilation up to 40% increase in grain yield | Simkin et al. (2017), Lefebvre et al. (2005), Simkin et al. (2015), Driever et al. (2017) |
| 2 | Cyanobacterial SBPase | Tobacco | Tissue-specific expression | More than 20% increase in the rate of photosynthetic $CO_2$ fixation 50% increase in final dry weight | Tamoi et al. (2006) |
| 3 | FBP aldolase | Arabidopsis | Tissue-specific expression | 46–80% increase in FBP aldolase activity, 31% increase in $CO_2$ assimilation 32% increase in dry weight, 35% increase in seed yield | Simkin et al. (2017) |
| 4 | Transketolase | Tobacco | Constitutive expression | 76–150% increase in transketolase activity, no increase in photosynthesis. Negative effect on plant growth resulting in leaf chlorosis | Khozaei et al. (2015) |
| 5 | Cyanobacterial SBP/FBPase | Soybean | Constitutive expression | 4–14% increase in $CO_2$ fixation rates in some lines. Under ambient $CO_2$, elevated temperature led to reductions in seed yield. Under elevated $CO_2$ and elevated temperature, seed yield was maintained while the WT showed 11% and 22% reductions | Köhler et al. (2017) |

(continued)

**Table 6.1** (continued)

| Sr. no. | Gene expressed | crop | Target trait | Percentage increase | Reference |
|---|---|---|---|---|---|
| 6 | Glycine decarboxylase H-protein | Arabidopsis | Tissue-specific expression | 19% increase in $CO_2$ assimilation and elevated photosynthetic electron transport rates compared with controls 50% increase in dry weight, no increase in seed yield | Simkin et al. (2017), Timm et al. (2012) |
| 7 | Algal Cyt $_{c6}$ | Tobacco | Constitutive expression. Higher photosynthetic/ electron transport rates and improved water use efficiency. Significant increases in chlorophyll and carotenoid content | Increased biomass | Yadav et al. (2018) |
| 8 | Cyanobacterial inorganic carbon transporter B | Rice | Constitutive expression | 18% increase in $CO_2$ assimilation 17.9% increase in biomass and increased plant height | Gong et al. (2015) |
| 9 | Rca[d] | Rice | Tissue-specific expression of the barley Rca | Reduction in Rubisco amount. Reduction in $CO_2$ assimilation increased rate of photosynthetic induction by light | Fukayama et al. (2012) |
| 10 | NADPH-dependent thioredoxin reductase (NTRC) | Arabidopsis | Constitutive expression of NTRC | 42–263% increase in dry weight. Increased starch. Increased photosynthesis. Enhanced tolerance to photo-oxidative and drought stresses | Toivola et al. (2013), Nikkanen et al. (2016), Kim et al. (2017) |

vapor pressure deficits in the air may rise, accelerating the process of transpiration from the plant's canopy. Complete summarization of the all consequence of high temperature on crop photosynthesis is still an area of discussion. The extreme increase in temperature can affect the enzyme degradation and can hamper the function of PSII, decrease electron transport rates, inhibit Rubisco activase (Rca), and decrease chlorophyll content (Allakhverdiev et al. 2008; Prasad and Djanaguiraman 2011). Increase in temperature can also results in the increased membrane permeability, which causes direct damage of the chloroplast thylakoid membranes, which leads to inhibition of light harvesting complex, electron transport rates, and ATP generation (Djanaguiraman et al. 2013; Pokharel et al. 2020). However, thermal liability of enzymes directly involved in photosynthesis remains the major cause of photosynthetic inhibition of $C_3$ and $C_4$ crops grown under elevated temperatures (Perdomo et al. 2015).

The intensity of light is also a considerable limiting factor of photosynthesis. The rate of the light-dependent reaction and photosynthesis generally increases linearly with increasing light intensity. Light is absorbed by the chlorophyll molecules present in mesophyll cells (Palit et al. 2020). As there is more number of photons of light that falling on the leaf, there will be more number of chlorophyll molecules that got excited and are ionized and more energy is produced in the form of ATP and NADPH. Change in temperature has no effect on the light dependent processes of photosynthesis. However, temperature has an effect on the light-independent reactions of photosynthesis since light independent reactions rely on enzymes. According to Wimalasekera (2019) when leaves are excessively exposed to light, the leaves try to dissipate the excess light energy and also protect the photosynthetic apparatus from photodamage. Dissipated light energy is based on irradiance, species, growth conditions, nutrition conditions, and temperature, and it has been observed that high light stress stops the metabolic processes of photosynthesis in rice. Photoinactivation of PSII complexes and photoprotection was observed in *Capsicum annuum* L. leaves when exposed to a high light intensity of 500 μmol photons $m^{-2}$ $s^{-1}$ (Lee et al. 2001).

## 6.5 Future Prospect

The higher predicted yields of physiological breeding over conventional breeding will improve the scope of physiological trait based breeding in future. For enhancing the productivity, photosynthesis-related traits are potentially promising and have to be desiccated further. Many research approaches have been followed by the scientific community to attain the actual photosynthetic potential, but still there is a scope of further improvement. Efforts are needed to develop smarter crop species with improved canopy architecture traits, updated photosynthetic system for maximizing the light harvesting for improving the biomass and ultimately the yield. Genetic engineering, and more recently genome editing, continues to improve our capability not only to alter the specific targets (cytochrome b/f) but also to introduce genes from other organisms like cyanobacteria. The desired manipulations can only be achieved with a multidisciplinary pathway which performed both lab- and field-based research approaches. Besides the high-throughput technologies, plant canopy architectural traits such as leaf posture, leaf angle, leaf area, and nonlaminar organs are the key determinants for improving photosynthesis. Hence, crop improvement breeding programs should be diverted from conventional breeding to trait based breeding for developing varieties which can help to improve farmers profit without additional cost.

## References

Acevedo-Siaca LG, Coe R, Wang Y, Kromdijk J, Quick WP, Long SP (2020) Variation in photosynthetic induction between rice accessions and its potential for improving productivity. New Phytol 227:1097–1108

Afzal M, Shabbir G, Ilyas M, Jan SSA, Jan SA (2018) Impact of climate change on crop adaptation: current challenges and future perspectives. Pure Appl Biol 7:965–972. https://doi.org/10.19045/bspab.2018.700115

Aigner H, Wilson RH, Bracher A, Calisse L, Bhat JY, Hartl FU, Hayer-Hartl M (2017) Plant RuBisCo assembly in E. coli with five chloroplast chaperones including BSD2. Science 358(6368):1272–1278

Allakhverdiev SI, Kreslavski VD, Klimov VV, Los DA, Carpentier R, Mohanty P (2008) Heat stress: an overview of molecular responses in photosynthesis. Photosynth Res 98:541–550

Atkinson N, Leitão N, Orr DJ, Meyer MT, Carmo-Silva E, Griffiths H, Smith AM, McCormick AJ (2017) Rubisco small subunits from the unicellular green alga Chlamydomonas complement Rubisco-deficient mutants of Arabidopsis. New Phytol 214(2):655–667

Baldocchi DD (2003) Assessing the eddy covariance technique for evaluating carbon dioxide exchange rates of ecosystems: past, present and future. Glob Chang Biol 9:479–492. https://doi.org/10.1046/j.1365-2486.2003.00629.x

Bassi D, Menossi M, Mattiello L (2018) Nitrogen supply influences photosynthesis establishment along the sugarcane leaf. Sci Rep 8(1):1–13

Batista-Silva W, da Fonseca-Pereira P, Martins AO, Zsögön A, Nunes-Nesi A, Araújo WL (2020) Engineering improved photosynthesis in the era of synthetic biology. Plant Commun 1(2): 100032

Bracher A, Whitney SM, Hartl FU, Hayer-Hartl M (2017) Biogenesis and metabolic maintenance of Rubisco. Annu Rev Plant Biol 68:29–60

Bykova NV, Keerberg O, Pärnik T, Bauwe H, Gardeström P (2005) Interaction between photorespiration and respiration in transgenic potato plants with antisense reduction in glycine decarboxylase. Planta 222(1):130–140

Carvalho JDF, Madgwick PJ, Powers SJ, Keys AJ, Lea PJ, Parry MA (2011) An engineered pathway for glyoxylate metabolism in tobacco plants aimed to avoid the release of ammonia in photorespiration. BMC Biotechnol 11(1):1–7

Chaudhary C, Sharma N, Khurana P (2021) Decoding the wheat awn transcriptome and overexpressing TaRca1β in rice for heat stress tolerance. Plant Mol Biol 105(1):133–146

Conlan B, Whitney S (2021) Preparing Rubisco for a tune up. Nat Plants 4(1):12–13

Conlan B, Birch R, Kelso C, Holland S, De Souza AP, Long SP, Beck JL, Whitney SM (2019) BSD2 is a Rubisco-specific assembly chaperone, forms intermediary hetero-oligomeric complexes, and is nonlimiting to growth in tobacco. Plant Cell Environ 42(4):1287–1301

Crafts-Brandner SJ, Salvucci ME (2002) Sensitivity of photosynthesis in a C4 plant, maize, to heat stress. Plant Physiol 129(4):1773–1780. https://doi.org/10.1104/pp.002170

Cui H (2021) Challenges and approaches to crop improvement through C3-to-C4 engineering. Front Plant Sci 12:715391. https://doi.org/10.3389/fpls.2021.715391
Dalal J, Lopez H, Vasani NB, Hu Z, Swift JE, Yalamanchili R, Dvora M, Lin X, Xie D, Qu R, Sederoff HW (2015) A photorespiratory bypass increases plant growth and seed yield in biofuel crop Camelina sativa. Biotechnol Biofuels 8(1):1–22
Dhanapal AP, Ray JD, Singh SK, Hoyos-Villegas V, Smith JR, Purcell LC, Fritschi FB (2016) Genome-wide association mapping of soybean chlorophyll traits based on canopy spectral reflectance and leaf extracts. BMC Plant Biol 16:174
Ding F, Wang M, Zhang S, Ai X (2016) Changes in SBPase activity influence photosynthetic capacity, growth, and tolerance to chilling stress in transgenic tomato plants. Sci Rep 6:32741
Djanaguiraman M, Prasad PVV, Schapaugh WT (2013) High day- or nighttime temperature alters leaf assimilation, reproductive success, and phosphatidic acid of pollen grain in soybean [*Glycine max (L.)* Merr.]. Crop Sci 53:1594–1604
Driever SM, Simkin AJ, Alotaibi S, Fisk SJ, Madgwick PJ, Sparks CA, Jones HD, Lawson T, Parry MA, Raines CA (2017) Increased SBPase activity improves photosynthesis and grain yield in wheat grown in greenhouse conditions. Philos Trans R Soc B Biol Sci 372(1730):20160384
Duveiller G, Cescatti A (2016) Spatially downscaling sun-induced chlorophyll fluorescence leads to an improved temporal correlation with gross primary productivity. Remote Sens Environ 182: 72–89. https://doi.org/10.1016/j.rse.2016.04.027
Ellis RJ (1979) The most abundant protein in the world. Trends Biochem Sci 4(11):241–244
Engel N, van den Daele K, Kolukisaoglu U, Morgenthal K, Weckwerth W, Parnik T, Keerberg O, Bauwe H (2007) Deletion of glycine decarboxylase in Arabidopsis is lethal under nonphotorespiratory conditions. Plant Physiol 144(3):1328–1335
Erb TJ, Zarzycki J (2018) A short history of RubisCO: the rise and fall (?) of Nature's predominant CO2 fixing enzyme. Curr Opin Biotechnol 49:100–107
Ermakova M, Arrivault S, Giuliani R, Danila F, Alonso-Cantabrana H, Vlad D, Ishihara H, Feil R, Guenther M, Borghi GL, Covshoff S (2021) Installation of C4 photosynthetic pathway enzymes in rice using a single construct. Plant Biotechnol J 19(3):575–588
Evans JR (2013) Improving photosynthesis. Plant Physiol 162(4):1780–1793
Evans LT, Bingham J, Jackson P, Sutherland J (1972) Effect of awns and drought on the supply of photosynthate and its distribution within wheat ears. Ann Appl Biol 70:67–76
Feng L, Wang K, Li Y, Tan Y, Kong J, Li H, Zhu Y (2007) Overexpression of SBPase enhances photosynthesis against high temperature stress in transgenic rice plants. Plant Cell Rep 26(9): 1635–1646
Flexas J, Barbour MM, Brendel O, Cabrera HM, Carriquí M, DíazEspejo A, Douthe C, Dreyer E, Ferrio JP, Gago J, Gallé A (2012) Mesophyll difusion conductance to $CO_2$: an unappreciated central player in photosynthesis. Plant Sci 193:70–84
Fukayama H, Ueguchi C, Nishikawa K, Katoh N, Ishikawa C, Masumoto C, Hatanaka T, Misoo S (2012) Overexpression of Rubisco activase decreases the photosynthetic CO2 assimilation rate by reducing Rubisco content in rice leaves. Plant Cell Physiol 53(6):976–986
Gebbing T, Schnyder H (1999) Pre-anthesis reserve utilization for protein and carbohydrate synthesis in grains of wheat. Plant Physiol 121:871–878
Gong HY, Li Y, Fang G, Hu DH, Jin WB, Wang ZH, Li YS (2015) Transgenic rice expressing IctB and FBP/SBPase derived from cyanobacteria exhibits enhanced photosynthesis and mesophyll conductance to $CO_2$. PLoS One 10:e0140928
Grundbacher FJ (1963) The physiological function of the cereal awn. Bot Rev 29:366–381
Gu J, Zhou Z, Li Z, Chen Y, Wang Z, Zhang H, Yang J (2017) Photosynthetic properties and potentials for improvement of photosynthesis in pale green leaf rice under high light conditions. Front Plant Sci 8:1082. https://doi.org/10.3389/fpls.2017.01082
Hammel A, Sommer F, Zimmer D, Stitt M, Mühlhaus T, Schroda M (2020) Overexpression of sedoheptulose-1,7-bisphosphatase enhances photosynthesis in *Chlamydomonas reinhardtii* and has no effect on the abundance of other Calvin-Benson cycle enzymes. Front Plant Sci 11:868. https://doi.org/10.3389/fpls.2020.00868

Hay WT, Bihmidine S, Mutlu N, Le Hoang K, Awada T, Weeks DP, Clemente TE, Long SP (2017) Enhancing soybean photosynthetic CO2 assimilation using a cyanobacterial membrane protein, ictB. J Plant Physiol 212:58–68

Heineke D, Bykova N, Gardeström P, Bauwe H (2001) Metabolic response of potato plants to an antisense reduction of the P-protein of glycine decarboxylase. Planta 212:880–887

Huang S, Sun L, Hu X, Wang Y, Zhang Y, Nevo E, Peng J, Sun D (2018) Associations of canopy leaf traits with SNP markers in durum wheat (*Triticum turgidum L. durum (Desf.)*). PLoS One 13(10):e0206226

IPCC (2014) Climate change 2014 synthesis report. In: Pachauri RK, Meyer Geneva LA (eds) Contribution of working groups I, II and III to the fifth assessment report of the intergovernmental panel on climate change. Intergovernmental Panel on Climate Change (IPCC)

IPCC (2018) "Global warming of 1.5°C," in An IPCC Special Report on the impacts of global warming of 1.5°C above pre-industrial levels and related global greenhouse gas emission pathways, in the context of strengthening the global response to the threat of climate change, sustainable development. Intergovernmental Panel on Climate Change (IPCC). Masson-Delmotte V, Zhai P, Pörtner HO, Roberts D, Skea J, Shukla PR, Pirani A, Moufouma-Okia W, Péan C, Pidcock R, Connors S, Matthews JBR, Chen Y, Zhou X, Gomis MI, Lonnoy E, Maycock T

Ishikawa C, Hatanaka T, Misoo S, Miyake C, Fukayama H (2011) Functional incorporation of sorghum small subunit increases the catalytic turnover rate of Rubisco in transgenic rice. Plant Physiol 156(3):1603–1611

Izumi M, Tsunoda H, Suzuki Y, Makino A, Ishida H (2012) RBCS1A and RBCS3B, two major members within the Arabidopsis RBCS multigene family, function to yield sufficient Rubisco content for leaf photosynthetic capacity. J Exp Bot 63(5):2159–2170

Jiang D, Wang F, Zhang H, Gao W, Tong X, Lv C, Chen G (2021) Formation of proto-Kranz in C3 rice induced by spike-stalk injection method. Int J Mol Sci 22(9):4305

Jiménez S, Fattahi M, Bedis K, Nasrolahpour-moghadam S, Irigoyen JJ, Gogorcena Y (2020) Interactional effects of climate change factors on the water status, photosynthetic rate, and metabolic regulation in peach. Front Plant Sci 11:43. https://doi.org/10.3389/fpls.2020.00043

Kaiser E, Morales A, Harbinson J (2018) Fluctuating light takes crop photosynthesis on a rollercoaster ride. Plant Physiol 176:977–989

Kate M, Giles NJ (2000) Review article. Chlorophyll fluorescence-a practical guide. J Exp Bot 51: 659–668

Kebeish R, Niessen M, Thiruveedhi K, Bari R, Hirsch HJ, Rosenkranz R, Peterhänsel C (2007) Chloroplastic photorespiratory bypass increases photosynthesis and biomass production in Arabidopsis thaliana. Nat Biotechnol 25(5):593–599

Khozaei M, Fisk S, Lawson T, Gibon Y, Sulpice R, Stitt M, Raines CA (2015) Overexpression of plastid transketolase in tobacco results in a thiamine auxotrophic phenotype. Plant Cell 27(2): 432–447

Kim MR, Khaleda L, Jung IJ, Kim JY, Lee SY, Cha JY, Kim WY (2017) Overexpression of chloroplast-localized NADPH-dependent thioredoxin reductase C (NTRC) enhances tolerance to photo-oxidative and drought stresses in Arabidopsis thaliana. J Plant Biol 60:175–180

Kirschbaum MU (2011) Does enhanced photosynthesis enhance growth? Lessons learned from $CO_2$ enrichment studies. Plant Physiol 155:117–124

Koester RP, Pignon CP, Kesler DC, Willison RS, Kang M, Shen Y, Priest HD, Begemann MB, Cook KA, Bannon GA, Oufattole M (2021) Transgenic insertion of the cyanobacterial membrane protein ictB increases grain yield in Zea mays through increased photosynthesis and carbohydrate production. PLoS One 16(2):e0246359

Köhler IH, Ruiz-Vera UM, VanLoocke A, Thomey ML, Clemente T, Long SP, Ort DR, Bernacchi CJ (2017) Expression of cyanobacterial FBP/SBPase in soybean prevents yield depression under future climate conditions. J Exp Bot 68:715–726

Kurek I, Chang TK, Bertain SM, Madrigal A, Liu L, Lassner MW, Zhu G (2007) Enhanced thermostability of Arabidopsis Rubisco activase improves photosynthesis and growth rates under moderate heat stress. Plant Cell 19(10):3230–3241

Lee HY, Hong YN, Chow WS (2001) Photoinactivation of photosystem II complexes and photoprotection by non-functional neighbours in *Capsicum annuum L.* leaves. Planta 212: 332–342

Lefebvre S, Lawson T, Fryer M, Zakhleniuk OV, Lloyd JC, Raines CA (2005) Increased sedoheptulose-1, 7-bisphosphatase activity in transgenic tobacco plants stimulates photosynthesis and growth from an early stage in development. Plant Physiol 138(1):451–460

Li X, Wang H, Li H, Zhang L, Teng N, Lin Q (2006) Awns play a dominant role in carbohydrate production during the grain-filling stages in wheat (*Triticum aestivum*). Physiol Plant 127:701–709

Lieman-Hurwitz J, Rachmilevitch S, Mittler R, Marcus Y, Kaplan A (2003) Enhanced photosynthesis and growth of transgenic plants that express ictB, a gene involved in $HCO_3^-$ accumulation in cyanobacteria. Plant Biotechnol J 1:43–50

Lin MT, Occhialini A, Andralojc PJ, Parry MA, Hanson MR (2014) A faster Rubisco with potential to increase photosynthesis in crops. Nature 513(7519):547–550

Long SP, Marshall-Colon A, Zhu XG (2015) Meeting the global food demand of the future by engineering crop photosynthesis and yield potential. Cell 161(1):56–66

López-Calcagno PE, Fisk S, Brown KL, Bull SE, South PF, Raines CA (2019 Jan) Overexpressing the H-protein of the glycine cleavage system increases biomass yield in glasshouse and field-grown transgenic tobacco plants. Plant Biotechnol J 17(1):141–151

Maier A, Fahnenstich H, Von Caemmerer S, Engqvist MK, Weber AP, Flügge UI, Maurino VG (2012) Transgenic introduction of a glycolate oxidative cycle into A. thaliana chloroplasts leads to growth improvement. Frontiers. Plant Sci 3:38

McDonald AE, Ivanov AG, Bode R, Maxwell DP, Rodermel SR, Hüner NP (2011) Flexibility in photosynthetic electron transport: the physiological role of plastoquinol terminal oxidase (PTOX). Biochim Biophys Acta 1807(8):954–967. https://doi.org/10.1016/j.bbabio.2010.10.024

Morales F, Pavlovič A, Abadía A, Abadía J (2018) Photosynthesis in poor nutrient soils, in compacted soils, and under drought. In: The leaf: a platform for performing photosynthesis. Springer, Cham, pp 371–399

Mott KA, Snyder GW, Woodrow IE (1997) Kinetics of Rubisco activation as determined from gas-exchange measurements in antisense plants of *Arabidopsis thaliana* containing reduced levels of Rubisco activase. Aust J Plant Physiol 24:811–818

Murchie EH, Kefauver S, Araus JL, Muller O, Rascher U, Flood PJ, Lawson T (2018) Measuring the dynamic photosynthome. Ann Bot 122(2):207–220

Nataraja KN, James J (1999) Clonal differences in photosynthesis in Heveabrasiliensis. Photosynthetica 36(1–2):89–98

Nikkanen L, Toivola J, Rintamäki E (2016) Crosstalk between chloroplast thioredoxin systems in regulation of photosynthesis. Plant Cell Environ 39:1691–1705

NOAA Mauna Loa Atmospheric Baseline Observatory (2019) Carbon dioxide levels hit record peak in May. Available at: https://research.noaa.gov/article/ArtMID/587/ArticleID/2461/Carbon-dioxide-levels-hit-record-peak-in-May

Occhialini A, Lin MT, Andralojc PJ, Hanson MR, Parry MA (2016) Transgenic tobacco plants with improved cyanobacterial Rubisco expression but no extra assembly factors grow at near wild-type rates if provided with elevated $CO_2$. Plant J 85(1):148–160

Ogawa S, Suzuki Y, Yoshizawa R, Kanno K, Makino A (2012) Effect of individual suppression of RBCS multigene family on Rubisco contents in rice leaves. Plant Cell Environ 35(3):546–553

Orr DJ, Pereira AM, da Fonseca Pereira P, Pereira-Lima ÍA, Zsögön A, Araújo WL (2017) Engineering photosynthesis: progress and perspectives. F1000Res 6:1891. https://doi.org/10.12688/f1000research.12181.1

Palit P, Kudapa H, Zougmore R, Kholova J, Whitbread A, Sharma M, Varshney RK (2020) An integrated research framework combining genomics, systems biology, physiology, modelling and breeding for legume improvement in response to elevated CO2 under climate change scenario. Curr Plant Biol 22:100149

Parazoo NC, Bowman K, Fisher JB, Frankenberg C, Jones DB, Cescatti A, Pérez-Priego Ó, Wohlfahrt G, Montagnani L (2014) Terrestrial gross primary production inferred from satellite fluorescence and vegetation models. Glob Chang Biol 20(10):3103–3121. https://doi.org/10.1111/gcb.12652

Perdomo JA, Cavanagh AP, Kubien DS, Galmés J (2015) Temperature dependence of in vitro Rubisco kinetics in species of Flaveria with different photosynthetic mechanisms. Photosynth Res 124(1):67–75

Pokharel M, Chiluwal A, Stamm M, Min D, Rhodes D, Jagadish SVK (2020) High night-time temperature during flowering and pod filling affects flower opening, yield and seed fatty acid composition in canola. J Agron Crop Sci 206:579–596

Pottier M, Gilis D, Boutry M (2018) The hidden face of Rubisco. Trends Plant Sci 23(5):382–392

Prasad PVV, Djanaguiraman M (2011) High night temperature decreases leaf photosynthesis and pollen function in grain sorghum. Funct Plant Biol 38:993–1003

Raven JA (2013) Rubisco: still the most abundant protein of earth? New Phytol 198(1):1–3

Richards RA (2000) Selectable traits to increase crop photosynthesis and yield of grain crops. J Exp Bot 51(1):447–458. https://doi.org/10.1093/jexbot/51.suppl_1.447

Robles-Zazueta CA, Molero G, Pinto F, Foulkes MJ, Reynolds MP, Murchie EH (2021) Field-based remote sensing models predict radiation use efficiency in wheat. J Exp Bot 72:3756–3773. https://doi.org/10.1093/jxb/erab115

Salazar-Parra C, Aranjuelo I, Pascual I, Aguirreolea J, Sánchez-Díaz M, Irigoyen JJ et al (2018) Is vegetative area, photosynthesis, or grape C uploading involved in the climate change-related grape sugar/anthocyanin decoupling in Tempranillo? Photosynth Res 138:115–128. https://doi.org/10.1007/s11120-018-0552-6

Salesse-Smith CE, Sharwood RE, Busch FA, Kromdijk J, Bardal V, Stern DB (2018) Overexpression of Rubisco subunits with RAF1 increases Rubisco content in maize. Nat Plants 4(10):802–810

Salter WT, Merchant AM, Richards RA, Trethowan R, Buckley TN (2019) Rate of photosynthetic induction in fluctuating light varies widely among genotypes of wheat. J Exp Bot 70:2787–2796

Salter WT, Li S, Dracatos PM, Barbour MM (2020) Identification of quantitative trait loci for dynamic and steady-state photosynthetic traits in a barley mapping population. AoB Plants 12(6):063

Salvucci ME, Crafts-Brandner SJ (2004) Relationship between the heat tolerance of photosynthesis and the thermal stability of Rubisco activase in plants from contrasting thermal environments. Plant Physiol 134(4):1460–1470

Sekhar KM, Kota VR, Reddy TP, Rao KV, Reddy AR (2021) Amelioration of plant responses to drought under elevated CO2 by rejuvenating photosynthesis and nitrogen use efficiency: implications for future climate-resilient crops. Photosynth Res 150(1):21–40

Sharwood RE (2017) Engineering chloroplasts to improve Rubisco catalysis: prospects for translating improvements into food and fiber crops. New Phytol 213(2):494–510

Shen BR, Wang LM, Lin XL, Yao Z, Xu HW, Zhu CH, Teng HY, Cui LL, Liu EE, Zhang JJ, He ZH (2019) Engineering a new chloroplastic photorespiratory bypass to increase photosynthetic efficiency and productivity in rice. Mol Plant 12(2):199–214

Simkin AJ, McAusland L, Headland LR, Lawson T, Raines CA (2015) Multigene manipulation of photosynthetic carbon assimilation increases CO2 fixation and biomass yield in tobacco. J Exp Bot 66:4075–4090

Simkin AJ, McAusland L, Lawson T, Raines CA (2017) Overexpression of the RieskeFeS protein increases electron transport rates and biomass yield. Plant Physiol 175:134–145

Simkin AJ, López-Calcagno PE, Raines CA (2019) Feeding the world: improving photosynthetic efficiency for sustainable crop *production*. J Exp Bot 70(4):1119–1140. https://doi.org/10.1093/jxb/ery445

Soleh MA, Tanaka Y, Kim SY, Huber SC, Sakoda K, Shiraiwa T (2017) Identification of large variation in the photosynthetic induction response among 37 soybean [*Glycine max (L.) Merr.*] genotypes that is not correlated with steady-state photosynthetic capacity. Photosynth Res 131:305–315

Sun Y, Gu L, Dickinson RE, Pallardy SG, Baker J, Cao Y, DaMatta FM, Dong X, Ellsworth D, Van Goethem D, Jensen AM (2014) Asymmetrical efects of mesophyll conductance on fundamental photosynthetic parameters and their relationships estimated from leaf gas exchange measurements. Plant Cell Environ 37:978–994

Takatani N, Ito T, Kiba T, Mori M, Miyamoto T, Maeda SI, Omata T (2014) Efects of high CO 2 on growth and metabolism of Arabidopsis seedlings during growth with a constantly limited supply of nitrogen. Plant Cell Physiol 55:281–292

Tambussi EA, Bort J, Guiamet JJ, Nogués S, Araus JL (2007) The photosynthetic role of ears in C3 cereals: metabolism, water use efficiency and contribution to grain yield. Crit Rev Plant Sci 26(1):1–6

Tamoi M, Nagaoka M, Miyagawa Y, Shigeoka S (2006) Contribution of fructose-1, 6-bisphosphatase and sedoheptulose-1, 7-bisphosphatase to the photosynthetic rate and carbon flow in the Calvin cycle in transgenic plants. Plant Cell Physiol 47(3):380–390

Taylor SH, Long SP (2017) Slow induction of photosynthesis on shade to sun transitions in wheat may cost at least 21% of productivity. Philos Trans R Soc B Biol Sci 372:20160543

Timm S, Florian A, Arrivault S, Stitt M, Fernie AR, Bauwe H (2012) Glycine decarboxylase controls photosynthesis and plant growth. FEBS Lett 586(20):3692–3697

Toivola J, Nikkanen L, Dahlström KM, Salminen TA, Lepistö A, Vignols HF, Rintamäki E (2013) Overexpression of chloroplast NADPHdependent thioredoxin reductase in Arabidopsis enhances leaf growth and elucidates in vivo function of reductase and thioredoxin domains. Front Plant Sci 4:389

Vico G, Way DA, Hurry V, Manzoni S (2019) Can leaf net photosynthesis acclimate to rising and more variable temperatures? Plant Cell Environ 42:1913–1928

Wang Y, Zhang YJ, Han JM, Li CH, Wang RJ, Zhang YL, Jia X (2019) Improve plant photosynthesis by a new slow-release carbon dioxide gas fertilizer. ACS Omega 4(6):10354–10361

Weigmann K (2019) Fixing carbon: to alleviate climate change, scientists are exploring ways to harness nature's ability to capture CO 2 from the atmosphere. EMBO Rep 20(2):e47580

Whitney SM, Sharwood RE (2008) Construction of a tobacco master line to improve Rubisco engineering in chloroplasts. J Exp Bot 59(7):1909–1921

Wimalasekera R (2019) Effect of light intensity on photosynthesis. In: Ahmad P, Ahanger MA, Alyemeni MN, Alam P (eds) Photosynthesis, productivity and environmental stress. Wiley, pp 65–73

Yadav SK, Khatri K, Rathore MS, Jha B (2018) Introgression of UfCyt c6, a thylakoid lumen protein from a green seaweed Ulva fasciata Delile enhanced photosynthesis and growth in tobacco. Mol Biol Rep 45:1745–1758

Yamori W, Masumoto C, Fukayama H, Makino A (2012) Rubisco activase is a key regulator of non-steady-state photosynthesis at any leaf temperature and, to a lesser extent, of steady-state photosynthesis at high temperature. Plant J 71(6):871–880

Zargar SM, Gupta N, Nazir M, Mahajan R, Malik FA, Sofi NR, Shikari AB, Salgotra RK (2017) Impact of drought on photosynthesis: molecular perspective. Plant Gene 11:154–159

Zhang Y, Guanter L, Berry JA, van der Tol C, Yang X, Tang J, Zhang F (2016) Model-based analysis of the relationship between sun-induced chlorophyll fluorescence and gross primary production for remote sensing applications. Remote Sens Environ 187:145–155

Zheng S, Ye C, Lu J, Liufu J, Lin L, Dong Z, Li J, Zhuang C (2021) Improving the rice photosynthetic efficiency and yield by editing OsHXK1 via CRISPR/Cas9 system. Int J Mol Sci 22(17):9554

Zhou Q, Yu Q, Wang Z, Pan Y, Lv W, Zhu L, Chen R, He G (2013) Knockdown of GDCH gene reveals reactive oxygen species-induced leaf senescence in rice. Plant Cell Environ 36(8): 1476–1489

Zhou X, Zeng Y, Tang Y, Huang Y, Lv F, Liu L, Wang S (2020) Artificial regulation of state transition for augmenting plant photosynthesis using synthetic light-harvesting polymer materials. Sci Adv 6(35):eabc5237

# Chapter 7
# Cuticular Waxes and Its Application in Crop Improvement

**Radha Sivarajan Sajeevan**

**Abstract** Cuticle and cuticular waxes form the first level of barrier between the land plants and their external environment. This hydrophobic layer protects the plant tissues from excessive non-stomatal water loss, controls exchange of gases and solutes, conferring tolerance to enormous abiotic and biotic challenges. The cuticular waxes synthesized in epidermal cells is a complex mixture of very long-chain fatty acids, their esters, and derivatives. Its biosynthesis, transport, and deposition involve multiple genes and are tightly coordinated by complex molecular networks, which in turn is regulated in response to various environmental factors. Past few decades of research evidences from model as well as from non-model systems greatly expanded our understanding and knowledge of the genes involved in cuticular wax biosynthesis and its regulation in plants. This chapter briefly summarizes on the significance of cuticular waxes, its biosynthesis, transport, and deposition. Further, focus has been given toward the transcription factors identified in wax biosynthesis, its positive and negative regulators, and the targeted manipulation of cuticular wax biosynthesis in *Arabidopsis* and different crop plants resulted in tolerance toward adverse conditions.

**Keywords** Cuticle · Waxes · Wax biosynthesis · Transcription factors · Abiotic and biotic stress tolerance

## 7.1 Introduction

Plant cuticle forms the first layer of resistance between all land plants and their surroundings. It performs multiple functions of which the most important is to restrict the non-stomatal water loss (Kerstiens 1996; Goodwin and Jenks 2005; Mamrutha et al. 2010; McFarlane et al. 2014). The cuticle mainly consists of

R. S. Sajeevan (✉)
Department of Plant Protection Biology, Swedish University of Agricultural Sciences (SLU), Alnarp, Sweden
e-mail: sajeevan.radha.sivarajan@slu.se

 
H. M. Mamrutha et al. (eds.), *Translating Physiological Tools to Augment Crop Breeding*, https://doi.org/10.1007/978-981-19-7498-4_7

cutin, lipid, intra-, and epi-cuticular waxes accumulated on the plant surface. The C16 and C18 oxygenated aliphatic monomers derived from fatty acids (FAs) and glycerol form the insoluble polymer cutin that can resist mechanical damage and act as structural support for cuticular waxes (Kolattukudy 1980; Pollard et al. 2008). The cuticular waxes are subdivided into intra- and epi-cuticular. These are generally complex mixtures of very long-chain (VLC) saturated FA derivatives (Borisjuk et al. 2014). The intra-cuticular waxes are mixture of amorphous lipids implanted in the cutin that links the cuticle with the cell wall matrix, and epi-cuticular waxes are the surface lipids forming various crystal like or smooth film structures (Jetter and Schaffer 2001; Kunst and Samuels 2003). Waxes are chemically complex mixtures of lipids consisting of very long-chain fatty acids (VLCFA), hydrocarbons, alkanes, alkenes, ketones, aldehydes, esters, primary alcohols, secondary alcohols, triterpenes, flavonoids, and sterols (Rashotte et al. 1997; Nawrath et al. 2013; Lee and Suh 2015a; Xue et al. 2017). The wax concentration and chemical composition highly vary among plant species, tissues, and developmental stages and contribute to wax crystal morphology, structure, and surface hydrophobicity (Mamrutha et al. 2010, 2017).

Cuticular waxes can play various roles in plant protection against stresses such as cold, salinity, drought, high temperature, ultraviolet (UV) radiations, and mechanical damage (Jenks et al. 1994; Long et al. 2003; Mamrutha et al. 2010; Sajeevan et al. 2017b), bacterial and fungal pathogens, and insects (Eigenbrode and Espelie 1995; Eigenbrode and Jetter 2002; Ziv et al. 2018; Zhang et al. 2019a, b, c; Wang et al. 2019; Kong et al. 2020). In addition to its protective roles, the cuticle is also involved in regulating the plant developmental processes (Ingram and Nawrath 2017). Both biotic and abiotic stresses can act as environmental clues and alter the concentration and composition of waxes. The *Arabidopsis thaliana* (*At*) plants under drought/dehydration stress showed altered cuticular wax biosynthesis and increased epi-cuticular wax deposition (Kosma et al. 2009; Yang et al. 2011). Similarly, drought stress-induced epi-cuticular wax deposition was reported in plants such as cotton, rose, peanut, and tree tobacco (Bondada et al. 1996; Jenks et al. 2001; Samdur et al. 2003; Cameron et al. 2006). A high correlation between improved drought tolerance and higher cuticular waxes was reported in oats (Bengtson et al. 1978), sorghum (Jordan et al. 1984), rice (Islam et al. 2009), alfalfa, and crested wheat grass (Jefferson et al. 1989). In biotic stress, cuticular waxes act as the plant's first physical barrier restricting pathogen entry. On the other hand, pathogens can exploit the cuticular waxes to initiate their pre-penetration and infection processes in regulating the plant–pathogen interactions (Skamnioti and Gurr 2007; Ju et al. 2017; Cui et al. 2019). With its diverse role in multiple abiotic and biotic stresses, cuticular waxes have gained increasing attention and considered to be an indispensable trait for crop improvement.

## 7.2 Cuticular Wax Biosynthesis in Plants

Through various forward and reverse genetics approaches in model plants like *A. thaliana* and tomato and crop plants such as rice and barley, a number of genes engaged in cuticular wax biosynthesis, transport, and deposition was identified and characterized. From the current knowledge, cuticular wax biosynthesis can be divided into three steps—a de novo synthesis of the C16 or C18 FAs followed by the extension to form VLCFAs. In the third step, the synthesis of various derivatives of VLCFAs such as aldehydes, alcohols, alkanes, ketones, esters, etc. via either the alcohol- or alkane-forming pathways. These VLCFA derivatives are further transported across plasma membrane and deposited as intra- and epi-cuticular waxes.

In short, the cuticular wax biosynthesis begins in endoplasmic reticulum (ER) by the addition of two carbons donated by malonyl CoA for the extension of C16 and C18 fatty acid (FA) precursors formed in plastid. This extension process is a sequential cycle that is facilitated by fatty acid elongase (FAE) complex results in the formation of VLCFAs consists of 20–36 carbons. The FA extension carries through a series of four consecutive reactions of condensation (β-ketoacyl-CoA synthase, KCS), reduction (β-ketoacyl-CoA reductase, KCR), dehydration (-β-hydroxyacyl-CoA dehydratase, HCD), and a second reduction (enoyl-CoA reductase, ECR), for each of two carbon atom extension, that are collectively called elongase (Ohlrogge et al. 1978). Mutation in one of the four extension enzymes (elongase) will result in pleiotropic effects and severe reduction in overall cuticular waxes, indicating the importance of FA extension is an important rate limiting step in cuticular wax synthesis (Beaudoin et al. 2009; Seo and Park 2011). These VLCFAs are further modified/processed to form a variety of cuticular wax components through two distinct pathways—decarbonylation pathway (alkane forming) and acyl reduction pathway (alcohol forming) (Li et al. 2008; Rowland et al. 2006; Rowland and Domergue 2012). In *Arabidopsis*, decarbonylation pathway is predominantly responsible for the production of major derivatives of cuticular waxes with chain length between 21 and 35C atoms such as aldehydes, alkanes, ketones, and secondary alcohols. On the other hand, acyl reduction pathway leads to the production of primary alcohols and wax esters (Bernard and Joubes 2013; Lee and Suh 2015a). A simplified schematic representation of plant cuticular wax biosynthesis pathways in ER is shown in Fig. 7.1.

## 7.3 Transporters of Cuticle Precursors

The cutin and wax precursors synthesized in ER are transported across the plasma membrane, cell wall, and the emerging cuticular membrane. To date, most of the steps involved in wax biosynthesis are well understood, but the mechanism of transport is poorly known. A close group of half transporters ABCG, an ATP binding cassette, are shown to be involved in the transport of both wax and cutin

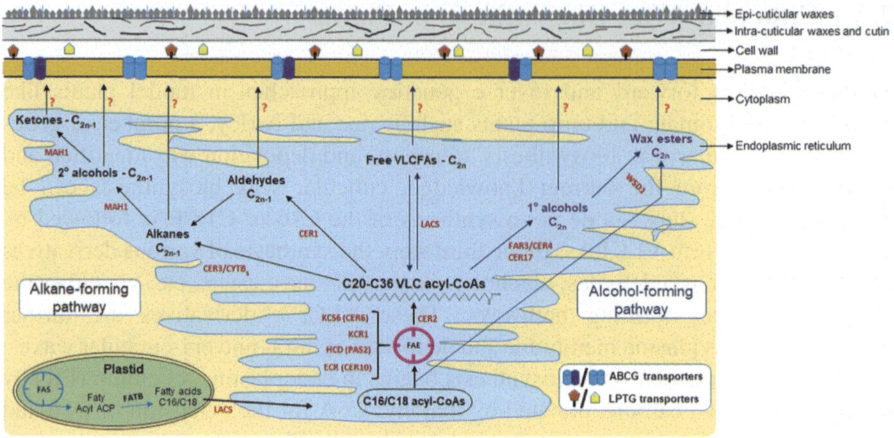


**Fig. 7.1** A simplified representation of biosynthesis of cuticular waxes in endoplasmic reticulum (ER). In the biosynthesis of cuticular waxes, fatty acid elongation (FAE) complex catalyzed four sequential consecutive reactions—condensation by KCS6, reduction by KCR1, dehydration by HCD, and a second reduction by ECR for the formation of VLCFAs. These elongated VLCFAs are then modify into different wax derivatives via alkane- and alcohol-forming pathways. Further, wax derivatives were transported to the plasma membrane (PM) by a fully unknown mechanism then to the extracellular matrix by ABC transporters and delivered to the cuticle by yet to be discovered machinery possibly implicating LTPG1 (CER—ECERIFERUM; LACS1—Long-Chain Acyl CoA Synthase1; MAH1—Mid chain Alkane Hydroxylase1; WSD1—Wax Synthase/Diacylglycerol acyltransferase1). The cuticular wax biosynthesis model is adapted from Bernard and Joubes (2013), Yeats and Rose (2008), and Lewandowska et al. (2020)

derivatives across the plasma membrane (Do et al. 2018). The *Arabidopsis* genome consists of four ABCG transporters—ABCG11, ABCG12, ABCG13, and an uncharacterized ABCG15 (Pighin et al. 2004; Bird et al. 2007; Panikashvili et al. 2011). The ABCG11 is a homodimer likely to export cutin precursors (Bird et al. 2007; Elejalde-Palmett et al. 2021) and ABCG11 and ABCG12/CER5 need to form heterodimer for wax secretion (Bird et al. 2007; McFarlane et al. 2010). The ABCG13, third half transporter, was reported to be involved in the cutin deposition in *Arabidopsis* flowers (Panikashvili et al. 2011).

A full transporter ABCG32 identified from *A. thaliana*, *Hordeum spontaneum*, and *Oryza sativa* is involved in cutin deposition (Bessire et al. 2011; Chen et al. 2011). More recently, another ABC transporter from rice (*OsABCG9*), an ortholog of *AtABCG11*, has been reported that specifically transport wax but not cutin (Nguyen et al. 2018). Despite having well-documented evidences to show the involvement of different ABC transporters in trafficking cuticular lipids, there is lack of evidences to demonstrate the substrate specificity in vitro. Till date, all the ABC transporters identified from different systems are members of the ABCG subfamily that are involved in the transport of lipids and hydrophobic compounds (Moitra et al. 2011). Studies have shown that ABC transporter mutants resulted in lipid embodiments intracellularly. This further supports the direct involvement of

ABC transporters in cuticular lipid transport (Pighin et al. 2004; Bird et al. 2007; Bessire et al. 2011).

Glycosyl phosphatidyl inositol (GPI)-anchored lipid transfer proteins (LTPs), LTPG1 and LTPG2, are plasma membrane bound that are involved in the transport of wax derivatives (Debono et al. 2009; Lee et al. 2009a; Kim et al. 2012). These LTPs are a unique class of soluble proteins that can bind to a variety of lipid substrates (Yeats and Rose 2008). It is proposed that the apoplastic LTPs are involved in the trafficking of wax derivatives, although genetic or biochemical evidences are clearly lacking (Yeats and Rose 2008). Recently, *Arabidopsis* mutant analysis demonstrated the involvement of gnom like1-1 (GNL1) and echidna (ECH)-dependent endo-membrane vesicle transport of waxes to plasma membrane-localized ATP-binding cassette transporters (McFarlane et al. 2014).

## 7.4 Transcriptional Regulation in Biosynthesis of Cuticular Waxes

Efforts to elucidate the biosynthesis of cuticular wax pathway and its players were mainly identified through mutants and concentrated with the model plant, *Arabidopsis*. A large number of genes involved in the biosynthesis of cuticular waxes has been identified, isolated, and characterized (Jetter et al. 2006; Jetter and Kunst 2008; Samuels et al. 2008). In *Arabidopsis*, more than 190 genes have been identified to be involved in the biosynthesis of cuticular waxes, its transport, or deposition (Li-Beisson et al. 2013). Among these, CER1, CER2, CER6/CUT1, KCS1, IDDLEHEAD (FDH), and WAX2 from *Arabidopsis*, GL1 and GL8 from maize encode wax synthesis and transport related enzymes (Aarts et al. 1995; Hansen et al. 1997; Todd et al. 1999; Fiebig et al. 2000; Chen et al. 2003; Zhang et al. 2005). A summary of genes identified from the model plant *Arabidopsis* that are involved in cuticular wax biosynthesis, transport, or deposition is detailed in Table 7.1.

The key regulators involved in the biosynthesis of waxes and cuticular components deposition are transcription factors (TFs). Different families of TFs belong to ethylene-responsive factors (ERFs), myeloblastosis family (MYB), and homeodomain-leucine zipper class IV (HD-Zip IV) factors identified as regulators of wax biosynthesis, of which ERFs gained more importance (Aharoni et al. 2004; Seo et al. 2011). Overexpression of these TFs lead to changes in wax biosynthesis, its accumulation, and changes in the chemical composition (Broun et al. 2004). It has also been demonstrated that overexpression of these TFs often resulted in increased stress tolerance (Broun et al. 2004; Javelle et al. 2010; Seo and Park 2011). However, despite their obvious positive effects on plant protection, it was also demonstrated that the ectopic expression could negatively affect plant growth, yield, and decreased stress tolerance (Aharoni et al. 2004; Zhang et al. 2005). A summary of TFs identified that play a role in the biosynthesis of cuticular waxes, targeted genes,

**Table 7.1** Genes identified in the biosynthesis of cuticular waxes and its transport from the model plant *Arabidopsis thaliana*

| Sl. No. | Gene abbreviation | Protein family name | Role | Function | Reference |
|---|---|---|---|---|---|
| 1. | *KCS1* | β-Ketoacyl-coenzyme A synthase | Extension of 24C FA | Biosynthesis | Todd et al. (1999) |
| 2. | *CUT1/CER6/KCS6* | β-Ketoacyl-coenzyme A synthase | Regulation of VLCFA biosynthesis/24C FA extension | Biosynthesis | Fiebig et al. (2000), Hooker et al. (2002) |
| 3. | *FATB* | Fatty acyl-ACP thioesterase B | Providing saturated FAs for wax biosynthesis | Biosynthesis | Bonaventure et al. (2003) |
| 4. | *CER10/ECR/ECR10* | Trans-2,3-Enoyl-coenzyme A reductase | VLCFA biosynthesis | Biosynthesis | Zheng et al. (2005) |
| 5. | *PASTICCINO2 (PAS2)/HCD* | β-Hydroxy-acyl-coenzyme A dehydratase | VLCFA synthesis in association with CER10, an enoyl-CoA reductase | Biosynthesis | Bach et al. (2008) |
| 6. | *KCR1* | β-Ketoacyl-coenzyme A reductase | VLCFA extension | Biosynthesis | Beaudoin et al. (2009) |
| 7. | *KCS20*; *KCS2/DAISY* | β-Ketoacyl-coenzyme A synthase | VLCFA extension to C22 | Biosynthesis | Franke et al. (2009), Lee et al. (2009b) |
| 8. | *LACS1/CER8*; *LCAS2*; *LCAS4* | Long-chain acyl CoA synthetase | Synthetase activity for VLCFAs C20-C30 | Biosynthesis | Lü et al. (2009), Jessen et al. (2011), Weng et al. (2010) |
| 9. | *ACC1* | Acetyl-coenzyme A carboxylase | Malonyl CoA substrate synthesis | Biosynthesis | Lü et al. (2011) |
| 10. | *CER2*; *CER2-Like1/2* | BAHD acyltransferase | FA extension beyond C28 | Biosynthesis | Haslam et al. (2012, 2015), Pascal et al. (2013) |
| 11. | *KCS9* | β-Ketoacyl-coenzyme A synthase | Extension of C22-C24 FAs | Biosynthesis | Kim et al. (2013) |
| 12. | *CER17 (ECERIFERUM1)* | Acyl-CoA desaturase like 4 | n-6 desaturation of VLC acyl-CoAs | Biosynthesis | Yang et al. (2017) |

(continued)

**Table 7.1** (continued)

| Sl. No. | Gene abbreviation | Protein family name | Role | Function | Reference |
|---|---|---|---|---|---|
| 13. | *WAX2/YRE/FLP1/CER3* | Aldehyde-generating acyl-CoA enzyme | Synthesis of aldehydes, alkanes, 2o-alcohols, and ketones; cuticular membrane biosynthesis | Biosynthesis—Alkane-forming pathway | Chen et al. (2003), Rowland et al. (2007), Bernard et al. (2012) |
| 14. | *CER1/CER22* | Aldehyde decarbonylase | Biosynthesis of VLC alkane | Biosynthesis—Alkane-forming pathway | Bourdenx et al. (2011), Bernard et al. (2012), Sakuradani et al. (2013) |
| 15. | *RST1—RESURRECTION1* | Aldo/keto reductase/cytochrome C/G-protein-coupled receptor family 1 | May act in reduction of acyl-CoAs to aldehydes | Biosynthesis—Alkane-forming pathway | Chen et al. (2005) |
| 16. | *CYTB5-B/C/D/E* | Cytochrome B5 | Redox-dependent synthesis of VLC alkanes | Biosynthesis—Alkane-forming pathway | Bernard et al. (2012) |
| 17. | *CYP96A15* (cytochrome P450 enzyme)/*MAH1* | Midchain alkane hydrolase 1 | Formation of 2o-alcohols and ketones | Biosynthesis—Alkane-forming pathway | Greer et al. (2007) |
| 18. | *CER4/FAR3* | FA CoA reductase | Formation of C24:0 and C26: 0 1o-alcohols | Biosynthesis—Alcohol-forming pathway | Rowland et al. (2006) |
| 19. | *WSD1* | Wax ester synthase/diacylglycerol acyltransferase | Wax ester biosynthesis | Biosynthesis—Alcohol-forming pathway | Li et al. (2008) |
| 20. | *ABCG12/CER5* | ATP-binding cassette (ABC) transporter | Cuticular waxes transport | Transport | Pighin et al. (2004) |
| 21. | *ABCG11/WBC11/DESPERADO* | ATP-binding cassette (ABC) transporter | Secretion of surface waxes in interaction with CER5 | Transport | Bird et al. (2007), Luo et al. (2007), Panikashvili et al. (2011) |

(continued)

**Table 7.1** (continued)

| Sl. No. | Gene abbreviation | Protein family name | Role | Function | Reference |
|---|---|---|---|---|---|
| 22. | *GLN1; ECH* | Vesicle trafficking | Vesicle trafficking | Transport | McFarlane et al. (2014) |
| 23. | *LTPG1; LTPG2* | GPI-anchored lipid transfer protein (LTPG) | Export or accumulation of cuticular waxes | Transport/ deposition | DeBono et al. (2009), Lee et al. (2009a). Kim et al. (2012) |

and cuticular composition affected identified through overexpression or down-regulation is detailed in Table 7.2.

### *7.4.1 APETALA2/Ethylene Responsive Factor*

The APETALA2/Ethylene Responsive Factor (AP2/ERF) superfamily is known to be one of the largest plant-specific families of TF involved in diverse plant physiological processes (Licausi et al. 2013). These TFs can regulate the gene expression transcriptionally and posttranslationally at different stages of plant growth and development, hormone signaling, and in response to various abiotic and biotic stresses (Elliott et al. 1996; Xu et al. 2011; Licausi et al. 2013). The AP2/ERF proteins were first identified from *Arabidopsis*, typically consists of a highly conserved AP2 domain of 40–70 amino acids in length (Jofuku et al. 1994). Based on the number of AP2 and other DNA binding domains, they are categorized into four different subfamilies—AP2, ERF, DREB (Dehydration Responsive Element Binding), and RAV (related to ABI3/VP1) (Mizoi et al. 2012). Members of AP2 subfamily consist of two AP2/ERF domains (Sakuma et al. 2002). The ERFs and DREB subfamilies contain single AP2 domain that usually binds to an ethylene responsive (AGCCGCC) cis-element designated as GCC-box (Ecker 1995; Eini et al. 2013). However, the RAV subfamily proteins are characterized by the presence of two different DNA binding domains, AP2/ERF and B3 (Kagaya et al. 1999).

The first TF identified and reported to be involved in cuticular wax biosynthesis was WAX INDUCER1/SHINE1 (WIN1/SHN1) from *Arabidopsis* simultaneously by two independent research groups designated as WAX INDUCER1 (WIN1) and SHINE1 (SHN1) (Aharoni et al. 2004; Broun et al. 2004). WIN1/SHN1 belonging to the subfamily of AP2/ERF TFs is a member of a clade of three close homolog proteins (SHN2 and SHN3) in *Arabidopsis* genome belonging to group V or B6 (Sakuma et al. 2002; Nakano et al. 2006). All three SHN clade genes exist with a single intron in nature, and no splice variants are reported. The *AtWIN1/SHN1* share 55% and 71% protein sequence homology with *AtSHN2* and *AtSHN3*, respectively (Aharoni et al. 2004). All three SHN proteins contain three conserved domains/

**Table 7.2** A list of TFs identified in biosynthesis of cuticular waxes, their target genes, and the chemical components

| Sl. No. | Plant | TF name | TF type | Target genes | Affected chemical components | Reference |
|---|---|---|---|---|---|---|
| 1. | *Arabidopsis thaliana* | WIN1/ SHN1 | AP2-EREBP | CYP86A7, CYP86A4, Lipase-like, HTH-like GPDHc1 NLM2, GPAT4, CER1, KCS1, CER2, FAE1 | C30/34 FA, C28/30 aldehyde, C27/C29/C33 alkane, C16: 0 and C18:1 ω-HFAs, C31 and C29 alkanes | Aharoni et al. (2004), Broun et al. (2004), Kannangara et al. (2007) |
| 2. | *Medicago truncatula* | WXP1 | ERF | FSE-like: MtTC79579, MtTC80406, MtTC87247, LCR-like: TC81689, TC84740, WAX2-like: TC82822, KAR (GL8-like) | C30/C28 primary alcohols, C29-C33 alkanes, C18: 1 DSA, C22/C24 FAs, C22/C23/C30 aldehydes, cholesterol, sitosterol | Zhang et al. (2005) |
| 3. | *Medicago truncatula* | WXP2 | ERF | – | C28/C32/ C18/C22 FAs, C32/C28 aldehyde C30/C34/C32 primary alcohols, C31/C23/ C25/C27 alkane, cholesterol, sitosterol | Zhang et al. (2007) |
| 4. | *Arabidopsis thaliana* | MYB41 | R2R3-type MYB | WIN1/SHN1 LACS2, ATT1, LTPs, GDSL-lipases, hydrolase α/β-fold family, AtEXP5 | – | Cominelli et al. (2008) |
| 5. | *Hordeum vulgare* | Nud | ERF | – | – | Taketa et al. (2008) |
| 6. | *Solanum lycopersicum* | MYB12 | R2R3-type MYB | 21 Phenylpropanoid/ flavonoidrelated transcripts, 8 genes related to FA metabolism | Mostly metabolites associated with a phenylpropanoid pathway | Adato et al. (2009) |

(continued)

**Table 7.2** (continued)

| Sl. No. | Plant | TF name | TF type | Target genes | Affected chemical components | Reference |
|---|---|---|---|---|---|---|
| 7. | *Zea mays* | OCL1 | HD-Zip IV | nsLTP, CYP78A6-like, ABC transporter, SEC14 | C25 alkane, C24, C26, and C28 alcohols, C48 ester, C28 and C30 aldehydes | Javelle et al. (2010) |
| 8. | *Arabidopsis thaliana* | SHN2/3 | AP2-EREBP | – | – | Aharoni et al. (2004), Shi et al. (2011) |
| 9. | *Arabidopsis thaliana* | MYB96 | R2R3-type MYB | RD22, some GH3 genes, KCR1, SER1, KCS1, KCS2, KCS6, PAS2, CER3, ESR, WBC11 and other ABC transporters, LTP | – | Seo et al. (2009, 2011) |
| 10. | *Oryza sativa* | CFL1 | WW domain | WIN1/SHN1, BDG, FDH | – | Wu et al. (2011) |
| 11. | *Arabidopsis thaliana* | HDG1 | HD-Zip IV | – | – | Wu et al. (2011) |
| 12. | *Eucalyptus gunnii* | CBF1a/b | DREB/CBF | – | – | Navarro et al. (2011) |
| 13. | *Arabidopsis thaliana* | WRI1/3/4 | AP2/EREBP | PKp2, MAT, KASI, KASIII, ENR, FATA, G3PDH, ROD1, BCCP2 | C16:0/C18:0/C18:1/C18:2 Data for triple wri1wri3 wri4 mutants: DSAs, C16:0/C18:1/C18:2 ω-OH and C16 10,16OH DCAs | Cernac and Benning (2004), Masaki et al. (2005), To et al. (2012) |
| 14. | *Arabidopsis thaliana* | NFXL2 | NFXL | BDG1, SHN1, SHN2, SHN3 | – | Lisso et al. (2012) |
| 15. | *Arabidopsis thaliana* | ANL2 | HD-Zip IV | – | C16:0, C18:0, C18:1, C18:2, and ω-OH DCAs, C27, C31, and C33 alkanes | Nadakuduti et al. (2012) |
| 16. | *Oryza sativa* | WR1 | ERF | OsKCS2, OsLACS1, | C16/C20/C24/C26/ | Wang et al. (2012) |

(continued)

**Table 7.2** (continued)

| Sl. No. | Plant | TF name | TF type | Target genes | Affected chemical components | Reference |
|---|---|---|---|---|---|---|
| | | | | OsCER3, OsCUT1, OsFDH1/2, OsKCS1, OsLACS1–2, OsCER1/2, OsFAE1-L | C30/C32 FAs, C32/C22 alcohols, C25/C27/ C29/C31 alkanes, C48 ester | |
| 17. | *Oryza sativa* | WR2/3/4 | ERF | – | – | Wang et al. (2012) |
| 18. | *Solanum lycopersicum* | CD2 | HD-Zip IV | – | Trans-CA, di OH C16:0 and ω-OH C18:1 Ph. C16:0, C20:0, and C22:0 DCAs, C27, C29, C30, and C31 alkanes CC18:0 FA | Nadakuduti et al. (2012) |
| 19. | *Solanum lycopersicum* | SHN3 | AP2-EREBP | SlGL2, SlMIXTA, SlCYP77A1, SlCYP86A8, SlCYP86A69 | BA, trans-CA, C16:0 and C16–9/10-H DFAs, C16–9/10, 16Di, C16-ω, C18:1–2 HFAs | Mintz-Oron et al. (2008), Shi et al. (2013) |
| 20. | *Arabidopsis thaliana* | MYB106/ NOK | R2R3-type MYB | WIN1/SHN1, FDH, LACS2, CYP84A4, CYP77A6, Atg04570, KCS1, CER1, CER2, LCR, LACS2 | – | Gilding and Marks (2010), Oshima et al. (2013) |
| 21. | *Arabidopsis thaliana* | MYB16 | R2R3-type MYB | WIN1/SHN1, FDH, LACS2, CYP84A4, CYP77A6, Atg04570, KCS1, CER1, CER2, LCR, LACS2 | – | Oshima et al. (2013) |
| 22. | *Eucalyptus grandis* | SHN1/2 | ERF | – | – | Marques et al. (2013) |
| 23. | *Solanum lycopersicum* | SHN1 | AP2-EREBP | GDSL, Enoyl-CoA reductase, | – | Al-Abdallat et al. (2014) |

(continued)

**Table 7.2** (continued)

| Sl. No. | Plant | TF name | TF type | Target genes | Affected chemical components | Reference |
|---|---|---|---|---|---|---|
| | | | | acyl-CoA synthase, Fiddle-head, HOT-HEAD-like | | |
| 24. | *Eutrema salsugineum* | WAX1 | R2R3-type MYB | *CER1*, *KCS2*, *KCR1*, *VTC1*, *GLDH*, *MIOX4* | – | Zhu et al. (2014) |
| 25. | *Arabidopsis thaliana* | MYB94, MYB96 | R2R3-type MYB | *KCR1*, KCS1, *KCS2/DAISY*, *KCS6*, *CER2*, *CER1*, *CER3*, *WSD1* | – | Lee and Suh (2015b), Lee et al. (2016) |
| 26. | *Arabidopsis thaliana* | WRI4 | AP2/EREBP | LACS1, KCR1, PAS2, ECR, WSD1, PKP1, PKP2, BCCP2, ENR1, PDH- E1a | C24 and C28 FAs, C27 alkanes, and C24, C26, C28 primary alcohols, C29 ketones | Park et al. (2016) |
| 27. | *Hordeum vulgare* | WIN1 | AP2/EREBP | KAS2, CYP86A2, CYP89A2, LACS2 | – | Kumar et al. (2016) |
| 28. | *Arabidopsis thaliana* | DEWAX2 | AP2/ERF | CER1, ACLA2, LACS1, LACS2, KCS12 | C29 and C31 alkanes, C28 primary alcohols, C29 ketone | Kim et al. (2018) |
| 29. | *Brassica napus* | WIN1/SHN1 | AP2/EREBP | BCCP1, GPAT9, LPAT5, DGAT2 *LACS2*, *KCS1*, *KCR1*, *CER1* | C29-alkanes, C31-alkanes, C28-alcohol, C29-alcohol | Liu et al. (2019) |
| 30. | *Jatropha curcas* | WRKY | WRKY | Cab40, Lhcb5, Rca1, WIN1 | FA, fatty alcohols, carboxylic acid, alkene, terpene, triterpenoid, aldehyde | More et al. (2019) |
| 31. | *Cucurbita moschata* | WIN1 | AP2/ERF | CER1, CER1-1, CER4, KCS1, ABC | Ester (C20: C22, C20: C24, C20: C26, C20: C28, C20: C24), alkanes | Zhang et al. (2019a) |

(continued)

**Table 7.2** (continued)

| Sl. No. | Plant | TF name | TF type | Target genes | Affected chemical components | Reference |
|---|---|---|---|---|---|---|
| | | | | | (C29 and C31) | |
| 32. | *Malus domestica* | SHINE2 | AP2/ EREBP | MYB30, MYB96, LACS2, CER1, CER3, CER6, KCS1, WIN1, DEWAX, SHINE3 | Alkanes, alcohols, aldehydes, FAs | Zhang et al. (2019b) |
| 33. | *Malus domestica* | MYB30 | R2R3-type MYB | *WRI1*, *WIN1*, *ACBP1*, *LACS2*, *SHINE2*, *SHINE3*, *KCS1* | C29 alkanes, C31 alcohols, C29 aldehydes, C16 FAs, C29 ketones, and C29 and C30 esters | Zhang et al. (2019c) |
| 34. | *Eustoma grandiflorum* | *MIXTA-like 1* | R2R3-type MYB | *CER3*, *CER6*, *CER10*, *KCS1*, *KCR1*, *CYP77A6*, *WIN1* | – | Wang et al. (2020) |
| 35. | *Arabidopsis thaliana* | RAP2.4 | AP2/ DREB | KCS2, CER1 | VLC-alkane, C27, C29, C31, C33 | Yang et al. (2020) |

motifs, AP2 domain at N-terminal, a middle, and C-terminal conserved motifs (Nakano et al. 2006).

The overexpression of WIN1/SHN1 showed increased accumulation of cutin/leaf epi-cuticular waxes and resulted in improved dehydration tolerance of transgenic *Arabidopsis* and downregulation leads to decreased cutin content in the outer parts of the plant (Aharoni et al. 2004; Broun et al. 2004; Kannangara et al. 2007). The WIN1/SHN1 overexpression also reflected in altering the structure of leaf epidermis and stomatal index, trichrome number, and branching (Aharoni et al. 2004). Therefore, it is possible that WIN1/SHN1 and other AP2 domain superfamily members not only involved in cuticle formation but also function in other metabolic pathways. Shi et al. (2011) showed that WIN1/SHN1 TF plays an important role in the metabolism of cell wall. The constitutive overexpression of WIN1/SHN1 TF leads to an upregulation of three downstream genes (*CER1*, *CER2*, and *KCS1*) that were initially identified and known to be involved in the biosynthesis of epidermal waxes (Broun et al. 2004). Transgenic *Arabidopsis* plants expressing WIN1/SHN1 resulted in altering the expression of 12 genes, of which 11 were upregulated and one with an unknown function strongly downregulated (Kannangara et al. 2007). Even though

WIN1/SHN1 overexpression altered cuticular wax load and resulted in improved stress tolerance, evidences for the direct activation of downstream genes by WIN1/SHN1 are still lacking. In the last few years, WIN/SHN-related members were identified and characterized from different crops like soybean, cotton, barley, wheat, etc. Overexpression resulted in altered cuticle properties and imparts tolerance to multiple abiotic stresses (Xu et al. 2016; Bi et al. 2018; Djemal et al. 2018; Djemal and Khoudi 2015, 2021).

### 7.4.2 Homologous of WIN/SHN

In rice, four homologous of *Arabidopsis* WIN/SHN were identified and designated as Wax Synthesis Regulatory genes 1–4 (*OsWR1–4*) (Wang et al. 2012). Sequence homology studies showed the *OsWR1* protein sequence is closely related to *AtWIN1/SHN1* protein. Transgenic rice plants overexpressing *OsWR1* resulted in decreased cuticle permeability, in contrast to the results exhibited in *AtWIN1/SHN1* overexpression studies by Aharoni et al. (2004), but an enhancement in drought tolerance has been reported in both the cases. The decreased cuticle permeability of *OsWR1* overexpression was due to alterations in long-chain FAs and alkanes. In addition, it was demonstrated that *OsWR1* could interact with wax-related genes, *OsLACS2* and *OsFAE1-L*, by direct binding to GCC and DRE elements present in the promoter region. The *OsWR1* overexpression in rice resulted in more than two-fold upregulation of 12 wax biosynthesis-related genes and four cutin biosynthesis genes. The overexpression also showed an increased expression of non-cuticle biosynthesis genes involved in membrane stabilization and reactive oxygen species (ROS) scavenging such as late embryogenesis abundant protein (LEA3), ascorbate peroxidase (APX1), superoxide dismutase (SOD), and catalase (Cat A and Cat B) that could independently contribute to improved drought tolerance. On the other hand, silencing of *OsWR1* by RNAi resulted in significant downregulation of many of those genes (Table 7.2) and partial silencing resulted in decreased transcript levels of Cat A and Cat B (Wang et al. 2012).

### 7.4.3 Negative Regulators of WIN/SHN

Two negative regulators, NUCLEAR FACTOR X LIKE2 (NFXL2) and SPINDLY (SPY) of *WIN/SHN* genes, were reported. The *Arabidopsis* NFXL2 mutant analysis showed difference in the composition of cutin, reduced stomatal aperture, and an increase in drought tolerance by regulating the expression of all three SHN genes (Lisso et al. 2012). Further analysis revealed that *NFXL2* gene could act as a negative regulator for WIN1/SHN1 and several others by directly interacting with the gene promoter region. Thus, NFXL2 protein modulated the cuticle components biosynthesis through a direct repression of *WIN1/SHN* gene (Lisso et al. 2012).

### 7.4.4 *WAX PRODUCTION1 (WXP1)*

WXP1, an AP2 domain containing TF from *Medicago truncatula* (*Mt*), is evidently distinct from other AP2/ERF TF family genes. The *MtWXP1* identified to be a homolog to ERFs has 53% amino acid sequence identity with RAP2.4. The WXP1 transcript was highly upregulated by abscisic acid (ABA) treatment in both shoot and root of seedlings, and upregulation was observed only in shoot under cold and drought stress. Under cold stress conditions, the upregulation was very rapid and could be detected within 30 minutes (Zhang et al. 2005). The constitutive overexpression of WXP1 in alfalfa resulted an increase in total leaf wax load to nearly 40% and enhanced drought stress tolerance (Zhang et al. 2005). The gas chromatography-mass spectrometry (GCMS) analysis of transgenic plants leaves showed a significant difference in multiple wax derivatives such as higher content of C30 alcohol moieties (25–35%) and elevated levels of other wax components (Table 7.2). The increase in wax content resulted in reduced water loss by decreasing water permeability, lower chlorophyll leaching, and showed better tolerance to drought stress. The WXP1 overexpression resulted in the upregulation of three FAE-like and two LACERATA (LCR, encoding cytochrome P450 monooxygenases) wax biosynthesis pathway genes (Zhang et al. 2005).

The constitutive overexpression of *MtWXP1* and its paralog WXP2 significantly enhanced the deposition of cuticular waxes due to the accumulation of specific wax components and their chain length distributions on leaves of *Arabidopsis* (Zhang et al. 2007). The WXP1 and WXP2 transgenic lines showed higher levels of n-alkanes. The primary alcohol levels were increased in WXP1 plants but showed an opposite trend in WXP2 as compared to their wild type plants (Table 7.2). The WXP1 plants did not show any changes in the cuticle permeability while WXP2 resulted in decreased levels. Surprisingly, detached leaves of WXP1 and WXP2 transgenic plants retained better water content and showed significantly enhanced survival under drought stress conditions. Under the low-temperature stress, WXP1 transgenic plants showed an improved tolerance while WXP2 was susceptible as compared to control plants. The *Arabidopsis* plants constitutively expressing WXP1 did not exhibit any negative effects on plant growth and development; however, slower plant growth was observed in WXP2 overexpression (Zhang et al. 2007).

### 7.4.5 *WRINKLED and CBF TFs*

The WRINKLED1 (*WRI1*) gene contain two AP2 domain was identified from *Arabidopsis* through the mutant analysis. The mature seeds of *wri1* mutants showed a wrinkled appearance and decreased content of water-insoluble oils (Cernac and Benning 2004). The overexpression of WRI1 resulted in 10–20% increased seed oil content without reducing the seed number (Cernac and Benning 2004). The WRI3 and WRI4 homologs of WRI1 were identified to be involved in gene expression of the synthesis of acyl chain and glycerol backbones that are main precursors of different lipid biosynthetic pathways (To et al. 2012). On the other hand, significant downregulation of most glycolytic and late FA biosynthetic genes were observed in *wri* triple mutants. The C-repeat binding factor genes (*CBF1a* and *1b*) associated with drought and cold tolerance were found to be involved in the regulation of deposition of cuticular waxes in *Eucalyptus gunii*, an Australian drought- and cold-tolerant tree species (Navarro et al. 2011). The eucalyptus transgenic plants exhibited a high accumulation of anthocyanins, decrease in the stomatal density, reduced growth, better water retention capacity with reduced leaf area, and increase in leaf thickness and leaf cell size as compared to the control plants. Also, transgenic plant leaves showed a higher density of oil glands, and amount of cuticular waxes were significantly higher (Navarro et al. 2011). Overexpression of *CBF4 TF* gene in grape vine resulted in enhanced freezing tolerance and decreased electrolyte leakage due to freezing. The mRNA expression profiling of transgenic line showed the expression of CBF4 targets the lipid metabolism, epi-cuticular wax formation, and cell wall structure-related genes (Tillett et al. 2012). So far, these are the only reports showing the involvement of *CBF* genes on cuticle wax deposition (Table 7.2).

### *7.4.6 Myeloblastosis Family (MYB)*

To date, many MYB TFs have been shown to be involved in the complex network that control cuticle biosynthesis, cell-wall modification, and cuticle deposition in the model plant *Arabidopsis*. A R2R3-type MYB TF in *Arabidopsis*, *MYB41* is reported to be involved in the cuticle biosynthesis and wax transport regulation (Cominelli et al. 2008). *AtMYB41* interacts with mitogen-activated protein kinase 6 (MPK6), a member of protein kinases family interacts with a number of signaling pathways involved in plant development and responses to stress (Hoang et al. 2012). It was demonstrated that *AtMYB41* can physically interact with MPK6 and get phosphorylated at residue Ser251, which can enhance MYB41 DNA binding capacity to the *LTP* gene promoter. This was further proved by wild type *AtMYB41* gene overexpression that showed improved tolerance to high salinity while overexpression of a mutated MYB41 (Serine 251 to alanine) resulted in decreased tolerance to salt stress (Hoang et al. 2012).

A R2R3-type MYB protein, MYB96, identified as a stress-responsive TF modulates the responses of drought stress by combining the auxin and ABA signals (Seo et al. 2009). The *Arabidopsis* mutant plants overexpressing MYB96 suppressed the lateral root growth but were resistant to drought stress, while the knockout mutants were highly sensitive to drought stress (Seo et al. 2009). The microarray results showed upregulation of a large group of genes encoding the wax biosynthetic enzymes by MYB96, specifically those of VLCFA condensing enzymes (Seo et al. 2011; Table 7.2). Most of the target genes of MYB96 were also upregulated under drought stress and ABA due to the presence of MYB-responsive cis-element "TAACTA/G" in their promoter. The transgenic *AtMYB96* plants showed increased epi-cuticular wax crystal deposition in leaves but reduced in stem and showed a slight change in the color of leaves. Also, these plants were significantly shorter with no characteristic "shiny" phenotype; however, no changes in epidermal development was observed (Seo et al. 2011). The *myb96* loss of function mutant was susceptible to drought stress due to the alteration in cuticular wax biosynthesis (Guo et al. 2013). A closely related *MYB94 TF* gene can effectively replace MYB96 in cuticular wax biosynthesis (Lee et al. 2016). The MYB94 and MYB96 TFs are closely related and can additively function in the biosynthesis of waxes under drought stress and well-watered conditions via an ABA-dependent pathway (Lee et al. 2016).

The role of *AtMYB96* in frost tolerance and response to biotic stresses were also reported (Guo et al. 2013; Seo and Park 2011). The *LTP3* gene overexpression resulted in increased freezing tolerance without showing an effect on CBF expression or their target cold regulated (COR) genes. The MYB96 directly binds to the *LTP3* gene promoter results in positive regulation of LTP3 expression results in enhanced freezing tolerance, consistent with MYB96 overexpressing transgenic plants (Guo et al. 2013). An inhibitor of the rust germ tube differentation1 (*irg1*) mutant showed complete loss of epi-cuticular wax crystals in the abaxial surface and consequent reduction in the surface hydrophobicity that conferred non-host resistance to biotrophic fungal pathogens. The abaxial leaf surface wax composition analysis of *irg1* mutant showed 90% reduction in primary alcohols (C30) and alkanes (C29 and C31) were increased compared to control (Table 7.2). It is proposed that IRG1 may be a direct or indirect regulator of *MtMYB96* transcription; however, there is no evidence to claim that IRG1 could regulate the cuticular wax biosynthesis-related genes directly or is performed only through MYB96.

MIXTA, an MYB-related TF, has been identified with the role in cuticular wax biosynthesis and epidermal cell shape formation. The *Arabidopsis* and *Torenia fournieri* MYB106 and MYB16, MIXTA-like TF genes can regulate the development of cuticle that coordinate with TF WIN1/SHN1 (Oshima et al. 2013; Table 7.2). The downregulation of *MYB106* and *MYB16 TF* genes resulted in cuticle deficiencies of flowering organs, organ adhesion, and decreased epi-cuticular wax crystals. Microarray results showed MYB106 and WIN1/SHN1 TFs regulate similar set of genes (Oshima et al. 2013; Table 7.2). Among these, the genes involved in the accumulation of waxes such as FDH, KCS1, and CER2 and cutin biosynthesis genes such as LACERATA and LONG-CHAIN ACYL COA SYNTHETASE2 were identified. The overexpression of MYB16 in *Arabidopsis* resulted in the accumulation of waxy substances on leaves, and both MYB106 and MYB16 downregulation by RNAi leads to reduced expression of cuticular wax biosynthesis genes *LACERATA* and *ECERIFERUM1* with severe permeable cuticle phenotype (Oshima and Mitsuda 2013; Oshima et al. 2013).

### 7.4.7 Homeodomain-Leucine Zipper Class IV Factors

The homeodomain leucine zipper IV (HD-Zip IV) TFs are predominantly expressed in epidermal cells with epidermis-related functions have been identified from number of plant systems (Javelle et al. 2011; Chew et al. 2013). Maize (*Zea Mays*) Outer Cell Layer 1 (*ZmOCL1*) gene is a member of HD-ZIP IV comes under the subclass of HD-ZIP homeodomain proteins, was detected in protoderm, floral organs, and developing leaves (Ingram et al. 1999, 2000). The transgenic maize plants overexpressing *ZmOCL1* gene had less effect on phenotype as compared to its control, but the transcriptome analysis revealed expression of many genes involved in the metabolism of lipids and its transport (Javelle et al. 2010). Some of the genes identified are carboxylesterase, type 2 LTP, phosphatidylinositol transport protein, three ABC transporters, and FA reductase (Table 7.2). The FA reductases responsible for the long-chain primary alcohol synthesis from FA precursors were closely related to CER4 protein in *Arabidopsis* (Rowland et al. 2006). The transgenic plants of *ZmOCL1* did not show significant changes in the wax layer structure or size as compared to the wild-type plants. However, wax chemical component analysis showed a significant increase in C32 alcohol content and decrease in C32 aldehydes in the young leaves of *ZmOCL1* transgenic. Few of the independent transgenic lines showed significant two- to threefold increase in C44 to C48 wax esters as compared to the control (Javelle et al. 2010).

### 7.4.8 Curly Flag Leaf1, a Negative Regulator of HD-Zip IV

The Curly Flag Leaf1 (*CFL1*) gene, a WW domain encoding protein, was reported as a negative regulator of cuticle development (Wu et al. 2011). The overexpression of *OsCFL1* and *AtCFL1* in transgenic *Arabidopsis* plants resulted in an organ fusion phenotype with decreased levels of epi-cuticular waxes and defective cuticles. Yeast two hybrid assay provided evidences for direct interaction of *AtCFL1* with HDG1, a HD-Zip IV protein (Wu et al. 2011). The *HDG1* gene suppression resulted in a defective cuticle phenotype in transgenic *Arabidopsis*, similar to that of the CFL1 overexpressing plants. The *AtCFL1* overexpression and HDG1 downregulation in transgenic *Arabidopsis* resulted in the downregulation of FIDDLEHEAD (FDH) and BODYGUARD (BDG), two cuticle biosynthesis-associated genes. The BDG encodes a member of the $\alpha/\beta$-hydrolase fold protein superfamily and FDH is also known as KCS10 (Kurdyukov et al. 2006; Wellesen et al. 2001; Yephremov et al. 1999). It was demonstrated that HDG1 could function as a positive regulator by directly binding to the L1 boxes in the promoters of *BDG* and *FDH* genes. The HDG1 function is negatively regulated by CFL1, thereby affecting the cuticle development (Wu et al. 2011; Table 7.2).

## 7.5 Cuticular Wax, a Multifunctional Trait

Plant cuticle and cuticular waxes play multifunctional role in crop protection and survival against various abiotic and biotic stresses like transpiration water loss, drought, high light intensity, salinity, invading pathogens, and insect herbivores (Lewandowska et al. 2020). It is well documented and demonstrated that drought stress induces wax production (Aharoni et al. 2004; Zhang et al. 2005; Cameron et al. 2006). Significant correlations were observed between the content of waxes, yield, water use efficiency (WUE), and drought tolerance in crops like rice, wheat, barley, and sorghum (Jordan et al. 1984; Richards et al. 1986; Febrero et al. 1998; Zhu and Xiong 2013). These evidences point toward the fact that as the wax content decreases, the crop plants will become more sensitive in general to desiccation and drought stress compared to more waxy ones (Guo et al. 2016). The role of cuticular waxes in imparting salinity stress tolerance is through controlling the residual transpiration, which is negatively correlated with wax content (Hasanuzzaman et al. 2017). Higher leaf surface wax containing genotypes generally have a cooler canopy temperature that helps to resist high temperature or heat stress (Awika et al. 2017). Similarly, higher cuticular waxes can protect from high light conditions such as excessive ultraviolet (UV) radiations, indicating these stresses can affect and alter the plant cuticular waxes (Fukuda et al. 2008; Xue et al. 2017; Lewandowska et al. 2020).

Infection with plant pathogens can also result in increased epi-cuticular wax load and change the cuticular properties. Infections with fungal pathogens *Colletotrichum gloeosporioides* and *C. acutatum* in tomato and citrus plants resulted in increased cuticular wax biosynthesis, deposition, and changes the cuticular structure (Alkan and Fortes 2015; Marques et al. 2016). The increase in epi-cuticular wax load and changes in chemical composition may not always necessarily result in plant resistance against biotic stresses. The epi-cuticular waxes can play divergent roles in different plants and for different pathogens. This was demonstrated through the functional studies of the *Arabidopsis DEWAX* gene, a negative regulator of wax biosynthesis. The *Arabidopsis dewax* mutant lines showed an increased epi-cuticular wax were susceptible to *Botrytis cinerea* and resistant to *Pseudomonas syringae*, fungal and bacterial pathogens, respectively (Ju et al. 2017). Overexpression of DEWAX in *Arabidopsis* and *Camelina* showed inverse defense regulation to *Botrytis* and *Pseudomonas* (Ju et al. 2017).

## 7.6 Attempts to Manipulate Cuticular Trait

Attempts have been made to improve crop plants by targeting the wax biosynthesis pathway and altering the cuticular properties by conventional and modern breeding as well as through transgenic approaches. The prerequisite for crop improvement through breeding or transgenic approaches is to have the prior knowledge about the genomic region/s and gene/s contributing for wax traits. This has been achieved to an extent through the loss- and gain-of-function mutants in either model or crop species. Over the domestication process of major crops like wheat, rice, corn, barley, soybean, and tomato, focus was on yield traits and the yield targeted breeding over generations resulted in reduced genetic diversity for other biotic and abiotic stressors in commercial varieties. A good source to regain the lost genetic diversity is to incorporate the wild relatives and landraces of crops plants in the breeding program. Multiple quantitative trait loci (QTL) regions involved in the biosynthesis of epi-cuticular waxes and its transport have been reported from multiple crops like rice, sorghum, cabbage, and pearl millet and can be used for marker-assisted breeding (MAS) programs (Srinivasan et al. 2008; Burow et al. 2009; Liu et al. 2018).

Considerable amount of work has been carried out in *Arabidopsis* to identify and characterize the functional and regulatory genes involved in cuticular wax biosynthesis (Aharoni et al. 2004; Kannangara et al. 2007; Seo et al. 2009; Shi et al. 2011; Yang et al. 2020). Many of these genes or its homologs identified from crop plants have been used for targeted engineering of wax biosynthetic pathway in crop plants that resulted in altered cuticle properties and showed multiple stress tolerance (Zhang et al. 2005, 2019a, b, c; Adato et al. 2009; Shi et al. 2013; Kumar et al. 2016; Sajeevan et al. 2017a; Liu et al. 2019; More et al. 2019; Wang et al. 2020). It has also been shown that ectopic expression of *Arabidopsis* or its homologs overexpression in biofuel crop, *Camelina sativa* and tree species like *Morus* and *Malus* resulted in altered total wax load, composition, structure, and contributed to drought tolerance (Lee et al. 2014; Sajeevan et al. 2017a; Zhang et al. 2019b, c). Alterations in the wax biosynthesis pathway is hampered due to the lack of clear knowledge in cuticular wax load, its chemical composition, and structural characteristics required to improve specific crops, and also to what extent these factors needs to be species- or tissue-specific.

## 7.7 Conclusion

Cuticle is a natural film covering the outer parts of the plant that consists of lipid polyesters covered and embedded with waxes that protect the tissues from multiple abiotic and biotic stresses. During the land plants evolution from aquatic to a more desiccating terrestrial environment, plants evolve to synthesize cuticular waxes as a fundamental morphological and physiological adaptation. There is a high level of compositional and structural differences exist in cuticular waxes among different crop plants and organs. These cuticular waxes are largely produced by two complex pathways controlled by the expression of different genes/enzymes in turn influenced by multiple environmental stresses. Although past decade advancement in genome sequencing technologies and through various forward and reverse genetics approaches allowed us to elucidate and understand the complex gene regulatory network involved in biosynthesis, transport, and deposition of cuticular waxes in model as well as different crop plants, to an extent. We still have long way to go towards fully understanding the regulatory mechanisms controlling the cuticular wax biosynthesis, compositional and structural differences, transport, and deposition in response to various stressors. In addition, a limited understanding of the role of plant cuticle components as signaling molecules that promote resistance or susceptibility to biotic stresses needs to be further investigated. Unraveling these mechanisms would aid in targeted manipulation of the trait using modern biotechnological applications for the development of crop cultivars with improved health thereby promoting sustainable agriculture.

## References

Aarts MGM, Keijzer CJ, Stiekema WJ, Pereira A (1995) Molecular characterization of the CER1 gene of *Arabidopsis* involved in epicuticular wax biosynthesis and pollen fertility. Plant Cell 7: 2115–2127

Adato A, Mandel T, Mintz-Oron S, Venger I, Levy D, Yativ M, Domínguez E, Wang Z, De Vos RC, Jetter R, Schreiber L, Heredia A, Rogachev I, Aharoni A (2009) Fruit–surface flavonoid accumulation in tomato is controlled by a *SlMYB12*–regulated transcriptional network. PLoS Genet 5:e1000777

Aharoni A, Dixit S, Jetter R, Thoenes E, van Arkel G, Pereira A (2004) The SHINE clade of AP2 domain transcription factors activates wax biosynthesis alters cuticle properties and confers drought tolerance when overexpressed in *Arabidopsis*. Plant Cell 16:2463–2480

Al-Abdallat AM, Al-Debei HS, Ayad JY, Hasan S (2014) Over-expression of *SlSHN1* gene improves drought tolerance by increasing cuticular wax accumulation in tomato. Int J Mol Sci 15:19499–19515

Alkan N, Fortes AM (2015) Insights into molecular and metabolic events associated with fruit response to postharvest fungal pathogens. Front Plant Sci 6:889

Awika HO, Hays DB, Mullet JE, Rooney WL, Weers BD (2017) QTL mapping and loci dissection for leaf epicuticular wax load and canopy temperature depression and their association with QTL for staygreen in *Sorghum bicolor* under stress. Euphytica 213:207

Bach L, Michaelson LV, Haslam R, Bellec Y, Gissot L, Marion J, Costa MD, Boutin JP, Miquel M, Tellier F, Domergue F, Markham JE, Beaudoin F, Napier JA, Faure JD (2008) The very- long-chain hydroxy fatty acyl-CoA dehydratase PASTICCINO2 is essential and limiting for plant development. Proc Natl Acad Sci U S A 105:14727–14731

Beaudoin F, Wu X, Li F, Haslam RP, Markham JE, Zheng H, Napier JA, Kunst L (2009) Functional characterization of the *Arabidopsis* β-Ketoacyl-Coenzyme A reductase candidates of the fatty acid elongase. Plant Physiol 150:1174–1191

Bengtson C, Larsson S, Liljenberg C (1978) Effects of water stress on cuticular transpiration rate and amount and composition of epicuticular wax in seedlings of six oat varieties. Physiol Plant 44:319–324

Bernard A, Joubes J (2013) *Arabidopsis* cuticular waxes: advances in synthesis export and regulation. Prog Lipid Res 52:110–129

Bernard A, Domergue F, Pascal S, Jetter R, Renne C, Faure JD, Haslam RP, Napier JA, Lessire R, Joubès J (2012) Reconstitution of plant alkane biosynthesis in yeast demonstrates that *Arabidopsis* ECERIFERUM1 and ECERIFERUM3 are core components of a very- long-chain alkane synthesis complex. Plant Cell 24:3106–3118

Bessire M, Borel S, Fabre G, Carraça L, Efremova N, Yephremov A, Cao Y, Jetter R, Jacquat AC, Métraux JP, Nawrath C (2011) A member of the Pleiotropic Drug Resistance family of ATP

binding cassette transporters is required for the formation of a functional cuticle in *Arabidopsis*. Plant Cell 23:1958–1970
Bi H, Shi J, Kovalchuk N, Luang S, Bazanova N, Chirkova L, Zhang D, Stepanenko A, Tricker P, Langridge P, Hrmova M, Lopato S, Borisjuk N (2018) Overexpression of the *TaSHN1* transcription factor in bread wheat leads to leaf surface modifications, improved drought tolerance, and no yield penalty under controlled growth conditions. Plant Cell Environ 41:2549–2566
Bird D, Beisson F, Brigham A, Shin J, Greer S, Jetter R, Kunst L, Wu X, Yephremov A, Samuels L (2007) Characterization of *Arabidopsis* ABCG11/WBC11 an ATP binding cassette (ABC) transporter that is required for cuticular lipid secretion. Plant J 52:485–498
Bonaventure G, Salas JJ, Pollard MR, Ohlrogge JB (2003) Disruption of the FATB gene in *Arabidopsis* demonstrates an essential role of saturated fatty acids in plant growth. Plant Cell 15:1020–1033
Bondada B, Oosterhuis DM, Murphy JB, Kim KS (1996) Effect of water stress on the epicuticular wax composition and ultrastructure of cotton (*Gossypium hirsutum* L.) leaf bract and boll. Environ Exp Bot 36:61–67
Borisjuk N, Hrmova M, Lopato S (2014) Transcriptional regulation of cuticle biosynthesis. Biotechnol Adv 2(2):526–540
Bourdenx B, Bernard A, Domergue F, Pascal S, Léger A, Roby D, Pervent M, Vile D, Haslam RP, Napier JA, Lessire R, Joubès J (2011) Overexpression of *Arabidopsis* ECERIFERUM1 promotes wax very-long-chain alkane biosynthesis and influences plant response to biotic and abiotic stresses. Plant Physiol 156:29–45
Broun P, Poindexter P, Osborne E, Jiang CZ, Riechmann JL (2004) WIN1 a transcriptional activator of epidermal wax accumulation in *Arabidopsis*. Proc Natl Acad Sci U S A 101: 4706–4711
Burow GB, Franks CD, Acosta-Martinez V, Xin Z (2009) Molecular mapping and characterization of BLMC, a locus for profuse wax (bloom) and enhanced cuticular features of sorghum (*Sorghum bicolor* (L.) Moench.). Theor Appl Genet 118:423–431
Cameron KD, Teece MA, Smart LB (2006) Increased accumulation of cuticular wax and expression of lipid transfer protein in response to periodic drying events in leaves of tree tobacco. Plant Physiol 140:176–183
Cernac A, Benning C (2004) WRINKLED1 encodes an AP2/EREB domain protein involved in the control of storage compound biosynthesis in *Arabidopsis*. Plant J 40:575–585
Chen X, Goodwin SM, Boroff VL, Liu X, Jenks MA (2003) Cloning and characterization of the WAX2 gene of *Arabidopsis* involved in cuticle membrane and wax production. Plant Cell 15: 1170–1185
Chen X, Goodwin SM, Liu X, Chen X, Bressan RA, Jenks MA (2005) Mutation of the RESURRECTION1 locus of *Arabidopsis* reveals an association of cuticular wax with embryo development. Plant Physiol 139:909–919
Chen G, Komatsuda T, Ma JF, Nawrath C, Pourkheirandish M, Tagiri A, Hu YG, Sameri M, Li X, Zhao X, Liu Y, Li C, Ma X, Wang A, Nair S, Wang N, Miyao A, Sakuma S, Yamaji N, Zheng X, Nevo E (2011) An ATP– binding cassette subfamily G full transporter is essential for the retention of leaf water in both wild barley and rice. Proc Natl Acad Sci U S A 108:12354–12359
Chew W, Hrmova M, Lopato S (2013) Role of Homeodomain Leucine Zipper (HD–Zip) IV transcription factors in plant development and plant protection from deleterious environmental factors. Int J Mol Sci 14:8122–8147
Cominelli E, Sala T, Calvi D, Gusmaroli G, Tonelli C (2008) Overexpression of the *Arabidopsis AtMYB41* gene alters cell expansion and leaf surface permeability. Plant J 53:53–64
Cui F, Wu W, Wang K, Zhang Y, Hu Z, Brosché M, Liu S, Overmyer K (2019) Cell death regulation but not abscisic acid signaling is required for enhanced immunity to *Botrytis* in *Arabidopsis* cuticle-permeable mutants. J Exp Bot 70:5971–5984

Debono A, Yeats TH, Rose JKC, Bird D, Jetter R, Kunst L, Samuels L (2009) *Arabidopsis* LTPG is a glycosylphosphatidyl inositol–anchored lipid transfer protein required for export of lipids to the plant s urface. Plant Cell 21:1230–1238

Djemal R, Khoudi H (2015) Isolation and molecular characterization of a novel WIN1/SHN1 ethylene-responsive transcription factor *TdSHN1* from durum wheat (*Triticum turgidum.* L. subsp. durum). Protoplasma 252:1461–1473

Djemal R, Khoudi H (2021) The barley SHN1-type transcription factor *HvSHN1* imparts heat, drought and salt tolerances in transgenic tobacco. Plant Physiol Biochem 16:44–53

Djemal R, Mila I, Bouzayen M, Pirrello J, Khoudi H (2018) Molecular cloning and characterization of a novel WIN1/SHN1 ethylene-responsive transcription *HvSHN1* in barley (*Hordeum vulgare* L.). J Plant Physiol 228:39–46

Do THT, Martinoia E, Lee Y (2018) Functions of ABC transporters in plant growth and development. Curr Opin Plant Biol 41:32–48

Ecker JR (1995) The ethylene signal transduction pathway in plants. Science 268:667–675

Eigenbrode SD, Espelie KE (1995) Effects of plant epicuticular lipids on insect herbivores. Annu Rev Entomol 40:171–194

Eigenbrode SD, Jetter R (2002) Attachment to plant surface waxes by an insect predator. Integr Comp Biol 42:1091–1099

Eini O, Yang N, Pyvovarenko T, Pillman K, Bazanova N, Tikhomirov N, Eliby S, Shirley N, Sivasankar S, Tingey S, Langridge P, Hrmova M, Lopato S (2013) Complex regulation by Apetala2 domain-containing transcription factors revealed through analysis of the stress-responsive *TdCor410b* promoter from durum wheat. PLoS One 8:e58713

Elejalde-Palmett C, Martinez San Segundo I, Garroum I, Charrier L, De Bellis D, Mucciolo A, Guerault A, Liu J, Zeisler-Diehl V, Aharoni A, Schreiber L, Bakan B, Clausen MH, Geisler M, Nawrath C (2021) ABCG transporters export cutin precursors for the formation of the plant cuticle. Curr Biol 31:2111–2123

Elliott RC, Betzner AS, Huttner E, Oakes MP, Tucker WQ, Gerentes D, Perez P, Smyth DR (1996) *AINTEGUMENTA* an APETALA2-like gene of *Arabidopsis* with pleiotropic roles in ovule development and floral organ growth. Plant Cell 8:155–168

Febrero A, Fernández S, Molina-Cano JL, Araus JL (1998) Yield, carbon isotope discrimination, canopy reflectance and cuticular conductance of barley isolines of differing glaucousness. J Exp Bot 49:1575–1581

Fiebig A, Mayfield JA, Miley NL, Chau S, Fischer RL, Preuss D (2000) Alterations in CER6 a gene identical to CUT1 differentially affect long-chain lipid content on the surface of pollen and stems. Plant Cell 12:2001–2008

Franke R, Höfer R, Briesen I, Emsermann M, Efremova N, Yephremov A, Schreiber L (2009) The DAISY gene from *Arabidopsis* encodes a fatty acid elongase condensing enzyme involved in the biosynthesis of aliphatic suberin in roots and the chalaza-micropyle region of seeds. Plant J 57:80–95

Fukuda S, Satoh A, Kasahara H, Matsuyama H, Takeuchi Y (2008) Effects of ultraviolet-B irradiation on the cuticular wax of cucumber (*Cucumis sativus*) cotyledons. J Plant Res 121: 179–189

Gilding EK, Marks MD (2010) Analysis of purified glabra 3-shapeshifter trichomes reveals a role for NOECK in regulating early trichome morphogenic events. Plant J 64:304–317

Goodwin SM, Jenks MA (2005) Plant cuticle function as a barrier to water loss. In: Jenks MA, Hasegawa PM (eds) Plant abiotic stress. Blackwell Publishing, Inc

Greer S, Wen M, Bird D, Wu X, Samuels L, Kunst L, Jetter R (2007) The Cytochrome P450 Enzyme CYP96A15 is the midchain alkane hydroxylase responsible for formation of secondary alcohols and ketones in stem cuticular wax of *Arabidopsis*. Plant Physiol 145:653–667

Guo L, Yang H, Zhang X, Yang S (2013) Lipid transfer protein 3 as a target of MYB96 mediates freezing and drought stress in *Arabidopsis*. J Exp Bot 64:1755–1767

Guo J, Xu W, Yu X, Shen H, Li H, Cheng D, Liu A, Liu J, Liu C, Zhao S, Song J (2016) Cuticular wax accumulation is associated with drought tolerance in wheat near-isogenic lines. Front Plant Sci 7:1809

Hansen JD, Pyee J, Xia Y, Wen TJ, Robertson DS, Kolattukudy PE, Nikolau BJ, Schnable PS (1997) The glossy1 locus of maize and an epidermis-specific cDNA from *Kleinia odora* define a class of receptor- like proteins required for the normal accumulation of cuticular waxes. Plant Physiol 113:1091–1100

Hasanuzzaman M, Davies NW, Shabala L, Zhou M, Brodribb TJ, Shabala S (2017) Residual transpiration as a component of salinity stress tolerance mechanism: a case study for barley. BMC Plant Biol 17:107

Haslam TM, Mañas-Fernández A, Zhao L, Kunst L (2012) *Arabidopsis* ECERIFERUM2 is a component of the fatty acid elongation machinery required for fatty acid extension to exceptional lengths. Plant Physiol 160:1164–1174

Haslam TM, Haslam R, Thoraval D, Pascal S, Delude C, Domergue F, Fernández AM, Beaudoin F, Napier JA, Kunst L, Joubès J (2015) ECERIFERUM2-LIKE proteins have unique biochemical and physiological functions in very-long-chain fatty acid elongation. Plant Physiol 167(3): 682–692

Hoang MH, Nguyen XC, Lee K, Kwon YS, Pham HT, Park HC, Yun DJ, Lim CO, Chung WS (2012) Phosphorylation by *AtMPK6* is required for the biological function of *AtMYB41* in *Arabidopsis*. Biochem Biophys Res Commun 422:181–186

Hooker TS, Millar AA, Kunst L (2002) Significance of the expression of the CER6 condensing enzyme for cuticular wax production in *Arabidopsis*. Plant Physiol 129:1568–1580

Ingram G, Nawrath C (2017) The roles of the cuticle in plant development: organ adhesions and beyond. J Exp Bot 68:5307–5321

Ingram GC, Magnard JL, Vergne P, Dumas C, Rogowsky PM (1999) *ZmOCL1* an HDGL2 family homeobox gene is expressed in the outer cell layer throughout maize development. Plant Mol Biol 40:343–354

Ingram GC, Boisnard-Lorig C, Dumas C, Rogowsky PM (2000) Expression patterns of genes encoding HD–zip IV homeo domain proteins define specific domains in maize embryos and meristems. Plant J 22:401–414

Islam MA, Du H, Ning J, Ye H, Xiong L (2009) Characterization of Glossy1-homologous genes in rice involved in leaf wax accumulation and drought resistance. Plant Mol Biol 70:443–456

Javelle M, Vernoud V, Depège-Fargeix N, Arnould C, Oursel D, Domergue F, Sarda X, Rogowsky PM (2010) Overexpression of the epidermis-specific homeodomain-leucine zipper IV transcription factor outer cell layer1 in maize identifies target genes involved in lipid metabolism and cuticle biosynthesis. Plant Physiol 154:273–286

Javelle M, Klein-Cosson C, Vernoud V, Boltz V, Maher C, Timmermans M, Depège-Fargeix N, Rogowsky PM (2011) Genome-wide characterization of the HD–ZIP IV transcription factor family in maize: preferential expression in the epidermis. Plant Physiol 157:790–803

Jefferson P, Johnson D, Rumbaugh M, Asay K (1989) Water stress and genotypic effects on epicuticular wax production of alfalfa and crested wheatgrass in relation to yield and excised leaf water loss rate. Can J Plant Sci 69:481–490

Jenks MA, Joly RJ, Peters PJ, Rich PJ, Axtell JD, Ashworth EN (1994) Chemically induced cuticle mutation affecting epidermal conductance to water vapor and disease susceptibility in *Sorghum bicolor* (L.) Moench. Plant Physiol 105:1239–1245

Jenks MA, Andersen L, Teusink RS, Williams MH (2001) Leaf cuticular waxes of potted rose cultivars as affected by plant development drought and paclobutrazol treatments. Physiol Plant 112:62–70

Jessen D, Olbrich A, Knüfer J, Krüger A, Hoppert M, Polle A, Fulda M (2011) Combined activity of LACS1 and LACS4 is required for proper pollen coat formation in *Arabidopsis*. Plant J 68: 715–726

Jetter R, Kunst L (2008) Plant surface lipid biosynthetic pathways and their utility for metabolic engineering of waxes and hydrocarbon biofuels. Plant J 54:670–683

Jetter R, Schaffer S (2001) Chemical composition of the Prunus laurocerasus leaf surface: dynamic changes of the epicuticular wax film during leaf development. Plant Physiol 126:1725–1734

Jetter R, Kunst L, Samuels AL (2006) Composition of plant cuticular waxes. In: Riederer M, Müller C (eds) Annual plant reviews 23: biology of the plant cuticle. Blackwell Publishing, Oxford, pp 145–181

Jofuku KD, den Boer BG, Van Montagu M, Okamuro JK (1994) Control of *Arabidopsis* flower and seed development by the homeotic gene APETALA2. Plant Cell 6:1211–1225

Jordan W, Shouse PJ, Blum A, Miller FR, Monk RL (1984) Environmental physiology of *sorghum*, II, epicuticular wax load and cuticular transpiration. Crop Sci 24:1168–1173

Ju S, Go YS, Choi HJ, Park JM, Suh MC (2017) DEWAX transcription factor is involved in resistance to *Botrytis cinerea* in *Arabidopsis thaliana* and *Camelina sativa*. Front Plant Sci 8: 1210

Kagaya Y, Ohmiya K, Hattori T (1999) RAV1, a novel DNA-binding protein, binds to bipartite recognition sequence through two distinct DNA-binding domains uniquely found in higher plants. Nucleic Acids Res 27(2):470–478

Kannangara R, Branigan C, Liu Y, Penfield T, Rao V, Mouille G, Hofte H, Pauly M, Riechmann JL, Broun P (2007) The transcription factor WIN1/SHN1 regulates cutin biosynthesis in *Arabidopsis thaliana*. Plant Cell 19(4):1278–1294

Kerstiens G (ed) (1996) Plant cuticles: an integrated functional approach. BIOS Scientific Publishers Limited, Oxford

Kim H, Lee SB, Kim HJ, Min MK, Hwang I, Suh MC (2012) Characterization of glycosylphosphatidyl inositol- anchored lipid transfer protein 2 (LTPG2) and overlapping function between LTPG/LTPG1 and LTPG2 in cuticular wax export or accumulation in *Arabidopsis thaliana*. Plant Cell Physiol 53:1391–1403

Kim J, Jung JH, Lee SB, Go YS, Kim HJ, Cahoon R, Markham JE, Cahoon EB, Suh MC (2013) *Arabidopsis* 3- Ketoacyl-Coenzyme A Synthase9 is involved in the synthesis of tetracosanoic acids as precursors of cuticular waxes, suberins, sphingolipids, and phospholipids. Plant Physiol 162:567–580

Kim H, Go YS, Suh MC (2018) DEWAX2 transcription factor negatively regulates cuticular wax biosynthesis in *Arabidopsis* leaves. Plant Cell Physiol 59(5):966–977

Kolattukudy P (1980) Cutin suberin and waxes. In: Stumpf PK (ed) The biochemistry of plants a comprehensive treatise, vol 4. Academic Press, New York, pp 571–641

Kong L, Zhi P, Liu J, Li H, Zhang X, Xu J, Zhou J, Wang X, Chang C (2020) Epigenetic activation of enoyl-CoA reductase by an acetyltransferase complex triggers wheat wax biosynthesis. Plant Physiol 183:1250–1267

Kosma DK, Bourdenx B, Bernard A, Parsons EP, Lü S, Joubès J, Jenks MA (2009) The impact of water deficiency on leaf cuticle lipids of *Arabidopsis*. Plant Physiol 151:1918–1929

Kumar A, Yogendra KN, Karre S, Kushalappa AC, Dion Y, Choo TM (2016) WAX INDUCER1 (*HvWIN1*) transcription factor regulates free fatty acid biosynthetic genes to reinforce cuticle to resist *fusarium* head blight in barley spikelets. J Exp Bot 67(14):4127–4139

Kunst L, Samuels AL (2003) Biosynthesis and secretion of plant cuticular wax. Prog Lipid Res 42: 51–80

Kurdyukov S, Faust A, Nawrath C, Bär S, Voisin D, Efremova N, Franke R, Schreiber L, Saedler H, Métraux JP, Yephremov A (2006) The epidermis-specific extracellular BODYGUARD controls cuticle development and morphogenesis in *Arabidopsis*. Plant Cell 18:321–339

Lee SB, Suh MC (2015a) Advances in the understanding of cuticular waxes in *Arabidopsis thaliana* and crop species. Plant Cell Rep 34:557–572

Lee SB, Suh MC (2015b) Cuticular wax biosynthesis is up-regulated by the MYB94 transcription factor in *Arabidopsis*. Plant Cell Physiol 56(1):48–60

Lee SB, Go YS, Bae HJ, Park JH, Cho SH, Cho HJ, Lee DS, Park OK, Hwang I, Suh MC (2009a) Disruption of glycosylphosphatidyl inositol-anchored lipid transfer protein gene altered cuticular lipid composition increased plastoglobules and enhanced susceptibility to infection by the fungal pathogen *Alternaria brassicicola*. Plant Physiol 150:42–54

Lee SB, Jung SJ, Go YS, Kim HU, Kim JK, Cho HJ, Park OK, Suh MC (2009b) Two *Arabidopsis* 3- ketoacyl CoA synthase genes, KCS20 and KCS2/DAISY, are functionally redundant in cuticular wax and root suberin biosynthesis, but differentially controlled by osmotic stress. Plant J 60:462–475

Lee SB, Kim H, Kim RJ, Suh MC (2014) Overexpression of *Arabidopsis* MYB96 confers drought resistance in *Camelina sativa* via cuticular wax accumulation. Plant Cell Rep 33:1535–1546

Lee SB, Kim HU, Suh MC (2016) MYB94 and MYB96 additively active cuticular wax biosynthesis in *Arabidopsis*. Plant Cell Physiol 57:2300–2311

Lewandowska M, Keyl A, Feussner I (2020) Wax biosynthesis in response to danger: its regulation upon abiotic and biotic stress. New Phytol 227:698–713

Li F, Wu X, Lam P, Bird D, Zheng H, Samuels L, Jetter R, Kunst L (2008) Identification of the wax ester synthase/acyl-coenzyme a: diacylglycerol acyltransferase WSD1 required for stem wax ester biosynthesis in *Arabidopsis*. Plant Physiol 148:97–107

Li-Beisson Y, Shorrosh B, Beisson F, Andersson MX, Arondel V, Bates PD, Baud S, Bird D, Debono A, Durrett TP, Franke RB, Graham IA, Katayama K, Kelly AA, Larson T, Markham JE, Miquel M, Molina I, Nishida I, Rowland O, Samuels L, Schmid KM, Wada H, Welti R, Xu C, Zallot R, Ohlrogge J (2013) Acyl- lipid metabolism. Arabidopsis Book 8:e0161

Licausi F, Ohme-Takagi M, Perata P (2013) APETALA2/Ethylene Responsive Factor (AP2/ERF) transcription factors: mediators of stress responses and developmental programs. New Phytol 199:639–649

Lisso J, Schroder F, Schippers JH, Mussig C (2012) NFXL2 modifies cuticle properties in *Arabidopsis*. Plant Signal Behav 7:551–555

Liu D, Dong X, Liu Z, Tang J, Zhuang M, Zhang Y, Lv H, Liu Y, Li Z, Fang Z, Yang L (2018) Fine mapping and candidate gene identification for wax biosynthesis locus, *BoWax1* in *Brassica oleracea* L. var. capitata. Front. Plant Sci 9:309

Liu N, Chen J, Wang T, Li Q, Cui P, Jia C, Hong Y (2019) Overexpression of WAX INDUCER1/ SHINE1 gene enhances wax accumulation under osmotic stress and oil synthesis in *Brassica napus*. Int J Mol Sci 20(18):4435

Long LM, Patel HP, Cory WC, Stapleton AE (2003) The maize epicuticular wax layer provides UV protection. Fun Plant Biol 30:75–81

Lü S, Song T, Kosma DK, Parsons EP, Rowland O, Jenks MA (2009) *Arabidopsis* CER8 encodes LONG-CHAIN ACYLCOA SYNTHETASE 1 (LACS1) that has overlapping functions with LACS2 in plant wax and cutin synthesis. Plant J 59:553–564

Lü S, Zhao H, Parsons EP, Xu C, Kosma DK, Xu X, Chao D, Lohrey G, Bangarusamy DK, Wang G, Bressan RA, Jenks MA (2011) The glossyhead1 allele of ACC1 reveals a principal role f or multidomain Acetyl- Coenzyme A Carboxylase in the biosynthesis of cuticular waxes by *Arabidopsis*. Plant Physiol 157:1079–1092

Luo B, Xue XY, Hu WL, Wang LJ, Chen XY (2007) An ABC transporter gene of *Arabidopsis thaliana*, *AtWBC11*, is involved in cuticle development and prevention of organ fusion. Plant Cell Physiol 48:1790–1802

Mamrutha HM, Mogili T, Lakshmi JK, Rama N, Kosma D, Udaya Kumar M, Jenks MA, Karaba NN (2010) Leaf cuticular wax amount and crystal morphology regulate post-harvest water loss in mulberry (*Morus* species). Plant Physiol Biochem 48:690–696

Mamrutha HM, Nataraja KN, Rama N, Kosma DK, Mogili T, Jhansi-Lakshmi K, Udaya Kumar M, Jenks MA (2017) Leaf surface wax composition of genetically diverse mulberry (*Morus* sp.) genotypes and its close association with expression of genes involved in wax metabolism. Curr Sci 112:759–766

Marques WL, Salazar MM, Camargo ELO, Lepikson-Neto J, Tiburcio RA, Costa do Nascimento L, Pereira GAG (2013) Identification of four *Eucalyptus* genes potentially involved in cell wall biosynthesis and evolutionarily related to SHINE transcription factors. Plant Growth Regul 69: 203–208

Marques JPR, Amorim L, Spósito MB, Appezzato-da-Glória B (2016) Ultrastructural changes in the epidermis of petals of the sweet orange infected by *Colletotrichum acutatum*. Protoplasma 253:1233–1242

Masaki T, Mitsui N, Tsukagoshi H, Nishii T, Morikami A, Nakamura K (2005) ACTIVATOR of Spomin::LUC1/WRINKLED1 of *Arabidopsis thaliana* transactivates sugar-inducible promoters. Plant Cell Physiol 46:547–556

McFarlane HE, Shin JJ, Bird DA, Samuels AL (2010) *Arabidopsis* ABCG transporters which are required for export of diverse cuticular lipids dimerize in different combinations. Plant Cell 22: 3066–3075

McFarlane HE, Watanabe Y, Yang W, Huang Y, Ohlrogge J, Samuels AL (2014) Golgi- and trans-Golgi network-mediated vesicle trafficking is required for wax secretion from epidermal cells. Plant Physiol 164(3):1250–1260

Mintz-Oron S, Mandel T, Rogachev I, Feldberg L, Lotan O, Yativ M, Wang Z, Jetter R, Venger I, Adato Aharoni A (2008) Gene expression and metabolism in tomato fruit surface tissues. Plant Physiol 147:823–851

Mizoi J, Shinozaki K, Yamaguchi-Shinozaki K (2012) AP2/ERF family transcription factors in plant abiotic stress responses. Biochim Biophys Acta 1819:86–96

Moitra K, Silverton L, Limpert K, Im K, Dean M (2011) Moving out: from sterol transport to drug resistance, the ABCG subfamily of efflux pumps. Drug Metabol Drug Interact 26:105–111

More P, Agarwal P, Joshi PS, Agarwal PK (2019) The *JcWRKY* tobacco transgenics showed improved photosynthetic efficiency and wax accumulation during salinity. Sci Rep 9:19617

Nadakuduti SS, Pollard M, Kosma DK, Allen JC, Ohlrogge JB, Barry CS (2012) Pleiotropic phenotypes of the sticky peel mutant provide new insight into the role of CUTIN DEFICIENT2 in epidermal cell function in tomato. Plant Physiol 159:945–960

Nakano T, Suzuki K, Fujimura T, Shinshi H (2006) Genome-wide analysis of the ERF gene family in *Arabidopsis* and rice. Plant Physiol 140:411–432

Navarro M, Ayax C, Martinez Y, Laur J, El Kayal W, Marque C, Teulières C (2011) Two *EguCBF1* genes overexpressed in *eucalyptus* display a different impact on stress tolerance and plant development. Plant Biotechnol J 9:50–63

Nawrath C, Schreiber L, Franke RB, Geldner N, Reina-Pinto JJ, Kunst L (2013) Apoplastic diffusion barriers in *Arabidopsis*. Arabidopsis Book 11:e0167

Nguyen VNT, Lee SB, Suh MC, An G, Jung KH (2018) *OsABCG9* is an important ABC transporter of cuticular wax deposition in rice. Front Plant Sci 9:960

Ohlrogge JB, Pollard MR, Stumpf PK (1978) Studies on biosynthesis of waxes by developing jojoba seed tissues. Lipids 13:203–210

Oshima Y, Mitsuda N (2013) The MIXTA-like transcription factor MYB16 is a major regulator of cuticle formation in vegetative organs. Plant Signal Behav 9:e26826

Oshima Y, Shikata M, Koyama T, Ohtsubo N, Mitsuda N, Ohme-Takagi M (2013) MIXTA-like transcription factors and WAX INDUCER1/SHINE1 coordinately regulate cuticle development in *Arabidopsis* and *Torenia fournieri*. Plant Cell 25:1609–1624

Panikashvili D, Shi JX, Schreiber L, Aharoni A (2011) The *Arabidopsis* ABCG13 transporter is required for flower cuticle secretion and patterning of the petal epidermis. New Phytol 190:113–124

Park CS, Go YS, Suh MC (2016) Cuticular wax biosynthesis is positively regulated by WRINKLED4, an AP2/ERF-type transcription factor, in *Arabidopsis* stems. Plant J 88(2):257–270

Pascal S, Bernard A, Sorel M, Pervent M, Vile D, Haslam RP, Napier JA, Lessire R, Domergue F, Joubès J (2013) The *Arabidopsis* cer26 mutant, like the cer2 mutant, is specifically affected in the very long-chain fatty acid elongation process. Plant J 73:733–746

Pighin JA, Zheng H, Balakshin LJ, Goodman IP, Western TL, Jetter R, Kunst L, Samuels AL (2004) Plant cuticular lipid export requires an ABC transporter. Science 306:702–704

Pollard M, Beisson F, Li Y, Ohlrogge JB (2008) Building lipid barriers: biosynthesis of cutin and suberin. Trends Plant Sci 13:236–246

Rashotte AM, Jenks MA, Nguyen TD, Feldmann KA (1997) Epicuticular wax variation in ecotypes of *Arabidopsis thaliana*. Phytochemistry 45(2):251–255

Richards RA, Rawson HM, Johnson DA (1986) Glaucousness in wheat: its development and effect on water use efficiency, gas exchange and photosynthetic tissue temperatures. Funct Plant Biol 13:465–473

Rowland O, Domergue F (2012) Plant fatty acyl reductases: enzymes generating fatty alcohols for protective layers with potential for industrial applications. Plant Sci 193–194:28–38

Rowland O, Zheng H, Hepworth SR, Lam P, Jetter R, Kunst L (2006) CER4 encodes an alcohol-forming fatty acyl-coenzyme a reductase involved in cuticular wax production in *Arabidopsis*. Plant Physiol 142:866–877

Rowland O, Lee R, Franke R, Schreiber L, Kunst L (2007) The CER3 wax biosynthetic gene from *Arabidopsis thaliana* is allelic to WAX2/YRE/FLP1. FEBS Lett 581:3538–3544

Sajeevan RS, Nataraja KN, Shivashankara KS, Pallavi N, Gurumurthy DS, Shivanna MB (2017a) Expression of *Arabidopsis* SHN1 in Indian mulberry (*Morus indica* L.) increases leaf surface wax content and reduces post-harvest water loss. Front Plant Sci 8:418

Sajeevan RS, Parvathi MS, Nataraja KN (2017b) Leaf wax trait in crops for drought and biotic stress tolerance: regulators of epicuticular wax synthesis and role of small RNAs. Indian J Plant Physiol 22:434–447

Sakuma Y, Liu Q, Dubouzet JG, Abe H, Shinozaki K, Yamaguchi-Shinozaki K (2002) DNA-binding specificity of the ERF/AP2 domain of *Arabidopsis* DREBs transcription factors involved in dehydration- and cold-inducible gene expression. Biochem Biophys Res Commun 290:998–1009

Sakuradani E, Zhao L, Haslam TM, Kunst L (2013) The CER22 gene required for the synthesis of cuticular wax alkanes in *Arabidopsis thaliana* is allelic to CER1. Planta 237:731–738

Samdur M, Manivel P, Jain VB, Chikani BM, Gor HK, Desai S, Misra JB (2003) Genotypic differences and water–deficit induced enhancement in epicuticular wax load in peanut. Crop Sci 43:1294–1299

Samuels L, DeBono A, Lam P, Wen M, Jetter R, Kunst L (2008) Use of *Arabidopsis* eceriferum mutants to explore plant cuticle biosynthesis. J Vis Exp 31:709

Seo PJ, Park CM (2011) Cuticular wax biosynthesis as a way of inducing drought resistance. Plant Signal Behav 6:1043–1045

Seo PJ, Xiang F, Qiao M, Park JY, Lee YN, Kim SG, Kim SG, Lee YH, Park WJ, Park CM (2009) The *MYB96* transcription factor mediates abscisic acid signaling during drought stress response in *Arabidopsis*. Plant Physiol 151:275–289

Seo PJ, Lee SB, Suh MC, Park MJ, Go YS, Park CM (2011) The MYB96 transcription factor regulates cuticular wax biosynthesis under drought conditions in *Arabidopsis*. Plant Cell 23: 1138–1152

Shi JX, Malitsky S, De Oliveira S, Branigan C, Franke RB, Schreiber L, Aharoni A (2011) SHINE transcription factors act redundantly to pattern the archetypal surface of *Arabidopsis* flower organs. PLoS Genet 7:e1001388

Shi JX, Adato A, Alkan N, He Y, Lashbrooke J, Matas AJ, Meir S, Malitsky S, Isaacson T, Prusky D, Leshkowitz D, Schreiber L, Granell AR, Widemann E, Grausem B, Pinot F, Rose JKC, Rogachev I, Rothan C, Aharoni A (2013) The tomato *SlSHINE3* transcription factor regulates fruit cuticle formation and epidermal patterning. New Phytol 197:468–480

Skamnioti P, Gurr SJ (2007) *Magnaporthe grisea* cutinase2 mediates aspersorium differentiation and host penetration and is required for full virulence. Plant Cell 19:2674–2689

Srinivasan S, Gomez SM, Kumar SS, Ganesh SK, Biji KR, Senthil A, Ranganathan C (2008) QTLs linked to leaf epicuticular wax, physio-morphological and plant production traits under drought stress in rice (*Oryza sativa* L.). Plant Growth Regul 56:245–256

Taketa S, Amano S, Tsujino Y, Sato T, Saisho D, Kakeda K, Nomura M, Suzuki T, Matsumoto T, Sato K, Kanamori H, Kawasaki S, Takeda K (2008) Barley grain with adhering hulls is controlled by an ERF family transcription factor gene regulating a lipid biosynthesis pathway. Proc Natl Acad Sci U S A 105:4062–4067

Tillett RL, Wheatley MD, Tattersall EA, Schlauch KA, Cramer GR, Cushman JC (2012) The *Vitis vinifera* C-repeat binding protein 4 (*VvCBF4*) transcriptional factor enhances freezing tolerance in wine grape. Plant Biotechnol J 10(1):105–124

To A, Joubès J, Barthole G, Lécureuil A, Scagnelli A, Jasinski S, Lepiniec L, Baud S (2012) WRINKLED transcription factors orchestrate tissue-specific regulation of fatty acid biosynthesis in *Arabidopsis*. Plant Cell 24:5007–5023

Todd J, Post BD, Jaworski JG (1999) KCS1 encodes a fatty acid elongase 3-ketoacyl-CoA synthase affecting wax biosynthesis in *Arabidopsis thaliana*. Plant J 17:119–130

Wang Y, Wan L, Zhang L, Zhang Z, Zhang H, Quan R, Zhou S, Huang R (2012) An ethylene response factor *OsWR1* responsive to drought stress transcriptionally activates wax synthesis related genes and increases wax production in rice. Plant Mol Biol 78:275–288

Wang X, Zhi P, Fan Q, Zhang M, Chang C (2019) Wheat CHD3 protein *TaCHR729* regulates the cuticular wax biosynthesis required for stimulating germination of *Blumeria graminis* f. sp. tritici. J Exp Bot 70:701–713

Wang L, Xue W, Li X, Li J, Wu J, Xie L, Kawabata S, Li Y, Zhang Y (2020) *EgMIXTA1*, a MYB-type transcription factor, promotes cuticular wax formation in *Eustoma grandiflorum* leaves. Front Plant Sci 11:524947

Wellesen K, Durst F, Pinot F, Benveniste I, Nettesheim K, Wisman E, Steiner-Lange S, Saedler H, Yephremov A (2001) Functional analysis of the LACERATA gene of *Arabidopsis* provides evidence for different roles of fatty acid omega-hydroxylation in development. Proc Natl Acad Sci U S A 98:9694–9699

Weng H, Molina I, Shockey J, Browse J (2010) Organ fusion and defective cuticle function in a lacs1 lacs2 double mutant of *Arabidopsis*. Planta 231:1089–1100

Wu R, Li S, He S, Wassmann F, Yu C, Qin G, Schreiber L, Qu LJ, Gu H (2011) CFL1 a WW domain protein regulates cuticle development by modulating the function of HDG1 a class IV homeodomain transcription factor in rice and *Arabidopsis*. Plant Cell 23:3392–3411

Xu Z-S, Chen M, Li L-C, Ma YZ (2011) Functions and application of the AP2/ERF transcription factor family in crop improvement. J Integr Plant Biol 53(7):570–585

Xu Y, Wu H, Zhao M, Wu W, Xu Y, Gu D (2016) Overexpression of the transcription factors *GmSHN1* and *GmSHN9* differentially regulates wax and cutin biosynthesis, alters cuticle properties, and changes leaf phenotypes in *Arabidopsis*. Int J Mol Sci 17:587

Xue D, Zhang X, Lu X, Chen G, Chen Z-H (2017) Molecular and evolutionary mechanisms of cuticular wax for plant drought tolerance. Front Plant Sci 8:621

Yang J, Zhao X, Liang L, Xia Z, Lei L, Niu X, Zou C, Zhang KQ (2011) Overexpression of a cuticle-degrading protease Ver112 increases the nematicidal activity of *Paecilomyces lilacinus*. Appl Microbiol Biotechnol 89:1895–1903

Yang X, Zhao H, Kosma DK, Tomasi P, Dyer JM, Li R, Liu X, Wang Z, Parsons EP, Jenks MA, Lü S (2017) The acyl desaturase CER17 is involved in producing wax unsaturated primary alcohols and cutin monomers. Plant Physiol 173:1109–1124

Yang SU, Kim H, Kim RJ, Kim J, Suh MC (2020) AP2/DREB transcription factor RAP2.4 activates cuticular wax biosynthesis in *Arabidopsis* leaves under drought. Front. Plant Sci 11: 895

Yeats TH, Rose JKC (2008) The biochemistry and biology of extracellular plant lipid-transfer proteins (LTPs). Protein Sci 17:191–198

Yephremov A, Wisman E, Huijser P, Huijser C, Wellesen K, Saedler H (1999) Characterization of the FIDDLE HEAD gene of *Arabidopsis* reveals a link between adhesion response and cell differentiation in the epidermis. Plant Cell 11:2187–2201

Zhang JY, Broeckling CD, Blancaflor EB, Sledge MK, Sumner LW, Wang ZY (2005) Overexpression of *WXP1* a putative *Medicago truncatula* AP2 domain-containing transcription factor gene increases cuticular wax accumulation and enhances drought tolerance in transgenic alfalfa (*Medicago sativa*). Plant J 42:689–707

Zhang JY, Broeckling CD, Sumner LW, Wang ZY (2007) Heterologous expression of two *Medicago truncatula* putative ERF transcription factor genes WXP1 and WXP2 in *Arabidopsis*

led to increased leaf wax accumulation and improved drought tolerance but differential response in freezing tolerance. Plant Mol Biol 64:265–278
Zhang J, Yang J, Yang Y, Luo J, Zheng X, Wen C, Xu Y (2019a) Transcription factor *CsWIN1* regulates pericarp wax biosynthesis in cucumber grafted on pumpkin. Front Plant Sci 10:1564
Zhang YL, Zhang CL, Wang GL, Wang YX, Qi CH, You CX, Li YY, Hao YJ (2019b) Apple AP2/EREBP transcription factor *MdSHINE2* confers drought resistance by regulating wax biosynthesis. Planta 249(5):1627–1643
Zhang YL, Zhang CL, Wang GL, Wang YX, Qi CH, Zhao Q, You CX, Li YY, Hao YJ (2019c) The R2R3 MYB transcription factor *MdMYB30* modulates plant resistance against pathogens by regulating cuticular wax biosynthesis. BMC Plant Biol 19:362
Zheng H, Rowland O, Kunst L (2005) Disruptions of the *Arabidopsis* Enoyl–CoA reductase gene reveal an essential role for very–long–chain fatty acid synthesis in cell expansion during plant morphogenesis. Plant Cell 17:1467–1481
Zhu X, Xiong L (2013) Putative megaenzyme DWA1 plays essential roles in drought resistance by regulating stress-induced wax deposition in rice. Proc Natl Acad Sci U S A 110:17790–17795
Zhu L, Guo J, Zhu J, Zhou C (2014) Enhanced expression of EsWAX1 improves drought tolerance with increased accumulation of cuticular wax and ascorbic acid in transgenic Arabidopsis. Plant Physiol Biochem 75:24–35
Ziv C, Zhao Z, Gao YG, Xia Y (2018) Multifunctional roles of plant cuticle during plant-pathogen interactions. Front Plant Sci 9:1088

# Chapter 8
# Radiation Use Efficiency (RUE) as Target for Improving Yield Potential: Current Status and Future Prospect

**Geetha Govind, Rajashekar Reddy, Chwan-Yang Hong, and B. T. Krishnaprasad**

**Abstract** Breeding and selection for harvest index to improve yield potentials have reached its biological ceiling limit. Recent efforts are on increasing yield potential by increasing biomass by optimizing radiation use efficiency (RUE). RUE is the efficiency of biomass production per unit light captured. It is essential to increase yield to meet the food demand of the growing population that is dependent on diminishing arable land. Improvements in RUE are possible by increasing photosynthetic efficiencies and its capacity. In this chapter, we describe the current understanding on RUE, the avenues that can be explored at leaf and canopy level to increase or optimize RUE, with major emphasis on improving photosynthetic efficiencies are discussed in detail.

**Keywords** Radiation use efficiency · Nonphotochemical quenching · Photorespiration · Rubisco · Smart canopy

G. Govind (✉)
Department of Crop Physiology, COA, Hassan, UASB, Bengaluru, Karnataka, India
e-mail: geetha.govind@uasbangalore.edu.in

R. Reddy
Department of Crop Physiology, GKVK, UASB, Bengaluru, Karnataka, India

C.-Y. Hong
Department of Agricultural Chemistry, National Taiwan University, Taipei, Taiwan

B. T. Krishnaprasad (✉)
Department of Biotechnology, COA, Hassan, UASB, Bengaluru, Karnataka, India
e-mail: krishnaprasad@uasbangalore.edu.in


H. M. Mamrutha et al. (eds.), *Translating Physiological Tools to Augment Crop Breeding*, https://doi.org/10.1007/978-981-19-7498-4_8

## 8.1 Introduction

By the year 2050, the world population is likely to exceed 9 billion (FAO 2009; Crist et al. 2017). Food shortage is of global concern and pressure on food production is increasing with increase in population. Four major crops: rice, wheat, maize, and soybean meet more than two thirds of the calories directly or indirectly. It is estimated that by 2050 the requirement is likely to increase by 87% more for these foods (FAO 2013; Ray et al. 2013). Reduction in availability of arable land along with negative impact of climate change and extreme weather conditions further adds to reduction in agricultural production and food supply. Climate change alone has resulted in 27% decline in yield potential of wheat, from 1990 to 2015 (Hochman et al. 2017). To meet the food demand of the increasing population, significant yield increase of major crops is essential. It is estimated that nearly 30% increase in yield of rice is required to be achieved by 2030, to meet the food demand. Similarly, in the next 20 years, demand for wheat will increase at the rate of 1.6% per year (Reynolds and Pfeiffer 2000). However, harvest index (HI) of many crops has reached or is approaching its biological ceiling limits. The genetic potential for HI in major crops like wheat and rice has reached almost 0.6 which is its maximum (Foulkes et al. 2011). Reduced height (Rht) genes played a major role in achieving higher HI in wheat. For the past 30 years, only 0.9% yield gain (genetic gain) per year has been achieved (Reynolds and Pfeiffer 2000).

With the existing improvement in yield potential maintained in the coming years, by 2050 we are likely to face 20–30% shortfall in major crops (rice, wheat, and soybean) (Ray et al. 2013). On the contrary, it is now apparent that the yield increase in the past decades is less than a quarter or near to zero of that during the Green Revolution years for major crops like rice and wheat (Long 2014; Kromdijk and Long 2016). The options left out to increase yield potential is by improving crop biomass, without sacrificing HI, that can be achieved by improving RUE via improving net carbon gain or net photosynthesis (Parry et al. 2011; Furbank et al. 2020).

Some of the strategies considered to accelerate breeding progress are to improve early generation selection tools for RUE traits—improved photosynthesis, biomass, and source–sink balance. Most breeding programs have fixed traits governed by additive effects. However, RUE is mostly controlled by genomic regions that display epistasis and has been difficult to fix them during breeding (Reynolds and Pfeiffer 2000). An improvement in photosynthesis so far, which is the conversion efficiency of the absorbed sunlight to biomass, is marginal (Long et al. 2006). One of the bottlenecks to increase yield potential is optimizing radiation use efficiency (RUE) to increase biomass. Although photosynthetic rate ($A_n$) is not always associated with high RUE, it is positively associated with yield. Therefore, selecting for traits regulating assimilation (photosynthetic rate, stomatal conductance, transpiration) can identify physiologically superior genotypes.

## 8.2 What Is RUE

Beers law is be used to express the radiation intercepted by a crop

$$\mathrm{RI} = \mathrm{RAD}\left(1 - e^{-k*\mathrm{LAI}}\right)$$

where, RAD: Global solar radiation incident above the canopy, *k*: Extinction coefficient, LAI: Leaf area index. [1-$e^{-k*\mathrm{LAI}}$]: Proportion of light intercepted by a crop.

The amount of crop biomass accumulated is linearly associated with the amount of radiation intercepted with slope depicting the radiation use efficiency (RUE g. $\mathrm{MJ}^{-1}$). Under nonlimiting conditions, the RUE ranges from 2.8 to 3.2 g.$\mathrm{MJ}^{-1}$ of PAR (Monteith 1977; Sinclair and Muchow 1999). The average *k* for erect and horizontal leaves is 0.6 and 0.8, respectively (Kemanian et al. 2004). Increase in LAI results in higher RI, but the rate of increase gradually diminishes. However, RI is larger for horizontal leaves compared to erect leaves for a given LAI. RUE is the sum total of efficiencies of process, viz. photosynthesis, respiration, synthesis of various molecules from the photosynthates (lipids, proteins, polysaccharides, etc.), transportation of photosynthates and metabolites which in turn depend on various internal and external factors. At canopy level, RUE can be extrapolated from net photosynthesis and respiration losses of a leaf to canopy (Hammer and Wright 1994).

Major plant process like photosynthesis and transpiration are affected by the amount and quality of light. The lower canopy of the plant usually only receives diffused light compared to upper canopy that majorly receives direct sunlight. Of the light received, generally $<50\%$ is used for photosynthesis in the upper canopy. The optical properties of the leaves differ in their absorption and transmission for PAR and NIR wavebands. In general, the absorptivity of a canopy is higher than absorptivity of a single leaf. Usually the lower canopy is enriched in NIR radiations, which is depended on the canopy structure, shape, and size (Shulski et al. 2004).

On an average, of the total solar energy (1.3 $\mathrm{kWm}^{-2}$) received only 5% is available or utilized by plants for production of carbohydrates. If the total solar energy received is considered 100 units, nearly 50 units of it is in wavelengths not absorbed by plants (non-PAR region). Of the 35 units from PAR (400–700 nm) region, the leaf absorbs 85–90% and the remaining 15 units is reflected or transmitted. Of the 35 units available, approximately 10 units is dissipated as heat, 17 units utilized for light reaction and 15 units utilized by the dark reaction. Of the remaining 6.5 units, only 5 units or 5% is utilized or converted to carbohydrates. Therefore, it is important to minimize the losses at various levels to increase radiation use efficiency (Fig. 8.1).

The maximum efficiency of photosynthesis is 5%, but is usually found to be lower RUE varies with photosynthetic metabolism type (C3, C4, and CAM) and seed composition (cereals, legumes, and oilseed). RUE is lower in C3 than C4 and much lower in oil and protein-rich crops. In general, C3 cereals have approximately 15% lower RUE than C4 plants. The lower RUE of C3 is due to lower

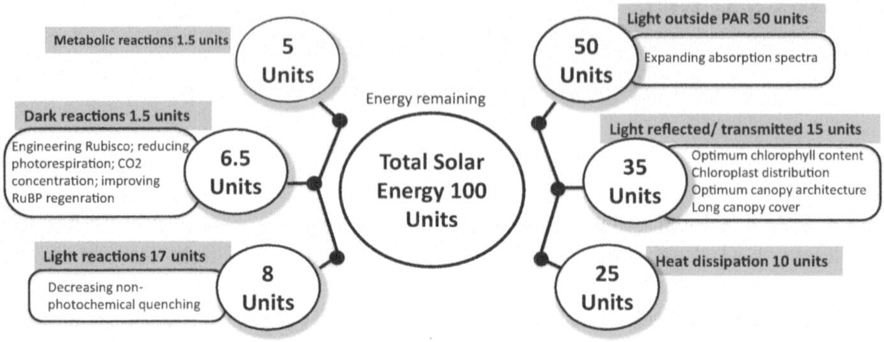


**Fig. 8.1** Total solar energy fixed by plants via photosynthesis after eliminating losses at various levels

photosynthetic efficiency that occurs because of photorespiratory $CO_2$ losses. Similarly, legumes and C3 oil crops have lower RUE than most crops, and it is approximately 15% lower than that of C3 cereals. The maximum RUE of C3 and C4 species is around 1.4 and 2.0 $g.MJ^{-1}$, respectively (Monteith 1977). Some C4 grasses like miscanthus and *Echinochloa polystachya* have reportedly higher RUE of around 2.3 $g.MJ^{-1}$. Among the C3 crops, maximum RUE is reported for crops that accumulate simple sugars, viz. sugarbeet and potato. Among the legumes, pea has relatively higher RUE of 1.5 $g.MJ^{-1}$ (Haro et al. 2007; Stöckle and Kemanian 2009). Environmental factors such as temperature, radiation, humidity, water, and nutritional status affect RUE.

## 8.3 Is It Possible to Improve RUE

The annual crops have $\leq$1% solar energy conversion of the incoming global solar radiation (de Groot 2008). To increase RUE, it is important to understand energy loss in the system to be able to fix them, to improve energy conversion efficiency. Decades of research have been carried out to understand and unravel the energy conversion efficiency of plants. Some of the routes to improve RUE can be taken care at leaf, canopy, and crop level. At single leaf level, improving the photochemical efficiency, photosynthetic capacity, decreasing photorespiration, and nonphotochemical quenching will result in increased RUE. Similarly at canopy level, optimizing canopy architecture, chlorophyll distribution with improved vertical distribution of photosynthetic resources will improve RUE. At crop level, increasing canopy cover duration achieved by early vigor and stay green types with simultaneous reduction in crop respiration loses will improve RUE (Yin and Struik 2015).

RUE can be improved by selecting for improved canopy photosynthesis, prolonged green leaf area duration and source–sink balance. Soluble carbohydrate (stem reserve) utilization and stay green (longer green leaf area duration) phenotype are two mutually exclusive events. However, improving both of these traits will improve RUE, and hence, there is a need to break the linkage associate between them. Genomic regions controlling stem reserve mobilization have been identified (Li et al. 2020). Stay green is an important trait that can be exploited, and selection for functional stay green has resulted in increased yield (Thomas and Ougham 2014; Munaiz et al. 2020). Candidate genes and genomic regions (QTLs) controlling stay green phenotype have been identified in rice (Singh et al. 2020; Habiba et al. 2021). Increasing photo-assimilation capacity and sink strength simultaneously will improve RUE. Altering spike morphology and sink capacity resulted in 13% increase in yield (Reynolds and Pfeiffer 2000).

GWAS analysis using 150 elite wheat genotypes identified the genetic basis (explained 7–17% of phenotypic variation) of biomass accumulation and RUE. Indicating, exotic populations are good source of genes for RUE that can be used to improve RUE and yield potentials (Molero et al. 2019). Translocation in 1BL/1RS chromosome in wheat is found to associate with high biomass and RUE (Foulkes et al. 2007). The QTL regions controlling RUE harbored genes coding for transcription factors (e.g., BTF-2 like) and functional proteins (e.g., ELIP, ALDH, CPT1) that regulate photosynthesis, photo protection, and stress response (Haga et al. 2005; Molero et al. 2019). RUE which relates yield to canopy photosynthesis is a relation between accumulation of biomass and interception of light by the crop (Hatfield 2014).

## 8.4 Approaches to Improve RUE

Only currently, breeders have shifted their focus to increase yield by increasing RUE/biomass in cereals and other crops of importance (Furbank et al. 2015). But, can increase in biomass increase yield potential, to meet the growing food demand? In this section, we explore the possibilities of improving RUE by targeting traits related to light interception, its utilization (photosynthetic efficiency and carbon assimilation) and minimizing energy loss for biomass production.

### *8.4.1 Light Harvesting and Electron Transport*

Maximizing light interception by leaves can maximize light harvesting by a canopy. Leaf is the fundamental photosynthetic functional unit that absorbs a great part of solar radiation. Using chlorophyll fluorescence, it is possible to estimate the efficiency of light harvesting. The inability of plants photosynthetic machinery to use all the mid-day high light intensities, strongly reduces the expected productivity. However, plants are efficient in using low light intensities.

Leaf angle, surface properties, thickness, chlorophyll concentration, leaf phyllotaxis, vertical stratification in addition to elevation of sun and distribution of radiation affect the radiation intercepted by a plant. Only 5% of total energy (100%) is efficiently converted to biomass (carbohydrates). Of the total energy, 60% is lost by the nonabsorbed wavelength, 8% by reflection and transmission, 8% is lost as heat, and metabolism losses 19%.

#### 8.4.1.1 Canopy Architecture

Leaf area index (LAI), a unit less measurement, is the leaf area ($m^2$) put forth by a plant per unit land area ($m^2$) it occupies. It is used to estimate the amount of light captured by the plant (Watson 1947). Plants that attain higher LAI early in stage can maximize light capture. So, plants with higher leaf area capture more light. Studies have identified genomic regions and genes controlling leaf size. In addition to small RNAs, candidate genes that code for proteins involved in phytohormone biosynthesis and signaling are shown to play role in regulating leaf size (Karamat et al. 2021).

Light interception also depends on leaf angle. At the canopy level, erectophile leaf canopies are believed to have high assimilation rates under high irradiation (Araus et al. 1993). Breeding for RUE is considerably successful by selecting for erectness in rice and wheat (Foulkes et al. 2009). However, canopy erectness is difficult to quantitatively measure, but digitalization tools have helped phenotype it (Deery et al. 2014). Leaf angle or the bending dynamics of the leaf blade is crucial in determining the plant architecture. Phenotyping has identified existence of genetic variation for erectness. Although only a few genes control leaf angle, the genetic control is found to be complex. Studies in rice, wheat, and other crops have identified genomic regions (QTL) and genes (e.g., D61/OsBRI1, ILI1, LC2, ILA1, RAV6, OsARF19, OsBCL2 and SLG) controlling leaf angle or erectness (Hirano et al. 2017; Liu et al. 2018, 2019a, b; Marone et al. 2020; Jang et al. 2021; Zhi et al. 2022). Phytohormones auxin, GA, and BR are shown to control leaf angle. Gain of function mutants in auxin (*OsGH3-1*: IAA amido synthetase) and brassinosteroid metabolism or signaling have shown to regulate leaf erectness. In addition, small RNAs and transcriptions factors—short vegetative phase group of MADS box, OsbHLH98, and OsWrKY108—also regulate leaf angle (Furbank et al. 2015; Guo et al. 2021; Wang et al. 2020). Plant with erect leaves (knockout lines) had significantly higher yield under dense planting (Wang et al. 2020). Similarly, maize and wheat plants with erect leaves had higher yields (Tian et al. 2019; Liu et al. 2019a, 2019b). Suggesting, manipulation of leaf angle, which is controlled by few genes, will improve RUE.

In addition to erectness, photosynthetic machinery distribution within the canopy regulates light penetration within the canopy. However, breeding or modeling of leaf level biochemistry (photosynthesis) along with parameters of architecture is not advanced (Sheehy and Mitchell 2015). Under high $CO_2$, improving RUE greatly increases the yield potential (Ainsworth and Long 2005).

#### 8.4.1.2 Expanding the Absorption Spectra

In addition, RUE can also be achieved by maximizing the light harvesting. Is it possible to expand the absorption spectra of the leaves to maximize light harvesting? Recent work with incorporation of chlorophylls other than *a* and *b* has provided proof that yes, it is possible to do so. Introduction of alternative pigments can be used to expand the wavelength of light absorbed by plants, optimize light interception and its use (Slattery and Ort 2021). Recent efforts, for example, incorporation of nanotubes or chlorophyll D increased the absorption spectrum (approximately 25 nm) in the near infrared region, resulting in improved light harvesting that is functional and effective (Giraldo et al. 2014; Elias et al. 2021). The Chl *f* pigment, discovered in 2010, was successfully incorporated into PSI complex of *Synechococcus* spp. PCc 7002 (Shen et al. 2019). Similar incorporation in crop plants will expand absorption into the far-red region up to 750–800 nm. Bacterial Chl*b* absorbs beyond 1000 nm which could also be a choice to expand in the far-red region (Blankenship et al. 1995). However, it is necessary to optimize far-red absorption, as its quantum efficiency is low (Kurashov et al. 2019; Tros et al. 2020). This strategy is highly promising for optimizing light usage as light in deeper layers of canopy is enriched in far-red and infrared photons (Ort et al. 2015).

In addition, the purple photosynthetic bacterium's reaction center consisting of bacterial chl*b* along with its cyclic electron system can be used to replace the PSI system. The *Blastochloris viridis* captures light up to 1075 nm using its reaction center-cytochrome $b_6f$ complex. Similarly, chl*a* in PSII can be replaced by chl*d* (cyanobacteria—*Acaryochloris marina*) to absorb light up to 730 nm, and the NADH can be reduced by quinone. If the two photosystems are developed such that they do not compete for similar wavelength of light, using the above-mentioned mechanism, the energetic efficiency would be twice than the current photosynthetic efficiency (Hanna and Nozik 2006; Blankenship et al. 2011).

#### 8.4.1.3 Maximizing Light Harvesting by Reducing Chlorophyll Content

The upper 20% of the canopy absorb 90% of the blue and red spectrum. One option to increase RUE is to reduce the amount of light absorbed under high light, such that the absorbed photons are converted to biomass with reduced or negligible loss of absorbed energy. Regulating the number of light harvesting pigments per reaction center can regulate the amount of light absorbed and or the number of reaction centers per chloroplast (Ort et al. 2011). Reducing chlorophyll is shown to increase RUE in computational studies, and experiment in algae and cyanobacteria displayed similar results (Nakajima and Itayama 2003; Kirst et al. 2014).

Pigment reduction can be achieved by mutation of antenna proteins, light harvesting complexes. Light attenuation is more gradual in chlorophyll-deficient mutants than wild type (Slattery et al. 2016). The photosynthetic and photoprotective capacity of antenna mutants reduced in their chlorophyll content are found to be on par with controls (Bielczynski et al. 2020). Mutants with reduced light antennae1 (tla1) (microalgae) and phycobilisome (cyanobacteria) have improved light penetration, photosynthetic efficiency, and yield (Kirst et al. 2014). Suppression of chlorophyllide-a conversion to chlorophyllide-b in green algae resulted in higher biomass (Negi et al. 2020). Similarly in higher plants like soybean, chlorophyll-deficient mutant (Y11y11) with 50–80% reduced chlorophyll has higher canopy photosynthesis (Slattery et al. 2017; Sakowska et al. 2018). Low chlorophyll rice mutants display higher light penetration to deeper layers of canopy and higher yields compared to wild type (Gu et al. 2017). In most cases, chlorophyll mutants compared to wild-type displayed higher RUE or photosynthetic rate under high light conditions at leaf level. Carboxylation rates of Rubisco and NUE were higher in a few chlorophyll deficient mutants (Gu et al. 2017; Slattery et al. 2017; Sakowska et al. 2018). Nevertheless, severe reduction in chl*b* resulted in inhibition of photosynthesis and photoprotection, indicating chl*a/b* ratio is crucial (Kim et al. 2009). Sometimes, mutations resulting in low chlorophyll display certain disadvantages. For example, low chlorophyll resulting from mutation in magnesium chelatase (MinnGold) is also known to regulate ABA signaling and stomatal movement, which could negatively affect yield under water limited condition (Slattery et al. 2017).

## 8.4.2 Improving Photosynthetic Efficiency

Generally, higher yield is associated with higher RUE, which is dependent on higher interception of solar radiation and photosynthetic efficiency (Liu et al. 2019a, b). Recent efforts are toward increasing RUE by targeting photosynthetic rate. Although high-throughput direct measurements of carbon fixation is difficult, advances in machine learning and optical sensing systems using spectral reflectance measurements (visible and NIR) have made it affordable to estimate traits (leaf N, phosphorous, mass per area, A-assimilation rate, J-potential electron transport rate, and $V_{cmax}$-Rubisco capacity) related to photosynthesis (Silva-Pérez et al. 2020; Furbank et al. 2020). Nevertheless, high-throughput estimation of photosynthetic electron transport and photosynthetic efficiency at leaf level is possible using chlorophyll fluorescence techniques (Murchie and Lawson 2013). Commercially available pulse amplitude-modulated chlorophyll fluorescence (PAM) systems estimates the light harvesting efficiency of PSII by calculating its photosynthetic electron transport rate and nonphotochemical quenching (NPQ) (Furbank et al. 2020).

#### 8.4.2.1 Reducing Energy Loss Via Nonphotochemical Quenching

Plants have evolved to be highly absorptive to capture low light occurring at dawn, dusk, or cloudy days. However, they capture more light in full sunlight, which is not used productively for photosynthesis, reducing solar conversion efficiencies. The excess energy should be carefully dissipated to avoid photooxidation. Thermal dissipation is the most commonly employed mechanism for safe disposal of excess energy, resulting in reduction in efficiency (Blankenship et al. 2011; Murchie and Niyogi 2011).

Various processes of energy dissipation like photoprotection of PSII from overexcitation under high light results in reduction of maximum photosynthetic quantum efficiency (Osmond and Forster 2008). Improvement in RUE can be achieved by optimizing light harvesting and photoprotection losses (Murchie and Niyogi 2011; Reynolds et al. 2012). Xanthophyll pigments in the thylakoid lumen play a major role in nonphotochemical energy dissipation. Zeaxanthin dissipates excess energy as heat, thereby reducing the efficiency of PSII. The level of zeaxanthin depends on conversion of violaxanthin (inactive in quenching) to zeaxanthin (active in quenching) by the enzyme violaxanthin de-epoxidase (*npq1*) and vice versa by zeaxanthin epoxidase (*npq2*). Mutants lacking *npq1* are unable to photoprotect themselves. Mutants of *npq2* (allelic to *aba1*) overproduce zeaxanthin, but were also found to be deficient in ABA (Niyogi et al. 1998). Benefits can be achieved by the precise manipulation of these genes in upper and lower leaves to maximize RUE. Enhanced thylakoid photoprotection by overexpression of PsbS, a central regulator of NPQ is found to increase RUE and grain yield. Indicating photoprotection can be a target for improving RUE. However, it is necessary to understand the tradeoffs with $CO_2$ assimilation (Hubbart et al. 2018). In addition, the mechanism of energy dissipation as heat by carotenoids is not clear. Recently, attempts have been made to understand the mechanism of chlorophyll to carotenoid energy transfer using nanodisc, comprising individual LHCII in a membrane disc, and ultrabroadband two dimensional electronic spectroscopy (Son et al. 2020).

#### 8.4.2.2 Engineering Rubisco and Reducing Energy Loss Via Photorespiration

Rubisco is a bi-functional enzyme that fixes carbon dioxide or oxygen to substrate RuBP. The kinetic properties of Rubisco and its activation state determine the rate of photosynthesis under saturating light. Rubisco is regarded as an inefficient catalyst that has poor specificity for $CO_2$. To sustain photosynthetic rates, plant invest large portion of its nitrogen (25% of leaf N) to synthesize high amounts of Rubisco. Therefore, increasing Rubisco further would increase the photosynthetic capacity under high light conditions, but it would reduce the NUE (Parry et al. 2007). There is a need to regulate the levels of Rubisco in different parts of the canopy (high levels in upper canopy and low levels in lower canopy) to increase RUE and NUE (Yabuta et al. 2008).

Carbon fixed by photosynthesis along with nitrogen is lost by photorespiration (oxygenation by Rubisco). C3 crops lose approximately one third of energy captured for photorespiration that results in reduced RUE. These losses are much higher under drought and high temperatures (Parry et al. 2011). However, substantial diversity exits for Rubisco in terrestrial C3, C4 plants and algae. Improving the Rubisco's catalytic parameters (Km for $CO_2$, carboxylation speed, and substrate specificity—$CO_2/O_2$) will improve photosynthetic efficiency (Orr et al. 2016; Heureux et al. 2017). Even with the well-resolved crystal structure and understanding of Rubisco, we have failed to improve its kinetic and catalytic (activity and substrate specificity) activity (Parry et al. 2007; Peterhansel et al. 2013).

Researchers were unsuccessful in eliminating or decreasing the oxygenation activity of Rubisco by site-directed mutagenesis. Nevertheless, mutated forms of Rubisco or introduction of Rubisco or its subunits from $C_4$ plants have been successful in reducing oxygenation. However, engineering Rubisco is complicated as the large subunit, and small subunits parts of it are encoded by chloroplast and nuclear genome, respectively (Ishikawa et al. 2011). Nevertheless, engineering Rubisco or its kinetics in $C_3$ would considerably increase RUE and improve photosynthesis (Parry et al. 2011). Decreasing photorespiration and introducing photorespiratory bypass are predicted to increase RUE (Kebeish et al. 2007). Some of the bypass mechanisms suggested are introduction of enzymes that can recycle photorespiratory products without releasing ammonia or release $CO_2$ in chloroplast instead of peroxisome to be fixed by Rubisco (Peterhansel et al. 2013; Carvalho et al. 2011).

Alternatively, reduction in photorespiration can be achieved by the introduction of synthetic pathways to metabolize glycolate, product of photorespiration. So far, three such pathways that metabolize glycolate are examined (Peterhansel et al. 2013; Xin et al. 2015). Few options are conversion of glycolate to glycerate, recycle glycolate to pyruvate without loss of $CO_2$ as occurs in few bacteria. Other options are introducing novel pathways to reduce phosphoglycolate to phosphoglycolaldehyde that can form xylulosebisphosphate on combining with dihydroxyacetonephosphate. The xylose bisphosphate formed can then be dephosphorylated to form xylose-5-phosophate, an intermediate of carbon reduction cycle (Shih et al. 2014). Bypassing Rubisco for carbon fixation by using oxygen insensitive carbon fixation could be another option; for example, using the 3-hydroxypropionate/4-hydroxybutryate or the 4-hydroxybutryate/dicarboxylate pathways that occur in aerobic and anerobic Archaea, respectively (Huber et al. 2008).

#### 8.4.2.3 Increasing $CO_2$ Uptake and Capture (CCM)

High concentrations of $CO_2$ around Rubisco decrease photorespiration and increase the rate of carboxylation, as Rubisco operates closer to its substrate ($CO_2$) saturation. The approaches to increase $CO_2$ uptake and capture are as follows.

##### 8.4.2.3.1 Expression of Algal or Photosynthetic Bacterial $CO_2$ and Bicarbonate Transporters in Chloroplast Membranes of C3

Mechanisms like presence of membrane localized inorganic carbon transporters (bicarbonate transporters) and NDH1-based $CO_2$ uptake systems in single-celled organisms of cyanobacteria and micro-algae have been well characterized. Bicarbonate transporters are either made up of single subunit (BicA and SbtA) or multisubunit (BCT1 and NDH1-based $CO_2$ uptake system) in single-celled organisms. Incorporation of these genes, especially the single subunit bicarbonate transporters that are simple to do, has not resulted in enhanced photosynthesis as expected (Price et al. 2013). Nevertheless, diffusion property of $CO_2$ across various plant cell membranes is not fully understood (Evans et al. 2009; Kaldenhoff 2012).

Modeling has shown that introduction of CCM would be energetically beneficially to crops. CCM would result in 60% increase in yield due to 60% increase in photosynthetic rate (Price et al. 2013). $CO_2$ concentrating mechanism (CCM) absent in $C_3$ plants has evolved multiple times in higher plants ($C_4$) and single-celled organisms (Price et al. 2013; von Caemmerer et al. 2012).

##### 8.4.2.3.2 Install the C4 Pathway in C3 Plants

Biochemical CCM in $C_4$ has evolved more than 60 times in a number of monocot and dicot plants. The pathway is usually a mixture of biochemical and anatomical specializations (Sage et al. 2012). Instead of using the membrane-based transporters or pumps, $C_4$ plants use PEP carboxylase to fix carbon as a 4-carbon compound (oxaloacetate) that is transported to specialized cell housing Rubisco. PEP carboxylase is also known as biochemical $CO_2$ pump. Rubisco present in chloroplast of bundle sheath cells fix the enzymatically released $CO_2$ from malate via calvin cycle similar to $C_3$ plants to form 3-phosohoglycerate. The increase in $CO_2$ in these specialized cells can reach up to 10-folds higher than the atmospheric concentration that results in almost elimination of photorespiration, thereby increasing RUE (Sheehy and Mitchell 2015). At present, attempts have been made to incorporate $C_3$ plants with $C_4$ pathways—with and without karnz anatomy in rice, which is funded by Bill and Melinda Gates foundation. Progress has been made in identifying the necessary genes of $C_4$ anatomy, NADPME-type $C_4$ mechanism, promoters, and factors (*cis* and *trans*) regulating cell-specific expression of genes in bundle sheath/ mesophyll cells (von Caemmerer et al. 2012; Furbank et al. 2015). Efforts to introduce CCM occurring in C4 plants into rice have given exciting results (von Caemmerer et al. 2012). Efforts are to understand $CO_2$ diffusion mechanisms across bundle sheath cells.

##### 8.4.2.3.3 Install Cyanobacterial or Algal CCM Characterized by Bicarbonate Transporters and Occurrence of Carboxysome or Pyrenoid-like Compartments That Contain Rubisco

Introduction of the well-characterized carboxysome or pyrenoid-like structures of single-celled organisms to harbor Rubisco in higher plants ($C_3$) could be beneficial in achieving CCM (Price et al. 2013). It is a challenging task to assemble proteins (>7) that form carboxysome structure. Aggregation of Rubisco in carboxysome-like structures in chloroplast was observed in plants expressing cyanobacterial carboxysome and Rubisco (Lin et al. 2014).

#### 8.4.2.4 Modulating RuBP Regeneration

Plants overexpressing sedoheptulose-1,7-bisphosphate or fructose 1,6-bisphosphate aldolase, the key rate limiting enzymes in regeneration of RuBP, displayed higher photosynthetic rate and biomass, under optimum conditions (Tamoi et al. 2006).

## 8.5 Future Prospect: Designing Smart Canopy

The angle of the sun, direction of planting, and plant characteristics (canopy structure, light absorption, and penetration) affect light interception by the canopy. The amount and quality of light, wind speed, and humidity vary within the canopy and with time of day (Li et al. 2021). Therefore, it is essential to construct efficient canopy structure by optimizing plant density to maximize light interception. An ideal LAI can make maximum use of PAR to increase biomass. In addition, it is also essential to maximize light distribution in the canopy at different leaf layers to improve RUE, biomass, and yield. Under optimum conditions, yield is directly proportional to the amount of light intercepted. Low RUE is an indication of lower efficiency of biomass production of the light intercepted. Hence, RUE increases as light interception increases at optimum conditions (Tian et al. 2020; Li et al. 2021).

We would have a smart canopy when the leaves in the canopy assemble themselves in a cooperative manner to maximize its light harvesting and biomass, thereby maximizing its RUE (Ort et al. 2015). Based on the amount of light received by the different layers of the canopy, the following modulations within the canopy can produce a smart canopy to maximize RUE (Fig. 8.2). A smart canopy must be able to regulate its leaf angle according the amount and quality of light it receives. The upper canopy receiving high light must have vertical leaves (acute leaf angle) while

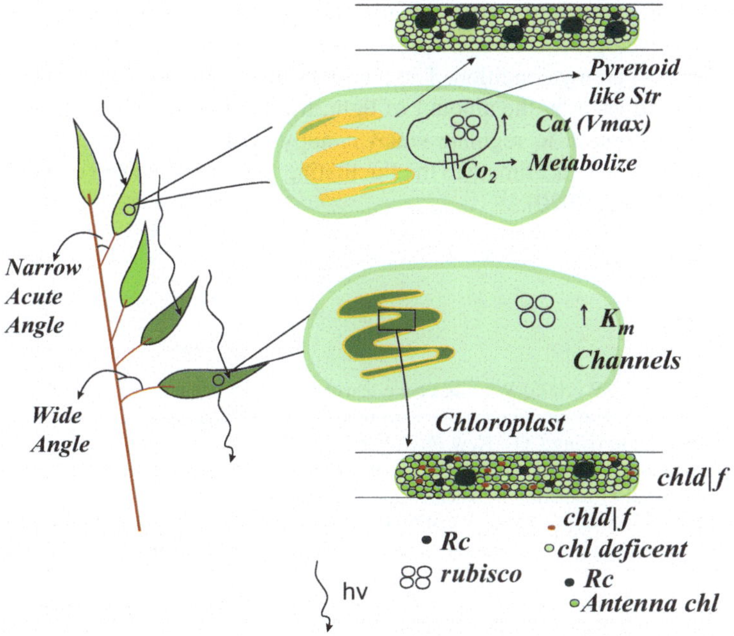


**Fig. 8.2** Model representing the regulation of leaf angle, pigment content, light harvesting, and carbon fixation that can lead to smart canopies with improved RUE

lower (deeper) canopy receiving low light (low R/FR ratio) must have horizontal leaves (right angle). Targeted manipulation of gene controlling leaf angle could be used to achieve smart canopy (Johnston et al. 2014). Similarly, Rubisco in the upper canopy must have high catalytic rate while in deeper canopy it must have high specificity, minimizing photorespiratory losses in deeper canopy. The upper canopy that receives high light should have small antenna systems that can feed many reaction centers while the deeper canopy should have larger antenna system with modified pigment composition that feed fewer reaction centers (Fig. 8.2). The modified pigments and larger antenna maximize light capture under low light conditions (Gilbert et al. 2001). In addition, reduction of chlorophyll content in upper canopy leaves can increase photosynthetic rate by decreasing photorespiratory losses and improve light distribution across canopy (Ort et al. 2011). However, lower chlorophyll under low light is responsible for reduced photosynthesis (Wu et al. 2021). Hence, it is important to regulate pigment content, such that reduction in chlorophyll is only restricted to upper canopy receiving high sunlight. Different pigments with absorption in different spectral range can be used to maximize light harvesting. In addition, smart canopies should have different photosynthetic machineries at different layers of the canopy. With the CCM mechanism present in upper canopy as it requires additional ATP while the $C_3$ pathway present in deeper canopy (Fig. 8.2).

Nevertheless, there is a need to identify molecular switches that can switch the system with development, to operate as upper canopy initially and with growth to switch to deeper canopy condition. The phytochromes sense the red to far-red ratios, which is least affected by the amount of light. Modulating the phytochrome signaling is one such option to regulate smart canopy (Ort et al. 2015). However, the canopy must be able to acclimate to the fluctuating light it receives to maximize RUE (Morales and Kaiser 2020).

## References

Ainsworth EA, Long SP (2005) What have we learned from 15 years of free-air $CO_2$ enrichment (FACE)? A meta-analytic review of the responses of photosynthesis, canopy properties and plant production to rising $CO_2$. New Phytol 165:351–371

Araus JL, Reynolds MP, Acevedo E (1993) Leaf posture, grain yield, growth, leaf structure and carbon isotope discrimination in wheat. Crop Sci 33:1273–1279

Bielczynski LW, Schansker G, Croce R (2020) Consequences of the reduction of the photosystem II antenna size on the light acclimation capacity of Arabidopsis thaliana. Plant Cell Environ 43: 866–879

Blankenship RE, Madigan MT, Bauer CE (eds) (1995) Anoxygenic photosynthetic bacteria. Kluwer Academic Publishers, Dordrecht

Blankenship RE, Tiede DM, Barber J, Brudvig GW, Fleming G, Ghirardi M, Gunner MR, Junge W, Kramer DM, Melis A, Moore TA, Moser CC, Nocera DG, Nozik AJ, Ort DR, Parson WW, Prince RC, Sayre RT (2011) Comparing photosynthetic and photovoltaic efficiencies and recognizing the potential for improvement. Science 6031:805–809

Carvalho J d FC, Madgwick PJ, Powers SJ, Keys AJ, Lea PJ, Parry MAJ (2011) An engineered pathway for glyoxylate metabolism in tobacco plants aimed to avoid the release of ammonia in photorespiration. BMC Biotechnol 11:111

Crist E, Mora C, Engelman R (2017) The interaction of human population, food production, and biodiversity protection. Science 356:260–264

de Groot H (2008) Harnessing solar energy for the production of clean fuels. In: van Keulen H, van Laar HH, Rabbinge R (eds) 40 years theory and model at Wageningen UR—on the occasion of the 40th anniversary of the inaugural address of C.T. de Wit in 1968. Wageningen University and Research Centre, pp 5–7

Deery D, Jimenez-Berni J, Jones H, Sirault X, Furbank R (2014) Proximal remote sensing buggies and potential applications for field-based phenotyping. Agronomy 4:349–379

Elias E, Liguori N, Saga Y, Schäfers J, Croce R (2021) Harvesting far-red light with plant antenna complexes incorporating chlorophyll d. Biomacromolecules 22(8):3313–3322

Evans JR, Kaldenhoff R, Genty B, Terashima I (2009) Resistances along the $CO_2$ diffusion pathway inside leaves. J Exp Bot 60:2235–2248

FAO (2009) How to feed the world: global agriculture towards 2050. www.fao.org/fileadmin/templates/wsfs/docs/Issues_papers/HLEF2050_Global_Agriculture.pdf

FAO (2013) FAO STAT. Food and Agriculture Organization of the United Nations, Rome

Foulkes MJ, Snape JW, Shearman VJ, Reynolds MP, Gaju O, Sylvester-Bradley R (2007) Genetic progress in yield potential in wheat: recent advances and future prospects. J Agric Sci 145:17–29

Foulkes MJ, Reynolds M, Sylvester-Bradley R (2009) Genetic improvement of grain crops: yield potential. In: Sadras VO, Calderini D (eds) Crop physiology applications for genetic improvement and agronomy. Elsevier, pp 235–256

Foulkes MJ, Slafer GA, Davies WJ, Berry PM, Sylvester-Bradley R, Martre P, Calderini DF, Griffiths S, Reynolds MP (2011) Raising yield potential of wheat. III. Optimizing partitioning to grain while maintaining lodging resistance. J Exp Bot 62:469–486

Furbank RT, Quick WP, Sirault X (2015) Improving photosynthesis and yield potential in cereal crops by targeted genetic manipulation: prospects, progress and challenges. Field Crop Res 182: 19–29

Furbank RT, Sharwood R, Estavillo GM, Silva-Perez V, Condon AG (2020) Photons to food: genetic improvement of cereal crop photosynthesis. J Exp Bot 71(7):2226–2238

Gilbert IR, Jarvis PG, Smith H (2001) Proximity signal and shade avoidance differences between early and late successional trees. Nature 411(6839):792–795

Giraldo JP, Landry MP, Faltermeier SM, Mc Nicholas TP, Everson NM, Boghossian AA, Reuel NF, Hilmer AJ, Sen F, Brew JA, Strano MS (2014) Plant nanobionics approach to augment photosynthesis and biochemical sensing. Nat Mater 13(4):400–408

Gu J, Zhou Z, Li Z, Chen Y, Wang Z, Zhang H (2017) Rice (*Oryza sativa* L.) with reduced chlorophyll content exhibit higher photosynthetic rate and efficiency, improved canopy light distribution, and greater yields than normally pigmented plants. F Crop Res 200:58–70

Guo JF, Li W, Shang LG, Wang YG, Yan P, Bai YH, Da XW, Wang K, Guo Q, Jiang RR et al (2021) *OsbHLH98* regulates leaf angle in rice through transcriptional repression of *OsBUL1*. New Phytol 230:1953–1966

Habiba, Xu J, Gad AG, Luo Y, Fan C, Uddin JBG, Ul Ain N, Huang C, Zhang Y, Miao Y, Zheng X (2021) Five OsS40 family members are identified as senescence-related genes in rice by reverse genetics approach. Front Plant Sci 12:701529

Haga K, Takano M, Neumann R, Iino M (2005) The Rice Coleoptile Phototropism1 gene encoding an ortholog of Arabidopsis NPH3 is required for phototropism of coleoptiles and lateral translocation of auxin. Plant Cell 17(1):103–115

Hammer GL, Wright GC (1994) A theoretical analysis of nitrogen and radiation effects on radiation use efficiency in peanut. Aust J Agric Res 45:575–589

Hanna MC, Nozik AJ (2006) Solar conversion efficiency of photovoltaic and photoelectrolysis cells with carrier multiplication absorbers. J Appl Phys 100:074510

Haro RJ, Otegui ME, Collino DJ, Dardanelli JL (2007) Environmental effects of seed yield determination of irrigated peanut crops: links with radiation use efficiency and crop growth rate. Field Crops Res 103:217–228

Hatfield J (2014) Radiation use efficiency: evaluation of cropping and management systems. Agron J 106:1820

Heureux AMC, Young JN, Whitney SM, Eason-Hubbard MR, Lee RBY, Sharwood RE, Rickaby REM (2017) The role of Rubisco kinetics and pyrenoid morphology in shaping the CCM of haptophyte microalgae. J Exp Bot 68:3959–3969

Hirano K, Yoshida H, Aya K, Kawamura M, Hayashi M, Hobo T, Sato-Izawa K, Kitano H, Ueguchi-Tanaka M, Matsuoka M et al (2017) SMALL ORGAN SIZE 1 and SMALL ORGAN SIZE 2/DWARF and LOW-TILLERING form a complex to integrate auxin and brassinosteroid signalling in rice. Mol Plant 10:590–604

Hochman Z, Gobbett DL, Horan H (2017) Climate trends account for stalled wheat yields in Australia since 1990. Glob Chang Biol 23:2071–2081

Hubbart S, Smillie IR, Heatley M, Swarup R, Foo CC, Zhao L, Murchie EH (2018) Enhanced thylakoid photoprotection can increase yield and canopy radiation use efficiency in rice. Commun Biol 1:22

Huber H, Gallenberger M, Jahn U, Eylert E, Berg IA, Kockelkorn D, Eisenreich W, Fuchs G (2008) A dicarboxylate/4-hydroxybutyrate autotrophic carbon assimilation cycle in the hyperthermophilicArchaeumIgnicoccushospitalis. Proc Natl Acad Sci U S A 105(22): 7851–7856

Ishikawa C, Hatanaka T, Misoo S, Miyake C, Fukayama H (2011) Functional incorporation of sorghum small subunit increases the catalytic turnover rate of Rubisco in transgenic rice. Plant Physiol 156:1603–1611

Jang S, Cho J-Y, Do G-R, Kang Y, Li H-Y, Song J, Kim H-Y, Kim B-G, Hsing Y-I (2021) Modulation of rice leaf angle and grain size by expressing OsBCL1 and OsBCL2 under the control of OsBUL1 promoter. Int J Mol Sci 22(15):7792. https://doi.org/10.3390/ijms22157792

Johnston R, Wang M, Sun Q, Sylvester AW, Hake S, Scanlon MJ (2014) Transcriptomic analyses indicate that maize ligule development recapitulates gene expression patterns that occur during lateral organ initiation. Plant Cell 26(12):4718–4732

Kaldenhoff R (2012) Mechanisms underlying $CO_2$ diffusion in leaves. Curr Opin Plant Biol 15(3): 276–281

Karamat U, Sun X, Li N, Zhao J (2021) Genetic regulators of leaf size in Brassica crops. Hort Res 8: 91

Kebeish R, Niessen M, Thiruveedhi K, Bari R, Hirsch HJ, Rosenkranz R, Stäbler N, Schönfeld B, Kreuzaler F, Peterhänsel C (2007) Chloroplastic photorespiratory bypass increases photosynthesis and biomass production in Arabidopsis thaliana. Nat Biotechnol 25:593–599

Kemanian AR, Stockle CO, Huggins D (2004) Variability of barley radiation-use efficiency. Crop Sci 44:1662–1672

Kim EH, Li XP, Razeghifard R, Anderson JM, Niyogi KK, Pogson BJ, Chow WS (2009) The multiple roles of light-harvesting chlorophyll a/b-protein complexes define structure and optimize function of Arabidopsis chloroplasts: a study using two chlorophyll b-less mutants. Biochim Biophys Acta 1787:973–984

Kirst H, Formighieri C, Melis A (2014) Maximizing photosynthetic efficiency and culture productivity in cyanobacteria upon minimizing the phycobilisome light-harvesting antenna size. Biochim Biophys Acta 1837:1653–1664

Kromdijk J, Long SP (2016) One crop breeding cycle from starvation? How engineering crop photosynthesis for rising $CO_2$ and temperature could be one important route to alleviation. Proc R Soc B 283:20152578. https://doi.org/10.1098/rspb

Kurashov V, Yang M, Shen G, Piedl K, Laremore TN, Bryant DA, Golbeck JH (2019) Energy transfer from chlorophyll f to the trapping center in naturally occurring and engineered Photosystem I complexes. Photosynth Res 141:151–163

Li M, Liu Y, Ma J, Zhang P, Wang C, Su J, Yang D (2020) Genetic dissection of stem WSC accumulation and remobilization in wheat (*Triticum aestivum* L.) under terminal drought stress. BMC Genet 21:50. https://doi.org/10.1186/s12863-020-00855-1

Li R, Zhang G, Xie R, Hou P, Ming B, Xue J, Wang K, Li S (2021) Optimizing row spacing increased radiation use efficiency and yield of maize. Agron J 113:4806–4818

Lin MT, Occhialini A, Andralojc JP, Devonshire J, Hines K, Parry MAJ, Hanson MR (2014) β-carboxysomal proteins assemble into highly organized structures in Nicotiana chloroplasts. Plant J 79:1–12

Liu X, Yang CY, Miao R, Zhou CL, Cao PH, Lan J, Zhu XJ, Mou CL, Huang YS, Liu SJ, Tian YL, Nguyen TL, Jiang L, Wan JM (2018) DS1/OsEMF1 interacts with OsARF11 to control rice architecture by regulation of brassinosteroid signaling. Rice 11(1):46

Liu K, Yang R, Lu J, Wang X, Lu B, Tian X, Zhang Y (2019a) Radiation use efficiency and source-sink changes of super hybrid rice under shade stress during grain-filling stage. Agron J 111(4): 1788–1798

Liu K, Cao J, Yu K, Liu X, Gao Y, Chen Q, Zhan W, Peng H, Du J, Xin M (2019b) Wheat TaSPL8 modulates leaf angle through auxin and brassinosteroid signaling. Plant Physiol 181:179–194

Long SP (2014) We need winners in the race to increase photosynthesis in rice, whether from conventional breeding, biotechnology or both. Plant Cell Environ 37:19–21

Long SP, Zhu X-G, Naidu SL, Ort DR (2006) Can improvement in photosynthesis increase crop yields? Plant Cell Environ 29:315–330

Marone D, Rodriguez M, Saia S, Papa R, Rau D, Pecorella I, Laidò G, Pecchioni N, Lafferty J, Rapp M, Longin FH, De Vita P (2020) Genome-wide association mapping of prostrate/erect growth habit in winter durum wheat. Int J Mol Sci 21(2):394

Molero G, Joynson R, Pinera-Chavez FJ, Gardiner LJ, Rivera-Amado C, Hall A, Reynolds MP (2019) Elucidating the genetic basis of biomass accumulation and radiation use efficiency in spring wheat and its role in yield potential. Plant Biotechnol J 17:1276–1288

Monteith JL (1977) Climate and the efficiency of crop production in Britain. Philos Trans R Soc Lond B Biol Sci 281(980):77–294

Morales A, Kaiser E (2020) Photosynthetic acclimation to fluctuating irradiance in plants. Front Plant Sci 11:268. https://doi.org/10.3389/fpls.2020.00268

Munaiz ED, Martínez S, Kumar A, Caicedo M, Ordás B (2020) The senescence (stay-green)—an important trait to exploit crop residuals for bioenergy. Energies 13:790. https://doi.org/10.3390/en13040790

Murchie EH, Lawson T (2013) Chlorophyll fluorescence analysis: a guide to good practice and understanding some new applications. J Exp Bot 64:3983–3998

Murchie EH, Niyogi KK (2011) Manipulation of photoprotection to improve plant photosynthesis. Plant Physiol 155(1):86–92

Nakajima Y, Itayama T (2003) Analysis of photosynthetic productivity of microalgal mass cultures. J Appl Phycol 15:497–505

Negi S, Perrine Z, Friedland N, Kumar A, Tokutsu R, Minagawa J, Berg H, Barry AN, Govindjee G, Sayre R (2020) Light regulation of light-harvesting antenna size substantially enhances photosynthetic efficiency and biomass yield in green algae. Plant J 103:584–603

Niyogi KK, Grossman AR, Bjorkman O (1998) Arabidopsis mutants define a cen- tral role for the xanthophylls cycle in the regulation of photosynthetic energy conversion. Plant Cell 10:1121–1134

Orr DJ, Alcântara A, Kapralov MV, Andralojc PJ, Carmo-Silva E, Parry MA (2016) Surveying Rubisco diversity and temperature response to improve crop photosynthetic efficiency. Plant Physiol 172:707–717

Ort DR, Zhu XG, Melis A (2011) Optimizing antenna size to maximize photosynthetic efficiency. Plant Physiol 155(1):79–85

Ort DR, Merchant SS, Alric J, Barkan A, Blankenship RE, Bock R, Croce R, Hanson MR, Hibberd JM, Long SP, Moore TA, Moroney JV, Niyogi KK, Parry MA, Peralta-Yahya P, Prince RC, Redding KE, Spalding MH, van Wijk KJ, Vermaas WF, von Caemmerer S, Weber AP, Yeates TO, Yuan JS, Zhu X (2015) Redesigning photosynthesis to sustainably meet global food and bioenergy demand. Proc Natl Acad Sci 112:8529–8536

Osmond B, Forster B (2008) Photoinhibition: then and now. In: Demmig-Adams B, Adams WW III, Mattoo AK (eds) Photoprotection, photoinhibition, gene regulation and environment. Springer, Dordrecht, pp 11–22

Parry MAJ, Madgwick PJ, Carvahlo JFC, Andralojc PJ (2007) Prospects for increasing photosynthesis by overcoming the limitations of rubisco. J Agric Sci 145:31–43

Parry MAJ, Reynolds M, Salvucci ME, Raines C, Andralojc PJ, Zhu XG, Price GD, Condon AG, Furbank RT (2011) Raising yield potential of wheat. II. Increasing photosynthetic capacity and efficiency. J Exp Bot 62:453–467

Peterhansel C, Blume C, Offermann S (2013) Photorespiratory bypasses: how can they work? J Exp Bot 64(3):709–715

Price GD, Pengelly JJL, Forster B, Du J, Whitney S, von Caemmerer S, Badger MR, Howitt SM, Evans JR (2013) The cyanobacterial CCM as a source of genes for improving photosynthetic $CO_2$ fixation in crop species. J Exp Bot 64(3):753–768

Ray DK, Mueller ND, West PC, Foley JA (2013) Yield trends are insufficient to double global crop production by 2050. PLoS One 8:e66428

Reynolds MP, Pfeiffer WH (2000) Applying physiological strategies to improve yield potential. In: Royo C, Nachit M, Di Fonzo N, Araus JL (eds) Durum wheat improvement in the Mediterranean region: new challenges. CIHEAM, Zaragoza, pp 95–103

Reynolds M, Foulkes J, Furbank R, Griffiths S, King J, Murchie E, Parry M, Slafer G (2012) Achieving yield gains in wheat. Plant Cell Environ 35:1799–1823

Sage RF, Sage TL, Kocacinar F (2012) Photorespiration and the evolution of $C_4$ photosynthesis. Annu Rev Plant Biol 63:19–47

Sakowska K, Rodeghiero M, Alberti G, Colombo R, Celesti M, Mevy JP, Benjamin W (2018) Leaf and canopy photosynthesis of a chlorophyll deficient soybean mutant. Plant Cell Environ 41(6): 1427–1437

Sheehy JE, Mitchell PL (2015) Calculating maximum theoretical yield in rice. Crop Past Res 182: 68–75

Shen G, Canniffe DP, Yang M, Vasily H, Art K, John VDE (2019) Characterization of chlorophyll f synthase heterologously produced in Synechococcussp. PCC 7002. Photosynth Res 140:77–92

Shih PM, Zarzycki J, Niyogi KK, Kerfeld CA (2014) Introduction of a synthetic $CO_2$-fixing photorespiratory bypass into a cyanobacterium. J Biol Chem 289(14):9493–9500

Shulski MD, Walter-Shea EA, Hubbard KG, Yuen GY, Horst G (2004) Penetration of photosynthetically active and ultraviolet radiation into alfalfa and tall fescue canopies. Agron J 96:1562–1571

Silva-Pérez V, De Faveri J, Molero G, Deery DM, Condon AG, Reynolds MP, Evans JR, Furbank RT (2020) Genetic variation for photosynthetic capacity and efficiency in spring wheat. J Exp Bot 71:2299–2311

Sinclair TR, Muchow RC (1999) Radiation use efficiency. Adv Agron 65:215–265

Singh UM, Sinha P, Dixit S, Abbai R, Venkateshwarlu C, Chitikineni A (2020) Unraveling candidate genomic regions responsible for delayed leaf senescence in rice. PLoS One 15(10): e0240591

Slattery RA, Ort DR (2021) Perspectives on improving light distribution and light use efficiency in crop canopies. Plant Physiol 185(1):34–48

Slattery RA, Grennan AK, Sivaguru M, Sozzani R, Ort DR (2016) Light sheet microscopy reveals more gradual light attenuation in light-green versus dark-green soybean leaves. J Exp Bot 67: 4697–4709

Slattery RA, Vanloocke A, Bernacchi CJ, Zhu XG, Ort DR (2017) Photosynthesis, light use efficiency, and yield of reduced-chlorophyll soybean mutants in field conditions. Front Plant Sci 8:1–19

Son M, Pinnola A, Gordon SC (2020) Observation of dissipative chlorophyll-to-carotenoid energy transfer in light-harvesting complex II in membrane nanodiscs. Nat Commun 11:1295

Stöckle CO, Kemanian AR (2009) Crop radiation capture and use efficiency: a framework for crop growth analysis. In: Sadras V, Calderini D (eds) Crop physiology. Academic Press, Elsevier Inc, pp 145–170

Tamoi M, Nagaoka M, Miyagawa Y, Shigeoka S (2006) Contribution of fructose-1,6-bisphosphatase and sedoheptulose-1,7-bisphosphatase to the photosynthetic rate and carbon flow in the Calvin cycle in transgenic plants. Plant Cell Physiol 47:380–390

Thomas H, Ougham H (2014) The stay-green trait. J Exp Bot 65(14):3889–3900

Tian J, Wang C, Xia J, Wu L, Xu G, Wu W, Li D, Qin W, Han X, Chen Q (2019) Teosinte ligule allele narrows plant architecture and enhances high-density maize yields. Science 365:658–664

Tian BJ, Zhu JC, Liu XW, Huang SB, Wang P (2020) Interacting leaf dynamics and environment to optimize maize sowing date in North China plain. J Integr Agric 19:1227–1240

Tros M, Bersanini L, Shen G, Ho MY, van Stokkum IHM, Bryant DA, Croce R (2020) Harvesting far-red light: functional integration of chlorophyll f into photosystem I complexes of Synechococcus sp. PCC 7002. Biochim Biophys Acta 1861:148206

von Caemmerer S, Quick WP, Furbank RT (2012) The development of $C_4$ rice: current progress and future challenges. Science 336:1671–1672

Wang R, Liu C, Li Q, Chen Z, Sun S, Wang X (2020) Spatiotemporal resolved leaf angle establishment improves rice grain yield via controlling population density. Science 23(9): 101489

Watson DJ (1947) Comparative physiological studies on the growth of field crops: I. variation in net assimilation rate and leaf area between species and varieties, and with and between years. Ann Bot 11:41–76

Wu X, Khan R, Gao H, Liu H, Zhang J, Ma X (2021) Low light alters the photosynthesis process in cigar tobacco via modulation of the chlorophyll content, chlorophyll fluorescence, and gene expression. Agriculture 11:755. https://doi.org/10.3390/agriculture11080755
Xin CP, Tholen D, Devloo V, Zhu XG (2015) The benefits of photorespiratory bypasses: how can they work? Plant Physiol 167(2):574–585
Yabuta Y, Tamoi M, Yamamoto K, Tomizawa K-I, Yokota A, Shigeoka S (2008) Molecular design of photosynthesis-elevated chloroplasts for mass accumula- tion of a foreign protein. Plant Cell Physiol 49:375–385
Yin X, Struik PC (2015) Constraints to the potential efficiency of converting solar radiation into phytoenergy in annual crops: from leaf biochemistry to canopy physiology and crop ecology. J Exp Bot 66(21):6535–6549
Zhi X, Tao Y, Jordan D, Borrell A, Hunt C, Cruickshank A, Potgieter A, Wu A, Hammer G, George-Jaeggli B, Mace E (2022) Genetic control of leaf angle in sorghum and its effect on light interception. J Exp Bot 73(3):801–816

# Chapter 9
# Application of Stable Isotopes in Crop Improvement

**S. Nadaradjan, R. Rajakumar, B. Aravind, and P. Boominathan**

**Abstract** Stable isotopes have always been useful to mankind in several archaeological explorations. Understanding the presence of these natural variants of elements has enabled us in using them in understanding several biological processes in several organisms. Apart from exploration of the presence of such isotopes in elucidating the possible life processes that has occurred, a little manipulation has been made to understand certain physiological processes as well. Such techniques have enabled plant scientists in understanding several plant physiological aspects, viz., water use efficiency (WUE), mean transpiration rate (MTR) etc. Estimating WUE and MTR has been very successful in crops like wheat and rice. This chapter compiles the exclusive information on the importance of stable isotopes, the different available forms of stable isotopes, their application in crop improvement programmes for estimating different physiological processes. Different ways for measuring stable isotopes and the success stories of their utilization in different crops for measuring different traits.

**Keywords** Stable isotopes · Water use efficiency · Mean transpiration rate · Carbon isotope discrimination · Isotopes of oxygen · Isotopes of carbon

S. Nadaradjan (✉) · R. Rajakumar
Pandit Jawaharlal Nehru College of Agricultural College and Research Institute, Karaikal, Puducherry, India

B. Aravind
Birla Institute of Technology and Science, Pilani, Hyderabad Campus, Secunderabad, Telangana, India

P. Boominathan
Forest College and Research Institute, Tamil Nadu Agricultural University, Mettupalyam, Tamil Nadu, India



H. M. Mamrutha et al. (eds.), *Translating Physiological Tools to Augment Crop Breeding*, https://doi.org/10.1007/978-981-19-7498-4_9

## 9.1 Introduction

Crop improvement has always been an important task for plant scientists because of the great need for food owing to an ever-increasing population and environmental issues that have a detrimental influence on cultivable land due to harsh circumstances. The improvement strategy is to breed for either high yield or a specific trait. Generally, breeding for environmental stress tolerance involves trait selective breeding approaches as plants tend to naturally adapt themselves towards the climatic stress factors with variations in their morphological, physiological, and other phenotypical parameters. In order to accelerate yield improvement for stress-influenced environments like drought, submergence, high and low temperature, salinity, low and high light conditions, weightage has to be given for the adaptive physiological traits. This comes with the challenge of screening larger populations precisely for all such traits. However, to overcome this challenge, analysing the stable isotopes can be a potential high-throughput screening methodology.

## 9.2 Stable Isotopes

Isotopes are different forms of atoms of the same element, which can be stable or unstable. Atoms become unstable, i.e., radioactive, if the nucleus has an excess of internal energy. For a given nonradioactive element, isotopes with the same number of protons but different numbers of neutrons are stable isotopes (Table 9.1).

Among 118 elements discovered so far, only 80 (among the 94 naturally occurring elements) accounts for a total of 254 stable isotopes. There are two stable isotopes of carbon: $^{12}C$ and $^{13}C$. $^{12}C$ contains six electrons, six protons, and six neutrons, whereas $^{13}C$ has an additional neutron. There are two stable isotopes of hydrogen: $^{1}H$, which has one electron and one proton, and $^{2}H$, which has one electron, one proton as well as a neutron. The specific notation for $^{13}C$ is

$$\delta^{13}C = \left(\frac{R_{\text{sample}}}{R_{\text{standard}}} - 1\right) \times 1000$$

where $R = {}^{13}C/{}^{12}C$.

**Table 9.1** Difference between unstable and stable isotopes

| S. no | Unstable isotopes | Stable isotopes |
|---|---|---|
| 1 | Atoms with unstable nuclei | Atoms with a stable nuclei |
| 2 | Shows radioactivity | Shows no radioactivity |
| 3 | Magic number: number of protons or number of electrons | Magic number: number of protons or number of neutrons |
| 4 | Half-life is generally short and can be easily computed | Half-life is either very long or does not have a half-life |

**Table 9.2** Average terrestrial occurrence of the isotopes of key elements used in environmental research

| Element | Isotope | Average terrestrial occurrence (%) |
|---|---|---|
| Hydrogen | $^{1}H$ | 99.98 |
| | $^{2}H$ | 0.015 |
| Carbon | $^{12}C$ | 98.89 |
| | $^{13}C$ | 1.110 |
| Oxygen | $^{16}O$ | 99.75 |
| | $^{17}O$ | 0.037 |
| | $^{18}O$ | 0.204 |
| Nitrogen | $^{14}N$ | 99.63 |
| | $^{15}N$ | 0.370 |
| Sulphur | $^{32}S$ | 95.00 |
| | $^{33}S$ | 0.760 |
| | $^{34}S$ | 4.220 |
| | $^{36}S$ | 0.014 |
| Strontium | $^{84}Sr$ | 0.560 |
| | $^{86}Sr$ | 9.860 |
| | $^{87}Sr$ | 7.020 |
| | $^{88}Sr$ | 82.56 |

This gives an interpretation of degree of sample lightness with decreasing $\delta^{13}C$ values. Stable isotopes are used in a broad variety of application like environmental studies, water and soil management, forensics, agriculture because of their unique properties. Among all the isotopes, the average terrestrial abundance of lighter stable isotope of the element is always higher the heavier isotope of the same element (Table 9.2) (Fritz and Fontes 1980).

## 9.3 Measuring Stable Isotopes

A mass spectrometer may be used to determine the sample's stable isotopic ratio. A mass spectrometer can distinguish between atoms/ions of various masses and hence determines their relative abundance. Isotopic molecules are ionized by electron bombardment and exposed to a high-voltage electric and magnetic field in a mass spectrometer. These fields separate light and heavy ions and collect them in distinct containers that can be carefully monitored and so the isotope ratios in the samples may be determined.

## 9.4 Utilization of Stable Isotopes in Different Field

Using naturally occurring stable isotopes of water and other chemicals, the origin, history, sources, sinks, and interactions in the water, carbon, and nitrogen cycles may all be tracked. Tracers, which are purposely placed into a system to be studied, such as agriculture or nutrition, can also be used with stable isotopes. For this, more advanced procedures such as mass spectrometry must be used to separate them. While the most regularly used stable isotope in nutrition studies is deuterium H-2, an isotope twice as heavy as hydrogen, nitrogen-15 is the most commonly used stable isotope in agriculture. Many more stable isotopes are currently being used more often.

Isotopes of the water molecule ($^{1}$H (protium), $^{2}$H (deuterium), $^{3}$H (tritium), $^{16}$O, and $^{18}$O) and carbon ($^{12}$C, $^{13}$C, and $^{14}$C) that exist in water as components of dissolved inorganic and organic compounds are the most often employed environmental isotopes for hydrological research. The use of stable H and O isotope ratios in ground water is mostly based on isotopic fluctuations in precipitation. Environmental isotope approaches aid in the study of recharge sources and mechanisms, ground water circulation and renewability, aquifer recharge regions and transit times, hydraulic interrelationships, and ground water pollution sources and mechanisms.

Because the light isotopes ($^{1}$H and $^{16}$O) evaporate more readily than the heavy isotopes ($^{2}$H and $^{18}$O), hydrogen and oxygen isotopes can be used to determine water losses by evaporation from the soil surface. Natural isotopic ratios of hydrogen ($^{2}$H/$^{1}$H) and oxygen ($^{18}$O/$^{16}$O) in soil water, water vapour inside a plant canopy, and plant leaves, which are frequently reported as delta units, can be used to estimate soil evaporation and plant transpiration. Irrigation and land management methods may be devised using this knowledge to reduce soil evaporation and channel water for agricultural production.

Isotopes can be used in studies to:

- Improve Fertilizer Use Efficiency (FUE) and sustainability of soil fertility, soil quality and water quantity and quality for adjustment or downstream communities (Bhat and Bhat 2010)
- Develop water and nutrient efficient crop varieties (Richards et al. 2002)
- Assess farm water and irrigation management techniques that increase WUE by crops (Ma and Song 2019)
- Assess soil erosion rates (Meusburger et al. 2013)
- Assess soil profile and soil formation rates (Suresh et al. 2013)
- Assess the recharge capacity of springs (Wang et al. 2020)
- Assess relative transport of solutes in the soil profile (Sprenger et al. 2015)
- Promote integrated natural resource management within agricultural catchment (Blake et al. 2012)

Greater abundances of $^{16}$O suggest warmer eras (more evaporation), whereas higher abundances of $^{18}$O indicate cooler periods (decreased evaporation) in ice cores and microorganism fossil remnants. As a result, isotope partitioning among substances during reactions or processes may be used to describe activities in the biological, geological, and hydrological domains, both past and present.

Within a watershed, stable isotopes may be utilised to characterize a range of water dynamics. Identifying probable water inputs (e.g., precipitation vs. groundwater), what happens to the water inside the system, and outflows (e.g., water lost to groundwater vs. streamflow) from the system, as well as identifying mixing and flow pathways of water with these applications.

Stable isotope analysis has shown to be a highly successful approach for revealing a wide range of ecological challenges, and stable isotope ecology has been the focus of several international conferences. Some examples of how stable isotopes can be used as biological tracers are as follows:

1. To pinpoint sources, such as the origins of basal carbon in a food web.
2. To determine if a breeding animal is utilizing local resources or its own reserves, or whether an animal migrates/disperses from one site to another.
3. To quantify relative inputs in a system, such as calculating the proportions of various prey items in a consumer's diet. When done effectively, stable isotope analysis has significant advantages over previous approaches and adds another tool to the ecologist's toolkit.

   (a) To get a better understanding of polar bear feeding behaviours, which may be used to predict future changes due to habitat degradation (Hilderbrand et al. 1996).
   (b) Look at dietary stress as a possible reason of Stellar sea lion population reduction (Scherer et al. 2015).
   (c) To determine the impact of a gillnet ban on bottlenose dolphins (Browning et al. 2014).
   (d) Before and after alien species invasions, recreate aquatic food network routes (Côté et al. 2013).
   (e) Use a shift in nutrition to predict the arrival timing of a long-distance migratory shorebird on arctic breeding grounds (Atkinson et al. 2005).
   (f) Elucidate the significance of marine-derived nutrients in the development of juvenile salmon. The findings of this study add to our understanding of the foraging ecology of species of management or conservation importance (Michener and Kaufman 2007).
   (g) Stable isotope ratios in fish tissues (other than otoliths) can give valuable information on fish ecology, especially physiological ecology (Trueman et al. 2012).

Good et al. (2015) employed deuterium, a stable isotope of hydrogen, to anticipate extratropical cyclones like Superstorm Sandy. Moisture evaporation, transport, and precipitation modify stable isotope ratios in cyclonic waters, resulting in spatiotemporal isotopic patterns that indicate synoptic-scale activities. The evolution of precipitation efficiency inside Superstorm Sandy is plotted across time utilizing a series of dispersed isotope collections based on this basic association between precipitation deuterium-excess and moisture source circumstances. These maps show a high-precipitation-efficiency zone in the storm's core, when severe rainfall rates likely outpaced moisture supplies, as well as outer rain-bands with reduced precipitation efficiency.

Stable isotopes have been recently utilized in measuring the status, bioavailability and bioefficacy of micronutrients/vitamins, viz. Fe, Zn, and vitamin A (Sheftel et al. 2018).

Isotope analysis is helpful in differentiating between shallow biogenic gas and deeper thermos-genic gas. Fertilizers labelled with stable isotopes like nitrogen-15 can be used to determine how much fertilizer is absorbed by the plant and how much is lost to the environment. Nitrogen-15 also gives a clear indication of how much nitrogen is being fixed from the environment in real-world situations. Although $^{15}N$ is the most commonly utilized stable isotope in soil/plant investigations, there are a plethora of other stable isotopes that are rapidly being employed in agricultural studies.

## 9.5 Stable Isotopes for Physiological Traits

### *9.5.1 Water Use Efficiency: An Important Trait for Drought*

Drought has a significant impact on agricultural output. Widawsky and O'Toole (1996), for example, estimated drought losses at 3.00 million tonnes from a rice-growing region of 24 million hectares in Eastern India, accounting for 22% of total losses due to technical restrictions. As a result, a lot of effort has gone into enhancing rice's genetic potential for root features based on the number of workers (Shen et al. 2001; Venuprasad et al. 2002). However, due to the challenges and complexity connected with measuring this critical attribute and further manipulating it for our purposes, little or no effort has been done to increase WUE.

The ratio of biomass produced over a period of time to water transpiration over the same period is known as water use efficiency (WUE). Increasing the WUE is thus critical for attaining long-term production under both water-limited and irrigated situations. Any characteristic that may be used to improve crops should have a high heritability, a low GxE interaction, and, on top of that, large genetic variability. In the instance of this critical feature, all of these criteria were satisfied. WUE is a possible physiological characteristic for crop improvement because of its minimal GxE interaction and high heredity (Ismail and Hall 1992, 1993).

According to Passioura's (1986) yield model (seed yield = WUE × $T$ × HI), a 0.1 unit increase in WUE would result in an increase in TDM production of 0.348 t/ha at an annual rainfall of 800 mm and around 45% of it being accessible for transpiration. Several writers have used various methodologies to document variance in water usage efficiency in a variety of crop species. Gravimetry, gas exchange investigations, and carbon isotope discrimination are some of the methods used to analyse WUE fluctuation.

#### 9.5.1.1 Gravimetric Approach

WUE is frequently measured using the gravimetric approach, in which the amount of water consumed by plants is meticulously estimated by weighing containers at regular intervals. Several attempts are being made to use a gravimetric approach to assess genetic variability in this trait in a variety of crop species (Table 9.3). Although reliable WUE measurement is achievable, this approach is time-consuming and does not allow for large-scale screening.

#### 9.5.1.2 Gas Exchange Studies

$CO_2$ fixation and transpiration rate can be used to assess physiological WUE. These tests are normally carried out on a single leaf for a short length of time (Caemerrer

**Table 9.3** Genetic variability in WUE across different crop species

| Species | Method | Genetic variability | References |
|---|---|---|---|
| *Cereals and millets* | | | |
| Rice | Gravimetry—pot | 3.94–5.47 g/kg | Nadaradjan (2004) |
| Rice | Gravimetry-pot | 2.50–5.4 g/kg | Impa et al. (2005) |
| Wheat | Gravimetry—pot | 4.60–5.27 g/kg | Condon et al. (1990) |
| *Pulses* | | | |
| Cow pea | Gravimetry—pot | 2.74–3.00 g/kg | Ismail and Hall (1992) |
| Cow pea | Field grown | 2.08–3.29 g/kg | Bindumadhava (2000) |
| Soybean | Minilysimeter | | White et al. (1996) |
| Irrigated<br>Water stress | | 1.66–2.44 g/kg<br>2.03–2.78 g/kg | |
| Navy bean | Minilysimeter—field | | Wright (1996) |
| Irrigated | | 2.26–3.57 g/kg | |
| Stressed | | 2.38–3.87 g/kg | |
| Chick pea | Minilysimeter—field | | Wright (1996) |
| Irrigated | | 1.61–2.23 g/kg | |
| Stressed | | 1.94–2.73 g/kg | |
| Chick pea | Gravimetry—pot | 1.41–2.67 g/kg | Devaraj (2000) |
| Chick pea | Gravimetry—pot | 1.29–2.14 g/kg | Gangadhara (1995) |
| *Oilseeds* | | | |
| Groundnut | Gravimetry—field | 1.57–2.66 g/kg | Hebbar (1990) |
| Ground nut | Minilysimeter—field | | Wright (1996) |
| Irrigated | | 2.46–3.71 g/kg | |
| Stressed | | 1.81–3.15 g/kg | |
| Ground nut | Gravimetry—pot | 1.41–3.30 g/kg | Shashidhar (2002) |
| Ground nut | Gravimetry—pot | 2.55–4.02 g/kg | Roy Stephen (1995) |
| *Others* | | | |
| Arabidopsis | Gravimetry—pot | 1.86–2.40 mg/g | Nienhuis et al. (1994) |

and Farquhar 1981; Farquhar and Sharkey 1982). Photosynthetic rate determines how much biomass is generated. As a result, WUE is defined as the carbon absorption rate (*A*) to transpiration rate (*T*) ratio at a particular leaf level (*T*). The intrinsic stomatal conductance and the existing leaf to air vapour pressure difference control the rate of transpiration (*V*). If the plants being investigated are cultivated under identical environmental conditions, the leaf to air vapour pressure should be similar, and the intrinsic stomatal conductance should be the primary factor determining transpiration ($g_s$). Therefore,

$$\mathrm{WUE} = \frac{A}{g_s}$$

A favourable correlation among WUE assessed by gravimetric and $A/g_s$ was seen, albeit it was not strong or statistically significant, suggesting that $A/g_s$ might be a helpful predictor of variance in whole plant WUE (Boominathan 2001). This gas exchange approach is quick, and it may be used for large-scale testing. However, because it is a time instantaneous measure, it does not account for diurnal fluctuations in the leaf gas exchange parameter.

#### 9.5.1.3 Carbon Isotope Discrimination

The revelation that plants discriminate between heavy isotopes of carbon gave a significant incentive for determining genetic variability in WUE (Farquhar and Richards 1984; Farquhar et al. 1989b), which was accompanied by a low GxE interaction. During photosynthesis, plants differentiate against the heavy carbon isotope ($^{13}$C), resulting in a decrease in $^{13}$C concentration in biomass (O'Leary 1981). The discriminating ($\Delta^{13}$C) of the carbon isotope ratio ($^{13}$C/$^{12}$C) of biomass from that of air is linked to the ratio of partial pressures of $CO_2$ inside the leaf to that in the ambient air (Pi/Pa) as follows (Farquhar et al. 1982, 1989a; Hubick and Farquhar 1989; O'Leary 1981):

$$\mathbf{\Delta^{13}C} = a + (b - a)\mathbf{Pi/Pa},$$

where *a* and *b* represent fractionation against $_{13}$C ($\delta^{13}$C) during stomatal diffusion and carboxylation by RuBisCO, respectively. Due to the fact that WUE is connected to $CO_2$ partial pressures, a high inter-relationship between $\Delta^{13}$C and WUE is predicted for a given VPD (Hubick and Farquhar 1989; O'Leary 1993).

$^{12}$C and $^{13}$C are two naturally available stable carbon isotopes. The majority of the carbon (98.9%) is $^{12}$C, with only 1.1% being $^{13}$C. The distribution of isotopes varies across and within compounds, and this variation can provide information about the physical, chemical, and metabolic processes that occur throughout carbon transformations. As indicated by the lower overall abundance of $^{13}$C relative to $^{12}$C in plant tissue than in atmospheric $CO_2$, carbon isotope discrimination occurs during the integration of $CO_2$ into plant biomass. Because carbon isotopes are stable and non-radioactive in nature, the information contained in the ratio of abundance of carbon isotopes, denoted by convention as $^{13}C/^{12}C$, is unchanged as long as carbon is not lost. The $\Delta^{13}C$ in plant samples is generally determined using a sophisticated analytical instrument called Isotope Ratio Mass Spectrometer (IRMS) specially designed for high precision measurements of the ratio *R*, defined as:

$$R = {}^{13}CO_2/{}^{12}CO_2$$

To evaluate the isotope composition, the plant material is transformed to $CO_2$. *R* is low in organic samples in general. In comparison to a typical PDB, atmosphere has a substantially greater fractionation value of roughly –7.8 per mil (‰) (Pee Dee Belemnite, from North Carolina, USA). The *R* value in this standard is 0.0124, whereas it is roughly 0.012 in various plant materials, implying very minimal differences in the *R* value, and so *R* in a sample may be compared to that of the standard and represented as $^{13}C$ in units parts per thousand or per mil (‰).

$$\delta^{13}C = R_{\text{sample}} - R_{\text{standard}}/R_{\text{standard}} \times 1000$$

Because the organic sample has a lower *R*-value than the standard, the $\delta^{13}C$ content of organic material is more negative, implying that there is less $^{13}C$ content and hence more discrimination (O'Leary 1984). Importance and biochemical basis of $\Delta^{13}C$ and the interaction of $\Delta$ with WUE have extensively been explored (Condon et al. 1990; Farquhar et al. 1982; Hubick et al. 1988; Read et al. 1991). The following table shows the range of $\Delta^{13}C$ levels for several crop species with distinct photosynthetic pathways:

#### 9.5.1.3.1 Range of $\delta^{13}C$ Composition

1. Atmosphere (air) = −6.4 to −7.0‰
2. $C_3$ plants = −22 to −44‰
3. $C_4$ plants = −9 to −19‰
4. CAM plants = −11‰ (approximate)

#### 9.5.1.3.2 Carbon Isotope Discrimination ($\Delta^{13}C$) at Different Steps During Photosynthesis

Several biological and physical mechanisms are involved in the fractionation of carbon isotopes during photosynthesis. These processes discriminate between $^{12}C$ and $^{13}C$ in different ways, and a plant's overall discrimination is a result of the $CO_2$ fixation mechanism it employs and the relative balance of the photosynthetic systems. During photosynthesis, $CO_2$ must go from the atmosphere to the

**Table 9.4** Fractionation of carbon isotope during photosynthesis

| Process | Discrimination (‰) | References |
|---|---|---|
| Diffusion of $CO_2$ in air through the stomatal pore | 4.4 | Craig (1953) |
| Diffusion of $CO_2$ in air through the boundary layer to the stomata | 2.9 | Farquhar (1983) |
| Diffusion of dissolved $CO_2$ through water | 0.7 | O'Leary (1984) |
| Fixation of gaseous $CO_2$ by RuBisCO | 29 | Guy and Hoering (1987) |

chloroplast stroma. Because $^{12}CO_2$ diffuses quicker than $^{13}CO_2$, many fractionation events occur along this diffusion path, resulting in a considerable reduction in $CO_2$ accessible at carboxylation sites in $^{13}C$ relative to the environment (Table 9.4) (Farquhar et al. 1989a; Bragnoli and Farquhar 1998).

The diffusion of $CO_2$ via the stomata and the carboxylation process mediated by RuBisCO are the two main components of photosynthesis that impact discrimination (O'Leary 1988, 1993). Regardless of the photosynthetic sub-component that determines discrimination, $^{13}C$ is connected to Pi in the following way:

$$\Delta = \{a + (b - a)\mathrm{Pi}/\mathrm{Pa} \quad - \quad d\}$$

where $a$ and $b$ are constants for discrimination against $^{13}CO_2$ during $CO_2$ diffusion into the leaf and carboxylation, respectively, and $d$ is a component supplied by respiration, dissolved $CO_2$ diffusion, and Pi and Pa are the intercellular and ambient $CO_2$ partial pressures.

#### 9.5.1.3.3 $\Delta^{13}C$ as a Surrogate for WUE

##### *9.5.1.3.3.1 $\Delta^{13}C$ and WUE Relationship*

During the photosynthetic process, plants discriminate against the heavy carbon isotope ($\Delta^{13}C$). However, because the degree of discrimination is determined by Pi, $\Delta^{13}C$ content in plant samples has emerged as a potential technique for determining Pi and WUE. According to some accounts, Pi and $\Delta^{13}C$ have a strong connection; hence, $\Delta^{13}C$ might represent a time-integrated approximation of Pi. Because Pi/Pa ratios are primarily responsible for fluctuations in WUE and $\Delta^{13}C$, the following equation may be used to predict and explain a significant link between $\Delta^{13}C$ and WUE (Farquhar et al. 1989b).

$$\mathbf{WUE} = \{(\mathbf{1} - \theta)(b - d - \mathbf{\Delta})\}/\mathbf{1.6}V(b - a)$$

where $V$ is the leaf–air vapour pressure gradient and $\theta$ is the fraction of fixed $CO_2$ lost in respiration. A positive link between Pi/Pa and $\Delta^{13}C$ (Meinzer et al. 1990; Richards and Tieszen 1993) and an inverse relationship between $A/g_s$ and $\Delta^{13}C$

**Table 9.5** Relationship between $\Delta^{13}C$ and WUE in crop species

| Species | Relationship | | | References |
|---|---|---|---|---|
| | *R* | *p*< | *N* | |
| *Cereals and millets* | | | | |
| Wheat | −0.750 | 0.01 | 12 | Farquhar and Richards (1984) |
| Wheat: Well-watered plant | −0.740 | 0.01 | 16 | Condon et al. (1990) |
| Wheat: Water stressed plant | −0.750 | 0.01 | 16 | |
| Wheat | 0.820 | 0.0001 | 16 | Liu et al. (2020) |
| Rice | −0.830 | 0.00001 | 34 | Boominathan (2001) |
| Rice | −0.470 | 0.01 | 39 | Nadaradjan et al. (2005) |
| Rice | −0.815 | 0.05 | 6 | Impa et al. (2005) |
| *Pulses* | | | | |
| Cowpea | −0.930 | 0.05 | 5 | Ismail and Hall (1992) |
| Soybean: Irrigated | −0.680 | – | – | White et al. (1996) |
| Soybean: Water stressed | −0.800 | – | – | |
| Navy bean | −0.800 | 0.05 | – | Wright (1996) |
| Chickpea | −0.830 | 0.001 | 12 | Gangadhara (1995) |
| Cowpea | −0.680 | 0.02 | 11 | Bindumadhava (2000) |
| *Oilseeds* | | | | |
| Peanut | −0.810 | 0.01 | 34 | Wright (1996) |
| Peanut | −0.640 | 0.05 | 8 | Roy Stephen (1995) |
| Peanut | −0.690 | 0.01 | 17 | Shashidhar (2002) |
| *Others* | | | | |
| Kentucky Bluegrass | −0.550 | 0.01 | 11 | Ebdon et al. (1998) |
| Crested wheat grass | −0.870 | 0.05 | 14 | Read et al. (1991) |

(Meinzer et al. 1990; Richards and Tieszen 1993) indicate that Pi determines the variability in $\Delta^{13}C$ (Hubick et al. 1988; Gutterrez and Meinzer 1994).

Although WUE and $\Delta^{13}C$ are connected via the Pi/Pa ratio and $A/g_s$ (Condon et al. 1990), these parameters will not provide an integrated estimate of WUE across time because of diurnal and seasonal changes in '*A*' and $g_s$ (Hall et al. 1993; Udayakumar and Prasad 1994). $^{13}C$ is a reliable statistic in this context since it reflects a time-integrated estimate of carbon uptake per unit transpiration, especially in $C_3$ plants. In pot grown sunflower, $\Delta^{13}C$ in whole plant dry matter appears to be a reliable indication of plant WUE, and a negative association was shown in structural carbon between these two features in both well-hydrated and drought circumstances (Johnson et al. 1993).

Even when plants were subjected to abiotic stressors, the association between $\Delta^{13}C$ and WUE in numerous crop species remained unchanged (Table 9.5). Because the relative ranking of genotypes in control and stress is maintained, genotype and environment interaction for WUE and $\Delta^{13}C$ is minimal, and broad sense heritability is high (Hubick et al. 1988; Wright et al. 1993). $\Delta^{13}C$ appears to be a particularly accurate indicator for identifying variability in WUE because of these specific benefits. As a result, various breeding initiatives to enhance WUE utilizing the carbon isotope discrimination approach were started (Hall et al. 1993; White 1993).

#### *9.5.1.3.3.2 Wheat*

In the northern wheat belt of Australia, a breeding effort for water use efficiency was done to introduce low carbon isotope discrimination (Δ) into the wheat variety Hartog (Rebetzke et al. 2002). Selection for low Δ led in higher grain yields in the Hartog backcross programme when compared to lines chosen for high Δ in nine different settings with varying seasonal rainfall. Furthermore, the biomass, harvest index, and kernel weight increased. The benefit of increased yield was shown to be greater associated with drier settings. Drysdale (October 2002) and Rees (October 2003) have been released as commercial wheat cultivars in eastern Australia as a result of these efforts (Condon et al. 2004).

#### *9.5.1.3.3.3 Rice*

Only when a drought tolerance trait is linked to crop growth rate, it is useful for agricultural development (Udayakumar and Prasad 1994). In this context, it is expected that a high ratio of carbon acquired per unit water lost via transpiration, referred to as high water use efficiency (WUE), will provide a fitness advantage during drought stress (Cohen 1970; Dudley 1996; McKay et al. 2001). Furthermore, the ability of the plant to harness water from a deeper soil profile associated with the roots, as well as superior water conservation strategies associated with epicuticular waxes, is frequently considered as potential traits that have relevance in crop improvement, particularly under water-scarce conditions. Under semi-irrigated aerobic conditions, the introduction of root and water usage efficiency into IR64 resulted in a significant yield benefit (Dharmappa et al. 2019). The surrogate 'Carbon isotope discrimination' was used to quantify the variability in water usage efficiency in this programme.

## *9.5.2 Mean Transpiration Rate (MTR)*

The mean transpiration rate is the rate of transpiration across the whole experimental time. This measure was calculated by dividing the total water transpired by the functional leaf area, and it is represented in grams or millilitres of water/dm$^2$ day.

### 9.5.3 Stable Isotopes of Oxygen

For more than half a century, scientists have known that natural oxygen and hydrogen isotopes exist. The first references for measuring $^{18}O$ in organic materials date back to the 1970s. However, because of the VPD effect, the oxygen isotope ratio of leaf water rises at the leaf (Ferhl and Letolle 1997). The amount of oxygen isotope enrichment in water on leaf surfaces is determined by a few physiological characteristics of the species, among other things. Enrichment of $^{18}O$ in water from CAM plants, for example, does not occur throughout the day, but it occurs in water from $C_3$ plants (Epstein et al. 1977).

Craig and Gordon (1965) created the first model of isotopic fractionation for the process of water evaporation from bodies of water. They observed that water molecules with the lighter isotope of oxygen ($H_2{}^{16}O$) spread and evaporate faster during evaporation than those with the heavier isoform ($H_2{}^{18}O$). They demonstrated that this resulted in an enrichment of $H_2{}^{18}O$ near the water's evaporating surface. They looked into the isotopic effects of oxygen in water as an answer.

Compositions of oxygen isotopes are stated in parts per thousand (per mil), much as carbon isotopes, i.e.,

$$\delta^{18}O = (R_{\text{sample}} - 1)/(R_{\text{standard}} \times 1000)$$

The molar ratio of the heavy to light isotope is $R$. The compositions of all water samples are indicated in terms of SMOW (Standard Mean Oceanic Water). The absolute ratio for SMOW used in the calculations were $^{18}O/^{16}O = 0.0020052$ (Ehleringer and Osmond 1989).

The rate of transpiration and stomatal conductance ($g_s$) in plants may be measured using $^{18}O$ signals. The stomatal conductance has a strong positive relationship with the $^{18}O$ content of leaf water, indicating that $g_s$ causes oxygenic isotopic enrichment in leaf water. On the day of leaf water extraction, the mean transpiration rate (MTR) calculated using gravimetric revealed a similar linear connection with leaf water $^{18}O$ (0.83: $p$ 0.005: $n = 7$). Through a metabolism that leads to cellulose formation, the enhanced $^{18}O$ makes its way into the cellulose. As a result, the $^{18}O$ composition of leaf biomass may be viewed as a reflection of the $^{18}O$ enrichment that happened throughout time (Bindumadhava et al. 1999).

Plants cultivated in low relative humidity have more $^{18}O$ than plants grown in high RH. Enhanced ABA concentration increased oxygen isotope enrichment of leaf cellulose ($^{18}Oc$) and entire leaf tissue ($^{18}Oi$) in each humidity condition (Barbour and Farquhar 2000). Some research conducted at the GKVK's Department of Crop Physiology in Bangalore suggests that $^{18}O$ and MTR have a positive link in various crop species (Table 9.6). Sheshshaye et al. (2005) found that the $^{18}O$ content of leaf biomass is a reasonable time-integrated proxy for plant mean transpiration rate.

**Table 9.6** Relationship between $\delta^{18}O$ and MTR in crop species

| Species | Relationship | | | References |
|---|---|---|---|---|
| | *R* | <*p* | *n* | |
| *Cereals and millets* | | | | |
| Rice | 0.68 | 0.02 | 11 | Impa (2002) |
| *Pulses* | | | | |
| Cowpea | 0.92 | 0.05 | 5 | Bindumadhava (2000) |
| *Oil seed crops* | | | | |
| Groundnut | 0.75 | 0.01 | 15 | Shashidhar (2002) |

## 9.6 Conclusion or Future Prospects

The purpose of this chapter was to demonstrate how stable isotope technology has aided in our understanding of plant–environment interactions. In terrestrial plants, stable C and N isotope ratios have the potential to reveal novel information about physiological processes and their interactions with the environment. The literature, we believe, clearly shows that collecting and analysing isotope data have resulted in significant breakthroughs in plant ecophysiology which aided in crop improvement programmes. Yet, several important issues demand more examination. Stable isotope techniques have the potential to increase our understanding of population and community dynamics when integrated with other data from modelling, molecular, and/or genetic data. This potential will be enhanced by ongoing study into environmental and physiological variables of Δ. This is especially critical when engineering plants with high water usage efficiency and photosynthetic capability. The idea of utilising Δ as an efficient selection index in plants to acquire genotypes with high WUE is found efficacious. Assessment of WUE and MTR and its combined usage might potentially assist physiologists and plant breeders in identifying crop management strategies and genotypes that are better suited to a variety of growing environments. These can be used as effective screening traits for abiotic stress tolerance breeding programmes.

## References

Atkinson PW, Baker AJ, Bevan RM, Clark NA, Cole KB, Gonzalez PM, Newton J, Niles LJ, Robinson RA (2005) Unravelling the migration and moult strategies of a long-distance migrant using stable isotopes: red knot Calidris canutus movements in the Americas. Ibis 147(4): 738–749

Barbour MM, Farquhar GD (2000) Relative humidity and ABA induced variation in carbon and oxygen isotope ratio of cotton leaves. Plant Cell Environ 23:473–483

Bhat MI, Bhat MA (2010) Applications of stable and radioactive isotopes in soil science. Curr Sci 98:1458–1471

Bindumadhava H (2000) Oxygen ($\Delta^{18}$ O) and Carbon ($\Delta^{13}$ C) isotope composition in plants—an approach to quantify transpiration and mesophyll factors associated with water use efficiency. Ph.D. thesis submitted to University of Agricultural Sciences, Bangalore

Bindumadhava H, Sheshshayee MS, Devendra A, Prasad TG, Udayakumar M (1999) Oxygen ($^{18}O$) isotopic enrichment in the leaves as a potential surrogate for transpiration and stomatal conductance. Curr Sci 76(11):1427–1428

Blake WH, Ficken KJ, Taylor P, Russell MA, Walling DE (2012) Tracing crop-specific sediment sources in agricultural catchments. Geomorphology 139:322–329

Boominathan P (2001) DNA polymorphism and genetic variability in physiological traits associated with water use efficiency in rice—an approach based on carbon isotope discrimination and RAPD profiles. M.Sc.(Ag) thesis submitted to Univ. of Agri. Sci., Bangalore

Bragnoli E, Farquhar GD (1998) Photosynthetic fractionation of carbon isotopes. In: Dekonchkovsky Y, Larher F (eds) Photosynthesis: physiology and metabolism, advances in photosynthesis. SFVP Universite de Renne. Springer, pp 399–434. Condon AC and Hall AC

Browning NE, McCulloch SD, Bossart GD, Worthy GAJ (2014) Fine-scale population structure of estuarine bottlenose dolphins (Tursiops truncatus) assessed using stable isotope ratios and fatty acid signature analyses. Marine Biol 161(6):1307–1317

Caemerrer SV, Farquhar GD (1981) Some relationship between the biochemistry of photosynthesis and the gas exchange of leaves. Planta 153:378–383

Cohen D (1970) The expected efficiency of water utilization in plants under different competition and selection regimes. Israel J Bot 19:52–54

Condon AG, Farquhar GD, Richards RA (1990) Genotypic variation in carbon isotope discrimination and transpiration efficiency in wheat. Leaf gas exchange and whole plant studies. Aust J Plant Physiol 17:9–22

Condon AG, Richards RA, Rebetzke GJ, Farquhar GD (2004) Breeding for high water use efficiency. J Exp Bot 55:2447–2460

Côté IM, Green SJ, Hixon MA (2013) Predatory fish invaders: insights from Indo-Pacific lionfish in the western Atlantic and Caribbean. Biol Conserv 164:50–61

Craig H (1953) The geochemistry of the stable carbon isotopes. Geochem Cosmochim Acta 3:53–92

Craig H, Gordon LI (1965) Deuterium and Oxygen-18 variations in the ocean and marine atmosphere. In: Tongiori E (ed) Stable isotopes in oceanographic studies and paleotemperatures, Spoleto, Italy. Lischi and Figli. Pisa

Devaraj A (2000) Relationship between a few root associated traits with growth and WUE under water stress and non-stress conditions in pigeon pea (Cajanus cajan, L) millip. Cultivars. M.Sc (Ag) thesis submitted to Univ. of Agri. Sci., Bangalore

Dharmappa PM, Doddaraju P, Malagondanahalli MV, Rangappa RB, Mallikarjuna NM, Rajendrareddy SH, Ramanjinappa R, Mavinahalli RP, Prasad TG, Udayakumar M, Sheshshayee SM (2019) Introgression of root and water use efficiency traits enhances water productivity: an evidence for physiological breeding in rice (Oryza sativa L.). Rice 12(1):1–14

Dudley SA (1996) Differing selection on plant physiological traits in responses to environmental water availability. A test of adaptive hypothesis. Evolution 50:92–102

Ebdon JS, Petrovic AM, Dawson TE (1998) Relationship between carbon isotope discrimination, water use efficiency and evapotranspiration in Kentucky Blue grass. Crop Sci 38(1):157–162

Ehleringer JR, Osmond CB (1989) Stable isotopes. In: Pearly RW, Ehleringer JR, Mooney HA, Rundel PW (eds) Plant physiological ecology, field methods and instrumentation. Chapman and Hall, London, pp 281–300

Epstein S, Thomson P, Yapp CJ (1977) Oxygen and hydrogen isotopic ratios in plant cellulose. Science 198:1209–1205

Farquhar GD (1983) On the nature of carbon isotope discrimination in C 4 species. Aust J Plant Physiol 10:281–292

Farquhar GD, Ehleringer JR, Hubick KT (1989a) Carbon isotope discrimination and photosynthesis. Annu Rev Plant Physiol Plant Mol Biol 40:503–537

Farquhar GD, Hubick KT, Condon AG, Richards RA (1989b) Carbon isotope fractionation and plant water use efficiency. In: Rundel PW, Ehleringer JR, Nagy KA (eds) Stable isotope in ecological research. Springer-Verlag, New York, pp 21–40

Farquhar GD, O'Leary MH, Berry JA (1982) On the relationship between carbon isotope discrimination and the intercellular $CO_2$ concentration in leaves. Aust J Plant Physiol 9:121–131

Farquhar GD, Sharkey TD (1982) Stomatal conductance and photosynthesis. Annu Rev Plant Physiol 33:317–345

Farquhar GD, Richards RA (1984) Isotopic composition of plant carbon correlates with water use efficiency of wheat cultivars. Aust J Plant Physiol 11:539–552

Ferhl C, Letolle TG (1997) Stable isotope composition of stem and leaf water: application to the study of plant water use. Funct Ecol 5:23–28

Fritz P, Fontes JC (eds) (1980) Handbook of environmental isotope geochemistry, vol I. The terrestrial environment. Elsevier, Amsterdam

Gangadhara GC (1995) Genotypic variations in WUE and related parameters in chickpea under stress and non-stress conditions. M.Sc. (Agri) thesis, submitted to University of Agricultural Sciences, Bangalore

Good SP, Mallia DV, Denis EH, Freeman KH, Feng X, Li S, Zegre N, Lin JC, Bowen GJ (2015) High frequency trends in the isotopic composition of Superstorm Sandy. In: Learning from the impacts of Superstorm Sandy. Academic, pp 41–55

Gutterrez MV, Meinzer FC (1994) Carbon isotope discrimination and photosynthetic gas exchange in coffee hedge rows during canopy development. Aust J Plant Physiol 21:207–209

Guy RD, Hoering TD (1987) Isotope fractionation during oxygen production and consumption by plants. In: Biggins J (ed) Progresses in photosynthetic research. III. Nijhoff, Dordrecht, pp 597–600

Hall AE, Ismail AM, Menendez C (1993) Implications in water use efficiency, carbon isotope discrimination and gas exchange. In: Ehleringer J (ed) Stable isotope and plant carbon/water relations. Academic, San Diego

Hebbar KB (1990) Studies on genotypic variation in WUE in groundnut and its relevance to productivity under field condition. M.Sc. (Agri) thesis submitted to Univ. of Agric. Sci, Bangalore

Hilderbrand GV, Farley SD, Robbins CT, Hanley TA, Titus K, Servheen C (1996) Use of stable isotopes to determine diets of living and extinct bears. Can J Zool 74(11):2080–2088

Hubick KT, Farquhar GD (1989) Carbon isotope discrimination and ratio of carbon gained to water lost in barley cultivars. Plant Cell Environ 12:795–804

Hubick KT, Shorter R, Farquhar GD (1988) Heritability and genotype × Environment interaction of carbon isotope discrimination and transpiration efficiency in peanut. Aust J Plant Physiol 15: 799–813

Impa SM (2002) Stable isotopic approach to identify gs and gm types in rice (Oryza sativa L.) and their genetic analysis using RAPD markers. M.Sc.(Ag) thesis submitted to Univ. of Agri. Sci, Bangalore

Impa SM, Nadaradjan S, Boominathan P, Shashidhar G, Bindhumadhava H, Sheshshayee MS (2005) Carbon isotope discrimination accurately reflects variability in WUE measured at a whole plant level in rice. Crop Sci 45:2517–2522

Ismail AM, Hall AE (1992) Correlation between water use efficiency and carbon isotope discrimination in diverse cowpea genotypes and isogenic lines cowpea. Crop Sci 32:7–12

Ismail AM, Hall AE (1993) Inheritance of carbon isotope discrimination and water use efficiency in cowpea. Crop Sci 33:498–503

Johnson DA, Asay KH, Read JJ (1993) Physiological ecology: a series of monographs, texts, and treatises. In: Ehleringer JR, Hall AE, Farquhar GD (eds) Stable isotopes and plant carbon/water relations. Academic, New York, pp 269–280

Liu Z, Ma FY, Hu TX, Zhao KG, Gao TP, Zhao HX, Ning TY (2020) Using stable isotopes to quantify water uptake from different soil layers and water use efficiency of wheat under long-term tillage and straw return practices. Agric Water Manag 229:1–9

Ma Y, Song X (2019) Applying stable isotopes to determine seasonal variability in evapotranspiration partitioning of winter wheat for optimizing agricultural management practices. Sci Total Environ 654:633–642

Mckay JK, Bishop JG, Lin JZ, Sala A, Richards JH, Mitchell-olds T (2001) Local adaptation across a climatic gradient despite small effective population size in the rare sapphire rockcress. Proc Royal Soc London Biol Sci 268:1715–1721

Meinzer FC, Rundel PW, Schultz LM (1990) Effects of leaf resin on stomatal behavior and gas exchange of Larrea tridenta (DC). Cov Funct Ecol 4:579–584

Meusburger K, Mabit L, Park JH, Sandor T, Alewell C (2013) Combined use of stable isotopes and fallout radionuclides as soil erosion indicators in a forested mountain site, South Korea. Biogeosciences 10(8):5627–5638

Michener RH, Kaufman L (2007) Stable isotope ratios as tracers in marine food webs: an update. In: Michener R, Lajtha K (eds) Stable isotopes in ecology and environmental science, vol 2. Wiley, pp 238–282

Nadaradjan S (2004) QTLs contributing to WUE and associated physiological traits in DHLs of rice. Ph.D. thesis submitted to Univ. of Agri. Sci., Bangalore

Nadaradjan S, Impa SM, Parsi SG, Sheshshayee MS, Udayakumar M, Prasad TG (2005) Overlapping QTLs for WUE and carbon isotope discrimination in DHLs of rice. J Plant Biol 32:1–8

Nienhuis J, Sills GR, Martin B, King G (1994) Variance for water use efficiency among ecotypes and recombinant inbred lines of Arabidopsis thaliana (Brassicaceae). Am J Bot 81(8):943–947

O'Leary MH (1981) Carbon isotope fractionation in plants. Phytochemistry 20:553–567

O'Leary MH (1984) Measurement of the isotope fractionation associated with diffusion of carbon dioxide in aqueous solution. J Phys Chem 88:823–825

O'Leary MH (1988) Carbon isotope in photosynthesis. Bio Sci 38:325–336

O'Leary MH (1993) Biochemical basis of carbon isotope fractionation. In: Ehleringer JR, Hall AE, Farquhar GD (eds) Stable isotopes and plant carbon/water relations. Academic, pp 19–20

Passioura JB (1986) Resistance to drought and salinity: avenues for improvement. Aust J Plant Physiol 13:191–201

Read JJ, Johnson DA, Tienzen C (1991) Carbon isotope discrimination, gas exchange and WUE in crested wheat grass clones. Crop Sci 31:1203–1208

Rebetzke GJ, Condon AG, Richards RA, Farquhar GD (2002) Selection for reduced carbon-isotope discrimination increases aerial biomass and grain yield of rainfed bread wheat. Crop Sci 42(3): 739–745

Richards RA, Rebetzke GJ, Condon AG, Van Herwaarden A (2002) Breeding opportunities for increasing the efficiency of water use and crop yield in temperate cereals. Crop Science 42(1): 111–121

Richards RA, Tieszen LL (1993) Carbon isotope discrimination, water relations and gas exchange in temperate grass species and accessions. In: Ehleriger JR, Hall AE, Farquhar GD (eds) Stable isotopes and plant carbon/water relations. Academic, pp 281–296

Scherer RD, Doll AC, Rea LD, Christ AM, Stricker CA, Witteveen B, Kline TC, Kurle CM, Wunder MB (2015) Stable isotope values in pup vibrissae reveal geographic variation in diets of gestating Steller sea lions Eumetopiasjubatus. Marine Ecol Prog Series 527:261–274

Shashidhar G (2002) Screening of diverse germplasm lines of groundnut (Arachis hypogaea L.) for genetic variability in water use efficiency and total dry matter based on stable isotopes and RAPD. M.Sc.(Agri) thesis submitted to the University of Agricultural Sciences, Bangalore

Sheftel J, Loechl C, Mokhtar N, Tanumihardjo SA (2018) Use of stable isotopes to evaluate bioefficacy of provitamin-A carotenoids, vitamin A status, and bioavailability of iron and zinc. Adv Nutr 9(5):625–636

Shen L, Courtois B, MCNally KL, Robin S, Li Z (2001) Evaluation of near-isogenic lines of rice introgressed with QTLs for root depth through marker-aided selection. Theor Appl Genet 103: 75–83

Sheshshaye MS, Bindumadhava H, Ramesh R, Prasad TG, Lakshminarayana MR, Udayakumar M (2005) Oxygen isotope enrichment ($\Delta^{18}O$) as a time averaged measure of transpiration rate. J Exp Bot 56:3033–3039

Sprenger M, Volkmann TH, Blume T, Weiler M (2015) Estimating flow and transport parameters in the unsaturated zone with pore water stable isotopes. Hydrol Earth Syst Sci 19(6):2617–2635
Stephen R (1995) Genotypic variation in water use efficiency in groundnut (Arachis hypogeae L.) and its relationship to carbon isotope discrimination and specific leaf area. M.Sc.(Agri.) thesis, submitted to the University of Agricultural Sciences, Bangalore, India
Suresh PO, Dosseto A, Hesse PP, Handley HK (2013) Soil formation rates determined from Uranium-series isotope disequilibria in soil profiles from the southeastern Australian highlands. Earth Planet Sci Lett 379:26–37
Trueman CN, MacKenzie KM, Palmer MR (2012) Identifying migrations in marine fishes through stable-isotope analysis. J Fish Biol 81(2):826–847
Udayakumar M, Prasad TG (1994) $^{13}$C isotope discrimination in plants—a potential technique to determine WUE. In: Rao RCN, Wright GC (eds) Selection for WUE in grain legumes, a report of workshop held at ICRISAT center, Andhra Pradesh, India. ICRISAT, pp 42–45
Venuprasad R, Shashidhar HE, Hittalmani S, Hemamalini GS (2002) Tagging quantitative trait loci associated with grain yield and root morphological traits in rice (Oryza sativa L.) under contrasting moisture regimes. Euphytica 128:293–300
Wang F, Chen H, Lian J, Fu Z, Nie Y (2020) Seasonal recharge of spring and stream waters in a karst catchment revealed by isotopic and hydrochemical analyses. J Hydrol 591:125595
White JW (1993) Implications of carbon isotope discrimination studies for breeding common bean under water deficits. In: Ehleringer JR, Hall AE, Farquhar GD (eds) Stable isotopes and plant carbon/water relations. Academic, pp 389–396
White DS, Bell MJ, Wright GC (1996) The potential to use carbon isotope discrimination as a selection tool to improve water use efficiency in soybean. In: Proc. of the 8th Australian agronomy conference, Toowoomba
Widawsky DA, O'Toole JC (1996) Prioritizing rice biotechnology research agenda for eastern India. In: Evenson RE, Herdd RW, Hossain M (eds) Rice research in Asia: progress and priorities. CAB International, Wallingford, pp 109–129
Wright GC (1996) Selection for water use efficiency in grain legume species. In: Proc. of the 8th Australian agronomy conference, Toowoomba
Wright GC, Hubick KT, Farquhar GD, Nageswara Rao RC (1993) Genetic and environmental variation in transpiration efficiency and its correlation with carbon isotope discrimination and specific leaf area in peanut. In: Ehleringer JR, Hall AE, Farquhar GD (eds) Stable isotope and plant carbon-water relations. Elsevier, p 226

# Chapter 10
# Root Phenotyping for Improved Resource Use Efficiency in Crops

**Jagadish Rane, P. S. Basavaraj, Krishna Kumar Jangid, Vinay Hegde, and Harohalli Masthigowda Mamrutha**

**Abstract** To ensure the conservation of natural resources and food security for future, utilization of different tools and techniques for improving resource use efficiency is mandatory. Roots are the key determinant for researcher and breeders for improving resource use efficiency. Crop root architecture improvements promise to increase water and nutrient resource utilization. The root phenotyping is challenging to discover root traits beneficial to crops and their incorporation into new cultivars through prebreeding. High precision and noninvasive methods are used mainly to measure the plant roots and structural traits. The nondestructive 2D techniques such as agar plates or rhizotrons have been integral to our understanding of root development while 3D imaging with a sophisticated platform that involves NMR and X-ray CT technologies are highly expensive and is evolving to meet the needs of field phenotyping. In this chapter, we summarize root response to resource deficient condition, needs of root phenotyping, methods of root phenotyping, integration of root traits into breeding program for improving resource use efficiency are discussed in detail.

**Keywords** Root · Phenotyping · Resource use efficiency · Mini-rhizotron · Image-based technique

## 10.1 Introduction

Many countries, including India, have traversed the journey from deficit to plenty of food during the Green Revolution triggered by agricultural technologies (Eliazer Nelson et al. 2019). During this phase of agriculture across the world, the genetic improvement of crops accompanied by technologies for production and protection

J. Rane (✉) · P. S. Basavaraj · K. K. Jangid · V. Hegde
ICAR-National Institute of Abiotic Stress Management, Baramati, Maharashtra, India
e-mail: jagadish.Rane@icar.gov.in

H. M. Mamrutha
ICAR-Indian Institute of Wheat and Barley Research, Karnal, Haryana, India



H. M. Mamrutha et al. (eds.), *Translating Physiological Tools to Augment Crop Breeding*, https://doi.org/10.1007/978-981-19-7498-4_10

substantially contributed to self-sufficiency in food production. However, there is evidence that the progress in food production occurred at the cost of environmental sustainability due to overuse of resources that have reduced productivity of land and quality of aquifers (John and Babu 2021; Smith et al. 2016). In addition, there is evidence that the nonjudicious use of nitrogen fertilizers can contribute to greenhouse gas such as nitrous oxide, which is a highly potential contributor to climate change (Sapkota et al. 2021). Concerted efforts to understand the climate change events have now conclusively proved that accelerated global warming is inevitable in the absence of precautionary and committed measures to mitigate greenhouse gases (IPCC 2021). Hence, the focus should be on adaptation and mitigation options at all possible scales and sectors, including agriculture. Nevertheless, such measures will compel the agriculture-input industries to adopt technologies that can add cost to their products, which can get reflected in increased cost of input and hence elevation in cost of cultivation of crop plants. In this context, resource use efficiency in crop plants assumes immense significance for reducing the cost to both the farmers and the environment.

Plants utilize natural resources, including solar radiation, water, carbon dioxide, and nutrients from the soil, for growth and development. Farmers provide additional inputs when these resources are limited due to the nature of agro-ecology and particular soil conditions. While resources that are to be acquired from the atmosphere are plenty during the growth of the plants, the efficiency of utilization of these resources are often limited by access to water and nutrients (Hodapp et al. 2019). However, crop intensification has resulted in a substantial reduction in the nutrient status of plants (Kopittke et al. 2019), and access to irrigation water is a constraint in drought-prone arid and semi-arid regions (Mitter and Schmid 2021). Further, predicted extremes in rainfall can aggravate the situation by exposing the crop plants to the deficit and excess soil moisture (IPCC 2021). Hence, the future crops should have the inherent capacity to withstand water stress caused by both the extremes of levels of soil moisture that, in turn, can determine plants' access to nutrients. In this context resource use efficiency can be defined as crop productivity per unit of resource including those provided by farmers and also those which are naturally exist in the agro-ecosystem (Hodapp et al. 2019). Resource use efficiency in broad sense includes radiation use efficiency (Kiniry et al. 1999), water use efficiency (Niu et al. 2011), nutrient use efficiency (Baligar et al. 2001), energy use efficiency etc. However, among the natural resources the soil moisture and the nutrient are highly critical for plant growth and development and are routed through plant root. In this regard, the inherent capacity of plants, as well as production technologies to maneuver the architecture of roots, is highly crucial in improving the resource use efficiency of crop plants and for ensuring sustainable food production in the events of amplified versions of abiotic stresses predicted due to climate change. Here we have made an attempt to focus on water and resource use efficiency and the need for phenotyping root responses to environmental stimuli.

## 10.2 How Can Resource Use Efficiency Be Enhanced?

Productive use of resources for growth and development is largely determined by those essential factors, which are limited though the other factors are in abundance (Briat et al. 2020; von Liebig 1855). Since much of these growth limiting factors, including nutrients and water are in the soil, their efficient acquisition and utilization are critical for the productivity of crop plants. The root system has to be established well before these resources are lost into the environment or transformed into an unavailable form. Hence, the efficiency with which this task is achieved can substantially determine the resource use efficiency of crop plants. In this regard, much of the research efforts were devoted to creating appropriate agronomic conditions to facilitate root development and their performance (Koevoets et al. 2016).

Consequently, there has been a substantial change in crop management practices that have resulted in enhanced food production. Now there has been enhanced focus on the inherent capacity of plants to take up and utilize the nutrient and water through deeper insights into the root system architecture responses and existing genetic variation in associated traits. Several studies on roots' responses to limitations imposed by soil moisture deficit and nutrients can hint about possible traits that have to be investigated for crop improvement (Kim et al. 2020) (Fig. 10.1).

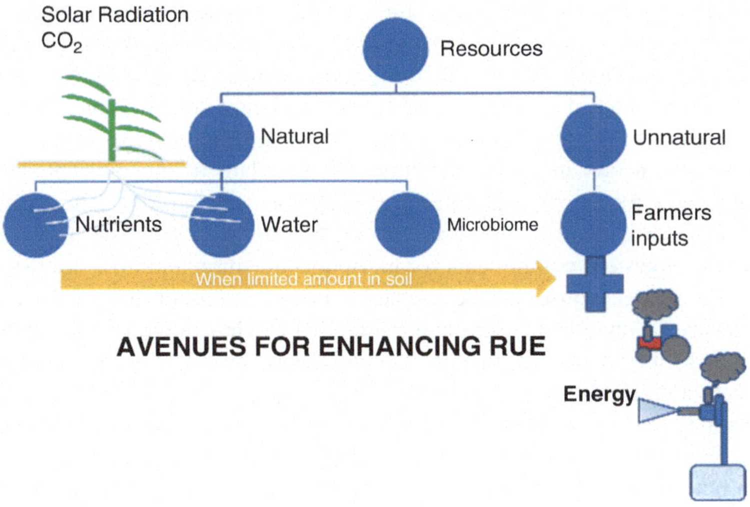


**Fig. 10.1** Avenues for enhancing resource use efficiency (RUE)

## 10.3 Root Responses to Resource Deficiency

The root system architecture of the plants is represented by its three-dimensional arrangement determined by the length and thickness of the primary root as in dicots or seminal roots in graminaceous crops and features of lateral roots such as density, length, thickness, and angles. Root system architecture at any point of time during the growth and development determines the soil space that a plant explores (Orman-Ligeza et al. 2013). In addition, the root hairs enable plants to intensively exploit resources such as water and nutrient from the soil space occupied by roots. Further, the temporal adjustment for resource acquisition is accomplished through the high degree of plasticity of root systems. Some of these aspects have been discussed in the following section concerning soil moisture and nutrients.

Crop plants adapt to soil moisture deficit through different mechanisms that involve an adjustment in the physiological process, such as osmosis or root morphology that includes root system architecture. At the same time, plants respond to the former mechanism during severe water deficits that allow them to survive. Changes in root system architecture enable plants to perform when soil moisture is relatively high, particularly in deeper soil layers. Under such conditions, increased penetration of roots, density, and proportion of biomass partitioned into roots help maintain the productivity of crops (Uga et al. 2013; Hsiao and Xu 2000). Signals sent by the root system for stomatal closure for tissue dehydration can be one of the adaptive mechanisms under drought stress (Medrano et al. 2002). In response to decreasing soil moisture levels, plants tend to form more lateral and thinner roots, which can increase the surface area for water and nutrient absorption with the optimum investment of assimilates for root development (Henry et al. 2012). Changes in anatomy manifested by changes in xylem vessel, the diameter of sclerenchyma cell, hydraulic conductivity, and changes in exudation of sap have also been reported as adaptive responses to soil moisture deficit (Henry et al. 2012).

During growth and development, plants have to take up soil nutrients, which are mostly heterogeneous in space and time. Plants adapt to the nutrient deficit by modifying their root system architecture to explore soil profiles efficiently. Such changes may be general or, to some extent, nutrient specific. The importance of individual root system components for nutrient acquisition and how developmental and physiological responses can be coupled to increase nutrient foraging by roots has been extensively reviewed (Giehl and Wirén 2014). Here some of them have been highlighted in the context of phenomics approaches for crop improvement through root system architecture.

Plants may respond to nutrient availability in the soil in a dose–response manner as nitrogen. During mild deficiency of nitrogen deficiency, root length tends to increase but not during severe deficiency (Gruber et al. 2013). On the other hand, plants may invest less in roots as nutrient availability reduces. Since soil is highly heterogeneous concerning available nutrients, plants have strategies to extend their root system as and when and where it is essential. For example, root hairs and lateral roots are developed only in the soil space where nutrients exist (Giehl et al. 2012).

The role of lateral roots in the acquirement of phosphorus, manganese, zinc, and copper is more significant than the acquisition of mobile nutrients such as sulfur and nutrient (Liu et al. 2018). The root system responses driven by in-built mechanisms to efficiently monitor and forage nutrient-rich patches in soils need to be revisited for developing root phenomics strategy for resource use efficiency. There is a need to explore natural genetic variation for the root system architecture response to nutrients (Ristova and Busch 2014).

## 10.4 Root Phenotyping Needs

Despite persistent efforts, our understanding of the root responses of plants to different environmental stimuli and management practices is limited mainly because they cannot be monitored so easily. Most of the studies have their interpretations based on the plant's capacity to produce roots through biomass investment under normal and stressed conditions (Paez-Garcia et al. 2015). Studies have revealed that plant capacity to spread their roots as reflected by root length density is critical rather than mere root biomass (Faye et al. 2019). However, it is essential to note that need-based investment in roots reflected by plasticity is essential to obtain nutrients and water from the soil to capture them before they are lost in the environment. Hence, the rate at which plant roots grow to acquire the resources can be one of the traits to be phenotyped to improve the resource use efficiency of crop plants. Other traits that need attention for phenotyping include the capacity of roots to modify the rhizosphere to convert immobilized nutrients into an available form, root capacity to facilitate beneficial microbe that contributes to nutrient acquisition, and plant growth through stimulating substances they produce at the rhizosphere (Backer et al. 2018). In addition, phenotyping root capacity to withstand excess soil moisture and salts in soil can contribute to genetic improvement in resource use efficiency. Deeper insights into the aspects contributing to gain in space and time by the root during plants growth and development can provide clues to phenotype roots for efficient use of resources for different agro-ecologies.

The gain in space by plant roots can be achieved through gain in mass, volume, density, and notable architecture, while the gain in time is achieved through the rate at which roots reach the resource-rich regions in the soil. Phenotyping efforts are needed to capture the genetic variation in these traits to improve plants' capacity to efficiently acquire the resources, particularly when plants adapt escape mechanisms to survive abiotic stresses such as drought and high temperature.

## 10.5 Demonstrated Efforts and Success of Root Phenotyping

There has been tremendous progress in phenotyping that starts from novel approaches of shavelomics (Trachsel et al. 2011) to 3D imaging with a sophisticated platform that involves NMR and X-ray CT technologies (Metzner et al. 2015). There has been a success in understanding the root system architecture with image-independent strategies as demonstrated in the rice deep root system (Uga et al. 2013), though they are labor-intensive and time-consuming. On the other hand, 3D image-based phenotyping is highly expensive and is evolving to meet the needs of field phenotyping. Some of the root phenotyping methods followed have been illustrated by Kim et al. (2020).

### *10.5.1 Perforated Basket Method*

Investigations on the genetic variation in the capacity of rice to grow deep roots led to the identification of the *DRO1* gene on chromosome 9 of rice (Uga et al. 2013). Advantage of this gene in drought tolerance has been successfully demonstrated by introgressing the gene in rice variety IR64 without yield penalty in sufficient-watered conditions. A phenotyping strategy for quantifying deep root traits involved perforated baskets that allow deep and shallow roots differentiation.

### *10.5.2 Soil Cylinder and Visual Scoring*

During the harvesting, roots excavated by removing a soil cylinder of 25-cm depth and 40 cm in diameter, with the plant base as the horizontal center of the soil cylinder. Roots are scored visually after washing (Trachsel et al. 2011). This method could enable quantification of root traits such as number and branching density of brace roots and number, angle, and branching density of crown root of maize plants.

### *10.5.3 Cylinder Containing Growth Medium*

In this method, germinated seeds of rice are transplanted in petri plates into glass growth cylinders containing 1.3 L of transparent gellan gum growth medium. The camera is located in front of the growth cylinder. Daily image sequences were captured for each plant root system grown in the growth medium, consisting of 40 silhouette images taken every 9° for the entire 360° of rotation. RootReader3D software was used to analyze the 3D root images (Clark et al. 2011). More than

27 root parameters were measured using this system. This system emerged as a powerful platform to measure root traits precisely. In addition, it enabled researchers to carry out a detailed study of the genetics of root system architecture.

### 10.5.4 Soil-Filled Root Chambers

This phenotyping platform consisted of 500 soil-filled root chambers (50, 45, 0.3 cm in size), made of transparent Perspex sheets placed in metal tubs and covered with polycarbonate sheets. After sowing, around 3 weeks later, first flush of nodal roots was visible. Roots were imaged in situ by an imaging box with digital cameras that were remotely controlled by two android tablets. Then free software (openGelPhoto.tcl) used for accurate measurement of various parameters including nodal root angle and other from the digital images. Root system architecture and nodal root angle could be measured by this method in sorghum plants (Joshi et al. 2017).

### 10.5.5 Buried Herbicide Method

This is a field-based method for large-scale screening of genotypes for deep, extensive roots by banding the herbicide metribuzin at a depth of 46, 64, and 76 cm. In this method, deep-rooted genotypes die more rapidly than shallow-rooted ones (Al-Shugeairy et al. 2014). This method is mainly for the phenotyping of rooting depth.

### 10.5.6 Agar Gel Method

Arabidopsis plant images were captured in the agarose gel condition contained in vertically arranged plates to allow roots to grow on the medium. Root system architecture—length, curvature, and stimulus response parameters (French et al. 2009). A study by Aziz et al. (2020) employed agar media to investigate root growth in black eye bean and Mungbean.

### 10.5.7 Grow Screen-Agar

Arabidopsis plants were grown in a petri dish filled with agar in this method. Images are captured either once or continuously. Imaging multiple sensors are deployed which operate simultaneously (Nagel et al. 2020).

### 10.5.8 NIR Image-Based Technique

This system measures the root system architecture (RSA) of maize, rapeseed, barley in high-throughput manner. For capturing images, transparent media was used, and NIR imaging was used to capture roots. Using this system's direction of root growth, root volume and other RSA traits were analyzed (Shi et al. 2018).

### 10.5.9 RGB Image-Based Technique

In this method, the root images were acquired using RGB camera after the staining of roots with powdered active charcoal. The images in jpg or tiff format were analyzed using ImageJ software. The method was compared with root system scanning with the help of specialized root scanner (STD4800 scanner) attached with WinRHIZO Pro software (Regent Instruments, Quebec, Canada) (Slota et al. 2016). In the destructive methods, total root length, root system surface, root volume, root diameter, and a number of tips were measured, while nondestructive methods could quantify root system depth, projected root surface, and the sum of the root lengths of spring barley.

### 10.5.10 Blotting Paper Method

Germinated seeds of maize were transferred to wet blotting paper in envelops. Root images were captured by the scanner and then analyzed by WinRHIZO software (Hund et al. 2009). This method could be studied for the root morphology of axile and lateral roots.

### 10.5.11 Backhoe-Assisted Monolith Method

This is a rapid method of collecting root samples from constant soil volume. In this method, soil samples with 20 cm width and 25 cm depth were excavated by the cylindrical steel monolith, from which root samples were then isolated. Root images were captured by the scanner and then analyzed by software (Teramoto et al. 2019). Different root systems such as shallow, intermediate, and deep roots can be quantified using this method.

### *10.5.12 Rhizoslides: Paper-Based Method*

Both the sides of the glasses are covered with germination paper for root growth in this method. Growing roots are captured through hyperspectral microscopy imaging and then analyzed by 'Smart Root', GiaRoots, and 'WhinRhizo'. WhinRhizo provided the most accurate results (Le Marié et al. 2014).

### *10.5.13 Field Mini-Rhizotron Method*

Field mini-rhizotrons were set up to investigate the roots of winter wheat. Transparent rhizotubes were inserted into the soil and images were captured with the help of the camera. The camera was situated using an indexing handle at 20 observation locations in the tubes (Cai et al. 2016). This method could help in root development and distribution, the number of total roots at different soil levels, and the number of roots per observation depths. Lu et al. (2019) deployed mini-rhizotron to phenotype the pepper root system. This enabled nondestructive measurement of root system architectural traits rapidly.

### *10.5.14 $^{15}N$ Tracer Method*

Deep rooting is a promising plant feature for enhancing crop productivity when water is scarce. This semi-field phenotyping method combines mini-rhizotrons to phenotype deep rooting in wheat (Chen et al. 2019).

### *10.5.15 X-Ray CT Method*

Sweet pea and sunflower seeds were planted on the surface and were grown for 30 days. An X-ray microtomography image was measured by high-resolution XMT beamline 8.3.2 at the Advanced Light Source (Lawrence Berkeley National Laboratory, USA). Transmitted X-ray light is converted to visible light using a CdWO4 single crystal scintillator, magnified by a Canon 2X lens, and imaged on a Cooke PCO 4000 CCD camera (Aravena et al. 2011). This technique assessed soil moisture flow from soil aggregate to the roots. Shao et al. (2021) combined the X-ray CT method with root pulling force, a high-throughput method of root extraction that offers an estimate of the root mass is linked to multiple 3D traits. This can be used to calibrate and interpret root pulling force measurements across a range of experimental contexts or scaled up as a stand-alone approach in extensive genetic studies of the root system. Teramoto et al. (2020) improved X-ray CT process flow to study regular RSA development in rice, enabling 4D RSA phenotyping.

### 10.5.16 *RhizoTubes*

This consists of an auto image capturing unit RhizoCab, which includes cameras and robots. This system was used to study the response of root traits to biotic and abiotic stresses in Medicagospp, rapeseed, etc. (Jeudy et al. 2016).

### 10.5.17 *MISIRoot: Minimally Invasive,* in Situ *Imaging System for Plant Root Phenotyping*

This robotics-based, nondestructive, in situ phenotyping method works under natural soil. This system consists of a camera for image capturing and image processing software for root recognition and feature extraction. It captures high-quality root images three-dimensionally and measures root structure in maize with an accuracy of 0.1 mm (Song et al. 2021).

### 10.5.18 *ChronoRoot*

This method is based on a deep learning technique which combines 3D-printed open-hardware with deep segmentation networks for high temporal resolution phenotyping of plant roots in agar medium. This method broadens the scope of root high-throughput phenotyping for genetics and natural variation studies, as well as the discovery of novel root features through the screening of clock-related mutations (Gaggion et al. 2021).

## 10.6 Way Forward

Though the plant root investigations have gained accelerations during the last couple of decades, much of the leads obtained through mechanisms involved or methods developed are yet to be translated into practical use for genetic improvement of resource use efficiency. Image-based root phenotyping platforms have opened up new avenues to assess genetic variation in root system architecture. Efforts are needed to optimize the method for phenotyping root plasticity in varying edaphic conditions that need simulation of natural method in a controlled environment or precise phenotyping tools for measurements in natural field conditions. A plant's capacity to proliferate root precisely in the nutrient-rich zone of soil space can be an interesting trait to discover for improving nutrient use efficiency. Phenotyping root traits influenced by microbe interaction in the rhizosphere can be another essential exploration avenue. Root phenotyping techniques have to be optimized for each of the agro-edaphic conditions as root responses are location-specific, resource-specific, and governed by developmental phases of the plant. Replacement or complementation of conventional destructive approaches with noninvasive tools can accelerate crop improvement efforts for an efficient root system that can allow plant growth with less nutrients and water, making a substantial proportion of resources used for its growth and development. This approach can help improve farmers' profit without compromising the sustainability of the agro ecosystems.

## References

Al-Shugeairy Z, Islam MS, Shrestha R, Al-Ogaidi F, Norton GJ, Price AH (2014) High throughput screening of rooting depth in rice using buried herbicide. Ann Appl Biol 165(1):96–107

Aravena JE, Berli M, Ghezzehei TA, Tyler SW (2011) Effects of root-induced compaction on rhizosphere hydraulic properties-X-ray microtomography imaging and numerical simulations. Environ Sci Technol 45(2):425–431

Aziz AA, Lim KB, Rahman EKA, Nurmawati MH, Zuruzi AS (2020) Agar with embedded channels to study root growth. Sci Rep 10(1):1–12

Backer R, Rokem JS, Ilangumaran G, Lamont J, Praslickova D, Ricci E, Smith DL (2018) Plant growth-promoting rhizobacteria: context, mechanisms of action, and roadmap to commercialization of biostimulants for sustainable agriculture. Front Plant Sci 9:1473

Baligar VC, Fageria NK, He ZL (2001) Nutrient use efficiency in plants. Commun Soil Sci Plant Anal 32(7–8):921–950

Briat JF, Gojon A, Plassard C, Rouached H, Lemaire G (2020) Reappraisal of the central role of soil nutrient availability in nutrient management in light of recent advances in plant nutrition at crop and molecular levels. Eur J Agron 116:126069

Cai G, Vanderborght J, Klotzsche A, van der Kruk J, Neumann J, Hermes N, Vereecken H (2016) Construction of minirhizotron facilities for investigating root zone processes. Vadose Zone J 15(9):1–13

Chen S, Svane SF, Thorup-Kristensen K (2019) Testing deep placement of an 15N tracer as a method for in situ deep root phenotyping of wheat, barley and ryegrass. Plant Methods 15(1): 1–12

Clark RT, MacCurdy RB, Jung JK, Shaff JE, McCouch SR, Aneshansley DJ, Kochian LV (2011) Three-dimensional root phenotyping with a novel imaging and software platform. Plant Physiol 156(2):455–465

Eliazer Nelson AR, Ravichandran K, Antony U (2019) The impact of the green revolution on indigenous crops of India. J Ethnic Foods 6(1):1–10

Faye A, Sine B, Chopart JL, Grondin A, Lucas M, Diedhiou AG, Gantet P, Cournac L, Min D, Audebert A, Kane A (2019) Development of a model estimating root length density from root impacts on a soil profile in pearl millet (Pennisetum glaucum (L.) R. Br). Application to measure root system response to water stress in field conditions. PLoS One 14(7):e0214182

French A, Ubeda-Tomás S, Holman TJ, Bennett MJ, Pridmore T (2009) High-throughput quantification of root growth using a novel image-analysis tool. Plant Physiol 150(4):1784–1795

Gaggion N, Ariel F, Daric V, Lambert É, Legendre S, Roulé T, Camoirano A, Milone DH, Crespi M, Blein T, Ferrante E (2021) ChronoRoot: high-throughput phenotyping by deep segmentation networks reveals novel temporal parameters of plant root system architecture. Giga Sci 10(7):giab052

Giehl RF, von Wirén N (2014) Root nutrient foraging. Plant Physiol 166(2):509–517

Giehl RF, Lima JE, von Wirén N (2012) Localized iron supply triggers lateral root elongation in Arabidopsis by altering the AUX1-mediated auxin distribution. Plant Cell 24(1):33–49

Gruber BD, Giehl RF, Friedel S, von Wirén N (2013) Plasticity of the Arabidopsis root system under nutrient deficiencies. Plant Physiol 163(1):161–179

Henry A, Cal AJ, Batoto TC, Torres RO, Serraj R (2012) Root attributes affecting water uptake of rice (Oryza sativa) under drought. J Exp Bot 63(13):4751–4763

Hodapp D, Hillebrand H, Striebel M (2019) "Unifying" the concept of resource use efficiency in ecology. Front Ecol Evol 233:49099

Hsiao TC, Xu LK (2000) Sensitivity of growth of roots versus leaves to water stress: biophysical analysis and relation to water transport. J Exp Bot 51(350):1595–1616

Hund A, Trachsel S, Stamp P (2009) Growth of axile and lateral roots of maize: I development of a phenotyping platform. Plant Soil 325(1):335–349

IPCC (2021) Climate change 2021: the physical science basis. In: Masson-Delmotte V, Zhai P, Pirani A, Connors SL, Péan C, Berger S, Caud N, Chen Y, Goldfarb L, Gomis MI, Huang M, Leitzell K, Lonnoy E, Matthews JBR, Maycock TK, Waterfield T, Yelekçi O, Yu R, Zhou B (eds) Contribution of working group I to the sixth assessment report of the intergovernmental panel on climate change. Cambridge University Press, Cambridge, 2391 pp

Jeudy C, Adrian M, Baussard C, Bernard C, Bernaud E, Bourion V, Busset H, Cabrera-Bosquet L, Cointault F, Han S, Lamboeuf M (2016) RhizoTubes as a new tool for high throughput imaging of plant root development and architecture: test, comparison with pot grown plants and validation. Plant Methods 12(1):1–18

John DA, Babu GR (2021) Lessons from the aftermaths of green revolution on food system and health. Front Sust Food Syst 5:644559

Joshi DC, Singh V, Hunt C, Mace E, van Oosterom E, Sulman R, Jordan D, Hammer G (2017) Development of a phenotyping platform for high throughput screening of nodal root angle in sorghum. Plant Methods 13(1):1–2

Kim Y, Chung YS, Lee E, Tripathi P, Heo S, Kim KH (2020) Root response to drought stress in rice (Oryza sativa L.). Int J Mol Sci 21(4):1513

Kiniry JR, Tischler CR, Van Esbroeck GA (1999) Radiation use efficiency and leaf CO2 exchange for diverse C4 grasses. Biomass Bioenergy 17(2):95–112

Koevoets IT, Venema JH, Elzenga JT, Testerink C (2016) Roots withstanding their environment: exploiting root system architecture responses to abiotic stress to improve crop tolerance. Front Plant Sci 7:1335

Kopittke PM, Menzies NW, Wang P, McKenna BA, Lombi E (2019) Soil and the intensification of agriculture for global food security. Environ Int 132:105078

Le Marié C, Kirchgessner N, Marschall D, Walter A, Hund A (2014) Rhizoslides: paper-based growth system for non-destructive, high throughput phenotyping of root development by means of image analysis. Plant Methods 10(1):1–16

Liu X, Dong X, Xue Q, Leskovar DI, Jifon J, Butnor JR, Marek T (2018) Ground penetrating radar (GPR) detects fine roots of agricultural crops in the field. Plant Soil 423(1):517–531

Lu W, Wang X, Wang F (2019) Adaptive minirhizotron for pepper roots observation and its installation based on root system architecture traits. Plant Methods 15(1):1–14

Medrano H, Escalona JM, Bota J, Gulías J, Flexas J (2002) Regulation of photosynthesis of C3 plants in response to progressive drought: stomatal conductance as a reference parameter. Ann Bot 89(7):895–905

Metzner R, Eggert A, van Dusschoten D, Pflugfelder D, Gerth S, Schurr U, Uhlmann N, Jahnke S (2015) Direct comparison of MRI and X-ray CT technologies for 3D imaging of root systems in soil: potential and challenges for root trait quantification. Plant Methods 11(1):1–11

Mitter H, Schmid E (2021) Informing groundwater policies in semi-arid agricultural production regions under stochastic climate scenario impacts. Ecol Econ 180:106908

Nagel KA, Lenz H, Kastenholz B, Gilmer F, Averesch A, Putz A, Heinz K, Fischbach A, Scharr H, Fiorani F, Walter A (2020) The platform GrowScreen-agar enables identification of phenotypic diversity in root and shoot growth traits of agar grown plants. Plant Methods 16(1):1–17

Niu S, Xing X, Zhang ZH, Xia J, Zhou X, Song B, Li L, Wan S (2011) Water-use efficiency in response to climate change: from leaf to ecosystem in a temperate steppe. Glob Chang Biol 17(2):1073–1082

Orman-Ligeza B, Parizot B, Gantet PP, Beeckman T, Bennett MJ, Draye X (2013) Post-embryonic root organogenesis in cereals: branching out from model plants. Trends Plant Sci 18(8):459–467

Paez-Garcia A, Motes CM, Scheible WR, Chen R, Blancaflor EB, Monteros MJ (2015) Root traits and phenotyping strategies for plant improvement. Plan Theory 4(2):334–355

Ristova D, Busch W (2014) Natural variation of root traits: from development to nutrient uptake. Plant Physiol 166(2):518–527

Sapkota TB, Jat ML, Rana DS, Khatri-Chhetri A, Jat HS, Bijarniya D, Sutaliya JM, Kumar M, Singh LK, Jat RK, Kalvaniya K (2021) Crop nutrient management using nutrient expert improves yield, increases farmers' income and reduces greenhouse gas emissions. Sci Rep 11(1):1–11

Shao MR, Jiang N, Li M, Howard A, Lehner K, Mullen JL, Gunn SL, McKay JK, Topp CN (2021) Complementary phenotyping of maize root system architecture by root pulling force and X-ray imaging. Plant Phenomics 2021:9859254

Shi R, Junker A, Seiler C, Altmann T (2018) Phenotyping roots in darkness: disturbance-free root imaging with near infrared illumination. Funct Plant Biol 45(4):400–411

Slota M, Maluszynski M, Szarejko I (2016) An automated, cost-effective and scalable, flood-and-drain based root phenotyping system for cereals. Plant Methods 12(1):1–15

Smith M, Cross K, Paden M, Laban P (2016) Spring—managing groundwater sustainably. IUCN, Gland

Song Z, Qiu W, Jin J (2021) MISIRoot: a robotic, minimally invasive, in situ imaging system for plant root phenotyping. Trans ASABE 64(5):1647–1658

Teramoto S, Kitomi Y, Nishijima R, Takayasu S, Maruyama N, Uga Y (2019) Backhoe-assisted monolith method for plant root phenotyping under upland conditions. Breed Sci 69:508–513

Teramoto S, Takayasu S, Kitomi Y, Arai-Sanoh Y, Tanabata T, Uga Y (2020) High-throughput threedimensional visualization of root system architecture of rice using X-ray computed tomography. Plant Methods 16(1):1–14

Trachsel S, Kaeppler SM, Brown KM, Lynch JP (2011) Shovelomics: high throughput phenotyping of maize (Zea mays L.) root architecture in the field. Plant Soil 341(1):75–87

Uga Y, Sugimoto K, Ogawa S, Rane J, Ishitani M, Hara N, Kitomi Y, Inukai Y, Ono K, Kanno N, Inoue H (2013) Control of root system architecture by DEEPER ROOTING 1 increases rice yield under drought conditions. Nat Genet 45(9):1097–1102

von Liebig JF. Die grundsätze der agricultur-chemiemitrücksicht auf die in englandangestelltenuntersuchungen. 1855.

# Chapter 11
# Root System Architecture and Phenotyping for Improved Resource Use Efficiency in Crops

**Anita Mann, Shoaib Mirza, Priyanka Chandra, Ashwani Kumar, Arvind Kumar, Sujata, Naresh Kumar, Pooja, S. K. Sanwal, Neeraj Kulshrestha, Bikram P. Banerjee, and Surya Kant**

**Abstract** Climate change, increasing population, devastating resources, and other environmental factors have many challenges for sustainable crop production globally. To pace with these future problems, plant researchers are underway to design robust cultivars with optimal performance under varying conditions to bridge the yield gap. Mostly, above-ground traits are focused for crop selection whereas root system is mostly affected by soil properties, creating yield gap. Furthermore, water and all nutrients distributed in the soil are absorbed and translocated by roots for plant growth. The root spatial arrangement in the soil along with its structural attributes defines the complete root system. Optimal absorption and translocation of water and essential nutrients under abiotic or biotic stresses are particular features of the robust root system. Hence, the overall plant development is actually a performance indicator of the root system. To explore the root architecture system and its structural phenes, therefore, holds great potential for sustainable crop productivity under sub-optimal conditions. Moreover, the advancement of third generation technologies like, high-throughput sequencing, omics, GWAS, and sensing, etc. have progressed over the last few years focusing on root traits in development of improved crop plants. Briefly, root structure and its components for better

---

A. Mann (✉) · P. Chandra · A. Kumar · A. Kumar · Sujata · N. Kumar · S. K. Sanwal · N. Kulshrestha
ICAR-Central Soil Salinity Research Institute, Karnal, Haryana, India
e-mail: Anita.mann@icar.gov.in

S. Mirza · B. P. Banerjee
Agriculture Victoria, Grains Innovation Park, Horsham, VIC, Australia

Pooja
ICAR-Suracane Breeding Institute, Regional Research Station, Karnal, Haryana, India

S. Kant
Agriculture Victoria, Grains Innovation Park, Horsham, VIC, Australia

School of Applied Systems Biology, AgriBio, La Trobe University, Bundoora, VIC, Australia

 
H. M. Mamrutha et al. (eds.), *Translating Physiological Tools to Augment Crop Breeding*, https://doi.org/10.1007/978-981-19-7498-4_11

phenotyping and the techniques involved for root phenotyping has been compiled in this chapter.

**Keywords** Root · Phenes · Phenotyping · Root structure architecture · Modeling

## 11.1 Introduction

Roots are the primary organ of the plant system, which allows the plant to anchor in the soil and supply the essential nutrients and water for plant growth and development (Lynch 2013). Roots provide the interface between plants and the complex soil environment, and their spatial distribution determines the source-sink relationship. The spatial distribution of the root system in the soil, such as root length, spread, number, and length of lateral roots, is referred to as root system architecture (RSA) that describes the morphological and structural organization of the root. The use of RSA as a trait to boost crop productivity is slow, even though it is an essential factor for overall plant growth. Although many studies are available for phenotyping of above-ground traits, research on root phenotyping still needs improvement.

Roots as well as shoots both are the equal growing axis for plant development. Although differentiation of roots and shoots occur in separate regimes but the specific plant architecture is based on the direct communication between these plant parts. Understanding the different parameters that define RSA and function is actually an expression of plant phenology and the surrounding root environment (Paez-Garcia et al. 2015; Tron et al. 2015). The root, generally defined as "hidden half of plants" show interspecific or intraspecific variations along with genotypic and/or environmental differences. Variations in the root structure acrhitecture are specific in different plant species as shown by land plants in terms of branching patterns with profuse to light for the effective performance. A specific root development program after the embryonic stage governs the development of RSA which offers flexible plasticity in root phenotypes under different environmental conditions. This phenomenon involves meristematic cell division in apical root with lateral cell division in pericycle further leading to root elongation through cell expansion. This cell division and expansion helps the plants to define their root system with best adaptive features under abiotic or biotic stress conditions. For better plant productivity, most of the variable edaphic resources should be accessed through RSA (de Dorlodot et al. 2007). During seed germination in cereals, emergence of radicle occurs through the coleorhiza and coleoptile follows after little extension of root (Shu et al. 2016; Ma et al. 2017). The newly developed roots will then acquire their signature RSA affecting further seedling establishment, growth, and crop productivity (Atkinson et al. 2015). Abiotic stresses affect crop productivity by interfering with the uptake and transport of essential nutrients along with water for optimum grain yield. One of the ways to dilute or buffer the adverse effects of stress on crop productivity is to select unique root phenes and underline the corresponding genes for RSA for better water and nutrient use efficiency. Plant ideotypes for root architecture possess interconnected multi-component traits. Lynch

(2013) model of steep-deep-cheap crop ideotype integrates root angel and root depth for increasing use efficiency of nitrogen (N) and water in some particular conditions. In spite of a peculiar RSA for each crop, some root classes have the conserved ontogeny across plants which can be used across multiple species. Advancement of gene identification for RSA has been explored in Arabidopsis and major cereals such as maize, wheat, and rice and few oilseeds such as soybean, but the genetic control of root system and traits/phenes needs elaborated research, specifically RSA under abiotic stress.

Root phenome of a plant constitutes a set of phenes where phenes are to phenotype similar to genes for genotype in a genome. RSA phenes are specific for resource uptake from soil, and their mechanisms at cellular and genetic level in crop breeding programs is not completely elucidated due to limited studies on functional utilities of root phenes in different environments. Crop yield is an extreme example of an aggregate phene, the main output of a plant phenotype. It consists of many basic and unique root phenes, interacting with different environments. Therefore, this article briefly describes the strategies for capturing root phenes of crop plants for better water and nutrients use efficiency.

## 11.2 Components of Root Structure Architecture (RSA)

RSA is a fundamental trait for plant water and other nutrients acquisition from different soil layers in different environmental conditions (Giehl et al. 2014). The efficient uptake of fertilizers/nutrients can be achieved by using genotypes with an efficient root system since deficiency of water and nutrients leads to a change in root–shoot growth ratio (Gruber et al. 2013). These changes in root growth and architecture in response to limited nutrients or water availability can highlight plant signal transduction and molecular mechanisms (Kellermeier et al. 2014). In conditions of water limitation or more salts, roots adapt by changing their morphology specifically root plasticity (Forde 2009). Therefore, for abiotic stress tolerance, root traits need to be identified for increasing the root capacity to forage deeper in soil at the minimal cost of respiration. A brief account of few important root phenes that can be used for crop production and improvement are described here.

### *11.2.1 Root Morphological Phenes*

Majorly, growing root angle, length and branching, and root hair defines the total root volume and surface area within the soil profiles with random distribution of water and nutrients (Lynch 2011). The fate of a plant is influenced by the environmental condition where the seed is sown, which further determines the root–shoot development, plant growth and yield. For root structure architecture, the potential metrics of a robust root system, such as, the average shape, size, and branching show

**Table 11.1** Spatial root characters of a standard root system

| Global root traits | Root traits derived from individual roots | Static root traits | Dynamic root traits |
|---|---|---|---|
| Total root length | • Root length<br>• Root number<br>• Root diameter<br>• Branching angle | Measured at a particular time of growth stage | Dynamic changes in root traits with time and space and different ecosystems |
| Spatial root distribution<br>• Root length density<br>• Rooting depth<br>• Root system width<br>• Area covered by roots | | | |

variation in their expression in different environments. Rellán-Álvarez et al. (2015) have observed suppressed lateral root growth and angle in the upper soil under water deficiency in *Arabidopsis* while decreased crown root growth was observed in *Setaria viridis* by Sebastian et al. (2016). Whereas, in rice, adventitious roots are developed under flooding while root elongation occurred in *Arabidopsis* (Lin and Sauter 2018). Less irrigation and use of deep stored water are required in future to face climate change. Nagel et al. (2012) have described the global, static, and dynamic root traits for root structure architecture (Table 11.1).

Root hairs are the first finite part of root architecture covering major ratio of root surface area whereby contributing for more than 50% water absorption. Nambiar, in 1976, compared root hairless mutant and wild-type of Arabidopsis plants depicting involvement of root hairs for water uptake in terms of sensitivity of mutant to salt, drought, and heat stress. Root hairs have also been reported to support root penetration for the establishment and P acquisition (Haling et al. 2013; Brown et al. 2013). A significantly higher plant biomass has been reported in Phaseolus vulgaris due to the combination of shallow base roots and long root hairs (Miguel et al. 2015). Molecular control of root hair growth has already been reported by Mendrinna and Persson (2015), which has led to the development of plants having long and highly branched root hairs (Velasquez et al. 2011; Yoo et al. 2012), but these changes regulate water or nutrient uptake is still a topic of research. Under varying N availability conditions, long and dense root hairs can improve N acquisition in maize (Saengwilai et al. 2021). Root hairs assists in forming a larger rhizosheath, that increases water availability in the root zone in dry condition. In water stress conditions, root hair enhanced the plant water status had lower leaf abscisic acid concentration in barley (Marin et al. 2021). In Arabidopsis, root hair also help in potassium (K) uptake in roots (Ahn et al. 2004) in addition to increased root exudation and rhizosphere extension (Holz et al. 2018), facilitating beneficial microbial communities. Quantitative trait loci (QTLs) of root hair length have been mapped in barley, maize, and wheat (Gong and McDonald 2017; Zhu et al. 2005a, b; Horn et al. 2016).

Root angle is also one of the important root phene for defining the stress adaptive features and better water or nutrient uptake in cereals. Horizontal or vertical distribution of root is based on root angle only since the degree of angle correlates with energy use during root penetration in soil under limited water supply or less available nutrients (Christopher et al. 2013; Uga et al. 2013; Meister et al. 2014; Oyiga et al. 2020). Some parts of the root are hidden in the soil, while a major part is visible in open media. The visible root portion further depends on the angle of growing media and root diameter, which can be correlated with total root length and root biomass. Genome-wide association in maize revealed threonine-protein kinase 15 (ZmCIPK15) gene (LOC100285495) influencing root angle (Schneider et al. 2021) where plants with steeper root angle had better N absorption and translocation. Under drought, steeper seminal and nodal root angle of maize genotype had a higher yield (Ali et al. 2015). In saline conditions, rice genotype with a shallower root growth angle had a higher yield (Kitomi et al. 2020). In P-deficient soils, shallow basal root growth angle (BRGA) improves P acquisition in bean as most of phosphorus is available in top soil, while in terminal drought conditions, steep BRGA improves water absorption and uptake (Ho et al. 2005). Influx and efflux of auxin transporters defines root auxin supply which indirectly controls differentiation, cell elongation, and bending of plant organs. Root bending is required for root angle and gravitropism, which determines the deep or shallow formation of RSA. Positive gravitropism leads to root growing downward and deep penetration while lateral roots, seminal roots, and crown roots may have different root angles and partial deviation from gravitropism (Koevoets et al. 2016).

Root length and depth in the soil are universal traits for all crop plants to acquire optimum water and nutrients in both favorable and unfavorable conditions. A small increase of 10–30 cm in root depth can increase approx. 0.5 t/ha yield in wheat (Kirkegaard et al. 2007). Soil properties are major factors for defining root length and depth in it where root anatomy also varies with available water or nutrients. Under drought, enlarged cortical cell size (CCS) has also been associated with deeper roots along with decreased cortical cell file number where these cortical cells reduce the respiration rate of roots. Hence, roots could penetrate deep at reduced metabolic costs with simultaneously explorating the soil (Chimungu et al. 2014a, b), imparting drought tolerance.

Root branching differentiates RSA in monocots and dicots. Dicot plants have one main primary root with several lateral roots. Under stress conditions, dicots can induce the formation of adventitious roots. On the other hand, monocot plants have axial roots divided into the embryonic (seminal) and non-embryonic (shoot borne) roots. Crown roots and nodal roots are important in monocots at an early stage, while shoot-borne roots play the role for nutrient uptake at later stages. Root density depends on available water and nutrients and soil texture, where contrasting effects on RSA has been observed due to phosphate and nitrate levels. In general, low nitrate inhibits plant growth (Linkohr et al. 2002) but does not inhibit root elongation and hence deeper root length (Gruber et al. 2013). Maize plants, having fewer crown roots, developed up to 45% deeper roots with higher N uptake and translocation (Saengwilai et al. 2014), on the other hand, lateral roots help the plants to adapt efficiently on P-deficit soils (Postma et al. 2014). Screening of maize with low and high phosphorus (P) levels identified root plasticity in maize roots. Plants with more lateral roots and dense root hairs had increased grain yield with plant biomass under P deficit conditions (Bayuelo-Jiménez et al. 2011). Deep roots with simple and less branching and lateral spreadings were correlated with higher yields in modern wheat genotypes than the varieties with primary seminal roots Zhu et al. (2019).

Many other roots morphological phene plays a significant role in crop resource use efficiency. In earlier studies, Singh and Das (1986) reported positive correlation between root density and soil water absorption and grain yield. During wet season, a longer seminal root axis was observed in wheat than in dry season, which depicted better adaptability (Morita and Okuda 1994). Mandal et al. (2003) observed root density and water use efficiency in wheat are significantly correlated. The efficacy of durum wheat under salt stress was evaluated by Soni et al. (2020, 2021), conferring their adaptability toward stress environments. Tomar et al. (2016) have found a correlation between root architecture and drought tolerance at seedling as well as reproductive stages in 158 diverse wheat genotypes of Indian and Australian origins. They reported compact and deeper roots in tolerant cultivars than sensitive ones alongwith higher absorptive root surface area. Mishra et al. (1997) reported that after 115 days of sowing, roots had sufficient length which is affected by water fluctuations in water table than water regime. The root length, root surface area, and root-to-shoot ratio were observed higher at maximum tillering and ear emergence stages in wheat genotypes under rainfed situations (Kalita et al. 2011). Root length and biomass decreased with combined stress of Boron and salinity in wheat genotypes compared to salt tolerant wheat variety KRL-210 (Lata et al. 2017). Biochemical constitutes like sugars and proline increased in wheat root with salinity depicting structural modifications and adaptations in saline environments (Kumar et al. 2018). An experiment on root phenotyping in field and hydroponic system identified translocation of factor 1RS coded for better root traits like root biomass and root length in wheat (Sharma et al. 2018). Genotypic variation among wheat germplasm lines collected from Australia, Canada, South Asia, Mexico, and Latin America revealed a positive correlation between root length, surface area, dry weight, and respective shoot traits (Narayanan et al. 2014).

### *11.2.2 Root Anatomical Phenes*

Plant productivity also depends on the proper use and balance of soil resources and property of roots traits to explore soil at the minimal cost in limited water or nutrient conditions. The proposed steep, deep, and cheap root ideotypes by Lynch (2013) refers steep and deep as root architectural phenes, whereas cheap accounts for phenes with lower cost of root metabolic energy for distribution in soil layers and maintaining root physiological activities. Root cortex possesses large intracellular spaces, known as, root cortical aerenchyma (RCA) which is also one abundant element of this ideotypes. Generally these large air spaces in root are formed due to abiotic stresses or nutrient deficiencies. Their role is to enhance oxygen translocation to the growing roots and improve the root metabolic efficiency (Saengwilai et al. 2014). Development of root cortical aerenchyma (RCA) has been reported to reduce root respiration rate to improve shoot growth under P deficit soils Fan et al. (2003). In maize, development of RCA under drought leads to deeper root length with improved leaf water status and significantly higher yields (Zhu et al. 2010). While comparing root anatomy of six major legume crops, groundnut (*Arachis hypogaea*), soybean (*Glycine max*), chickpea (*Cicer arietinum*), cowpea (*Vigna unguiculata*), pigeon pea (*Cajanus cajan*), and common bean (*Phaseolus vulgaris*), Purushothaman et al. (2015) observed that xylem vessel size and the numbers were the most discriminating traits for their drought adaptation strategies. In cereals, drought tolerance is positively correlated with size of root xylem vessels. The metaxylem regulates the root axial hydraulic conductivity (Kadam et al. 2015, 2017) and thus, water uptake per unit area of root increases due to reduced xylem vessel diameter under drought stress (Giuliani et al. 2005; Comas et al. 2013) and hence, overcome the stress. Root cortical and stele ratio is significantly affected by water availability in the root system. Other anatomical traits like cell density, size, number, and configuration are involved in various pathways for translocation of essential nutrients and water (Marschner et al. 1996; Burton et al. 2013).

Some root anatomical adaptations facilitate roots to penetrate deeper into the soil and increase water uptake. In maize and wheat, root penetration in compacted soils has been deepened due to multiseriate cortical sclerenchyma (Schneider et al. 2021). Some anatomical root phene also enable microbial symbiosis, thus indirectly facilitate increased nutrition and water uptake. For instance, in maize, larger root diameters and larger aerenchyma lacunae facilitate mycorrhizal colonization (Galindo-Castañeda et al. 2019). Furthermore, root plasticity has also been reported as the best adaptive feature for efficient uptake of water and nutrients under water limited conditions (Schneider et al. 2020a, b; Sandhu et al. 2016; Kadam et al. 2015) in different crops.

The intraspecific variation of root system of *Arenaria tetraquetra* subsp. Amabilis was examined with specific reference to plant architecture, leaf functional phenes, and anatomy of stem-xylem in varying aspects of elevation at Sierra Nevada mountains, Southern Spain García-Cervigón et al. (2021). They observed that nurse cushioned conditions in different elevations and aspects were correlated with xylem and architectural traits with plants having less leaves per branch and a high compact canopy.

## 11.3 Rhizospheric Components Defining Root Traits

The immediate growing environment of roots, rhizosphere, helps to shape the root structure and ultimately the plant functioning. Rhizosphere interplays between soil microbes and plant roots to stimulate water and nutrient absorption to host plants in addition to protection against plant pathogens. It has rather been postulated that belowground microbial diversity in soil is an insurance to maintain plant health and productivity in various climatic scenarios (Wagg et al. 2011). Rhizosphere was first described by German agronomist Lorentz Hiltner in 1904 to define the vicinity of plant root colonized by unique population of microorganisms which can interact with plant roots through chemicals released by them (Walker et al. 2003). It has also been observed that land plants can also preferentially modulate rhizospheric microbiome as per their requirement through selective stimulation of certain beneficial microbes (Cook et al. 1995; Hu et al. 2018; Jacoby et al. 2021). The annual plants export 30–60% of net fixed carbon through roots into rhizosphere, called rhizodeposition, which may change the physical, biological, and chemical properties of rhizospheric soil drastically (Kuzyakov et al. 2007). Rhizo deposition directs the plant–soil–microbial interactions of the soil ecology (Fig. 11.1). Plant species types, root system including root architecture, shoot-to-root ratio, presence of carbon, photosynthesis, nodulation, and mycorrhizal colonization are the biotic factors which affect rhizo deposition along with abiotic factors like moisture, humidity, temperature, depth of root and also soil properties such as texture and nutrient availability. Rhizo deposits mainly include bulky polysaccharides, predominantly

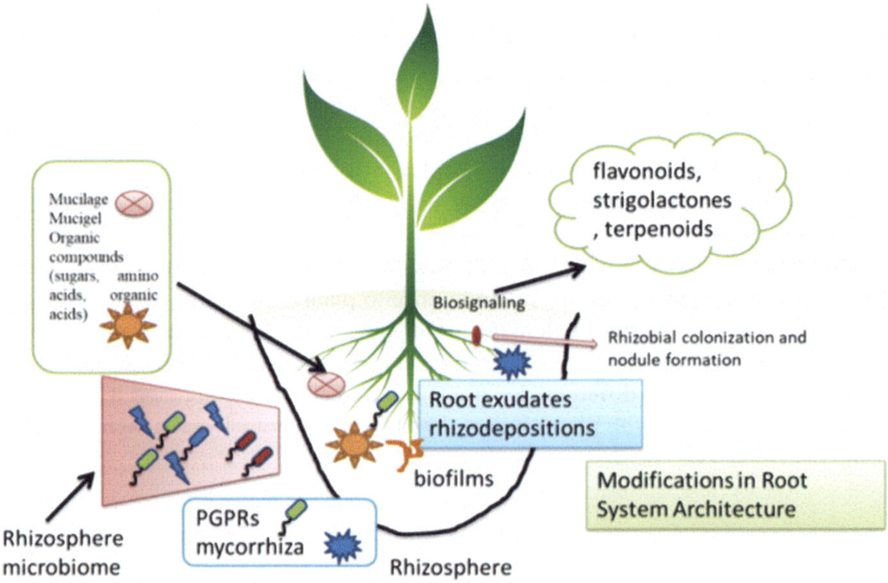


**Fig. 11.1** Plant-soil-microbial interactions in soil ecology

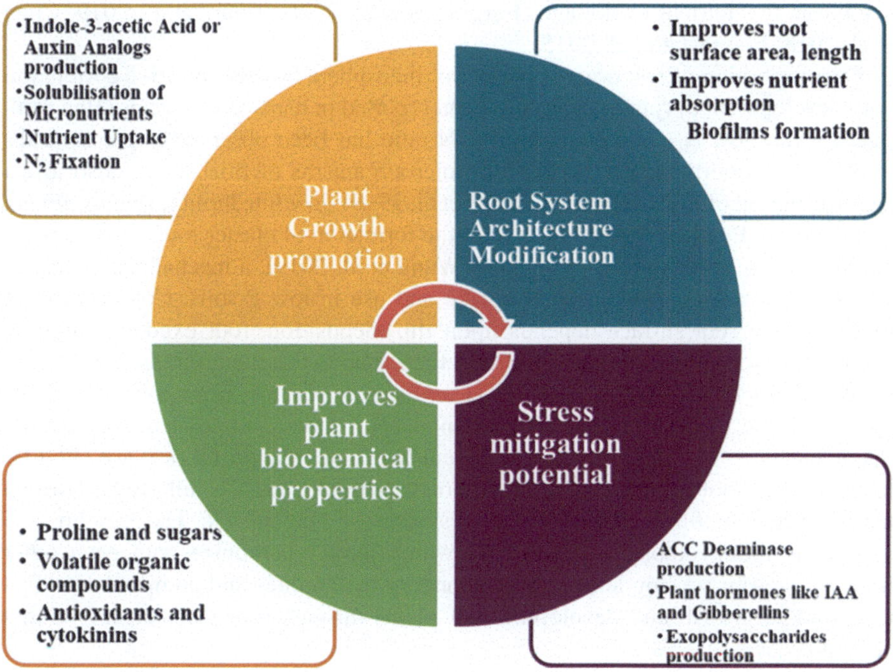


**Fig. 11.2** Mechanisms of plant growth promoting rhizobacteria (PGPR) in enhancing plant growth

mucilages which is around 12% of total rhizodeposition, in addition to several organic compounds including organic acids, sugars, enzymes, amino acids, vitamins, fatty acids, and growth factors (Hassan et al. 2019; Dey and Sengupta 2020). The host root system secretes flavonoid compounds into the rhizosphere, activating the bacterial nodulation (nod) genes. After the triggering of nod factors, a series of developmental processes follow in root system of host plant, allowing rhizobium to enter into the host cell and forming nodules where the bacteria dwell (Wang et al. 2018). Plant Growth Promoting Rhizobacteria (PGPR), dwelling in the rhizosphere of host plant, provide benefits to the host plant through several mechanisms as shown in Fig. 11.2. PGPR colonize plant roots and further improve plant growth by releasing plant hormones such as auxins or cytokinins. They also have the N fixation capability and also produce secondary metabolites such as HCN, siderophore, and ammonia which inhibit the presence of pathogens in the rhizosphere potential (Chandra et al. 2021). Various reporter genes for different mechanisms in the rhizosphere specifically including responses of bacteria with temperature, water potential, carbon, N and P availability in the soil have been reported (Kragelund et al. 1997; Herron et al. 2010; Mendes et al. 2013). Vice versa, plant root traits like root number, diameter, and specific root length are the boosting factors for rhizosphere fungal population (Sweeney et al. 2021). A "collaboration gradient" is developed between plant strategies for tissue building and fungal symbiosis for nutrient exploitage (Bergmann et al. 2020; Zhang et al. 2019; Chen et al. 2018; Eissenstat et al. 2015).

Decomposition of soil organic matter in rhizosphere has also positive correlation with development of root traits, as has been reported in hardwood species (Han et al. 2020) where low N content and high C/N ratio has been observed in roots. Additionally, phytohormones released from microorganisms of rhizosphere also affect the distribution of root architecture (Pitts et al. 1998). In white lupin (*Lupinus albus*), nodulation in P-deficit conditions lead to the formation of cluster roots at the cost of nodule resources for root development (Wang et al. 2019), a mechanism generally adopted by plants for enhanced P uptake and use in low P-soils. Colonization of PGPR on the root surface depends upon rhizodeposition, root exudates, nutrient availability, soil properties, etc. PGPRs get attracted to the chemo-attractants of root exudates which force them to get fixed on the root surface. Root hairs, cap cells, epidermal cell junctions, sites of emerging lateral roots, and axial groves are the major locations where bacterial attachment and colonization takes place. PGPRs after utilizing root exudates secrete exo-polysaccharides extra-cellularly and form a matrix called biofilms around the root system of host plants. The biofilm is a systematic micro colony of PGPRs in which they synchronize with each other through quorum sensing and release secondary metabolites and compounds which are promote growth and development of plants through root mechanisms (Gupta et al. 2017).

The role of microbes in enhanced uptake of P in roots from the soil has been in literature since very old times (Barber 1966) and N-fixers and P-acquiring microbial strains have now been identified and characterized using advanced molecular technologies. After adding organic matter, nutrient availability in soil is altered by microbial activity (Gichangi et al. 2009; Khan and Joergensen 2009). Nitrogen and phosphorus uptake gets increased with high availability of carbon in soil, thereby, creating a competition for nutrients between microbes and plants. Simultaneously, P-uptake by microbes restricts N leaching and fixation of P in soil colloids (Gichangi et al. 2009; Khan and Joergensen 2009; Ayaga et al. 2006). In that case, microbes slowly release the stored fertilizers (Docampo et al. 2010) to plant roots acting as source of nutrition.

## 11.4 Genetics of Root Phenes

High-throughput screening can expedite the selection of novel traits for crop improvement in plant breeding (Richard et al. 2015; Adeleke et al. 2019), however, high throughput screening of root traits is often limited by the lack of suitable phenotyping growth systems (Joshi et al. 2017). Root architecture ideotypes are comprised of complex, multi-component, and interconnected traits. The steep-deep-cheap ideotype, proposed by Lynch (2013) to increase N and water use efficiency for crops grown under certain conditions, integrates root angles suitable for nitrate recovery, deep rooting for water and nutrient acquisition. Several RSA traits have been used for selection of bean in Africa and latin America. Acquisition of water, N and P is significantly correlated with RSA traits in different crop plants (Henry et al. 2011; Manschadi et al. 2008; Hirel et al. 2007). When β-expansin gene GmEXPB2 was overexpressed in soybean, it improved RSA with increased P efficiency (Zhou et al. 2014; Guo et al. 2011), similarly overexpression of TaEXPB23 gene in tobacco (Han et al. 2015), as evidenced through more number of lateral roots. Yield benefit under water stress has been observed through cloning of major gene DEEPER ROOTING 1(DRO1) in rice (Uga et al. 2013). QUICK ROOTING 1 & 2 have also been mapped with DRO1 for development of arsenal root ideotypes under specific environments (Kitomi et al. 2018). In wheat and barley, the flowering gene VERNALIZATION1 is modulating root architecture (Voss-Fels et al. 2018). At early seedling stage, a significant variation for root phenotypes was seen in bread wheat genotypes (Chen et al. 2020) although the genetic determinants for basal root number are still unknown. In cereals, nodal roots, which originate from shoot, form the major root system after few weeks of root development. It has been reported in maize mutants where initiation of specific roots has been distinct indicating genetic control of different root classes. Modifications in the development of lateral root could affect the overall root structure through competition for carbohydrates, which further depends on other root phenes like number of axial roots (Trachsel et al. 2011). Few reports are available exploring genetic control of root system and further how it helps to overcome drought stress for better yields (Polania et al. 2017; Koevoets et al. 2016; Uga et al. 2015; Lynch 2011; de Dorlodot et al. 2007). Molecular mechanism of root gravity sensing and signaling through auxin, cytoskeleton, and cell wall has also been characterized (Shin et al. 2005; Blancaflor 2013; Sato et al. 2015).

QTLs linked to seminal roots and lateral branching, root hairs, and P-acquisition affecting crop yield under water scarcity and flooding situations (Hochholdinger and Tuberosa 2009) have also been identified. Lynch and Brown (2012) have used greater basal root whorl number (BRWN) trait to introgress into bean lines through direct root phenotyping. BRWN is an important root trait of RSA for better soil acquisition. Cai et al. (2012) mapped 30 QTLs in maize BC4F3 population where on chromosome 6, QTLs for root traits were located at stage I, and on chromosome 10 for number of axial root traits across all developmental stages. Many studies in maize have been conducted showing significant genetic variation for RSA contributing toward tolerance against lodging and drought along with better nutrient efficiency (Hebert et al. 1992; Tuberosa et al. 2003; Chun et al. 2005) and relevant QTLs have been identified (Liu et al. 2008; Zhu et al. 2005a, b, 2016; Hund et al. 2011). Tuberosa et al. (2003) identified root traits as good candidates for QTL prediction, and can be exploited by integrating it with crop improvement. QTLs for root length and depth, root angle and seminal root number were identified with significant phenotypic variations in spring barley by Jia et al. (2019). Root plasticity has also been reported as a potential root phene for stress adaptation through identification of genetic loci for root phenotypic plasticity in cereal crops (Schneider and Lynch 2020; Schneider et al. 2020a, b; Kadam et al. 2017; Sandhu et al. 2016). In wheat, SNP/DArt platform was used for QTLs analysis for root traits like root length, angle, and number and total 50 QTLs were identified. Furthermore, root spreading angle and root system depth were found to be important root phenes for grain enhancement. In water limited conditions, QTLs for root morphology and anatomical phenes were identified through GWAS (Oyiga et al. 2020) in barley. Alahmad et al. (2019) suggested deploying traits of root architecture in durum wheat cultivars for enhanced yield stability in arid or semi-arid like conditions. They quantified seven marker-trait associations for seminal root angle through GWAS clustered on chromosome6A. Root depth, length, and root density are the main components of RSA affecting water absorption from deep soils (Carvalho et al. 2014; Asif and Kamran 2011; King et al. 2003). El Hassouni et al. (2018) identified durum wheat genotypes having 35% enhanced grain yield with deep root systems along with 9% increased 1000 kernel weight under water stress than the genotypes having shallow root systems.

The molecular characterization of root structure architecture and its related genes has although been explored up to QTL identification in wheat, rice, and maize and soybean but still this requires detailed deep scaling. Despite a large gap in illustration of the genetic control of RSA, many studies are successful in altering plant based root traits through phenotypic or molecular abbreviations. Rice leads other crops in translating various molecular factors using transcription factors or genes such as OsNAC5/9, OsMYB2, OsEXPA8, OsRAA1, a cell wall extension gene, etc. for improved RSA. Other genes for root traits in rice include ARL1; CRL1, coding for adventitious and crown roots and some conserved domain of LOB TFs (Kitomi et al. 2011). Improved tolerance under drought conditions has been observed through overexpression of GmbZIP1 along with enhanced root–shoot growth of wheat (Gao et al. 2011). Also, aquaporin (AQP) has been reported for enhancing water uptake in plant roots (Javot et al. 2003). Knockout experiment in Arabidopsis, for AQP (PIP2.2) reduced root hydraulic conductivity by 25–30%. Root plasticity has been reported as a potential root phene for stress adaptation through identification of genetic loci for root phenotypic plasticity in cereal crops (Sandhu et al. 2016; Kadam et al. 2017; Schneider et al. 2020a, b; Schneider and Lynch 2020).

## 11.5 High-Thoroughput Techniques for Root Phenotyping

The need to evaluate a large number of samples in the least time scale and minimal human effort has led to the concept of "high-throughput phenotyping." High-throughput phenotyping techniques automate monitoring plant traits over different growth stages from sowing to harvest, in controlled or field environments, through data capturing, storage and image analysis pipelines. Developing high-throughput phenotyping approaches for roots is challenging due to the obstruction imposed by soil or growth media. A straight forward practice is to uproot the plant for ex situ sampling of root phenes at a selected time point. In situ phenotyping is an alternative approach that saves roots to capture the desired phenes in live media regularly. The phenotyping environment for a particular phene of interest should be artefactual for other phenes as well. Field-based and lab-based traits can be compiled together for the above and below ground plant systems to correlate changes in RSA with above-ground selection. Advancements in root phenotyping systems have taken place from germination paper, coring, shovelomics to imaging and scanning, a few are briefly described here.

### *11.5.1 Trenching*

Given by John Weaver (1919), this is one of the oldest methods used to characterize root traits. In this method, roots were excavated in different horizontal layers and then manually measured for different root traits, e.g., root depth, width, etc. The limitation of this method is indirect calculation of root numbers and biomass. Secondly, for breeding, in addition to being more laborious, it is difficult to trench plant roots in field plots since it will not give accurate correlation with yield.

### *11.5.2 Shovelomics*

This method simply involves uprooting soil with a shovel and further analysis using different platforms for root phenotyping in the field (Trachsel et al. 2011). Initially, plant root crowns are excavated at a depth of ~20–25 cm (crop specific), after washing, cleaning, and air drying, different root parameters like root length and diameter, root angle, number of roots, and branching, etc. can be analyzed manually or using various available softwares, WinRhizo (Bauhus and Messier 1999), RootReader (Clark et al. 2011), GiaRoots (Galkovskyi et al. 2012), DIRT (Das et al. 2015), RhizoVision Explorer (Seethepalli et al. 2021). This phenotyping technique has been used widely for maize, wheat, common bean, soybean, cluster/ broad bean, and other crops. The main limitations of this technique are being laborious destructive.

### *11.5.3 Core Cutting*

Soil core sampling is also the classical phenotyping method where soil cores are collected at 1–2 cm top soil layer. After washing, roots are characterized for various traits (Kücke et al. 1995). But the disadvantage of soil coring is under estimation of root depth and distribution since partial segments of roots are collected.

## *11.5.4 Rhizotron*

Rhizotron have been a good success in root phenotyping for a large number of crops being nondestructive and capturing 2D images, secondly it produced best assessment of roots in situ. This includes insertion of glass tubes at an angle of 45° in the planting fields before crop sowing which are connected with live scanners and cameras to capture root traits with 360° view (Bragg et al. 1983). The only limitation of this technique is missing root information which is away from camera and scanner.

Jia et al. (2019) have designed Rhizoboxes for root phenotyping in green house conditions to capture the natural structural variation of root architectural system in a diverse panel of spring barley (Fig. 11.3). Nagel et al. (2012) have designed a phenotyping platform (GROWSCREEN-Rhizo) for automatic imaging of complete plant growing in rhizotrons (soil-filled, 18 L volume) with a sped output of 60 rhizotrons in 1 h as shown in Fig. 11.3. Capture of minor alterations in RSA induced by a moderate change in soil compaction is the advantage of this method and analyzed using GROWSCREEN-Root software but it can measure root images up to 90 cm length and shoot from 10 to 80 cm.

**Fig. 11.3** Setup view of different phenotyping structures. (Source: Nagel et al. 2012; Jia et al. 2019; Qiao et al. 2019)

## 11.6 Sensing Technologies in High-Throughput Root Phenotyping

Presently, sensors used in high-throughput systems play an important role in imaging root architecture and associated traits. Sensing technologies provide routine data capture and analytics standardization, which are essential for scientific rigor. Additionally, multi-modal sensors such as optical imaging, hyperspectral, X-ray, magnetic resonance imaging (MRI), microwave, ground penetrating radar (GPR), etc., bring different dimensionalities in capturing RSA. Broadly, sensors used in root phenotyping could be classified into two categories: (1) surface scanning systems, i.e. optical and spectral imaging systems requiring a direct view of the root system, and (2) sub-surface scanning systems, i.e. X-ray, MRI, microwave, can penetrate the soil of growth substrate.

### *11.6.1 Surface Scanning Systems: Optical and Spectra Imaging*

RGB cameras, hyperspectral, and fluorescence sensors are surface imaging tools. These sensors measure the light reflected off the root surface to decipher the rhizosphere's complex root architecture and chemistry. Therefore, the system requires specialized designs of plant growth environment, including different growth media or substrates to suit surface scanning of a root system. Clark et al. (2011) used 3D imaging and a software platform to quantify root structural traits in rice using a gellan gum system analyzing with Root Reader 3D. A germination paper-based moisture replacement system was used to capture root traits of wheat seedlings (Adeleke et al. 2019). Few others have designed the root box-pin board method (Kono et al. 1987) to record complete root distribution, but it is not feasible for large scale root phenotyping. Hydrogel-based transparent soil has also been used for phenotyping of roots (Ma et al. 2019) being economic and nondestructive. The root phenes observed with hydrogel are parallel with plants grown in soil or in hydroponics. Qiao et al. (2019) used a semi-hydroponic like phenotypic platform for root traits in maize (Fig. 11.3). Raab and Martin (2001) visualized the organic chemistry of rhizosphere in legumes using Fourier-transform infrared (FTIR) microscope spectroscopy in the mid-infrared (MIR) region. The experimentation mapped the lipid, lignin, and sugars levels in root epidermal tissue through spectral analytical approaches. Ilhardt et al. (2019) demonstrated laser-induced breakdown spectroscopy (LIBS) for imaging of both organic and inorganic components of soil-root-rhizosphere systems. The diagnostic spectral signatures detected trace elements and carbon accumulations within the rhizosphere region, which identified few host sites for microbial accumulation in the form of residues (detritusphere). Arnold et al. (2016) reported using near-infrared (NIR) hyperspectral to discriminate plant root from soil using spectral signatures in rhizoboxes. The system was used in drought-resistance research to map water distribution in plant roots and soil. An image processing setup with the highest spatial resolution through hyperspectral imaging and automated segmentation for plant roots grown in soil has been established to define spectral root signatures based on root zone properties (Bodner et al. 2018). Rahman et al. (2020) fabricated a mini-minirhizotron with optical and multispectral imaging capabilities for in-situ imaging of root systems. The system provides offline image-processing software to eliminate motion blur and geometric distortion for reconstructing 360° panoramic views.

### 11.6.2 *Sub-surface Scanning Systems: X-Ray, MRI, Microwave, and GPR*

X-Ray, MRI, Microwave, and GPR are sub-surface scanning tools. These sensors have been used widely for scientific research in nanosciences, health and medicine, and material research. In recent years, these scanning tools have been developed significantly to find applications in plant based research. Sub-surface scanning systems emit electromagnetic radiation that can penetrate to a certain depth of the soil substrate. The surface penetration ability of the sensor system depends on the dielectric property of the material, which varies between soil or substrate, water, and root, enabling precise differentiation of RSA. Images collected from X-ray based computed tomography (CT) have been used to analyze segmentation of plant roots, i.e., separation of root from non-root matter, reconstruction of complete root system, and identification of higher-level phenotypes (Perret et al. 2007; Xu et al. 2018). The promoted uptake of P in poplar through endophytes was captured through different techniques like proteomics, spectroscopy, and imaging through X-ray Varga et al. (2020). Plant samples inoculated with phosphate solubilizing bacteria strains were having contracting root architecture along with other root traits. Yang et al. (2017) integrated explicitly-resolved root structure from X-Ray CT scans and pore-scale computational fluid dynamics (CFD) simulations for calculating root-soil-ground-water system. The coupled imaging-modeling was found to be a realistic solution to investigate rhizosphere flow processes. Van Dusschoten et al. (2016) reported an imaging, visualization, and analysis method combining MRI and NM Rooting toolbox both. The research identified the most approximate all the root traits, root length and diameter, root angle (in two-dimensional polar coordinates), root mass, and tip number along with spatial distribution. Pflugfelder et al. (2017) investigated in the suitability of six commercially available soil equivalent substrates used in MRI for root phenotyping using barley as a model plant. Liu et al. (2014) tested different concentrations of Gd-DTPA, an MRI contrasting agent for imaging roots in rice at different growth stages. X-ray CT and MRI (Koteen et al. 2015; Fisher and Sozzani 2016) are also opted for root phenotyping for some crops. However, these had their disadvantages since the plants need to be transferred to the instrument, the plant/pot size is required to be kept small. Moreover, it cannot capture all the roots (Zappala et al. 2013; Metzner et al. 2015). Krzyzaniak et al. (2021) have compared cylindric rhizotrons with hydroponics and plane rhizotrons for root architectural traits in grapevine and captured 2D (hydroponics and rhizotron) and 3D (neutron tomography) images. The method included segmentation of captured image, further skeletonization and then detection of adventitious root skeleton along with adventitious root reconstruction. Ground penetrating radars have also been explored for infield root imaging (Araus and Cairns 2014).

Along with root architectural phene phenotyping, progress has also been made in root exudate collection and identification of exuded metabolites (Williams and Vries 2020). To identify a specific component of exudates approaches like gas or liquid chromatography, spectrophotometry, tandem mass spectrometry (GC–MS/MS, LC-MS/MS), ion chromatography (IC) or high-pressure liquid chromatography (HPLC) were used, whereas for non-targeted metabolite analysis nuclear magnetic resonance (NMR), hydrophilic interaction liquid chromatography (HILIC), mass spectrometry: reversed-phase liquid chromatography (RPLC), were used (Oburger and Jones 2018).

In brief, whatever the root phenotyping may be adopted, it should reveal the reproducible quantification of growth and architecture of roots, elucidate dynamic establishment of roots temporally and spatially and should correlate the interaction of root traits with above-ground traits in all types of growth media from artificial to soil in varying environmental conditions.

## 11.7 Root Modeling

A number of variations in root traits are observed with time and space which accounts for constraining root phenes in different species and ecosystems. This demands identification of most variable specific root traits across time and space and adopt some defined protocols for correlation between root trait variation and the ecosystem. Phene-integrated structural-functional modeling and its simulation in breeding approaches can offer a better understanding of non-linear signaling processes adopted by roots under abiotic stresses for better water and nutrient use efficiency (Chickarmane et al. 2010). Few formulations have been defined over years, for example, Fine-Root Ecology Database (FRED, www.roots.ornl.gov) is providing hypothesized variations among various root traits within different species with age of root across environmental gradients. These hypotheses are based on already reported fine-root traits with coefficient of variation. With more exploration of root traits and testing, the specific variations among them as well as within same plant species can be helpful to design more advanced models to correlate the best root trait with a positive correlation with varying environmental condition in a competitive way (McCormack et al. 2017).

The modeling approach for maize yield (Hammer et al. 2009) in the U.S. Corn Belt correlated the continuous increase with structural and functional modifications in root architecture. The Steep, Deep, and Cheap model of Lynch (2013) was proposed for enhanced N acquisition in maize based on rhizoeconomics and spatio-temporal availability of soil N. But the ratio of structural input to that of metabolic cost of root function may, sometime, be more than half of net daily photosynthesis rate (Lambers et al. 2006). Thus, the complete economic benefit–cost ratio of root system is necessary for crop improvement programs in deficient conditions (Lynch 2007). Moreover, a particular ideotype might not be universally applicable. In low rainfall areas where water only stays in the upper soil layer "Wide, Shallow and Fine" ideotype (Rao et al. 2021) might be more appropriate in harnessing water.

Models have been designed for better uptake and translocation of nutrients based on root structure and function, e.g., maize lines with better RCA had improved N-acquisition under limited N availability (Saengwilai et al. 2014). Likewise, the plant model SimRoot also correlated RCA formation with NPK use efficiency by decreasing the soil exploration cost (Postma and Lynch 2011; Postma et al. 2017). The 3D water and nutrient transport model was developed based on Rhizospheric interactions with RSA (Mai et al. 2019). Few other root models such as ROOTMAP (Dunbabin et al. 2002), ArchiSimple (Pagès et al. 2012), R-SWMS (Schröder et al. 2012), DigR (Barczi et al. 2018), and CRootBox (Schnepf et al. 2018) have also been designed for study of diversity of root architectural system in various environmental conditions and the basic data required for root modeling (Takahashi and Pradal 2021) for water and nutrient uptake efficiency. A model has also been developed for efficient water uptake in drought, flooded, and average conditions using rainfall data, root, and soil hydraulic conductance integrating the basic root architecture and distribution in soils. Draye et al. (2010) have reported a root architectural view based on root and soil hydraulics for efficient water uptake.

## 11.8 Future Prospects

Abiotic and biotic stresses, decreasing soil fertility, changing climate, and ecosystems are the major challenges being faced globally. For sustainable crop productivity, in addition to the plant based traits, the hidden part, root traits, also need to be explored for enhancing the source-sink relationship. For maximum water and nutrient uptake and translocation, exploration of root phenes which are capable of adaption under varying edaphic conditions is required. The lab based identification of traits should be reproducible with field assays and contribute to actual yield enhancement. Therefore, identification of important root phenes, their translation to new improved crop varieties could add one step into cop improvement programs. With the advancement of technologies and softwares, phenotyping of roots traits will advance for enhanced water and nutrient uptake along with disease resistance and incorporation of important root phenes in development of new germplasm would be a continuous process in changing era of climate change for agricultural sustainability.

## References

Adeleke E, Millas R, McNeal W, Faris J, Taheri A (2019) Assessing root system architecture of wheat seedlings using a high-throughput root phenotyping system. Bio Rxiv 677955

Ahn SJ, Shin R, Schachtman DP (2004) Expression of KT/KUP genes in Arabidopsis and the role of root hairs in $K^+$ uptake. Plant Physiol 134(3):1135–1145

Alahmad S, El Hassouni K, Bassi FM, Dinglasan E, Youssef C, Quarry G, Aksoy A, Mazzucotelli E, Juhász A, Able JA, Christopher J, Voss-Fels KP, Hickey LT (2019) A major root architecture QTL responding to water limitation in Durum wheat. Front Plant Sci 10:436

Ali ML, Luetchens J, Nascimento J, Shaver TM, Kruger GR, Lorenz AJ (2015) Genetic variation in seminal and nodal root angle and their association with grain yield of maize under water-stressed field conditions. Plant Soil 397(1):213–225

Araus JL, Cairns JE (2014) Field high-throughput phenotyping: the new crop breeding frontier. Trends Plant Sci 19(1):52–61

Arnold T, Leitner R, Bodner G (2016) Application of NIR hyperspectral imaging for water distribution measurements in plant roots and soil. IEEE Sensors, pp 1–3

Asif M, Kamran A (2011) Plant breeding for water-limited environments. Crop Sci 51(6):2911

Atkinson JA, Wingen LU, Griffiths M, Pound MP, Gaju O, Foulkes MJ, Le Gouis J, Griffiths S, Bennett MJ, King J, Wells DM (2015) Phenotyping pipeline reveals major seedling root growth QTL in hexaploid wheat. J Exp Bot 66(8):283–2292

Ayaga G, Todd A, Brookes PC (2006) Enhanced biological cycling of phosphorus increases its availability to crops in low-input sub-Saharan farming systems. Soil Biol Biochem 38(1):81–90

Barber SA (1966) The role of root interception, mass flow and diffusion in regulating the uptake of ions by plants from soil. In: International Atomic Energy Agency (ed) Limiting steps in ion uptake by plants from soil. Proceedings. FAO/IAEA, Vien, pp 39–45

Barczi JF, Rey H, Griffon S, Jourdan C (2018) DigR: a generic model and its open-source simulation software to mimic three-dimensional root-system architecture diversity. Ann Bot 121(5):1089–1104

Bauhus J, Messier C (1999) Evaluation of fine root length and diameter measurements obtained using RHIZO image analysis. Agron J 91(1):142–147

Bayuelo-Jiménez JS, Gallardo-Valdéz M, Pérez-Decelis VA, Magdaleno-Armas L, Ochoa I, Lynch JP (2011) Genotypic variation for root traits of maize (*Zea mays* L.) from the Purhepecha plateau under contrasting phosphorus availability. Field Crops Res 121(3):350–362

Bergmann J, Weigelt A, van Der Plas F, Laughlin DC, Kuyper TW, Guerrero-Ramirez N, Valverde-Barrantes OJ, Bruelheide H, Freschet GT, Iversen CM, Kattge J (2020) The fungal collaboration gradient dominates the root economics space in plants. Sci Adv 6(27):3756

Blancaflor EB (2013) Regulation of plant gravity sensing and signaling by the actin cytoskeleton. Am J Bot 100(1):143–152

Bodner G, Nakhforoosh A, Arnold T, Leitner D (2018) Hyperspectral imaging: a novel approach for plant root phenotyping. Plant Methods 14(1):1–7

Bragg PL, Govi G, Cannell RQ (1983) A comparison of methods, including angled and vertical minirhizotrons, for studying root growth and distribution in a spring oat crop. Plant Soil 73(3): 435–440

Brown LK, George TS, Dupuy LX, White PJ (2013) A conceptual model of root hair ideotypes for future agricultural environments: what combination of traits should be targeted to cope with limited P availability? Ann Bot 112(2):317–330

Burton AL, Brown KM, Lynch JP (2013) Phenotypic diversity of root anatomical and architectural traits in *Zea* species. Crop Sci 53(3):1042–1055

Cai H, Chen F, Mi G, Zhang F, Maurer HP, Liu W, Reif JC, Yuan L (2012) Mapping QTLs for root system architecture of maize (*Zea mays* L.) in the field at different developmental stages. Theor App Genet 125(6):1313–1324

Carvalho P, Azam-Ali S, Foulkes MJ (2014) Quantifying relationships between rooting traits and water uptake under drought in Mediterranean barley and durum wheat. J Integr Plant Biol 56(5): 455–469

Chandra P, Wunnava A, Verma P, Chandra A, Sharma RK (2021) Strategies to mitigate the adverse effect of drought stress on crop plants—influences of soil bacteria: a review. Pedosphere 31(3): 496–509

Chen W, Eissenstat DM, Koide RT (2018) Root diameter predicts the extramatrical hyphal exploration distance of the ectomycorrhizal fungal community. Ecosphere 9(4):e02202

Chen Y, Palta J, Prasad PV, Siddique KH (2020) Phenotypic variability in bread wheat root systems at the early vegetative stage. BMC Plant Biol 20(1):1–6

Chickarmane V, Roeder AH, Tarr PT, Cunha A, Tobin C, Meyerowitz EM (2010) Computational morphodynamics: a modeling framework to understand plant growth. Ann Rev Plant Biol 61: 65–87

Chimungu JG, Brown KM, Lynch JP (2014a) Reduced root cortical cell file number improves drought tolerance in maize. Plant Physiol 166(4):1943–1955

Chimungu JG, Brown KM, Lynch JP (2014b) Large root cortical cell size improves drought tolerance in maize. Plant Physiol 166(4):2166–2178

Christopher J, Christopher M, Jennings R, Jones S, Fletcher S, Borrell A, Manschadi AM, Jordan D, Mace E, Hammer G (2013) QTL for root angle and number in a population developed from bread wheats (Triticum aestivum) with contrasting adaptation to water-limited environments. Theor App Genet 126(6):1563–1574

Chun L, Mi G, Li J, Chen F, Zhang F (2005) Genetic analysis of maize root characteristics in response to low nitrogen stress. Plant Soil 276(1):369–382

Clark RT, MacCurdy RB, Jung JK, Shaff JE, McCouch SR, Aneshansley DJ, Kochian LV (2011) Three-dimensional root phenotyping with a novel imaging and software platform. Plant Physiol 156(2):455–465

Comas L, Becker S, Cruz VM, Byrne PF, Dierig DA (2013) Root traits contributing to plant productivity under drought. Front Plant Sci 4:442

Cook RJ, Thomashow LS, Weller DM, Fujimoto D, Mazzola M, Bangera G, Kim DS (1995) Molecular mechanisms of defense by rhizobacteria against root disease. Proc Natl Acad Sci 92(10):4197–4201

Das A, Schneider H, Burridge J, Ascanio AK, Wojciechowski T, Topp CN, Lynch JP, Weitz JS, Bucksch A (2015) Digital imaging of root traits (DIRT): a high-throughput computing and collaboration platform for field-based root phenomics. Plant Methods 11(1):1–2

de Dorlodot S, Forster B, Pagès L, Price A, Tuberosa R, Draye X (2007) Root system architecture: opportunities and constraints for genetic improvement of crops. Trends Plant Sci 12(10): 474–481

Dey S, Sengupta S (2020) Classification of organic compounds present in the rhizosphere. Agric Food 2(4):266–269. e-Newsletter

Docampo R, Ulrich P, Moreno SN (2010) Evolution of acidocalcisomes and their role in polyphosphate storage and osmoregulation in eukaryotic microbes. Philos Trans R Soc B Biol Sci 365(1541):775–784

Draye X, Kim Y, Lobet G, Javaux M (2010) Model-assisted integration of physiological and environmental constraints affecting the dynamic and spatial patterns of root water uptake from soils. J Exp Bot 61(8):2145–2155

Dunbabin VM, Diggle AJ, Rengel Z, Van Hugten R (2002) Modelling the interactions between water and nutrient uptake and root growth. Plant Soil 239(1):19–38

Eissenstat DM, Kucharski JM, Zadworny M, Adams TS, Koide RT (2015) Linking root traits to nutrient foraging in arbuscular mycorrhizal trees in a temperate forest. New Phytol 208(1): 114–124

El Hassouni K, Alahmad S, Belkadi B, Filali-Maltouf A, Hickey LT, Bassi FM (2018) Root system architecture and its association with yield under different water regimes in durum wheat. Crop Sci 58(6):2331–2346

Fan M, Zhu J, Richards C, Brown KM, Lynch JP (2003) Physiological roles for aerenchyma in phosphorus-stressed roots. Funct Plant Biol 30(5):493–506

Fisher AP, Sozzani R (2016) Uncovering the networks involved in stem cell maintenance and asymmetric cell division in the *Arabidopsis* root. Curr Opin Plant Biol 29:38–43

Forde BG (2009) Is it good noise? The role of developmental instability in the shaping of a root system. J Exp Bot 60(14):3989–4002

Galindo-Castañeda T, Brown KM, Kuldau GA, Roth GW, Wenner NG, Ray S, Schneider H, Lynch JP (2019) Root cortical anatomy is associated with differential pathogenic and symbiotic fungal colonization in maize. Plant Cell Environ 42(11):2999–3014

Galkovskyi T, Mileyko Y, Bucksch A, Moore B, Symonova O, Price CA, Topp CN, Iyer-Pascuzzi AS, Zurek PR, Fang S, Harer J (2012) GiA roots: software for the high throughput analysis of plant root system architecture. BMC Plant Biol 12(1):116

Gao SQ, Chen M, Xu ZS, Zhao CP, Li L, Xu HJ, Tang YM, Zhao X, Ma YZ (2011) The soybean GmbZIP1 transcription factor enhances multiple abiotic stress tolerances in transgenic plants. Plant Mol Biol 75(6):537–553

García-Cervigón AI, García-López MA, Pistón N, Pugnaire FI, Olano JM (2021) Co-ordination between xylem anatomy, plant architecture and leaf functional traits in response to abiotic and biotic drivers in a nurse cushion plant. Ann Bot 127(7):919–929

Gichangi EM, Mnkeni PN, Brookes PC (2009) Effects of goat manure and inorganic phosphate addition on soil inorganic and microbial biomass phosphorus fractions under laboratory incubation conditions. Soil Sci Plant Nutr 55(6):764–771

Giehl RF, Gruber BD, von Wirén N (2014) It's time to make changes: modulation of root system architecture by nutrient signals. J Exp Bot 65(3):769–778

Giuliani S, Sanguineti MC, Tuberosa R, Bellotti M, Salvi S, Landi P (2005) Root-ABA1, a major constitutive QTL, affects maize root architecture and leaf ABA concentration at different water regimes. J Exp Bot 56(422):3061–3070

Gong X, McDonald G (2017) QTL mapping of root traits in phosphorus-deficient soils reveals important genomic regions for improving NDVI and grain yield in barley. Theor App Genet 130(9):1885–1902

Gruber BD, Giehl RF, Friedel S, von Wirén N (2013) Plasticity of the Arabidopsis root system under nutrient deficiencies. Plant Physiol 63(1):161–179

Guo W, Zhao J, Li X, Qin L, Yan X, Liao H (2011) A soybean β-expansin gene GmEXPB2 intrinsically involved in root system architecture responses to abiotic stresses. Plant J 66(3): 541–552

Gupta G, Snehi SK, Singh V (2017) Role of PGPR in biofilm formations and its importance in plant health. Biofilms Plant Soil Health 27:27–40

Haling RE, Brown LK, Bengough AG, Young IM, Hallett PD, White PJ, George TS (2013) Root hairs improve root penetration, root–soil contact, and phosphorus acquisition in soils of different strength. J Exp Bot 64(12):3711–3721

Hammer GL, Dong Z, McLean G, Doherty A, Messina C, Schussler J, Zinselmeier C, Paszkiewicz S, Cooper M (2009) Can changes in canopy and/or root system architecture explain historical maize yield trends in the US corn belt? Crop Sci 49(1):299–312

Han Y, Chen Y, Yin S, Zhang M, Wang W (2015) Over-expression of TaEXPB23, a wheat expansin gene, improves oxidative stress tolerance in transgenic tobacco plants. J Plant Physiol 173:62–71

Han M, Sun L, Gan D, Fu L, Zhu B (2020) Root functional traits are key determinants of the rhizosphere effect on soil organic matter decomposition across 14 temperate hardwood species. Soil Biol Biochem 151:108019

Hassan MK, McInroy JA, Kloepper JW (2019) The interactions of rhizodeposits with plant growth-promoting rhizobacteria in the rhizosphere: a review. Agriculture 9(7):142

Hebert Y, Barriere Y, Bertholeau JC (1992) Root lodging resistance in forage maize: genetic variability of root system and aerial part. Maydica 37(2):173–183

Henry A, Gowda VR, Torres RO, McNally KL, Serraj R (2011) Variation in root system architecture and drought response in rice (Oryza sativa): phenotyping of the OryzaSNP panel in rainfed lowland fields. Field Crops Res 120(2):205–214

Herron PM, Gage DJ, Cardon ZG (2010) Micro-scale water potential gradients visualized in soil around plant root tips using microbiosensors. Plant Cell Environ 33(2):199–210

Hirel B, Le Gouis J, Ney B, Gallais A (2007) The challenge of improving nitrogen use efficiency in crop plants: towards a more central role for genetic variability and quantitative genetics within integrated approaches. J Exp Bot 58(9):2369–2387

Ho MD, Rosas JC, Brown KM, Lynch JP (2005) Root architectural tradeoffs for water and phosphorus acquisition. Funct Plant Biol 32(8):737–748

Hochholdinger F, Tuberosa R (2009) Genetic and genomic dissection of maize root development and architecture. Curr Opin Plant Biol 12(2):172–177

Holz M, Zarebanadkouki M, Kuzyakov Y, Pausch J, Carminati A (2018) Root hairs increase rhizosphere extension and carbon input to soil. Ann Bot 121(1):61–69

Horn R, Wingen LU, Snape JW, Dolan L (2016) Mapping of quantitative trait loci for root hair length in wheat identifies loci that co-locate with loci for yield components. J Exp Bot 67(15): 4535–4543

Hu L, Robert CA, Cadot S, Zhang X, Ye M, Li B, Manzo D, Chervet N, Steinger T, Van Der Heijden MG, Schlaeppi K (2018) Root exudate metabolites drive plant-soil feedbacks on growth and defense by shaping the rhizosphere microbiota. Nat Commun 9(1):1–3

Hund A, Reimer R, Messmer R (2011) A consensus map of QTLs controlling the root length of maize. Plant Soil 344(1):143–158

Ilhardt PD, Nuñez JR, Denis EH, Rosnow JJ, Krogstad EJ, Renslow RS, Moran JJ (2019) High-resolution elemental mapping of the root-rhizosphere-soil continuum using laser-induced breakdown spectroscopy (LIBS). Soil Biol Biochem 131:119–132

Jacoby RP, Koprivova A, Kopriva S (2021) Pinpointing secondary metabolites that shape the composition and function of the plant microbiome. J Exp Bot 72(1):57–69

Javot H, Lauvergeat V, Santoni V, Martin-Laurent F, Güçlü J, Vinh J, Heyes J, Franck KI, Schaffner AR, Bouchez D, Maurel C (2003) Role of a single aquaporin isoform in root water uptake. Plant Cell 15(2):509–522

Jia Z, Liu Y, Gruber BD, Neumann K, Kilian B, Graner A, Von Wirén N (2019) Genetic dissection of root system architectural traits in spring barley. Front Plant Sci 10:400

Joshi DC, Singh V, Hunt C, Mace E, van Oosterom E, Sulman R, Jordan D, Hammer G (2017) Development of a phenotyping platform for high throughput screening of nodal root angle in sorghum. Plant Methods 13(1):1–2

Kadam NN, Yin X, Bindraban PS, Struik PC, Jagadish KS (2015) Does morphological and anatomical plasticity during the vegetative stage make wheat more tolerant of water deficit stress than rice? Plant Physiol 167(4):1389–1401

Kadam NN, Tamilselvan A, Lawas LM, Quinones C, Bahuguna RN, Thomson MJ, Dingkuhn M, Muthurajan R, Struik PC, Yin X, Jagadish SK (2017) Genetic control of plasticity in root morphology and anatomy of rice in response to water deficit. Plant Physiol 174(4):2302–2315

Kalita P, Kalita R, Das R, Rabha BK (2011) Growth and yield response of wheat genotypes to rainfed condition of Assam. Indian J Plant Physiol 16(1):10–18

Kellermeier F, Armengaud P, Seditas TJ, Danku J, Salt DE, Amtmann A (2014) Analysis of the root system architecture of Arabidopsis provides a quantitative readout of crosstalk between nutritional signals. Plant Cell 26(4):1480–1496

Khan KS, Joergensen RG (2009) Changes in microbial biomass and P fractions in biogenic household waste compost amended with inorganic P fertilizers. Bioresour Technol 100(1): 303–309

King J, Gay A, Sylvester-Bradley RO, Bingham IA, Foulkes J, Gregory P, Robinson D (2003) Modelling cereal root systems for water and nitrogen capture: towards an economic optimum. Ann Bot 91(3):383–390

Kirkegaard JA, Lilley JM, Howe GN, Graham JM (2007) Impact of subsoil water use on wheat yield. Aust J Agric Res 58(4):303–315

Kitomi Y, Ito H, Hobo T, Aya K, Kitano H, Inukai Y (2011) The auxin responsive AP2/ERF transcription factor CROWN ROOTLESS5 is involved in crown root initiation in rice through the induction of OsRR1, a type-A response regulator of cytokinin signaling. Plant J 67(3): 472–484

Kitomi Y, Itoh JI, Uga Y (2018) Genetic mechanisms involved in the formation of root system architecture. In: Rice genomics, genetics and breeding. Springer, Singapore, pp 241–274

Kitomi Y, Hanzawa E, Kuya N, Inoue H, Hara N, Kawai S, Kanno N, Endo M, Sugimoto K, Yamazaki T, Sakamoto S (2020) Root angle modifications by the DRO1 homolog improve rice yields in saline paddy fields. Proc Natl Acad Sci 117(35):21242–21250

Koevoets IT, Venema JH, Elzenga JT, Testerink C (2016) Roots withstanding their environment: exploiting root system architecture responses to abiotic stress to improve crop tolerance. Front Plant Sci 7:1335

Kono Y, Yamauchi A, Nonoyama T, Tatsumi J, Kawamura N (1987) A revised experimental system of root-soil interaction for laboratory work. Environ Control Biol 25(4):141–151

Koteen LE, Raz-Yaseef N, Baldocchi DD (2015) Spatial heterogeneity of fine root biomass and soil carbon in a California oak savanna illuminates plant functional strategy across periods of high and low resource supply. Ecohydrology 8(2):294–308

Kragelund L, Hosbond C, Nybroe O (1997) Distribution of metabolic activity and phosphate starvation response of lux-tagged Pseudomonas fluorescens reporter bacteria in the barley rhizosphere. Appl Environ Microbiol 63(12):4920–4928

Krzyzaniak Y, Cointault F, Loupiac C, Bernaud E, Ott F, Salon C, Laybros A, Han S, Héloir MC, Adrian M, Trouvelot S (2021) In situ phenotyping of grapevine root system architecture by 2D or 3D imaging: advantages and limits of three cultivation methods. Front Plant Sci 12:638688

Kücke M, Schmid H, Spiess A (1995) A comparison of four methods for measuring roots of field crops in three contrasting soils. Plant Soil 172(1):63–71

Kumar A, Sharma SK, Lata C, Kulshrestha N, Krishnamurthy SL, Singh K, Yadav RK (2018) Impact of water deficit (salt and drought) stress on physiological, biochemical and yield attributes on wheat (*Triticum aestivum*) varieties. Indian J Agri Sci 88(10):1624–1632

Kuzyakov Y, Hill PW, Jones DL (2007) Root exudate components change litter decomposition in a simulated rhizosphere depending on temperature. Plant Soil 290(1):293–305

Lambers H, Shane MW, Cramer MD, Pearse SJ, Veneklaas EJ (2006) Root structure and functioning for efficient acquisition of phosphorus: matching morphological and physiological traits. Ann Bot 98(4):693–713

Lata C, Kumar A, Sharma SK, Singh J, Sheokand S, Mann A, Rani B (2017) Tolerance to combined boron and salt stress in wheat varieties: biochemical and molecular analyses. Ind J Exp Bot 55: 321–328

Lin C, Sauter M (2018) Control of adventitious root architecture in rice by darkness, light, and gravity. Plant Physiol 176(2):1352–1364

Linkohr BI, Williamson LC, Fitter AH, Leyser HO (2002) Nitrate and phosphate availability and distribution have different effects on root system architecture of Arabidopsis. Plant J 29(6): 751–760

Liu J, Li J, Chen F, Zhang F, Ren T, Zhuang Z, Mi G (2008) Mapping QTLs for root traits under different nitrate levels at the seedling stage in maize (Zea mays L.). Plant Soil 305(1):253–265

Liu Z, Qian J, Liu B, Wang Q, Ni X, Dong Y, Zhong K, Wu Y (2014) Effects of the magnetic resonance imaging contrast agent Gd-DTPA on plant growth and root imaging in rice. PLoS One 9(6):e100246

Lynch JP (2007) Rhizoeconomics: the roots of shoot growth limitations. HortSci 42:1107–1109

Lynch JP (2011) Root phenes for enhanced soil exploration and phosphorus acquisition: tools for future crops. Plant Physiol 156(3):1041–1049

Lynch JP (2013) Steep, cheap and deep: an ideotype to optimize water and N acquisition by maize root systems. Ann Bot 112(2):347–357

Lynch JP, Brown KM (2012) New roots for agriculture: exploiting the root phenome. Philos Trans R Soc Lond Ser B Biol Sci 367(1595):1598–1604
Ma Z, Bykova NV, Igamberdiev AU (2017) Cell signaling mechanisms and metabolic regulation of germination and dormancy in barley seeds. Crop J 5(6):459–477
Ma L, Shi Y, Siemianowski O, Yuan B, Egner TK, Mirnezami SV, Lind KR, Ganapathysubramanian B, Venditti V, Cademartiri L (2019) Hydrogel-based transparent soils for root phenotyping in vivo. Proc Natl Acad Sci 116(22):11063–11068
Mai TH, Schnepf A, Vereecken H, Vanderborght J (2019) Continuum multiscale model of root water and nutrient uptake from soil with explicit consideration of the 3D root architecture and the rhizosphere gradients. Plant Soil 439(1):273–292
Mandal KG, Hati KM, Misra AK, Ghosh PK, Bandyopadhyay KK (2003) Root density and water use efficiency of wheat as affected by irrigation and nutrient management. J Agric Phys 3(1): 49–55
Manschadi AM, Hammer GL, Christopher JT, Devoil P (2008) Genotypic variation in seedling root architectural traits and implications for drought adaptation in wheat (*Triticum aestivum* L.). Plant Soil 303(1):115–129
Marin M, Feeney DS, Brown LK, Naveed M, Ruiz S, Koebernick N, Bengough AG, Hallett PD, Roose T, Puértolas J, Dodd IC (2021) Significance of root hairs for plant performance under contrasting field conditions and water deficit. Ann Bot 128(1):1–6
Marschner H, Kirkby EA, Cakmak I (1996) Effect of mineral nutritional status on shoot—root partitioning of photoassimilates and cycling of mineral nutrients. J Exp Bot 47:1255–1263
McCormack ML, Guo D, Iversen CM, Chen W, Eissenstat DM, Fernandez CW, Li L, Ma C, Ma Z, Poorter H, Reich PB (2017) Building a better foundation: improving root-trait measurements to understand and model plant and ecosystem processes. New Phytol 215(1):27–37
Meister R, Rajani MS, Ruzicka D, Schachtman DP (2014) Challenges of modifying root traits in crops for agriculture. Trends Plant Sci 19(12):779–788
Mendes R, Garbeva P, Raaijmakers JM (2013) The rhizosphere microbiome: significance of plant beneficial, plant pathogenic, and human pathogenic microorganisms. FEMS Microbiol Rev 37(5):634–663
Mendrinna A, Persson S (2015) Root hair growth: it's a one way street. F1000Prime Rep 7:23
Metzner R, Eggert A, van Dusschoten D, Pflugfelder D, Gerth S, Schurr U, Uhlmann N, Jahnke S (2015) Direct comparison of MRI and X-ray CT technologies for 3D imaging of root systems in soil: potential and challenges for root trait quantification. Plant Methods 11(1):1–1
Miguel MA, Postma JA, Lynch JP (2015) Phene synergism between root hair length and basal root growth angle for phosphorus acquisition. Plant Physiol 167(4):1430–1439
Mishra HS, Rathore TR, Pant RC (1997) Root growth, water potential, and yield of irrigated rice. Irrig Sci 17(2):69–75
Morita S, Okuda H (1994) Effects of soil water condition on root development of wheat seedling with special reference to branching of primary seminal root. Jpn J Crop Sci 63(3):418–422
Nagel KA, Putz A, Gilmer F, Heinz K, Fischbach A, Pfeifer J, Faget M, Blossfeld S, Ernst M, Dimaki C, Kastenholz B (2012) GROWSCREEN-Rhizo is a novel phenotyping robot enabling simultaneous measurements of root and shoot growth for plants grown in soil-filled rhizotrons. Funct Plant Biol 39(11):891–904
Narayanan S, Mohan A, Gill KS, Prasad PV (2014) Variability of root traits in spring wheat germplasm. PLoS One 9(6):e100317
Oburger E, Jones DL (2018) Sampling root exudates–mission impossible? Rhizosphere 6:116–133
Oyiga BC, Palczak J, Wojciechowski T, Lynch JP, Naz AA, Léon J, Ballvora A (2020) Genetic components of root architecture and anatomy adjustments to water-deficit stress in spring barley. Plant Cell Environ 43(3):692–711
Paez-Garcia A, Motes CM, Scheible WR, Chen R, Blancaflor EB, Monteros MJ (2015) Root traits and phenotyping strategies for plant improvement. Plan Theory 4(2):334–355

Pagès L, Moreau D, Sarlikioti V, Boukcim H, Nguyen C (2012) ArchiSimple: a parsimonious model of the root system architecture. In: IEEE 4th International symposium on plant growth modeling, simulation, visualization and applications. 297–303

Perret JS, Al-Belushi ME, Deadman M (2007) Non-destructive visualization and quantification of roots using computed tomography. Soil Biol Biochem 39(2):391–399

Pflugfelder D, Metzner R, van Dusschoten D, Reichel R, Jahnke S, Koller R (2017) Non-invasive imaging of plant roots in different soils using magnetic resonance imaging (MRI). Plant Methods 13(1):1–9

Pitts RJ, Cernac A, Estelle M (1998) Auxin and ethylene promote root hair elongation in Arabidopsis. Plant J 16(5):553–560

Polania J, Poschenrieder C, Rao I, Beebe S (2017) Root traits and their potential links to plant ideotypes to improve drought resistance in common bean. Theor Exp Plant Phys 29(3):143–154

Postma JA, Lynch JP (2011) Root cortical aerenchyma enhances the growth of maize on soils with suboptimal availability of nitrogen, phosphorus, and potassium. Plant Physiol 156(3): 1190–1201

Postma JA, Dathe A, Lynch JP (2014) The optimal lateral root branching density for maize depends on nitrogen and phosphorus availability. Plant Physiol 166(2):590–602

Postma JA, Kuppe C, Owen MR, Mellor N, Griffiths M, Bennett MJ, Lynch JP, Watt M (2017) OpenSimRoot: widening the scope and application of root architectural models. New Phytol 215(3):1274–1286

Purushothaman R, Thudi M, Krishnamurthy L, Upadhyaya HD, Kashiwagi J, Gowda CL, Varshney RK (2015) Association of mid-reproductive stage canopy temperature depression with the molecular markers and grain yields of chickpea (*Cicer arietinum* L.) germplasm under terminal drought. Field Crops Res 174:1–1

Qiao S, Fang Y, Wu A, Xu B, Zhang S, Deng X, Djalovic I, Siddique KH, Chen Y (2019) Dissecting root trait variability in maize genotypes using the semi-hydroponic phenotyping platform. Plant Soil 439(1):75–90

Raab TK, Martin MC (2001) Visualizing rhizosphere chemistry of legumes with mid-infrared synchrotron radiation. Planta 213(6):881–887

Rahman G, Sohag H, Chowdhury R, Wahid KA, Dinh A, Arcand M, Vail S (2020) SoilCam: a fully automated minirhizotron using multispectral imaging for root activity monitoring. Sensors 20(3):787

Rao S, Armstrong R, Silva-Perez V, Tefera AT, Rosewarne GM (2021) Pulse root ideotype for water stress in temperate cropping system. Plan Theory 10(4):692

Rellán-Álvarez R, Lobet G, Lindner H, Pradier PL, Sebastian J, Yee MC, Geng Y, Trontin C, LaRue T, Schrager-Lavelle A, Haney CH (2015) GLO-roots: an imaging platform enabling multidimensional characterization of soil-grown root systems. elife 4:e07597

Richard CA, Hickey LT, Fletcher S, Jennings R, Chenu K, Christopher JT (2015) High-throughput phenotyping of seminal root traits in wheat. Plant Methods 11(1):1–1

Saengwilai P, Nord EA, Chimungu JG, Brown KM, Lynch JP (2014) Root cortical aerenchyma enhances nitrogen acquisition from low-nitrogen soils in maize. Plant Physiol 166(2):726–735

Saengwilai P, Strock C, Rangarajan H, Chimungu J, Salungyu J, Lynch JP (2021) Root hair phenotypes influence nitrogen acquisition in maize. Ann Bot 128(7):849–858

Sandhu N, Raman KA, Torres RO, Audebert A, Dardou A, Kumar A, Henry A (2016) Rice root architectural plasticity traits and genetic regions for adaptability to variable cultivation and stress conditions. Plant Physiol 171(4):2562–2576

Sato EM, Hijazi H, Bennett MJ, Vissenberg K, Swarup R (2015) New insights into root gravitropic signalling. J Exp Bot 66(8):2155–2165

Schneider HM, Lynch JP (2020) Should root plasticity be a crop breeding target? Front Plant Sci 11:546

Schneider HM, Klein SP, Hanlon MT, Nord EA, Kaeppler S, Brown KM, Warry A, Bhosale R, Lynch JP (2020a) Genetic control of root architectural plasticity in maize. J Exp Bot 71(10): 3185–3197

Schneider HM, Klein SP, Hanlon MT, Kaeppler S, Brown KM, Lynch JP (2020b) Genetic control of root anatomical plasticity in maize. Plant Genome 13(1):e20003

Schneider HM, Strock CF, Hanlon MT, Vanhees DJ, Perkins AC, Ajmera IB, Sidhu JS, Mooney SJ, Brown KM, Lynch JP (2021) Multiseriate cortical sclerenchyma enhance root penetration in compacted soils. Proc Natl Acad Sci 118(6):e2012087118

Schnepf A, Leitner D, Landl M, Lobet G, Mai TH, Morandage S, Sheng C, Zörner M, Vanderborght J, Vereecken H (2018) CRootBox: a structural–functional modelling framework for root systems. Ann Bot 121(5):1033–1053

Schröder N, Javaux M, Vanderborght J, Steffen B, Vereecken H (2012) Effect of root water and solute uptake on apparent soil dispersivity: a simulation study. Vadose Zone J 11(3):vzj2012-0009

Sebastian J, Yee MC, Viana WG, Rellán-Álvarez R, Feldman M, Priest HD, Trontin C, Lee T, Jiang H, Baxter I, Mockler TC (2016) Grasses suppress shoot-borne roots to conserve water during drought. Proc Natl Acad Sci 113(31):8861–8866

Seethepalli A, Dhakal K, Griffiths M, Guo H, Freschet GT, York LM (2021) RhizoVision explorer: open-source software for root image analysis and measurement standardization. AoB Pleants 13(6):plab056

Sharma A, Sheikh I, Kumar R, Kumar K, Vyas P, Dhaliwal HS (2018) Evaluation of end use quality and root traits in wheat cultivars associated with 1RS. 1BL translocation. Euphytica 214(4):1–9

Shin H, Shin HS, Guo Z, Blancaflor EB, Masson PH, Chen R (2005) Complex regulation of Arabidopsis AGR1/PIN2-mediated root gravitropic response and basipetal auxin transport by cantharidin-sensitive protein phosphatases. Plant J 42(2):188–200

Shu K, Liu XD, Xie Q, He ZH (2016) Two faces of one seed: hormonal regulation of dormancy and germination. Mol Plant 9(1):34–45

Singh CJ, Das DK (1986) Influence of varying levels of nitrogen and water supply on root growth of wheat. Ann Agric Res 7:234–244

Soni S, Kumar A, Sehrawat N, Kumar N, Meena BL, Mann A (2020) Study of salt tolerance in wheat varieties through growth and development approaches. Bhartiya Krishi Anushandhan Patrika 35(1&2):71–74

Soni S, Kumar A, Sehrawat N, Kumar A, Kumar N, Lata C, Mann A (2021) Effect of saline irrigation on plant water traits, photosynthesis and ionic balance in durum wheat genotypes. Saudi J Biol Sci 28(4):2510–2517

Sweeney CJ, de Vries FT, van Dongen BE, Bardgett RD (2021) Root traits explain rhizosphere fungal community composition among temperate grassland plant species. New Phytol 229(3): 1492–1507

Takahashi H, Pradal C (2021) Root phenotyping: important and minimum information required for root modeling in crop plants. Breed Sci 71(1):109–116

Tomar RS, Tiwari S, Naik BK, Chand S, Deshmukh R, Mallick N, Singh S, Singh NK, Tomar SM (2016) Molecular and morpho-agronomical characterization of root architecture at seedling and reproductive stages for drought tolerance in wheat. PLoS One 11(6):e0156528

Trachsel S, Kaeppler SM, Brown KM, Lynch JP (2011) Shovelomics: high throughput phenotyping of maize (Zea mays L.) root architecture in the field. Plant Soil 341(1):75–87

Tron S, Bodner G, Laio F, Ridolfi L, Leitner D (2015) Can diversity in root architecture explain plant water use efficiency? A modeling study. Ecol Model 312:200–210

Tuberosa R, Salvi S, Sanguineti MC, Maccaferri M, Giuliani S, Landi P (2003) Searching for quantitative trait loci controlling root traits in maize: a critical appraisal. Plant Soil 255(1):35–54

Uga Y, Sugimoto K, Ogawa S, Rane J, Ishitani M, Hara N, Kitomi Y, Inukai Y, Ono K, Kanno N, Inoue H (2013) Control of root system architecture by DEEPER ROOTING 1 increases rice yield under drought conditions. Nat Genet 45(9):1097–1102

Uga Y, Kitomi Y, Ishikawa S, Yano M (2015) Genetic improvement for root growth angle to enhance crop production. Breed Sci 65(2):111–119

Van Dusschoten D, Metzner R, Kochs J, Postma JA, Pflugfelder D, Bühler J, Schurr U, Jahnke S (2016) Quantitative 3D analysis of plant roots growing in soil using magnetic resonance imaging. Plant Physiol 170(3):1176–1188

Varga T, Hixson KK, Ahkami AH, Sher AW, Barnes ME, Chu RK, Battu AK, Nicora CD, Winkler TE, Reno LR, Fakra SC (2020) Endophyte-promoted phosphorus solubilization in Populus. Front Plant Sci 11:1585

Velasquez SM, Ricardi MM, Dorosz JG, Fernandez PV, Nadra AD, Pol-Fachin L, Egelund J, Gille S, Harholt J, Ciancia M, Verli H (2011) O-Glycosylated cell wall proteins are essential in root hair growth. Science 332(6036):1401–1403

Voss-Fels KP, Robinson H, Mudge SR, Richard C, Newman S, Wittkop B, Stahl A, Friedt W, Frisch M, Gabur I, Miller-Cooper A (2018) VERNALIZATION1 modulates root system architecture in wheat and barley. Mol Plant 11(1):226–229

Wagg C, Jansa J, Schmid B, van der Heijden MG (2011) Belowground biodiversity effects of plant symbionts support aboveground productivity. Ecol Lett 14(10):1001–1009

Walker TS, Bais HP, Grotewold E, Vivanco JM (2003) Root exudation and rhizosphere biology. Plant Physiol 132(1):44–51

Wang Q, Liu J, Zhu H (2018) Genetic and molecular mechanisms underlying symbiotic specificity in legume-rhizobium interactions. Front Plant Sci 9:313

Wang Z, Yu K, Lv S, Niklas KJ, Mipam TD, Crowther TW, Umaña MN, Zhao Q, Huang H, Reich PB (2019) The scaling of fine root nitrogen versus phosphorus in terrestrial plants: a global synthesis. Funct Ecol 33(11):2081–2094

Weaver JE (1919) The ecological relations of roots. Carnegie Institution of Washington, p 286

Williams A, de Vries FT (2020) Plant root exudation under drought: implications for ecosystem functioning. New Phytol 225(5):1899–1905

Xu Z, Valdes C, Clarke J (2018) Existing and potential statistical and computational approaches for the analysis of 3D CT images of plant roots. Agronomy 8(5):71

Yang X, Varga T, Liu C, Scheibe TD (2017) What can we learn from in-soil imaging of a live plant: X-ray computed tomography and 3D numerical simulation of root-soil system. Rhizosphere 3: 259–262

Yoo CM, Quan L, Blancaflor EB (2012) Divergence and redundancy in CSLD2 and CSLD3 function during *Arabidopsis thaliana* root hair and female gametophyte development. Front Plant Sci 3:111

Zappala S, Mairhofer S, Tracy S, Sturrock CJ, Bennett M, Pridmore T, Mooney SJ (2013) Quantifying the effect of soil moisture content on segmenting root system architecture in X-ray computed tomography images. Plant Soil 370(1):35–45

Zhang J, Wang P, Xue K, Hao Y, Wang Y, Cui X (2019) Trait complementarity between fine roots of Stipa purpurea and their associated arbuscular mycorrhizal fungi along a precipitation gradient in Tibetan alpine steppe. J Mountain Sci 16:542–547

Zhou J, Xie J, Liao H, Wang X (2014) Overexpression of β-expansin gene GmEXPB2 improves phosphorus efficiency in soybean. Physiol Plant 150(2):194–204

Zhu J, Kaeppler SM, Lynch JP (2005a) Mapping of QTL controlling root hair length in maize (*Zea mays* L.) under phosphorus deficiency. Plant Soil 270(1):299–310

Zhu J, Kaeppler SM, Lynch JP (2005b) Mapping of QTLs for lateral root branching and length in maize (*Zea mays* L.) under differential phosphorus supply. Theor Appl Genet 111(4):688–695

Zhu J, Brown KM, Lynch JP (2010) Root cortical aerenchyma improves the drought tolerance of maize (*Zea mays* L.). Plant Cell Environ 33(5):740–749

Zhu J, Mickelson SM, Kaeppler SM, Lynch JP (2016) Detection of quantitative trait loci for seminal root traits in maize (*Zea mays* L.) seedlings grown under differential phosphorus levels. Theor App Genet 113(1):1–10

Zhu YH, Weiner J, Yu MX, Li FM (2019) Evolutionary agroecology: trends in root architecture during wheat breeding. Evol Appl 12(4):733–743

# Chapter 12
# Harnessing Root Associated Traits and Rhizosphere Efficiency for Crop Improvement

**Krishnappa Rangappa, Rumi Narzari, Jayanta Layek, Prabha Moirangthem, Burhan U. Choudhury, Bijoya Bhattacharjee, Ramesh Thangavel, and Vinay K. Mishra**

**Abstract** Crop production under stressful and fragile agro-ecologies is really formidable challenge. Under changing climate, the development of crop stress resilience is further challenging owing to its intricacies and linkage among abiotic and biotic stress responses and unprecedented prevalence of climatic vulnerabilities. Root mediated stress modulation and rhizosphere engineering is one of the novel approaches for improved crop growth and productivity under resource poor stressful environments. Holistic understanding and harnessing/tapping of inherent root associated traits including versatile root architecture and complex plant–microbe interactions in consonance with enhanced root exudation potential is one of the frontier areas of crop science research. Crop plants with wider root adaptability and root growth plasticity is quite essential for scavenging more quantities of essential nutrients under marginal and degraded lands. But the extent of employing root adaptability with differential exudation of organic compounds to drive selective plant–microbe interactions is merely exploited as source for crop resilience particularly under imminent multiple stresses. "Root microbiome" being integral reservoir of microbes is controlled by stress and cultivar specific root exudation and reported to evoke various cellular, biochemical, and molecular processes to improve the metabolic capability of crop under stressful environment. A systematic study of structure and function of root microbiome in accordance with expression of related candidate genes as directed by crop adaptability and environmental factors is critical for devising reliable crop production strategies to improve productivity. Conventional efforts to manage unabated crop stresses and to attain potential crop yields are not withstanding and long lasting. Comprehensive understanding of root adaptive

K. Rangappa (✉) · R. Narzari · B. Bhattacharjee
Division of Crop Sciences, ICAR Research Complex for North Eastern Hill Region, Umiam, Meghalaya, India

J. Layek · P. Moirangthem · B. U. Choudhury · R. Thangavel · V. K. Mishra
Division of System Research and Engineering, ICAR Research Complex for North Eastern Hill Region, Umiam, Meghalaya, India

 
H. M. Mamrutha et al. (eds.), *Translating Physiological Tools to Augment Crop Breeding*, https://doi.org/10.1007/978-981-19-7498-4_12

mechanisms coupled with genetic and modern biotechnological approaches besides integrated and innovative fertilization practices are the need of the hour to yield economically feasible and environmentally sustainable results. In this backdrop this chapter deals with various root associated traits such as root architecture, root exudation potential, and root microbiome which are majorly important for developing crop stress resilience.

**Keywords** Crop stresses · Nutrient uptake · Root architecture · Root exudation · Root microbiome · Rhizosphere efficiency

## 12.1 Introduction

The task of meeting the food supply for burgeoning population of 7.6 billion which is further estimated to rise 9.5–10 billion by 2050 is a herculean and formidable challenge for the research community across the globe (Kumar and Dubey 2020). Unstable and stressful agro-environments with increasing climate fluctuation under changing climate scenario and limited land and resource availability are some of the major constraints that are severely impacting global crop production (Kumar and Dubey 2020; Lesk et al. 2016). Among crop stresses drought, salinity, soil pH, temperatures, and nutrient availability are cardinal factors responsible for major loss in crop yields. Globally, various abiotic stress and their extent of impacts suggest that drought is the major factor impacting >60% of land productivity followed by cold (57%), 15% by low pH, 9% by nutrient deficiencies, and about 6% by salinity (Cramer et al. 2011; Mittler 2006). To overcome such limitations and sustain food supply to rapidly growing human and cattle population, yields of essential crops have to be increased invariably which could be achieved only via adoption of appropriate crop production technologies with the application of adequate water, nutrient, and soil amendments (Tester and Langridge 2010; White and Brown 2010).

Crop Production in problematic soils like calcareous and acid soils are majorly influenced by the availability of Cu, Zn and Fe (White and Broadley 2009) and P, K, Ca, Mg (von Uexküll and Mutert 1995) and undue accumulation of Mn, protons and Al(White and Greenwood 2013; Fageria et al. 2011) respectively. Recent observations suggest that even after increased application of fertilizers the total crop yield has declined severely in intensive farming system. Over use of chemical fertilizers usually results in elevated accumulation of nutrients in root zone, which inhibits the rhizosphere activities and root growth (Mi et al. 2010; Zhang et al. 2010). Hence, to exploit the efficiency of rhizosphere with holistic understanding of root associated traits and root environment it is important to synchronize nutrient supply in the root zone with crop requirement which ultimately results in more nutrient acquisition and mobilization and thereby increase resource use efficiency under constrained and stressful environments.

The word “rhizosphere” was originally conceived by Lorenz Hiltner in 1904. This narrow region of contact among root and soil in real time is of great significance because of its phenomenal effect on nutrient alteration, mobilization, acquisition, and efficient utilization by most of crop (Adl 2016; Dessaux et al. 2016). Apart from regulating morphological characters to acclimatize to soil environment, plant roots also actively alter rhizosphere activities via their inherent physiological actions, such as exudation, release of signaling substances, phosphatases, proton, and regular modifications in redox potentials (Zhang et al. 2010; Hinsinger et al. 2009). Since rhizosphere abundantly influences the crop production and sustainability, its manipulation might be an effective tactic to enhance the effectiveness of nutrient utilization and crop yield (Zhang et al. 2010).

Rhizosphere management strategies usually emphasize more on maximization of root efficiency and rhizosphere processes through plants instead of being reliant on chemical fertilizers. However, efficiency of rhizosphere is very reliant on soil inherent fertility and nutrient supply status of the soil. Low nutrient content of soil adversely affects root growth and expansion. The real strategies of rhizosphere management hover around (1) manipulation of root growth (both physiological and morphological characters); (2) intensification of rhizosphere activities with regard to increased root exudation for favored chemical environment; and (3) introducing or enhancing root zone microbiome population (Chen et al. 2011; Zhang et al. 2010).

Plants have continuously progressed with sophisticated adaptive mechanisms to withstand the multiple abiotic stresses to which they are invariably exposed in challenging under fragile agro-ecologies (Lamers et al. 2020). This chapter keenly examines the important root traits and their assessment and their viable association with crop stresses and crop growth that improve the acquisition of essential nutrients and many research studies that are conducted repeatedly in various agricultural and horticultural crops to minimize the input cost and maximize overall resource use efficiency. In this chapter we will be more focusing on the potentiality and importance of root associated traits such as root exudation and rhizosphere acidification, root architecture alterations, and rhizosphere microbiome profiling to enhance rhizosphere efficiency for improvement of crop production.

## 12.2 Root Exudation

Many crop plants are known to secret various compounds to their surrounding soil media from various organs, such as shoot or roots. These secretions could be in viscous solid, mucilaginous liquids, or in gaseous forms. These compounds released into the rhizosphere that actively contribute to nutrient mobilization and acquisition (Jones and George 2002). Roots exude numerous possibly valuable low molecular weight organic complexes (LMWO) into the root zone are usually known as root exudates, are inclusive of various bioactive metabolites accountable for suppression of weeds, infiltration of plant pathogen, improved root environment for maximized nutrient efficiency, and favored microbial community structure (Krishnappa et al. 2014). Plants can modify the chemical configuration of exudates to alter soil properties to guarantee better adaptability of plant towards hostile conditions.

They can employ various approaches to do, these are: (1) Solubilization of nutrients to assimilable forms via alteration of soil pH, (2) Chelates formation of toxic compounds, (3) alluring advantageous microbiota, or (4) discharging compounds lethal for pathogens, etc. Approximately 11% to 18% of the net fixed C is transuded into the rhizosphere via roots under various stressful environments (Nakayama and Tateno 2018; Jones and George 2002; Bais et al. 2006).

It is an amalgamation of varied components, comprising both primary and secondary metabolites (Table 12.1). Primary metabolites include amino acids, carbohydrates, and organic acids, while the later consists of glucosinolates, flavonoids, auxins, etc. (Badri and Vivanco 2009; Nguyen 2009). They act as a reserve of energy and carbon for rhizobacteria, the close vicinity of their catabolic pathways is vital for economical establishment of roots and annihilation of disease (Lugtenberg and Kamilova 2009; Kamilova et al. 2005). Various researchers have established that root exudates also comprise of plentiful signal molecules, however, its significance is still under investigation (Badri and Vivanco 2009; Walker et al. 2003).

Both abiotic and biotic factors are responsible for influencing the volume and chemical composition of organic compounds of exudates (Matilla et al. 2010). Composition of root exudates can be altered due to various soil phenomenons such as chelation, acidification, redox reaction, and precipitation. Change in the nutritional status also triggers release of exudates from the roots, it is more pronounce in case of P, zinc (Zn), or iron (Fe) deficiencies. The redox potential in the root zone can be altered by the plants via direct release of $CO_2$ into the soil and promoting microbial respiration mediated by providing readily available carbon. Thus, the disparity in redox potential between root and bulk soil affects the redox reaction of contaminants including iron (Fe), arsenic (As), manganese (Mn), selenium (Se), and chromium (Cr). Apart from this, the composition and amount of exudation in dependent on photoperiod and temperature since they tend to influence plants respiration, translocation and photosynthesis. Exudation rate is also dependent on plant growth state for instance the rate of exudation of some organic acids is twice or more times higher during vegetative and reproductive stages with elevated levels of temperature and photoperiod.

The disparity anion and cation absorption directly influence the root exudation of organic acids. For example: if excess number of cations are taken up by the plants more precisely $K^+$, then to maintain the electronic equilibrium negatively charged that organic acids are released. Accumulation of organic acids in the rhizosphere of plant grown in $NH_4^-$ fertilized area is lower than those in $NO_3^-$ fertilized fields. The chemical makeup of root exudates diverges based on various aspects which have been recorded in Table 12.2.

## 12.3 Root Architecture

Hodge et al. (2009) stress upon the importance of plant root system for its growth and various other functions such as providing anchorage, moisture and nutrient, and aiding nutrient absorption efficiency through symbiotic association with

**Table 12.1** Compositions of root exudates by crop plants

| Compound class | Components identified |
|---|---|
| Organic acids | Aconitic, acetic, ascorbic, benzoic, caffeic, aldonic, butyric, citric, erythronic, p-coumaric, fumaric, formic, ferulic, glycolic, glutaric, malonic, oxalacetic, malic, lactic, glyoxilic, p-hydroxybenzoic, piscidic, pyruvic, propionic, oxalic, syringic, succinic, tetronic, tartaric, vanillic, valeric |
| Amino acids | α-Aminoadipic, γ-aminobutyric, α-alanine, aspartic, β-alanine, arginine, asparagine, citrulline, cysteine, cystathionine, methionine, cystine, 3-epihydroxymugineic, deoxymugineic, glutamic, glycine, homoserine, glutamine, histidine, leucine, isoleucine, lysine, mugineic, phenylalanine, ornithine, serine, proline, tryptophan, threonine, valine, tyrosine |
| Fatty acids | Oleic, linoleic, palmitic, linolenic, stearic |
| Carbohydrates | Mucilages of various compositions, Arabinose, galactose, mannose, glucose, maltose, fructose, xylose, oligosaccharides, rhamnose, ribose, raffinose, deoxyribose, sucrose |
| Sterols | Stigmasterol, sitosterol, campesterol, cholesterol |
| Growth factors and vitamins | Biotin, p-amino benzoic acid, inositol, N-methyl nicotinic acid, strigolactones, niacin, choline, pantothenate, pyridoxine riboflavin, pathothenic, thiamine |
| Enzymes | Amylase, invertase, peroxidase, phenolase, acid/alkaline phosphatase, polygalacturonase, protease |
| Proteins and enzymes | Lectins, hydrolases, PR proteins, peroxidases, acid phosphatases, proteases, lipase |
| Flavonoids | Coumarine, flavones, flavanones, chalcone, flavonones, flavonols, isoflavones |
| Coumarins | Umbelliferone |
| Aurones | Benzyl auronessynapates, sinapoyl choline |
| Anthocyanins | Delphinidin, cyanidin, pelargonidin and their substitutes with sugar molecules |
| Glucosinolates | Desuphoguconapin, cyclobrassinone, desulphonapoleiferin, desulphoprogoitrin, desulphoglucoalyssin |
| Indole compounds | Brassitin, indole-3-acetic acid, sinalexin, methyl indole carboxylate, brassilexin, camalexinglucoside |
| Allomones | DIBOA, Jugulone, 5,7,4′-trihydroxy-3′,5′-dimethoxyflavone, sorgoleone, DIMBOA |
| Nucleotides/purines | Adenine, guanine, uridine/cytidine |
| Flavonols | Kaempferol, naringin, naringenin, myricetin, rutin, quercitin, genistein, strigolactone and their substitutes with sugars |
| Lignins | Nicotinic acid, pyroglutamic acid, benzoic acid, catechol, phloroglucinol, gallic acid, cinnamic acid, ferulic acid, sinapoyl aldehyde, coumaric acid, chlorogenic acid, vanillin, sinapyl alcohol, syringic acid, quinic acid |
| Miscellaneous | Alcohols, glucosinolates, alkyl sulfides, camalexin, auxins, dihydroquinone, glucosides, ethanol, glycinebetaine, inorganic ions and gaseous molecules (e.g., $H_2$, $CO_2$, $HCO_3$, $OH^-$, $H^+$), hydrocyanic acid, Al-induced polypeptides, unidentified ninhydrin positive compounds, isothiocyanates, reducing compounds, unidentifiable soluble proteins, sorgoleone, scopoletin, strigolactones |

**Table 12.2** Factors affecting root exudation process in most crop plants

| Category | Factors | Effects |
|---|---|---|
| Edaphic | Moisture | Moisture stress elevates exudation |
| | pH | Influence exudates composition |
| | Texture | Large quantity of exudates is produced in sandy soil |
| | Salinity | Low salinity causes release of high amount of organic and amino acids |
| | $NO_3^-/NH_4^+$ | Roots under $NH_4^+$ have lower organic acid concentrations |
| | $O_2$ availability | Compositions are different under aerobic and anaerobic conditions |
| Abiotic | Light | High light intensity increases exudation |
| | Temperature | Acts as a regulator in secretion of root exudates usually high temperature stimulates exudation |
| | Nutrient | Nutrient deficiencies increase exudation |
| | Stress | Influence exudates' composition |
| Biotic | Microorganism | Exudation increases based on the presence of microorganisms. However, it is dependent on the species of microbes colonizing the root zone |
| | Plant species | Unlike plants and cultivars have dissimilar exudate compositions, amount, and range |
| | Plant age | Mostly higher number of exudates are produced during the initial stages of plant growth |
| | Root damage | Any physical or chemical root damage increases root exudation |
| Others | Supporting media | Composition of exudates may vary with the media in which the roots are grown |

advantageous microflora. Root architecture means to the three-dimensional arrangement of the root system, i.e., the explicit geometric deployment of root axes across the soil profile. Root architecture in plant productivity is important owing to irregular dissemination of soil resources, its depletion, so the spatial arrangement of the root system will ensure capability of a plant to acquire those resources. Disparity in soil vital parameters including availability of oxygen and nutrient, moisture status, soil temperature, bulk density, and pH happen with soil depth over a scale of centimeters. These variations usually create both limitations and opportunities in terms of resource procurement; for instance, top soils compared to subsoil contains more nutrients, less moisture and subject to more temperature extremes. In this regard root system architecture (RSA) has its own mechanism to overcome issues such as crop nutrient efficiency; environmental pollution due to high fertilizer inputs; increasing cost of fertilizer inputs; and poor yield under low-input conditions (Lynch 1995). Thus, optimization of root traits has emerged as an economically viable option for improving the efficiency of nutrients (Ahmadi et al. 2014).

For instance, traits like higher root density in subsoil, deep roots, and elevated length of root hair enhance nutrient and water procurement (Lynch 2013; Wasson et al. 2012). Scientist postulated that the "steep, cheap, and deep" ideotype might actually help in better acquisition of nutrient water from soil through refining root exploration of subsoil (Lynch 2013; Mi et al. 2010). According to Lynch and Ho

(2005) architectural phenes like brace roots and growth angles of crown which improves the depth of soil exploration is referred as "Steep" whereas phenes which helps in the reduction of metabolic cost for soil exploration is termed as "Cheap." Postma and Lynch (2011) stated that high amount of aerenchyma tissues in the root cortex is a key factor in the efficient acquisition of N with minimal carbon utilization during root growth. Apart from that it also helps in effective uptake of deep water which in turn increases resilience drought stress (Hund et al. 2009). However, shallow root system with specialized adventitious roots is crucial for phosphorus uptake due to its higher abundance in surface soil strata (Lynch 2007, 2011). Thus, to qualify as an effective genotype with ideotype root architecture a plant must have the capability to merge diverse root functions especially seizure and utilization of various soil resources, both spatially and temporally. The nutrient content of the soil determines the size of the roots whereas localized supply of nutrients determine its distribution and proliferation.

There is tremendous scope to improvise root plasticity and its adaptability to withstand an array of abiotic stresses comprising apparent water and nutrient stresses and thereby to increase crop productivity in degraded lands owing to its unpresented genetic variability and foraging strategies (Lynch and Brown 2012; Lynch 2007).

## 12.4 Rhizosphere Microbiome

The term microbiome refers to the total microbial population residing as a community. Microbiome (bacteria, fungi, archaea, algae, nematode, viruses, and protozoa) observed in the root zone is often referred as plant's second genome due to its influence on plants well-being (Buée et al. 2009; Berendsen et al. 2012). Plant's rhizobiome may have positive (such as symbiosis, mutualistic, etc.), negative (such as pathogens and predators), or no effect on plant community (Bever et al. 2012). However, irrespective of its nature of response the rhizobiome strongly influences plants metabolic processes and its tolerance towards stress which is the main driver behind its increasing popularity (Mendes et al. 2013) (Fig. 12.1).

Rhizobiome can impact the plant's growth and health both directly via production of phytohormones which enhances phosphate solubilization and nutrient uptake; and indirectly through providing protection against plant pathogens mediated by releasing of siderophore, antibiotics, and cell wall-degrading enzymes (Berendsen et al. 2012). Rhizobiome helps in nutrient uptake by releasing certain siderophores or organic acid anions that leads to the formation of chelate ion of trace elements which is then transferred to the cell surface (Mendes et al. 2013). However, a multiple number of studies confirm the fact that a consortium of microorganisms is responsible for such an effect rather than a single microbe. For instance, during downy mildew pathogen attack in *Arabidopsis thaliana* three bacterial genera, *Xanthomonas*, *Microbacterium*, and *Stenotrophomonas* were observed in the root zone (Fig. 12.2).

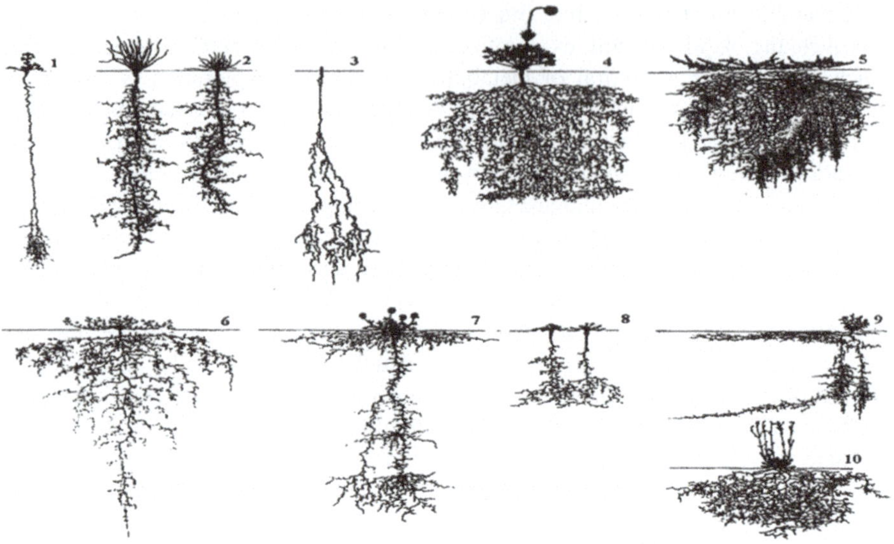


**Fig. 12.1** Variation in root architecture across different crop plants. *1*, *eryngium campestre*; *2*, *Scorzonera villosa*; *3*, *Chondrilla juncea*; *4*, *Pulsatilla pratensis*; *5*, *Genista germanica*; *6*, *Trigonella balansae*; *7*, *Trifolium trichocephalum*; *8*, *Carum caucasicum*; *9*, *Onosma arenarium*; *10*, *Silene otites*

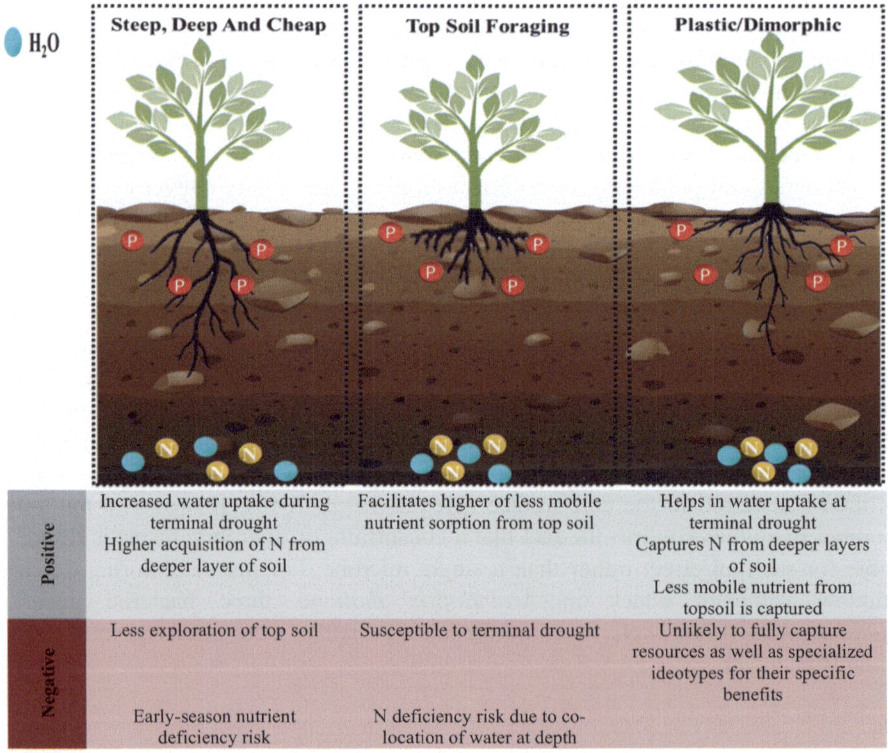


| | Steep, Deep And Cheap | Top Soil Foraging | Plastic/Dimorphic |
|---|---|---|---|
| Positive | Increased water uptake during terminal drought<br>Higher acquisition of N from deeper layer of soil | Facilitates higher of less mobile nutrient sorption from top soil | Helps in water uptake at terminal drought<br>Captures N from deeper layers of soil<br>Less mobile nutrient from topsoil is captured |
| Negative | Less exploration of top soil<br>Early-season nutrient deficiency risk | Susceptible to terminal drought<br>N deficiency risk due to co-location of water at depth | Unlikely to fully capture resources as well as specialized ideotypes for their specific benefits |

**Fig. 12.2** Root ideotypes and potential trade-offs

A wide variety of microflora have been reported to support plant growth and helps in mitigation of abiotic stresses owing to its genetic and intrinsic metabolic capabilities (Meena et al. 2017). Naveed et al. (2014) stated that wheat when inoculated with *Burkholderia phytofirmans* Ps JN showed higher chlorophyll content, elevated rate of photosynthesis, and grain yield even under water stress condition. Ahmad et al. (2015) observed that fungus treatment improvise uptake of nutrient and production of antioxidants and osmolytes while Na+ acquisition was reduced in *Brassica juncea* under saline environments. Other studies revealed that tolerance towards salinity was enhanced in beet root in *Arabidopsis Bacillus subtilis* GB03 control the expression of HKT1 (high-affinity K+ transporter 1) which imparts tolerance towards salt stress (Zhang et al. 2008). Srivastava et al. (2008) observed that chickpea in association with *Pseudomonas putida* strain NBR10987 was able to evade moisture stress due to the production of exo-polysaccharides with water holding features. Theocharis et al. (2012) suggest that rhizobiome helps in the rapid and elevated levels of stress-related metabolites and proteins which provides shield low non-freezing temperatures. Jha et al. (2012) reported new stress resistant bacteria such as *Zhihengliuella* sp., *Brachybacterium saurashtrense*, and *Brevibacterium casei*. However, there are certain bacteria (*Actinobacteria*, *Pseudomonas*, *Verruco microbia*, and *Microbacterium*) which can be utilized for rhizoremediation due to their ability to modify movement and bioavailability of metals, resulting in their enhanced uptake by plants (Yang et al. 2016).

The rhizosphere is a region characterized by lower species diversity of microorganisms but an intense zone abundance and activity. The microbiome assemblage in the rhizosphere is strongly dependent upon the plant-derived metabolite or in other words root exudates, which are crucial constituents engaged in the advent and successive seminal of the rhizosphere (Haney et al. 2015). There are numerous factors which determine the composition of rhizobiome such as plant species, genotype, and soil constituents thereby making the microbe's biome specific which was observed in case of *Bacillus subtilis* by malic acid (Fig. 12.3) (Lakshmanan et al. 2014; Mendes et al. 2013).

Rhizosphere microbiome configuration is affected by complex interactions between microbe, hosts, and environmental aspects such as temperature, cultivation practices, soil, and climate as listed in Table 12.3.

## 12.5 Mechanisms of Adaptation for Root Traits

### *12.5.1 Mechanism of Exudation*

Mechanisms involved with plant exudations are chiefly of two types of passive and active transports. The passive process is mediated through three diverse pathways: diffusion, ionic channels, and vesicles transport (Fig. 12.4) (Baetz and Martinoia 2014). The pathways followed for the transport is determined by the chemical properties of exuded.

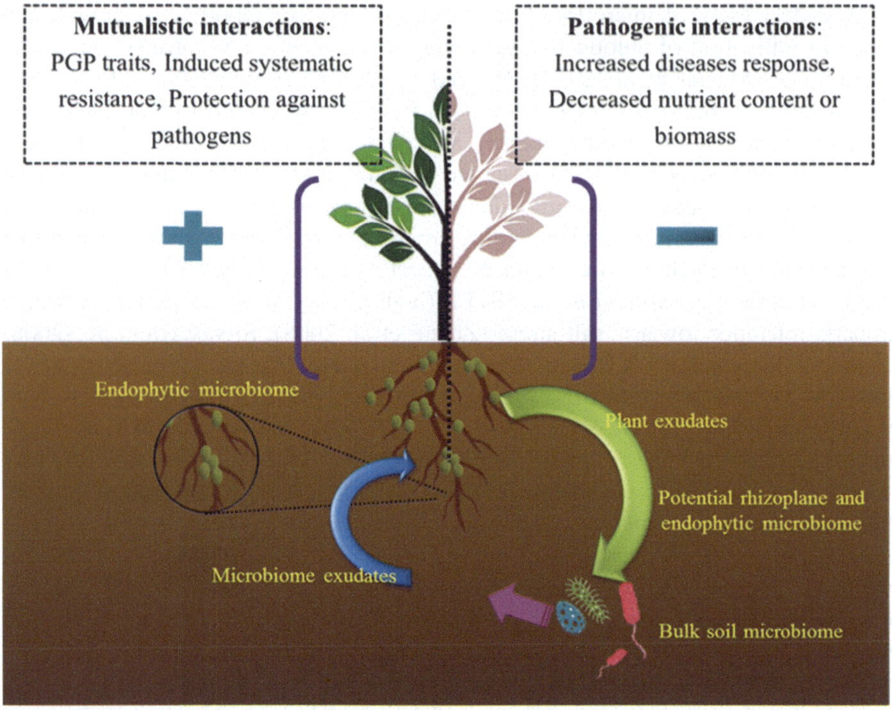


**Fig. 12.3** Schematic representation of overall rhizosphere microbiome and its effect on plant

Metabolites with low molecular weight (such as amino acids, phenolics, sugars, and carboxylic acids) are released through concentration gradient generated between cytoplasm and plasma membrane of root cells and the rhizosphere, which in turn affects permeability of root membrane, root cells integrity, and the polarity which leads to release of exudate through passive diffusion (Badri and Vivanco 2009). In case of passive transport through ionic channels proteins are employed for the release of certain carboxylates (such as malate and oxalate) and carbohydrates at high amounts. Basically, the ionic channels can be described in two ways: SLow Anion Channels (SLACs), which have an activation time period of several seconds; and QUick Anion Channels (QUACs), with an activation period of a few milliseconds (Dreyer et al. 2012). Exudation of organic acid with the help of Aluminum-activated-malate transporters (ALMT) leads to the formation of Al chelate which inactivate $Al^{3+}$ ions, thereby providing tolerance to plant against aluminum toxicity (Sharma et al. 2016). Similarly, malate is secreted during phosphorus stress condition through AtALMT3 (Maruyama et al. 2019). In addition to this, overexpression of AtALMT1 coupled with Aluminum toxicity tolerance facilitates colonisation of rhizobacteria which improvise plant immunity. Lastly, metabolites of high molecular weight are packaged and transported via vesicles secretion also known as exocytosis

**Table 12.3** Drivers of rhizosphere microbiome composition in plants

| Category | Factors | Effects | References |
|---|---|---|---|
| Biotic | Type of plant | Specific exudates are responsible for definite microbial community | Wei et al. (2018) |
| | Pathogenic incursion | Negatively affects the population and microbes | Wei et al. (2018) |
| | Free-living Protists | Due to its predation of certain bacteria and fungi it causes reorientation of microbial composition | Gao et al. (2019) |
| Edaphic | Elevated temperature | Upsurges alpha group diversity later | van der Voort et al. (2016) |
| | Soil sterilization | Encourages advantageous microbe | Li et al. (2019) |
| | Soil salinization | Supremacy of Actinobacteria | Mukhtar et al. (2018) |
| | | Exhibited 55% of bacteroidetes, proteobacteria, and actinobacteria | Ibekwe et al. (2017) |
| | Non-saline soil | Supremacy of proteobacteria | Mukhtar et al. (2018) |
| | Elevated nitrogen level | Subdue microbial respiration | Maghnia et al. (2019) |
| Agricultural practices | Change in land use pattern | Amplified quantity of Acidobacteria and Verrucomicrobia | Maghnia et al. (2019) |
| | Tillage | Alters phylogenetic configuration of the rhizobium | Wang et al. (2020) |
| | | Enhances and stabilizes plant growth-promoting rhizobacteria population as it causes better aeration and moisture | Wang et al. (2017) |
| | Irrigation with wastewater | Changes bacteria composition | Cui et al. (2008) |

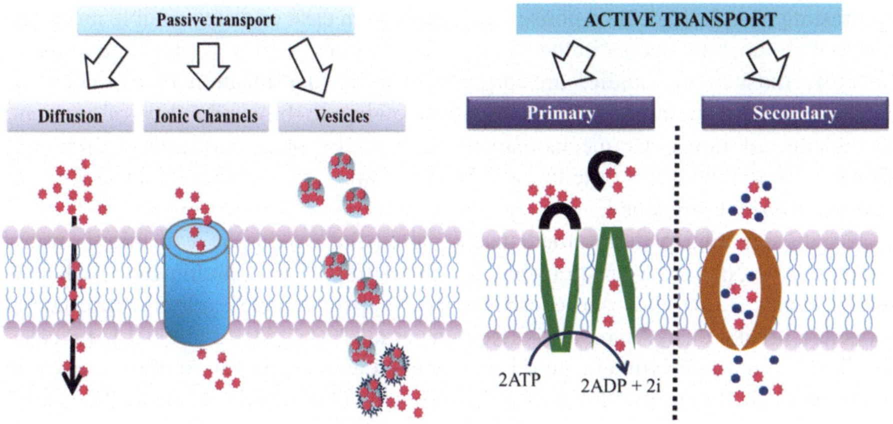


**Fig. 12.4** Root exudation mechanisms in plants

(Badri and Vivanco 2009). These metabolites are released from Golgi apparatus or endoplasmic reticulum and provides defense against pathogens (Weston et al. 2012).

Proteins situated in the root plasmatic membrane are responsible for active transport of metabolites (Baetz and Martinoia 2014). Two types of membrane transporters are employed in active transport, viz., ABC (ATP-Binding Cassette) and MATE (Multidrug and toxic compound extrusion) are involved in case of active transport (Kang et al. 2019). Proteins mediated exudation generally materializes under the following circumstance: (1) secretion of different metabolites by transporters; (2) ability of different membrane transporter to produce metabolites into the rhizosphere; and (3) exudation of compound by a unique transporter. Utilization of ATP (adenosine triphosphate) hydrolysis for translocation of solutes and qualifies ABC transporters as primary transporters (Jones and George 2002). This transporter is also capable of modifying soil microbiota, due to higher exudation of phenolic compounds compared to sugars (Badri et al. 2009). MATE transporters family have 58 members that serves as secondary active transporters which regulates the transfer of the compounds based on electrochemical gradient (Weston et al. 2012). Transportation of flavonoids to the vacuole is mediated by both ABC and MATEs transporters (Yazaki 2005).

### *12.5.2 Mechanisms of Root Architecture Modification*

Roots developmental factors such as diameters of primary (PR) and lateral roots (LR), root length, angle, development of root hair, and its number are determined by the nutrient concentrations. Thus, it cannot be denied that the root architecture shows nutrient-dependent patterns. For instance, elongation of primary roots is restricted whereas enhanced LR formation was observed in plants grown under P deficit condition (Williamson et al. 2001), which leads to the development of high root surface area with a shallow root morphology. Phosphate generally accrues on the top layer of soil where it immobilizes (Jobbagy and Jackson 2001). Therefore, in order to acquire P more efficiently plant deploys more roots towards the P. However, a contrasting morphological response was observed in case of water-soluble nutrients, due to their leaching deeper in the soil layer deep root system was prevalent in them. Recently most of the studies are targeted towards identification of distinct gene accountable for alteration in root structure owing to bioavailability of nutrient. Knowledge of molecular mechanism through which plant can modify their root structure based on the nutrient availability would aid genetic enhancements to improve nutrient efficiency.

Nitrate signaling pathway mediated by auxin in association with nitrate transporter/sensor NRT1.1 is responsible for alteration of root architecture, whereas NRT2.1 have shown dual purposes with respect growth and development of lateral roots and uptake of in Arabidopsis (Remans et al. 2006). Ma et al. (2014) revealed that TAR2 (auxin biosynthetic gene) is associated with reorientation of root structure in Arabidopsis due to low level of N in soil. This observation confirms the role of auxin in management of root architecture towards N accessibility. Though, there are ample amount of literary evidence to support numerous signaling pathways concerning soil N availability and root architecture hypothesis (Forde 2014). However, there is a dearth of studies which have identified the genes responsible for this. Ubiquitin ligase (EL5) in rice is responsible for avoiding death of root meristematic cell due modification in root triggered by N (Mochizuki et al. 2014). Apart from this, microRNAs play a crucial part in plant reactions towards nutrient stress. Two microRNAs, namely miR393 and miR167 in Arabidopsis regulates genes AFB3 and ARF8 which controls the modification of root architecture according to the availability of N (Gifford et al. 2008). Some of the transcription factors identified which regulated root architecture modification in response to P stress are: in case of rice, it's OsPHR2/OsMYB2P-1 (Dai et al. 2012), for Arabidopsis it's SIZ1/YRKY75 (Devaiah et al. 2007), and ZmPTF1 for maize (Li et al. 2011). Both elongation of roots and Fe accumulation were found to be controlled by OsARF12 (Qi et al. 2012). However, the working principal of these genes is yet not known.

### *12.5.3 Mechanism Adopted by Microbiome Community*

As mentioned before the microbial community residing in the root zone affect plant productivity and composition directly or indirectly (van der Heijden et al. 1998). Three different mechanisms have been suggested to elucidate plant health enhancement adopted by plant are: (1) manipulation plants' hormonal signaling (Verbon and Liberman 2016); (2) preventing or surpassing pathogenic strains (Mendes et al. 2013); and (3) elevating soil nutrients bioavailability.

By targeting the hormonal balance in the plant rhizomicrobiome can encourage plant growth. They either releases small secondary metabolite which emulates plant growth hormone or manipulate signaling of plant defense hormone for effective colonization in plants tissues (Patkar and Naqvi 2017). Various fungi and bacterial species associated with plant are well adapted to generate indole-3-acetic acid (IAA) or auxin-mimicking molecules which directly regulates growth and development of plants. Some phytohomones or molecules similar to phytohormoes including cytokinins, gibberellins, jasmonic acid (JA)-isoleucine or salicylic acid (SA) helps in the successful colonization of rhizosphere microbiome (Stringlis et al. 2018). Most of the beneficial rhizosphere microbes tend to produce 1-aminocyclopropane-1-carboxylate (ACC) deaminase which cleaves ACC that acts as a precursor of ethylene (ET), and thereby enhance plant development by suppressing the ET levels for plant development under stress conditions (Stringlis et al. 2018; Glick 2014).

One of the mechanisms of root microbiome to facilitate plant productivity is via elevation of plant defense potential acknowledged as induced systemic resistance (ISR) that enable plant resistant against numerous infective threats (Pieterse et al. 2009). Plant defense system is activated by network of hormone signaling pathways that mediate the transfer of information to the different organs of plants (Pieterse et al. 2009). This phenomenon was first identified in Pseudomonas, and the principal behind it was derived from "systemic acquired resistance" (SAR) induced due to pathogenic outbreak. Various plant growth-promoting fungi (PGPF) and bacteria (PGPR) have also been recognized to exhibit ISR which is distinguished by the kind of secondary metabolite secreted by the microbes (Stringlis et al. 2018). However, defense response in ISR is activated upon pathogenic attack thus saving energy. Such kind of defense response is also referred as priming which silently alert plants defense system against an insect or pathogen when attacked. After the attack the plant deploys all its responses faster and/or stronger for effective and efficient defense.

Due to association of nutrients such as P and N with organic molecules under natural ecosystems, their bioavailability is restricted and not adequate for plants. Therefore, to acquire these nutrients plants need metabolic machinery to depolymerize and mineralize nutrient thus they require help from the microbes such as fungi and bacteria. The microbe releases its contents by protozoic or cell lysis which causes liberation of inorganic forms of nutrients into the soil, including phosphate, ammonium, sulfate, and nitrate, making them bioavailable to plants. Thus, these entire process acts as a rate-limiting step for ecosystem productivity (Schimel and Bennett 2004).

## 12.6 State-of-the Art Analytical Tools Available for Assessment of Root Traits

### *12.6.1 Root Exudate Analysis*

Owing to their wide variability in quality, configuration and quantity and their dependence on root environment, genotype and type of cultivar, research studies on plant root exudates is more complicated (Badri and Vivanco 2009). Thus, the scientific community heavily relies on a combination of various analytical techniques and methods to understand and explore the various aspects of these metabolites (Zhang et al. 2012). Non-targeted metabolic investigation is gaining popularity in the recent times (Fuhrer and Zamboni 2015). To explore and improvise understanding regarding interaction between plant and microbial community the dependence on the modern methods and analytical techniques is inevitable to interpret the intricacy of root exudate. Isolation, quantification, and characterization of root exudate metabolites can be made possible owing to the rapid advancement in metabolomic examination (Zhang et al. 2012). For comprehensive understanding of numerous compounds various platforms such as ultra-performance liquid chromatography mass spectrometry (UPLC-MS) and gas chromatography mass spectrometry (GC/MS) are utilized for detail profiling of root exudation. Though, the protocols followed for root exudate assessment are usually familiar; collection methods employed for its collection may vary considerably owing to unapproachability of the root system (Oburger et al. 2018).

The extent of root exudation and consequent rhizosphere acidification were assessed and harnessed through various approaches. Simple assays like agar media plating with pH sensitive dyes at seedling level are better suited for low root systems. These assays adequately help the researchers to identify the genotypes having the higher inherent potential for root exudation with better root environment and thereby result in enhanced nutrient use efficiency. At vegetative stage other approaches like spectrophotometric and chromatographic analytical tools play important role after appropriate collection, condensation, purification, and processing of root exudates. Some simple organic acids like citric acid and malic acid could be quantified through spectrophotometric methods and novel compounds in the root exudates were identified through HPLC (high-performance liquid chromatography) and mass spectrometer-based assays.

### *12.6.2 Root Architecture Study*

Multiple root parameters including branching angle, root width, length, depth, distribution of root volume, and ratio of root depth to width across soil profile has been described. These traits are becoming very essential to precisely model, measure, and quantify root architecture to enhance crop productivity, improvise crop sustainability, and understand the key interaction between plant–microbes. Hallé and Oldeman (1970) were the first to conduct qualitative investigation on crowns of tropical trees architecture and coin the term "architectural analysis." Excavation of root is an invariable and tedious part of root architectural study owing to its underground growth habit. However, this procedure might cause some damage to RSA reflected in loss of fine root structures if the finer roots are not properly handled. In the recent decade advancement in the technologies results in improvised and easy methods to record and evaluate root systems without causing any visible or destruction. For example, *WinRhizo*, *Minirhizotrons*, *X-ray computed tomography* (*CT*), *and magnetic resonance imaging* (*MRI*) are the recently developed user-friendly software used for finer imaging of intact roots (Table 12.4).

Both geometric and topological characteristic of the root encompasses the three-dimensional (3D) root structure of plant. Topology deals with the corporeal associations among plant apparatus, while geometry deals with the orientation, size, shape, and spatial location of the root components. Geometry is related to exchanges between plant and environment, anchorage, and acquisition of resources, while biological sequences present is the axes which acts as the base for flux of mass, energy, and information falls under topology. During excavation or washing in traditional topology study most of the time the root system is destroyed. Therefore, utmost care and precaution must be taken while removing the soil around the roots, which permits the removal of the root from the ground; it should be washed smoothly with smooth flush of water. However, it is labour intensive and lengthy approach and due to the damaging nature of soil coring method, sampling of finest roots is difficult at most of the time.

**Table 12.4** Overview of analytical techniques employed for root exudates analysis

| Techniques | | Advantages | Disadvantages | References |
|---|---|---|---|---|
| Analytical | Mass spectrometry (MS) | • Thorough examination of variations in exudation patterns of compounds specific to plant species<br>• High sensitivity and cover wide range of metabolites | • Insufficient data about the remaining C compounds secreted<br>• Co-suppression of compounds along with low ionization efficiencies | Fuhrer and Zamboni (2015) |
| | Gas chromatography mass spectrometry (GC-MS) | • Detection of known and unknown analytes and signals independent of hypothetical targets among the vast range of metabolites<br>• Steady retention times<br>• Outstanding separation resolution | • GC-MS is limited by compound volatilization<br>• Often requires chemical derivatization | Pantigoso et al. (2021) |
| | Liquid chromatography mass spectrometry (LC-MS) | • Appropriate for an extensive array of compounds<br>• Unsophisticated sample preparation | • Not suitable for highly polar compounds | Casas and Matamoros (2021) |
| | Nuclear magnetic resonance (NMR) | • Minimal sample preparation<br>• A reduction in loss of unknown compounds<br>• Rapid analysis | • Low sensitivity<br>• Hard to combine with separation techniques<br>• Costly | van Dam and Bouwmeester (2016) |
| | FTIR | • Non-destructive nature<br>• Sample preparation does not require reagents<br>• Rapid and inexpensive<br>• Portable | • Low sensitivity<br>• Quality of spectra may be affected by sample<br>• Dehydration of sample vital | Pantigoso et al. (2021) |
| | Raman | • Non-destructive nature<br>• Sample preparation does not require reagents<br>• Rapid and inexpensive<br>• Fresh plant material can be used<br>• Portable | • Low sensitivity | Casas and Matamoros (2021) |

(continued)

**Table 12.4** (continued)

| Techniques | | Advantages | Disadvantages | References |
|---|---|---|---|---|
| | High-performance liquid chromatography (HPLC) | • Wide range of analytes amenable<br>• Short sample runs<br>• Various mechanisms of separation<br>• Structure confirmation | • Potential suppression of ionization efficiency<br>• Eluent compatibility required | Dell'mour (2010) |
| Isotope based | $^{13}C/^{14}C/^{11}C$ pulse labeling | • Gives data on distribution of photosynthate at a specific developmental stage | • Does not denote the circulation of total C | Kuzyakov and Domanski (2000) |
| | Continuous $^{13}C$ labeling | • Provides information on total amount of assimilated C and total C distribution | • Expensive<br>• High sample processing time | Oburger and Jones (2018) |
| | $^{13}C$ natural abundance approach | • Do not require extensive equipment or large quantities of isotopes | • Rough estimate of rhizodeposition owing to poor availability of suitable soil-plant pairs and variability in $\delta^{13}C$ values | Oburger and Jones (2018) |
| Imaging | Positron emitting tracer imaging system (PETIS) | • Non-destructive<br>• Produces quantitative real-time images | – | Suzuki et al. (2008) |
| | Autoradiography phosphor imaging | • Provides dynamics images (absorption, translocation and distribution) of mineral | • Destructive<br>• Time consuming | Ferrari et al. (2016) |
| | Zymography | – | – | Oburger and Schmidt (2016) |

Root architecture can also be indicated by various sub-traits like fractal dimensions reflecting the area occupied by the roots, or linkage among the basal root and total root system. Topologically some model studies the root system with set of links connecting two points of root branch (interior link) or tip of root branch (external link) deriving additional parameters like the magnitude and altitude. This trait ultimately signifies the actual carbon cost, absorption cum exploitation efficiency which implicates the actual differences of complex but contrast branching root systems.

### *12.6.3 Rhizosphere Microbiome*

Various studies have established the fact that the community structure and abundance of the rhizobiome depends upon the developmental stage of the plant generating a rendering adequate number of replicates which is very important for a prerequisite for comprehensive statistical understanding of sequencing data. Apart from this, stochastic effects (timing of arrival of species) may also affect species distribution.

Identification and understanding of microbial dynamics and structure have been made possible with the advancement in high-throughput sequencing integrated with an array of "omics." The advanced sequencing techniques have enabled the researcher to obtain detailed information regarding the identity and relative abundance of the rhizosphere microbes. Cultivation of microbial isolates are not required due to the generation of sequences directly from environmental sample. Prior to selection of sequencing method detailed information on the types of expected biases and errors must be considered. Table 12.5 provides a quick summary of available techniques for identification soil microbial community.

## 12.7 Case Studies for Rhizosphere Management

Many research studies inclined on the exploration of methods to optimize and integrate various rhizospheric processes for intensification crop productivity in a justifiable way via increasing nutrient use efficiency, improve crop yields, mineral fertilizer inputs, and achieve sustainable crop production. Most of the work is primarily based on plants acquisition, mobilization, or utilization of P from soils, which is essential for their growth. Plant roots influence the processes occurring at rhizosphere zone through morphology, physiological metabolic activities or microbiome to improve the bioavailability of soil P. Density of root hair and its lateral branching are some of the useful trait for efficient use of phosphorus. In Arabidopsis and beans (Lynch and Brown 2012) higher secondary branching of roots and lateral root growth was observed over primary root. Phosphorus capture capacity of the plant was enhanced due to enhanced root length and root hair densities. Richardson et al. (2009) detected that in case of maize and common bean, the axial root length increased instead of lateral branching in quest of soil patches enriched with phosphorus (Richardson et al. 2009). Deng et al. (2014) established that maize maintained optimal root efficiency in terms of mycorrhizal infection, root surface area, and root growth vitality at a topsoil (0–20 cm) Olsen-P of 5–10 mg/kg, and at the same time maintained the maximum yield (Deng et al. 2014). The length of root hair and root hair density are negatively correlated. This may be due to the plant's policy to conserve carbon input which will otherwise utilize lot of carbon to maintain both the root traits. According to Hao et al. (2010), nutrient capture capacity of root was increased by several folds by modification of root hair

**Table 12.5** Molecular methods employed to characterize soil microbial communities

| Technique | Advantages | Limitations | References |
|---|---|---|---|
| Chloroform fumigation extraction | • Measurements of microbial biomass can be done in recently added and freshly decomposed substrates | • Clay soils may need to be corrected for the amount of chloroform C added to assess the concentration of biomass C | Rincon-Florez et al. (2013) |
| Phospholipid fatty acid analysis | • Sensitive detection and accurate quantification of different microbial groups<br>• Rapid and efficient<br>• Useful information on the dynamics of viable bacteria<br>• Reproducible | • Time taking<br>• Small number of samples can be treated at the same time | Kaur et al. (2005) |
| Quantitative PCR | • Quick, accurate and highly sensitive method for sequence quantification that can also be used to quantify microbial groups<br>• Relatively cheap and easy to implement<br>• Specific amplification can be confirmed by melting curve analysis | • Can only be used for targeting of known sequences.<br>• DNA impurities and artifacts may create false-positives or inhibit amplification | Smith and Osborn (2009) |
| Denaturing gradient gel electrophoresis/ temperature gradient gel electrophoresis | • Sensitive to variation in DNA sequences<br>• Bands can be excised, cloned and sequenced for identification | • Time consuming—multiple bands for a single species can be generated due to micro-heterogeneity<br>• Can be used only for short fragments<br>• Complex communities may appear smeared due to a large number of bands<br>• Difficult to reproduce (gel to gel variation) | Kirk et al. (2004) |
| Single-strand conformation polymorphism | • Community members can be identified<br>• Screening of potential variations in sequences<br>• Helps to identify new mutations | • Short fragments<br>• Lack of reproducibility<br>• Several factors like mutation and size of fragments can affect the sensitivity of the method | Konstantinos et al. (2008) |
| Terminal restriction fragment length polymorphism fingerprinting | • Enables analyses of a wide array of microbes<br>• Highly reproducible<br>• Convenient way to store data and compare between different samples | • Artifacts might appear as false peaks<br>• Distinct sequences sharing a restriction site will result in one peak.<br>• Unable to retrieve sequences | Rincon-Florez et al. (2013) |

(continued)

**Table 12.5** (continued)

| Technique | Advantages | Limitations | References |
|---|---|---|---|
| Automated ribosomal intergenic spacer analysis | • High resolution when detecting microbial diversity<br>• Quick and sensitive | • More than one peak could be generated for a single organism<br>• Similar spacer length in unrelated organisms may lead to underestimations of community diversity | Kirk et al. (2004) |
| Length-heterogeneity PCR | • Results are reproducible<br>• Easy and rapid<br>• Efficient and reliable | • Limited by the bacterial species known in public databases<br>• Not enough information is available for fragment length on databases to compare LH-PCR lengths with environmental microorganisms | Rincon-Florez et al. (2013) |
| Random amplified polymorphic DNA | • Suitable for unknown genomes<br>• Requires low quantities of DNA<br>• Efficient, fast, and low cost | • Low reproducibility<br>• Sensitive to reaction conditions | Rincon-Florez et al. (2013) |
| Amplified ribosomal DNA restriction analysis | • Highly useful for detection of structural changes in simple microbial communities<br>• No special equipment required | • More applicable to environments with low complexity<br>• Several restrictions are needed for adequate resolution<br>• Labor- and time-intensive<br>• Different bands can belong to the same group | Nocker et al. (2007) |
| Fluorescence in situ hybridization | • Allows detection and spatial distribution of more than one samples at the same time | • Autofluorescence of microorganisms<br>• Accuracy and reliability is highly dependent on specificity of probe(s) | Moter and Göbel (2000) |
| DNA Array | • Analyzes a vast amount of genetic information simultaneously | • Requires the construction of an array and access to a scanner<br>• Issues with specificity/cross hybridization<br>• Requires normalization<br>• Sensitivity and reproducibility can be problematic<br>• Limited by the | Everett et al. (2010) |

(continued)

**Table 12.5** (continued)

| Technique | Advantages | Limitations | References |
|---|---|---|---|
| | | presence of probes on the array | |
| Next generation sequencing (16S rRNA amplicon sequencing) | • Rapid method to assess biodiversity and abundance of many species/organizational taxonomic units simultaneously and at a considerable depth compared to the methods that have been available so far | • Relatively expensive<br>• Replication and statistical analysis are essential<br>• Computationally intensive<br>• Challenging in terms of data analysis | Rincon-Florez et al. (2013) |
| Next generation sequencing (metagenomics) | • Biodiversity can be studied in more detail<br>• Captures polymorphism in microbial communities<br>• Reveals the presence of thousands of microbial genomes simultaneously<br>• Provides information about the functions of microbial communities in a given environment | • High cost<br>• Data analysis is challenging and time consuming<br>• Difficult to use for low-abundance communities<br>• The high biodiversity in soil leads to many incomplete genomes<br>• Current sequencing methods and computing power still in its infancy to the high biodiversity in soil | Rincon-Florez et al. (2013) |
| Fluorescein diacetate | • Low-cost, easy and fast method to measure microbial activity for soil samples | • The measurement of soil microbes by FDA can be contaminated by external sources, e.g., plant matter | Manichanh et al. (2008) |
| Stable-isotope probing | • High sensitivity<br>• Provides evidence on the function of microorganisms in a controlled experimental setup | • Incubation and cycling of the stable isotope might cause biases within the microbial communities | Dumont and Murrell (2005) |
| Functional gene arrays | • Analyzes a vast amount of genetic information simultaneously | • Requires the construction of an array and access to a scanner<br>• Issues with specificity/cross hybridization<br>• Requires normalization<br>• Insufficient sensitivity and reproducibility can be problematic<br>• Limited by the presence of probes on the array | Everett et al. (2010) |

(continued)

**Table 12.5** (continued)

| Technique | Advantages | Limitations | References |
|---|---|---|---|
| | | • Issues with RNA extraction from soil | |
| Next generation sequencing (Metatranscriptomics) | • Allows rRNA and/or mRNA profiling and quantification without prior knowledge of sequence<br>• Provides a snapshot of microbial transcripts at the time of sampling that may allow deduction of microbial ecosystem function<br>• Helps to understand the response of microbial communities to changes in their environment | • Many issues with isolation of RNA from soil<br>• mRNA isolation and often amplification are required for gene expression analyses<br>• Current sequencing methods, data bases and computing power are not sufficient yet to cover the high biodiversity in soil | Carvalhais et al. (2013) |
| Metabolomics | • Allows profiling the metabolites microbiota produce<br>• Semi quantitative | • Unknown metabolite origin | Galloway-Peña and Hanson (2020) |
| Metaproteomics | • Combines metagenomics and proteomics and provides for the verification of metagenomics studies by protein data | • Insufficient databases for soil protein identification | Lagos et al. (2015) |

Adapted from Rincon-Florez et al. (2013)

density and growth angle of lateral roots. Wissuwa et al. (2006) found that P deficiency enthused root elongation in rice, whereas low phosphorus conditions was responsible for high root number, length, surface for tolerant rice cultivars. Brassica cultivars with high phosphorus-use efficiency exhibited efficient translocation of phosphate metabolically active nonmature plant parts (Akhtar et al. 2011). Under high phosphorus condition most of the phosphorus content (around 75%) in plant may be present vacuole as orthophosphate (Akhtar et al. 2011).

Apart from root architecture root exudation also showed a potential to improvise plant productivity. Exudation in the form of phytosiderophores from gramineous plant roots under Zn-deficient conditions facilitate insoluble Fe mobilization along with Zn and Cu. Rhizosphere interaction via root exudation plays an important role in improving Fe nutrition of peanut (*Arachis hypogaea* L.). Radish (*Raphanus sativus* L.) exuded tartaric acid to mobilize Fe–P and Al–P, while malic and citric acids were the dominant organic acids exuded by rape (*Brassica napus* L.). Mucilage exuded by roots helps in the formation of rhizosheath under drought conditions which assembles soil particles and stabilize soil aggregates around roots. This provides plant protection against drought stress such as maize, oat, sorghum, barley, and wheat (Ndour et al. 2020). P deficiency induces exudation of caffeic and protocatechuic acid which facilitate desorption of P for plant uptake (Hu et al. 2005). Acid phosphatases (APases) secreted into the rhizosphere hydrolyze P from both extracellular and intracellular organic P pool before they can be absorbed by roots of wheat, maize, and rice (Hurley et al. 2010; Tadano and Sakai 1991). Some of the cases studied in regard to the modification of microbes in the rhizosphere for enhanced crop productivity or plant growth have been enumerated below (Table 12.6).

## 12.8 Maximization of Rhizosphere Efficiency for Higher Crop Productivity

Nutrient transformation, mobilisation, efficient uptake and utilization is greatly influenced by rhizospheric processes which regulate and drives rhizosphere biology and chemistry and thereby coordinating the plant soil interactions. In order to improvise plant productivity and nutrient use efficiency alteration in the rhizosphere processes by employing microbes might be a suitable and sustainable approach. This approach may be utilized to increase the efficiency of nutrient procurement and utilization at the same time reducing dependence on chemical fertilizers. Optimum supply of nutrient may also aid to enhance rhizosphere efficiency. Zhang et al. (2010) suggested that the inefficacy of the rhizosphere can be suppressed either by supplying excessive amount of nutrient or by subjecting plant to severe nutrient stress condition. By controlling the root development, exudation of carboxylates, protons as well as activity of acid phosphatase can be regulated which in turn would help in manipulation of rhizosphere processes and efficiency?

Optimization of the amount of nutrient supplied to the high yielding cropping system is very vital along with evaluation of the biological potential of the root system. Based on the nutrient demand by the specific crop at specific growth stage the supply or rate of chemical fertilizer application can be regulated (Ju et al. 2009). Management of nutrient supply of the plant at root zone can leads to better nutrient capture from soil by the plant thereby diminishing chemical fertilizer application. As an ideal root zone nutrient management strategy stresses upon application of nutrient in the root zone within a limit in accordance to the requirement of crop, time, and development stage of crop growth, especially in case of N management (Cui et al. 2008).

Plant roots are not only influenced and regulated by morphological traits to adapt to soil environmental conditions but it also significantly modify rhizosphere processes through their physiological activities. Rhizospheric processes are the linkages between plant processes and soil processes to some extent determining the bioavailability of soil nutrient and thus affect the crop production. The rhizosphere management approaches completely stress upon maximizing the efficacy of root rhizosphere in nutrient procurement towards high yield of crops by optimizing nutrient input in rooting zone, regulating root growth, root architecture, release of exudates, and

**Table 12.6** Application of microbes in the rhizosphere for stimulation of plant growth

| Rhizosphere | Microbes utilized | Effect | References |
|---|---|---|---|
| Potato | *Brevundimonas spp., Azospirillum spp.* | N-fixation, IAA production, P solubilization, N-fixation | Naqqash et al. (2016) |
| Potato, rice, wheat, maize, soybean | *Serratia spp.* | AHLs, IAA production, phytase activity | Hanif et al. (2020) |
| Pea | *Azospirillium, Agrobacterium tumefaciens* | N-fixation, P-solubilization | Ejaz et al. (2020) |
| Rice | *Acinetobacter soli, Bacillus sp., Pseudomonas putida, Pseudomonas mosselii, Arthrobacter woluwensis* | P-solubilization, ACC deaminase activity, Siderophores production | Xiao et al. (2020) |
| Cabbage | *B. subtilis* | Gibberellins production | Kang et al. (2019) |
| Arabidopsis thaliana | *Bacillus amyloliquefaciens* | Phytohormone production, Lipopeptide | Asari et al. (2017) |
| Tomato | *Bacillus pumilus, Bacillus amyloliquefaciens* | HCN, Siderophore, N-fixation, IAA, P-solubilization | Xiaohui et al. (2017) |
| Soybean | *Bacillus cereus* | Siderophores, IAA, P-solubilization, EPS | Arif et al. (2017) |
| Mungbean | *Rhizobium, Bradyrhizobium, Bacillus cereus, B. drentensis, B. pumilus, B. subtilis, Enterobacter cloacae, Pseudomonas putida, Ochrabactrum* | N-fixation, P-solubilization, IAA production, ACC deaminase activity | Mahmood et al. (2016), Akhtar and Ali (2011) |
| Chickpea | *Mesorhizobium ciceri, Ochrabactrum ciceri, S. marcescens* | P-solubilization, IAA production, N-fixation | Imran et al. (2015) |
| Wheat | *Enterobacter sp., A. chlorophenolicus, S. marcescens, B. megaterium* | P-solubilization, IAA, HCN, N-fixation, Gibberellin, Siderophores | Kumar et al. (2015) |
| Maize | *Achromobacter xylosoxidans, Azospirillum brasilense, Bacillus subtilis, Bacillus megaterium, Pseudomonas stutzeri, Rhodococcus rhodococcus* | IAA production, Zn and P-solubilization | Zahid (2015) |
| Cotton | *Bacillus amyloliquefaciens, Bacillus spp.* | N-fixation, production of phytohormones, P-solubilization, antibiotic activity | Fahimi et al. (2014) |

manipulating rhizospheric interactions. Successful integration of technical advances and rhizosphere mechanisms (root exudation, root architecture, and microbial) would surely help in maximization of nutrient acquisition, transformation, and utilization, thereby increasing crop yield in a sustainable way.

Selective microbial simulation in the rhizosphere is greatly influenced by root exudates. Plants have a significant consequence on soil microbiological properties owing to the release of various nutrients and organic compounds into the soil. The microbial diversity has been found to be more in rhizosphere compared to bulk soil. Soil microorganisms help in maintaining soil quality and as well as nutrient cycling in the soil of P, N, and S by interfering in soil hydrolytic reactions. Microbes facilitate approximately 80–90% of the processes in the soil. The microbial community structure and population in the rhizosphere is maintained by root shedding of root cells and exudation. Majority of plant in association with fungi forms mycorrhizae which effects microbial population in the rhizosphere. Many bacteria are intimately associated with plant roots; the microbial population in the rhizosphere differs in their species distribution and abundance in the surrounding soil. It is generally accepted that microbial extracellular enzyme-production abides by a cost-effective rule of economy, i.e., microbes will maximize the energy, C, and nutrient returns by minimizing resource utilization. The quantity and quality of the root exudates can be modulated by activities of microbiome by affecting nutrient supply and root. Rhizospheric microbes have the ability to mobilize and immobilize plant nutrients as well production of both phytotoxins and growth-promoting phytohormones. The presence of certain antimicrobial compounds in the roots of plant tissues may have antagonistic effect on the soil microbial diversity in the rhizosphere.

Crop having better root architecture for efficient acquisition of nutrients and moisture from the soil can be developed in breeding programs. This kind of crops is expected to decrease cost of production with enhanced resource use efficiency with remarkable reduction in environmental pollution and, would improve food security. In this regard it has been observed that plants have deeper root system have higher carbon sequestration as well as climate resistant ability. Such root system with better N and moisture absorption capacity are often facilitated by steeper root growth angles, reduced lateral branching, fewer axial roots, and anatomical phenotypes which lowers the metabolic cost. Hypoxia, mechanical resistance, and elemental toxicity are some of the constraints faced during exploration of subsoil. Shallow root system accompanied with high lateral branching, axial roots and shallow root growth angles aids the efficiency of plant to obtain resources (such as P and K) present in the top soil as its metabolically cheaper. Parsimonious root phenotypes with emphasis on acquisition of moisture might prove to be beneficial under high-input agricultural systems. Optimization of root ideotypes for any one function may not be possible but it has the ability to provide resistance against a wide range of both biotic and abiotic stress in low-input systems. Root hairs, reduced metabolic cost, and developmental regulation of plasticity may be useful in all environments. Understanding and harnessing rhizosphere processes for crop improvement appears to be interesting and challenging transdisciplinary opportunity to address global food and nutritional security.

## 12.9 Challenges and Future Prospects

Appropriate exploration and harnessing of rhizosphere efficiency to improvise plant productivity is not as simple as it seems. Various limitation and challenges associated with rhizosphere manipulation such as stability of root exudates, sample collection, analytical tools, successful establishment, and survival of inoculants are matter of great concern. Even after increased identification and characterization of molecules involved, the metabolic profiling is major constraint. Hurdles associated with the extraction and analysis of organic compounds is one of the prime reasons for difficulty in exploring root exudation. Most of the studies regarding root chemical environment conducted crucially pave the way for conducting many real-time in situ assays which helps to know the dynamic role in plant roots. Root architectural phenotyping is really challenging with lack of information or underdeveloped understanding regarding the underlying mechanism which limits the application and exploitation of these beneficial traits to their full potential. Root biology, with possible interactions being complex, real challenge lies in proper microbial screening, synthesis, and promotion under different environmental stresses with their optimum mode of action.

Next generation tools like "Metabolomics" tools with inventory of molecules helps in greater understanding of rhizodeposition patterns both quantitatively and qualitatively. Modern tools like metaproteomics, nano-technology, and rhizo-engineering provide appropriate utility of plant growth-promoting rhizobacteria (PGPR) inoculants and their ecological and physiological acceptance.

## 12.10 Conclusions

Plants growing under stressful and challenging agro-ecosystems are potentially adapt and grow with dynamic, divine, and complex changes in rhizosphere that influences the health and productivity of crop production. The roles of root exudates, root architecture, and rhizosphere microbial population cannot be denied for imparting biotic and abiotic stress tolerance. Crop root and their efficiency can be improved by breeding or genetic engineering. As rhizosphere microbes are intricately linked to various edaphic factors for determining soil fertility and nutrient recycling and to overcome stress alleviations based on its environmental adaptations and genetic traits. With the application of metagenomics approaches, plant beneficial bacteria can be utilized for improvement of crop production. Plant growth-promoting (PGP) microorganisms maybe used as beneficial products for the higher productivity of crops.

Since the importance of understanding the underlying mechanism is rising, research efforts would go long way for unraveling the complexity of the rhizosphere processes. Microbial interactions with varied root chemical environment and root morphology are not fully characterized. Multidisciplinary approaches are needed for possibly allowing appropriate rhizosphere exploration, which could be promising for development of climate resilient crops for enhanced crop productivity.

## References


Abràmoff MD, Magalhães PJ, Ram SJ (2004) Image processing with image. J Biophotonics Int 11(7):36–42

Adl S (2016) Rhizosphere, food security, and climate change: a critical role for plant-soil research. Rhizosphere 1:1–3

Ahmad P, Hashem A, Abd-Allah EF, Alqarawi AA, John R, Egamberdieva D, Gucel S (2015) Role of *Trichoderma harzianum* in mitigating NaCl stress in Indian mustard (*Brassica juncea L*) through antioxidative defense system. Front Plant Sci 6:868

Ahmadi N, Audebert A, Bennett MJ, Bishopp A, de Oliveira AC, Courtois B, Diedhiou A, Diévart A, Gantet P, Ghesquière A, Guiderdoni E (2014) The roots of future rice harvests. Rice 7(1):1–9

Akhtar S, Ali B (2011) Evaluation of rhizobacteria as non-rhizobial inoculants for mung beans. Aust J Crop Sci 5(13):1723–1729

Arif MS, Muhammad RI, Shahzad SM, Yasmeen T, Shafaqat AL, Akhtar MJ (2017) Phosphorus-mobilizing rhizobacterial strain Bacillus cereus GS6 improves symbiotic efficiency of soybean on an Aridisol amended with phosphorus-enriched compost. Pedosphere 27(6):1049–1061

Armengaud P, Zambaux K, Hills A, Sulpice R, Pattison RJ, Blatt MR, Amtmann A (2009) EZ-Rhizo: integrated software for the fast and accurate measurement of root system architecture. Plant J 57(5):945–956

Arsenault JL, Poulcur S, Messier C, Guay R (1995) WinRHIZO™, a root-measuring system with a unique overlap correction method. Hort Sci 30(4):906D

Asari S, Tarkowská D, Rolčík J, Novák O, Palmero DV, Bejai S, Meijer J (2017) Analysis of plant growth-promoting properties of Bacillus amyloliquefaciens UCMB5113 using Arabidopsis thaliana as host plant. Planta 245(1):15–30

Badri DV, Vivanco JM (2009) Regulation and function of root exudates. Plant Cell Environ 32(6): 666–681

Baetz U, Martinoia E (2014) Root exudates: the hidden part of plant defense. Trends Plant Sci 19(2):90–98

Bais HP, Weir TL, Perry LG, Gilroy S, Vivanco JM (2006) The role of root exudates in rhizosphere interactions with plants and other organisms. Annu Rev Plant Biol 57:233–266

Barton CV, Montagu KD (2004) Detection of tree roots and determination of root diameters by ground penetrating radar under optimal conditions. Tree Physiol 24(12):1323–1331

Berendsen RL, Pieterse CM, Bakker PA (2012) The rhizosphere microbiome and plant health. Trends Plant Sci 17(8):478–486

Bever JD, Platt TG, Morton ER (2012) Microbial population and community dynamics on plant roots and their feedbacks on plant communities. Annu Rev Microbiol 66:265–283

Bonser AM, Lynch J, Snapp S (1996) Effect of phosphorus deficiency on growth angle of basal roots in Phaseolus vulgaris. New Phytol 132(2):281–288

Buee M, De Boer W, Martin F, Van Overbeek L, Jurkevitch E (2009) The rhizosphere zoo: an overview of plant-associated communities of microorganisms, including phages, bacteria, archaea, and fungi, and of some of their structuring factors. Plant Soil 321(1–2):189–212

Carvalhais LC, Dennis PG, Tyson GW, Schenk PM (2013) Rhizosphere metatranscriptomics: challenges and opportunities. Mol Microb Ecol Rhizosphere 2:1137–1144

Casas ME, Matamoros V (2021) Analytical challenges and solutions for performing metabolomic analysis of root exudates. Trends Environ Anal Chem 31:e00130

Chen L, Yang X, Raza W, Li J, Liu Y, Qiu M, Zhang F, Shen Q (2011) Trichoderma harzianum SQR-T037 rapidly degrades allelochemicals in rhizospheres of continuously cropped cucumbers. Appl Microbiol Biotechnol 89(5):1653–1663

Cramer GR, Urano K, Delrot S, Pezzotti M, Shinozaki K (2011) Effects of abiotic stress on plants: a systems biology perspective. BMC Plant Biol 11(1):1–14

Cui Z, Chen X, Miao Y, Zhang F, Sun Q, Schroder J, Zhang H, Li J, Shi L, Xu J, Ye Y (2008) On-farm evaluation of the improved soil Nmin-based nitrogen management for summer maize in North China Plain. Agron J 100(3):517–525

Dai X, Wang Y, Yang A, Zhang WH (2012) OsMYB2P-1, an R2R3 MYB transcription factor, is involved in the regulation of phosphate-starvation responses and root architecture in rice. Plant Physiol 159(1):169–183

van Dam NM, Bouwmeester HJ (2016) Metabolomics in the rhizosphere: tapping into belowground chemical communication. Trends Plant Sci 21(3):256–265

Dell'mour M (2010) Development of novel methods for the analysis of root exudates. University of Natural Resources and Life Sciences, Vienna

Deng Y, Chen K, Teng W, Zhan A, Tong Y, Feng G, Cui Z, Zhang F, Chen X (2014) Is the inherent potential of maize roots efficient for soil phosphorus acquisition? PLoS One 9(3):e90287

Dessaux Y, Grandclément C, Faure D (2016) Engineering the rhizosphere. Trends Plant Sci 21(3): 266–278

Devaiah BN, Karthikeyan AS, Raghothama KG (2007) WRKY75 transcription factor is a modulator of phosphate acquisition and root development in Arabidopsis. Plant Physiol 143(4): 1789–1801

Dreyer I, Gomez-Porras JL, Riaño-Pachón DM, Hedrich R, Geiger D (2012) Molecular evolution of slow and quick anion channels (SLACs and QUACs/ALMTs). Front Plant Sci 3:263

Dumont MG, Murrell JC (2005) Stable isotope probing—linking microbial identity to function. Nat Rev Microbiol 3(6):499–504

Ejaz S, Batool S, Anjum MA, Naz S, Qayyum MF, Naqqash T, Shah KH, Ali S (2020) Effects of inoculation of root-associative Azospirillum and Agrobacterium strains on growth, yield and quality of pea (Pisum sativum L.) grown under different nitrogen and phosphorus regimes. Sci Hortic 270:109401

Everett KR, Rees-George J, Pushparajah IP, Janssen BJ, Luo Z (2010) Advantages and disadvantages of microarrays to study microbial population dynamics a minireview. N Z Plant Protect 63: 1–6

Evers JB, Vos J, Fournier C, Andrieu B, Chelle M, Struik PC (2005) Towards a generic architectural model of tillering in Gramineae, as exemplified by spring wheat (Triticum aestivum). New Phytol 166(3):801–812

Fageria NK, Moreira A, Coelho AM (2011) Yield and yield components of upland rice as influenced by nitrogen sources. J Plant Nutr 34(3):361–370

Fahimi A, Ashouri A, Ahmadzadeh M, Hoseini Naveh V, Asgharzadeh A, Maleki F, Felton GW (2014) Effect of PGPR on population growth parameters of cotton aphid. Arch Phytopathol Plant Protect 47(11):1274–1285

Fang Y, Yabusaki SB, Ahkami AH, Chen X, Scheibe TD (2019) An efficient three-dimensional rhizosphere modeling capability to study the effect of root system architecture on soil water and reactive transport. Plant Soil 441(1):33–48

Ferrari A, Hagedorn F, Niklaus PA (2016) Experimental soil warming and cooling alters the partitioning of recent assimilates: evidence from a 14C-labelling study at the alpine treeline. Oecologia 181(1):25–37

Forde BG (2014) Nitrogen signalling pathways shaping root system architecture: an update. Curr Opin Plant Biol 21:30–36

Fuhrer T, Zamboni N (2015) High-throughput discovery metabolomics. Curr Opin Biotechnol 31: 73–78

Galloway-Peña J, Hanson B (2020) Tools for analysis of the microbiome. Dig Dis Sci 65(3): 674–685

Gao Z, Karlsson I, Geisen S, Kowalchuk G, Jousset A (2019) Protists: puppet masters of the rhizosphere microbiome. Trends Plant Sci 24(2):165–176
Gärtner H, Denier C (2006) Application of a 3D laser scanning device to acquire the structure of whole root systems—a pilot study. TRACE–tree rings in archaeology. Climatol Ecol 4:288–294
Gärtner H, Wagner B, Heinrich I, Denier C (2009) 3D-laser scanning: a new method to analyze coarse tree root systems. Forest Snow Lands Res 82(1):95–106
Gifford ML, Dean A, Gutierrez RA, Coruzzi GM, Birnbaum KD (2008) Cell-specific nitrogen responses mediate developmental plasticity. Proc Natl Acad Sci 105(2):803–808
Glick BR (2014) Bacteria with ACC deaminase can promote plant growth and help to feed the world. Microbiol Res 169(1):30–39
Godin C, Costes E, Sinoquet H (2005) Plant architecture modelling: virtual plants and complex systems. In: Turnbull CGN (ed) Plant architecture and its manipulation. vol 17. Blackwell, Oxford, pp 238–287. Annual Plant Reviews
Gregory PJ, Hutchison DJ, Read DB, Jenneson PM, Gilboy WB, Morton EJ (2003) Non-invasive imaging of roots with high resolution X-ray micro-tomography. In: Roots: the dynamic interface between plants and the earth. Springer, Dordrecht, pp 351–359
Hallé F, Oldeman RA (1970) Essai sur l'architecture et la dynamique de croissance des arbres tropicaux (Vol. 6). Masson, Paris
Haney CH, Samuel BS, Bush J, Ausubel FM (2015) Associations with rhizosphere bacteria can confer an adaptive advantage to plants. Nat Plants 1(6):1–9
Hanif MK, Malik KA, Hameed S, Saddique MJ, Fatima K, Naqqash T, Majeed A, Iqbal MJ, Imran A (2020) Growth stimulatory effect of AHL producing Serratia spp. from potato on homologous and non-homologous host plants. Microbiol Res 238:126506
Hao S, Xue J, Guo D, Wang D (2010) Earliest rooting system and root: shoot ratio from a new Zosterophyllum plant. New Phytol 185(1):217–225
Hinsinger P, Bengough AG, Vetterlein D, Young IM (2009) Rhizosphere: biophysics, biogeochemistry and ecological relevance. Plant Soil 321(1):117–152
Hodge A, Berta G, Doussan C, Merchan F, Crespi M (2009) Plant root growth, architecture and function. Plant Soil 321(1):153–187
Hu H, Tang C, Rengel Z (2005) Role of phenolics and organic acids in phosphorus mobilization in calcareous and acidic soils. J Plant Nutr 28(8):1427–1439
Hund A, Ruta N, Liedgens M (2009) Rooting depth and water use efficiency of tropical maize inbred lines, differing in drought tolerance. Plant Soil 318(1):311–325
Hurley BA, Tran HT, Marty NJ, Park J, Snedden WA, Mullen RT, Plaxton WC (2010) The dual-targeted purple acid phosphatase isozyme AtPAP26 is essential for efficient acclimation of Arabidopsis to nutritional phosphate deprivation. Plant Physiol 153(3):1112–1122
Ibekwe AM, Ors S, Ferreira JF, Liu X, Suarez DL (2017) Seasonal induced changes in spinach rhizosphere microbial community structure with varying salinity and drought. Sci Total Environ 579:1485–1495
Imran A, Mirza MS, Shah TM, Malik KA, Hafeez FY (2015) Differential response of kabuli and desi chickpea genotypes toward inoculation with PGPR in different soils. Front Microbiol 6:859
Iyer-Pascuzzi AS, Symonova O, Mileyko Y, Hao Y, Belcher H, Harer J, Weitz JS, Benfey PN (2010) Imaging and analysis platform for automatic phenotyping and trait ranking of plant root systems. Plant Physiol 152(3):1148–1157
Jahnke S, Menzel MI, Van Dusschoten D, Roeb GW, Bühler J, Minwuyelet S, Blümler P, Temperton VM, Hombach T, Streun M, Beer S (2009) Combined MRI–PET dissects dynamic changes in plant structures and functions. Plant J 59(4):634–644
Jha B, Gontia I, Hartmann A (2012) The roots of the halophyte Salicornia brachiata are a source of new halotolerant diazotrophic bacteria with plant growth-promoting potential. Plant Soil 356(1): 265–277
Jobbagy EG, Jackson RB (2001) The distribution of soil nutrients with depth: global patterns and the imprint of plants. Biogeochemistry 53(1):51–77

Johnson MG, Tingey DT, Phillips DL, Storm MJ (2001) Advancing fine root research with minirhizotrons. Environ Exp Bot 45(3):263–289
Jones PM, George AM (2002) Mechanism of ABC transporters: a molecular dynamics simulation of a well characterized nucleotide-binding subunit. Proc Natl Acad Sci 99(20):12639–12644
Ju XT, Xing GX, Chen XP, Zhang SL, Zhang LJ, Liu XJ, Cui ZL, Yin B, Christie P, Zhu ZL, Zhang FS (2009) Reducing environmental risk by improving N management in intensive Chinese agricultural systems. Proc Natl Acad Sci 106(9):3041–3046
Kaestner A, Schneebeli M, Graf F (2006) Visualizing three-dimensional root networks using computed tomography. Geoderma 136(1–2):459–469
Kamilova F, Validov S, Azarova T, Mulders I, Lugtenberg B (2005) Enrichment for enhanced competitive plant root tip colonizers selects for a new class of biocontrol bacteria. Environ Microbiol 7(11):1809–1817
Kang SM, Hamayun M, Khan MA, Iqbal A, Lee IJ (2019) Bacillus subtilis JW1 enhances plant growth and nutrient uptake of Chinese cabbage through gibberellins secretion. J Appl Bot Food Qual 92:172–178
Kaur A, Chaudhary A, Kaur A, Choudhary R, Kaushik R (2005) Phospholipid fatty acid—a bioindicator of environment monitoring and assessment in soil ecosystem. Curr Sci 89:1103–1112
Kirk JL, Beaudette LA, Hart M, Moutoglis P, Klironomos JN, Lee H, Trevors JT (2004) Methods of studying soil microbial diversity. J Microbiol Methods 58(2):169–188
Konstantinos KV, Panagiotis P, Antonios VT, Agelos P, Argiris NV (2008) PCR–SSCP: a method for the molecular analysis of genetic diseases. Mol Biotechnol 38(2):155–163
Krishnappa R, Aftab Hussain IS (2014) Phosphorus acquisition from deficient soil: involvement of organic acids and acid phosphatase in Pigeon pea. Indian J Plant Physiol (12). https://doi.org/10.1007/s40502-014-0101-z
Kumar A, Dubey A (2020) Rhizosphere microbiome: engineering bacterial competitiveness for enhancing crop production. J Adv Res 24:337–352
Kumar A, Maurya BR, Raghuwanshi R (2015) Characterization of bacterial strains and their impact on plant growth promotion and yield of wheat and microbial populations of soil. Afr J Agric Res 10(12):1367–1375
Kuzyakov Y, Domanski G (2000) Carbon input by plants into the soil. Rev J Plant Nutr Soil Sci 163(4):421–431
Lagos L, Maruyama F, Nannipieri P, Mora ML, Ogram A, Jorquera MA (2015) Current overview on the study of bacteria in the rhizosphere by modern molecular techniques: a mini-review. J Soil Sci Plant Nutr 15(2):504–523
Lakshmanan V, Selvaraj G, Bais HP (2014) Functional soil microbiome: belowground solutions to an aboveground problem. Plant Physiol 166(2):689–700
Lamers J, Van Der Meer T, Testerink C (2020) How plants sense and respond to stressful environments. Plant Physiol 182(4):1624–1635
Lang AR (1973) Leaf orientation of a cotton plant. Agric Meteorol 11:37–51
Lesk C, Rowhani P, Ramankutty N (2016) Influence of extreme weather disasters on global crop production. Nature 529(7584):84–87
Li K, DiLegge MJ, Minas IS, Hamm A, Manter D, Vivanco JM (2019) Soil sterilization leads to re-colonization of a healthier rhizosphere microbiome. Rhizosphere 12:100176
Li Z, Gao Q, Liu Y, He C, Zhang X, Zhang J (2011) Overexpression of transcription factor ZmPTF1 improves low phosphate tolerance of maize by regulating carbon metabolism and root growth. Planta 233(6):1129–1143
Lugtenberg BJ, Dekkers L, Bloemberg GV (2011) Molecular determinants of rhizosphere colonization by Pseudomonas. Annu Rev Phytopathol 39(1):461–490
Lugtenberg B, Kamilova F (2009) Plant-growth-promoting rhizobacteria. Annu Rev Microbiol 63: 541–556
Lynch J (1995) Root architecture and plant productivity. Plant Physiol 109(1):7
Lynch JP (2007) Roots of the second green revolution. Aust J Bot 55(5):493–512

Lynch JP (2011) Root phenes for enhanced soil exploration and phosphorus acquisition: tools for future crops. Plant Physiol 156(3):1041–1049

Lynch JP (2013) Steep, cheap and deep: an ideotype to optimize water and N acquisition by maize root systems. Ann Bot 112(2):347–357

Lynch JP, Brown KM (2012) New roots for agriculture: exploiting the root phenome. Philosop Trans Royal Soc B Biol Sci 367(1595):1598–1604

Lynch JP, Ho MD (2005) Rhizoeconomics: carbon costs of phosphorus acquisition. Plant Soil 269(1):45–56

Ma W, Li J, Qu B, He X, Zhao X, Li B, Fu X, Tong Y (2014) Auxin biosynthetic gene TAR 2 is involved in low nitrogen-mediated reprogramming of root architecture in Arabidopsis. Plant J 78(1):70–79

Maghnia FZ, Abbas Y, Mahé F, Prin Y, El Ghachtouli N, Duponnois R, Sanguin H (2019) The rhizosphere microbiome: a key component of sustainable cork oak forests in trouble. For Ecol Manag 434:29–39

Mahmood S, Daur I, Al-Solaimani SG, Ahmad S, Madkour MH, Yasir M, Hirt H, Ali S, Ali Z (2016) Plant growth promoting rhizobacteria and silicon synergistically enhance salinity tolerance of mung bean. Front Plant Sci 7:876

Manichanh C, Chapple CE, Frangeul L, Gloux K, Guigo R, Dore JA (2008) Comparison of random sequence reads versus 16S rDNA sequences for estimating the biodiversity of a metagenomic library. Nucleic Acids Res 36(16):5180–5188

Maruyama H, Sasaki T, Yamamoto Y, Wasaki J (2019) AtALMT3 is involved in malate efflux induced by phosphorus deficiency in Arabidopsis thaliana root hairs. Plant Cell Physiol 60(1): 107–115

Matilla MA, Ramos JL, Bakker PA, Doornbos R, Badri DV, Vivanco JM, Ramos-González MI (2010) Pseudomonas putida KT2440 causes induced systemic resistance and changes in Arabidopsis root exudation. Environ Microbiol Rep 2(3):381–388

van Noordwijk ME, de Jager AN, Floris J (1985) A new dimension to observations in minirhizotrons: a stereoscopic view on root photographs. Plant Soil 86(3):447–453

Meena KK, Sorty AM, Bitla UM, Choudhary K, Gupta P, Pareek A, Singh DP, Prabha R, Sahu PK, Gupta VK, Singh HB (2017) Abiotic stress responses and microbe-mediated mitigation in plants: the omics strategies. Front Plant Sci 8:172

Mendes R, Garbeva P, Raaijmakers JM (2013) The rhizosphere microbiome: significance of plant beneficial, plant pathogenic, and human pathogenic microorganisms. FEMS Microbiol Rev 37(5):634–663

Mi G, Chen F, Wu Q, Lai N, Yuan L, Zhang F (2010) Ideotype root architecture for efficient nitrogen acquisition by maize in intensive cropping systems. Sci China Life Sci 53(12): 1369–1373

Mittler R (2006) Abiotic stress, the field environment and stress combination. Trends Plant Sci 11(1):15–19

Mochizuki S, Jikumaru Y, Nakamura H, Koiwai H, Sasaki K, Kamiya Y, Ichikawa H, Minami E, Nishizawa Y (2014) Ubiquitin ligase EL5 maintains the viability of root meristems by influencing cytokinin-mediated nitrogen effects in rice. J Exp Bot 65(9):2307–2318

Moter A, Göbel UB (2000) Fluorescence in situ hybridization (FISH) for direct visualization of microorganisms. J Microbiol Methods 41(2):85–112

Mukhtar MS, Liu X, Somssich IE (2018) Elucidating the role of WRKY27 in male sterility in Arabidopsis. Plant Signal Behav 13(11):e1363945

Nakayama M, Tateno R (2018) Solar radiation strongly influences the quantity of forest tree root exudates. Trees 32(3):871–879

Naqqash T, Hameed S, Imran A, Hanif MK, Majeed A, van Elsas JD (2016) Differential response of potato toward inoculation with taxonomically diverse plant growth promoting rhizobacteria. Front Plant Sci 7:144

Naveed M, Hussain MB, Zahir ZA, Mitter B, Sessitsch A (2014) Drought stress amelioration in wheat through inoculation with Burkholderia phytofirmans strain PsJN. Plant Growth Regul 73(2):121–131

Ndour PM, Heulin T, Achouak W, Laplaze L, Cournac L (2020) The rhizosheath: from desert plants adaptation to crop breeding. Plant Soil 456(1):1–13

Nguyen C (2009) Rhizodeposition of organic C by plant: mechanisms and controls. In: Lichtfouse E, Navarrete M, Debaeke P, Véronique S, Alberola C (eds) Sustainable agriculture. Springer, Dordrecht, pp 97–123

Nigris S, Baldan E, Tondello A, Zanella F, Vitulo N, Favaro G, Guidolin V, Bordin N, Telatin A, Barizza E, Marcato S (2018) Biocontrol traits of Bacillus licheniformis GL174, a culturable endophyte of *Vitis vinifera* cv Glera. BMC Microbiol 18(1):1–16

Nocker A, Burr M, Camper AK (2007) Genotypic microbial community profiling: a critical technical review. Microb Ecol 54(2):276–289

Oburger E, Schmidt H (2016) New methods to unravel rhizosphere processes. Trends Plant Sci 21(3):243–255

Oburger E, Jones DL (2018) Sampling root exudates–mission impossible? Rhizosphere 6:116–133

Pantigoso HA, He Y, DiLegge MJ, Vivanco JM (2021) Methods for root exudate collection and analysis. Methods Mol Biol 2232:291–303

Patkar RN, Naqvi NI (2017) Fungal manipulation of hormone-regulated plant defense. PLoS Pathog 13(6):e1006334

Pieterse CM, Leon-Reyes A, Van der Ent S, Van Wees SC (2009) Networking by small-molecule hormones in plant immunity. Nat Chem Biol 5(5):308–316

Postma JA, Lynch JP (2011) Root cortical aerenchyma enhances the growth of maize on soils with suboptimal availability of nitrogen, phosphorus, and potassium. Plant Physiol 156(3): 1190–1201

Qi Y, Wang S, Shen C, Zhang S, Chen Y, Xu Y, Liu Y, Wu Y, Jiang D (2012) OsARF12, a transcription activator on auxin response gene, regulates root elongation and affects iron accumulation in rice (Oryza sativa). New Phytol 193(1):109–120

Remans T, Nacry P, Pervent M, Filleur S, Diatloff E, Mounier E, Tillard P, Forde BG, Gojon A (2006) The Arabidopsis NRT1. 1 transporter participates in the signaling pathway triggering root colonization of nitrate-rich patches. Proc Natl Acad Sci 103(50):19206–19211

Richardson K, Steffen W, Schellnhuber HJ, Alcamo J, Barker T, Kammen DM, Leemans R, Liverman D, Munasinghe M, Osman-Elasha B, Stern N (2009) Synthesis report. Climate change: global risks, challenges and decisions, Copenhagen, Denmark, 10–12. University of Copenhagen

Rincon-Florez VA, Carvalhais LC, Schenk PM (2013) Culture-independent molecular tools for soil and rhizosphere microbiology. Diversity 5(3):581–612

Schäfer P, Pfiffi S, Voll LM, Zajic D, Chandler PM, Waller F, Scholz U, Pons-Kühnemann J, Sonnewald S, Sonnewald U, Kogel KH (2009) Manipulation of plant innate immunity and gibberellin as factor of compatibility in the mutualistic association of barley roots with Piriformospora indica. Plant J 59(3):461–474

Schimel JP, Bennett J (2004) Nitrogen mineralization: challenges of a changing paradigm. Ecology 85(3):591–602

Sharma T, Dreyer I, Kochian L, Piñeros MA (2016) The ALMT family of organic acid transporters in plants and their involvement in detoxification and nutrient security. Front Plant Sci 7:1488

Smith CJ, Osborn AM (2009) Advantages and limitations of quantitative PCR (Q-PCR)-based approaches in microbial ecology. FEMS Microbiol Ecol 67(1):6–20

Sorty AM, Meena KK, Choudhary K, Bitla UM, Minhas PS, Krishnani KK (2016) Effect of plant growth promoting bacteria associated with halophytic weed (*Psoralea corylifolia* L) on germination and seedling growth of wheat under saline conditions. Appl Biochem Biotechnol 180(5): 872–882

Srivastava S, Yadav A, Seem K, Mishra S, Chaudhary V, Nautiyal CS (2008) Effect of high temperature on Pseudomonas putida NBRI0987 biofilm formation and expression of stress sigma factor RpoS. Curr Microbiol 56(5):453–457

Stringlis IA, Zhang H, Pieterse CM, Bolton MD, de Jonge R (2018) Microbial small molecules–weapons of plant subversion. Nat Prod Rep 35(5):410–433

Suzuki M, Tsukamoto T, Inoue H, Watanabe S, Matsuhashi S, Takahashi M, Nakanishi H, Mori S, Nishizawa NK (2008) Deoxymugineic acid increases Zn translocation in Zn-deficient rice plants. Plant Mol Biol 66(6):609–617

Tadano T, Sakai H (1991) Secretion of acid phosphatase by the roots of several crop species under phosphorus-deficient conditions. Soil Sci Plant Nutr 37(1):129–140

Tester M, Langridge P (2010) Breeding technologies to increase crop production in a changing world. Science 327(5967):818–822

Theocharis A, Bordiec S, Fernandez O, Paquis S, Dhondt-Cordelier S, Baillieul F, Clément C, Barka EA (2012) *Burkholderia phytofirmans* PsJN primes *Vitis vinifera* L. and confers a better tolerance to low nonfreezing temperatures. Mol Plant-Microbe Interact 25(2):241–249

Tracy SJ (2010) Qualitative quality: eight "big-tent" criteria for excellent qualitative research. Qual Inq 16(10):837–851

Van der Heijden MG, Boller T, Wiemken A, Sanders IR (1998) Different arbuscular mycorrhizal fungal species are potential determinants of plant community structure. Ecology 79(6): 2082–2091

Verbon EH, Liberman LM (2016) Beneficial microbes affect endogenous mechanisms controlling root development. Trends Plant Sci 21(3):218–229

Von Uexküll HR, Mutert E (1995) Global extent, development and economic impact of acid soils. Plant Soil 171(1):1–15

van der Voort M, Kempenaar M, van Driel M, Raaijmakers JM, Mendes R (2016) Impact of soil heat on reassembly of bacterial communities in the rhizosphere microbiome and plant disease suppression. Ecol Lett 19(4):375–382

Walker TS, Bais HP, Halligan KM, Stermitz FR, Vivanco JM (2003) Metabolic profiling of root exudates of *Arabidopsis thaliana*. J Agric Food Chem 51(9):2548–2554

Wang Z, Li T, Li Y, Zhao D, Han J, Liu Y, Liao Y (2020) Relationship between the microbial community and catabolic diversity in response to conservation tillage. Soil Tillage Res 196: 104431

Wang Z, Liu L, Chen Q, Wen X, Liu Y, Han J, Liao Y (2017) Conservation tillage enhances the stability of the rhizosphere bacterial community responding to plant growth. Agron Sustain Dev 37(5):1–11

Wasson AP, Richards RA, Chatrath R, Misra SC, Prasad SS, Rebetzke GJ, Kirkegaard JA, Christopher J, Watt M (2012) Traits and selection strategies to improve root systems and water uptake in water-limited wheat crops. J Exp Bot 63(9):3485–3498

Wei Y, Zhao Y, Shi M, Cao Z, Lu Q, Yang T, Fan Y, Wei Z (2018) Effect of organic acids production and bacterial community on the possible mechanism of phosphorus solubilization during composting with enriched phosphate-solubilizing bacteria inoculation. Bioresour Technol 247:190–199

Weston LA, Ryan PR, Watt M (2012) Mechanisms for cellular transport and release of allelochemicals from plant roots into the rhizosphere. J Exp Bot 63(9):3445–3454

White PJ, Broadley MR (2009) Biofortification of crops with seven mineral elements often lacking in human diets– iron, zinc, copper, calcium, magnesium, selenium and iodine. New Phytol 182(1):49–84

White PJ, Brown P (2010) Plant nutrition for sustainable development and global health. Ann Bot 105(7):1073–1080

White PJ, Greenwood DJ (2013) Properties and management of cationic elements for crop growth. In: Gregory PJ, Nortcliff S (eds) Soil conditions and plant growth. Blackwell, Oxford, pp 160–194

Williamson LC, Ribrioux SP, Fitter AH, Leyser HO (2001) Phosphate availability regulates root system architecture in Arabidopsis. Plant Physiol 126(2):875–882

Wissuwa M, Ismail AM, Yanagihara S (2006) Effects of zinc deficiency on rice growth and genetic factors contributing to tolerance. Plant Physiol 142:731–741

Wu P, Wang XM (2008) Role of OsPHR2 on phosphorus homoestasis and root hairs development in rice (Oryza sativa L.). Plant Signal Behav 3(9):674–675

Xiao AW, Li Z, Li WC, Ye ZH (2020) The effect of plant growth-promoting rhizobacteria (PGPR) on arsenic accumulation and the growth of rice plants (*Oryza sativa* L.). Chemosphere 242: 125136

Xiaohui FA, Zhang S, Xiaodan MO, Yuncong LI, Yuqing FU, Zhiguang LI (2017) Effects of plant growth-promoting rhizobacteria and N source on plant growth and N and P uptake by tomato grown on calcareous soils. Pedosphere 27(6):1027–1036

Yang Y, Liang Y, Han X, Chiu TY, Ghosh A, Chen H, Tang M (2016) The roles of arbuscular mycorrhizal fungi (AMF) in phytoremediation and tree-herb interactions in Pb contaminated soil. Sci Rep 6(1):1–4

Yazaki K (2005) Transporters of secondary metabolites. Curr Opin Plant Biol 8(3):301–307

Zahid M (2015) Isolation and identification of indigenous plant growth promoting rhizobacteria from Himalayan region of Kashmir and their effect on improving growth and nutrient contents of maize (Zea mays L.). Front Microbiol 6:207

Zhang H, Kim MS, Sun Y, Dowd SE, Shi H, Paré PW (2008) Soil bacteria confer plant salt tolerance by tissue-specific regulation of the sodium transporter HKT1. Mol Plant-Microbe Interact 21(6):737–744

Zhang F, Shen J, Zhang J, Zuo Y, Li L, Chen X (2010) Rhizosphere processes and management for improving nutrient use efficiency and crop productivity: implications for China. In: Advances in agronomy, vol 107. Academic, pp 1–32

Zhang A, Sun H, Wang P, Han Y, Wang X (2012) Modern analytical techniques in metabolomics analysis. Analyst 137(2):293–300

Zhu Z, Brilakis I (2009) Comparison of optical sensor-based spatial data collection techniques for civil infrastructure modeling. J Comput Civ Eng 23(3):170–177

# Chapter 13
# High-Throughput Phenomics of Crops for Water and Nitrogen Stress

**Rabi N. Sahoo, C. Viswanathan, Mahesh Kumar, Swati Bhugra, Sourabh Karwa, Tanuj Misra, R. Dhandapani, and Sudhir Kumar**

**Abstract** While both water and nitrogen (N) are necessary for crop output and quality, excessive N application increases production costs and environmental degradation. Recent years have seen the development of NUE and WUE assessment tools, which have shown value. This chapter discusses the properties of next-generation phenomics that are critical for recognising rice genotypes during periods of water scarcity. In addition to the newly proposed one, the experiment featured non-imaging hyperspectral remote sensing, thermal imaging, and colour and multi-spectral imaging sensors from the ground and aerial platforms. Numerous multivariate models for the non-invasive evaluation of rice plants' relative water content (RWC) and sugar content were examined using spectral reflectance data gathered in the 350–2500 nm spectral region. Differentiating rice genotypes was accomplished using spectral data. A crop water stress score developed from thermal imaging of selected rice genotypes may be used to identify rice with high drought resistance and low drought sensitivity. The researchers used multispectral and RGB sensors mounted on a drone for field remote sensing and heat map mapping to record the distinct responses of various genotypes and characteristics. The procedures established are rapid, low-cost, and non-invasive, providing a viable alternative to traditional approaches. These techniques are now being used to do high-throughput plant phenotyping in water scarcity and nutrient deficiency conditions.

**Keywords** Sensors · Spectral · Water use efficiency (WUE) · Nitrogen use efficiency (NUE) · High throughput · Phenotyping

---

R. N. Sahoo (✉)
Division of Agricultural Physics, ICAR-Indian Agricultural Research Institute, Delhi, India

C. Viswanathan · M. Kumar · S. Karwa · R. Dhandapani · S. Kumar
Division of Plant Physiology, ICAR-Indian Agricultural Research Institute (ICAR-IARI), New Delhi, India

S. Bhugra
Indian Institute of Technology, Delhi, India

T. Misra
Rani Lakshmi Bai Central Agricultural University, Jhansi, India

 
H. M. Mamrutha et al. (eds.), *Translating Physiological Tools to Augment Crop Breeding*, https://doi.org/10.1007/978-981-19-7498-4_13

## 13.1 Introduction

The primary goal for agricultural specialists is to double global food production by 2050. The increase in output must be accomplished by increasing agricultural productivity per unit area. Rice, wheat, maize, and soybean yield growth rates are 1.0%, 0.9%, 1.6%, and 1.3% per year, respectively, which is much less than the 2.4% annual growth rate required to double food output and assure food security by 2050 (Ray et al. 2013). Competition for land and water resources, as well as the severity of biotic and abiotic stresses, will intensify this dilemma in the future. Thus, genetic improvement in terms of production, nutritional quality, resilience to biotic and abiotic stress, and decrease of environmental imprint is vital for food and nutritional security. Water and nitrogen (N) availability continue to be the primary constraints on crop development on a worldwide scale. Due to rising population, increased demand for food, we will need to boost the efficiency with which we use water and nitrogen for crops. Water and nitrogen control have been crucial in closing the gap in the production of minor grain crops. The combined response of plant growth to water and nitrogen is larger than the combined response to each element individually, meaning that methods to maximise plant development require an equal distribution of both resources.

Collecting trustworthy genotypic data for a diverse range of germplasm and breeding populations has become more accessible, cost-effective, and time efficient as a result of next-generation sequencing (NGS) and automated high-throughput SNP genotyping technologies. However, one disadvantage of using genomic information for crop growth is the high cost and time required to comprehend phenotypes arising from genotype × environment interactions (Kumar et al. 2016). Traditional phenotyping procedures are laborious, repetitive, and frequently provide only end-point phenotypes, making them less exact in manual scoring situations. As a result, phenotyping is the primary impediment to exploiting the abundance of germplasm resources and the genomics potential for identifying and applying new genotypes, performing high-resolution linkage mapping, genome-wide association mapping, and training genomic selection models for crop development.

High-throughput phenomics can rapidly collect vast amounts of data on a large number of plants and plots using non-destructive techniques including remote sensing. Phenomics is an interdisciplinary discipline that use sensor technologies to acquire and analyse high-dimensional phenotypic data at an organismal scale in a non-destructive, high-throughput manner. This approach is capable of elucidating the inner workings of living plants and bridging the phenotype-genotype divide (Fiorani and Schurr 2013). According to the authors, robotized transportation of plants to imaging sensors is standard practise in controlled environment phenomics platforms, whereas sensors are brought to the plants in field phenomics platforms. Visual imaging, hyperspectral imaging, infrared thermography, near-infrared image analysis, chlorophyll fluorescence imaging, bioluminescence imaging, and fluorescence imaging are all examples of non-invasive sensors that are frequently employed in non-destructive automated plant phenomics facilities. It is feasible to capture a large variety of phenotypic data on an entire plant throughout its life cycle using phenomics technology, which is not attainable using conventional phenotyping techniques (Kumar et al. 2016). Additional sensors such as light detection and ranging (LIDAR) and laser triangulation are used to assess plant growth, shoot biomass, leaf angle distributions, and canopy structure, while magnetic resonance imaging (MRI) is used to obtain spatial information about a plant's root system architecture. Phenomics is now being applied in both controlled and natural environments.

Crop output is affected by agronomic characteristics and environmental factors such as available water and temperature (Awika 2011). The most sensitive variable affecting plant growth and agricultural productivity is water scarcity or drought (Mishra and Singh 2010). To address these possible climate change issues, the creation of elite germplasm capable of handling these abiotic stresses, as well as the technological advancements necessary to generate climate smart germplasm, would be a priority. Researchers must do comprehensive research on crops' reactions to phenotypic changes in these environments and decipher the processes and crucial regulatory components that enable these plants to thrive in these abiotic environments. Numerous genetic approaches, including marker-assisted breeding (MAB), marker-assisted backcross (MABC), and marker-assisted recurrent selection (MARS), enable in the production of climate smart germplasm for enhancing agricultural potential under these variable climatic conditions (Varshney et al. 2014).

In comparison, there is a significant, continuous gap between genomic and practical agricultural advancements because of crop phenotyping instability, which results in insufficient selection of essential QTL or gene for genomics-assisted breeding. Phenomics is a new age of analysis that enables researchers to compare plant phenotypic data to genomes under stress situations and aids in the discovery of accurate and related genes and quantitative trait loci (QTL) for desired stress conditions. Due to the sophisticated and rapid nature of these phonemics algorithms, they provide a great volume of field data that assists in the selection of elite lines in high-stress settings. A comprehensive approach to plant phenotyping advances for understanding of environmental-influenced characteristics is crucial. The advancements in phonemics enable scientists to efficiently describe germplasm that is resistant to abiotic stress. This chapter will discuss several high-throughput phenotyping methods that may aid in our understanding of the likely molecular components of stressful circumstances.

## 13.2 HTP Sensors for Water and Nitrogen Stress/Use Efficiency

Conveyor belts are employed in high-throughput phenotyping (HTP) technology to automate the collection of plant biometric data. HTP imaging employs a variety of sensors and technologies, including RGB cameras, fluorescence cameras, thermal infrared (IR), near-infrared (NIR), and hyperspectral imaging. Numerous studies have been conducted on the response of numerous crops to abiotic conditions, notably drought stress. The NIR intensity of pictures acquired by various sensors on HTP platforms may be utilised to assess the water content (Seelig et al. 2008). Near-infrared methods can be used to determine the biomass of plants. When plants are stressed by drought, infrared thermography can offer critical data regarding their temperature data. Additionally, the temperature differential will benefit in research on photosynthetic efficiency, drought tolerance, and other relevant areas (Jones et al. 2009). Plants that are stressed by a lack of water have a higher-than-normal temperature in the leaf blade and sheath (Jones et al. 2009; James and Sirault 2012). When an item emits light of a certain wavelength while concurrently absorbing light of a different wavelength, a fluorescence picture is formed. This technology enables the photosynthesis process and the plant's health to be monitored. The chlorophyll fluorescence method is used to evaluate the effect of various genes or environmental conditions on photosynthesis efficiency. When scanning with visible light, a shift in hue can be utilised to determine a plant's or leaf's senescence. In a dose-response experiment, high-throughput phenotyping was used to examine sorghum's reaction to nitrogen and water limitation, as evaluated by growth, composition, and shape. Senescence of growing leaves represents a plant's attempt to escape or avoid water stress, whereas genotypes with remain-green leaves may continue photosynthesis and are thus categorised as resistant.

Precision biomass estimation has been demonstrated in a variety of crops, including rice, wheat, barley, sorghum, and other grains, using digital images (Yang et al. 2013). Magnification resonance imaging (MRI) is a technique for examining the roots of plants. Magnetic fields and radio waves are employed to acquire photographs of the roots. By using MRI, it is possible to create roots with a 3D shape equal to that of a genuine plant growing in soil. A non-transparent surface's spectral reflectance is defined as the proportion of light reflected by the surface. Researchers can use spectral reflectance to detect plants that have been drought-stressed prior to the manifestation of apparent disease symptoms. A hyperspectral camera detects and records all wavelengths of light reflected or absorbed by a plant.

When the soil is moist, nitrogen may easily move through it; however, when the soil is dry, nitrogen mobility is restricted. Water stress and nitrogen stress can be utilised in combination in HTP systems due to the control over water supply. As a result, it is used to identify genotypes that are vulnerable to varying amounts of nitrogen and water stress. High-throughput platforms (HTPs) are capable of identifying germplasm that is N responsive under Mediterranean field conditions, which are unique to wheat and barley growing regions. As electromagnetic radiation interacts with molecules in the leaf, spectral reflectance measurements enable accurate and non-destructive assessment of the leaf's chemistry without causing damage to the leaf. Leaves or canopy spectrometry is a flexible approach that has been demonstrated to properly quantify nitrogen in wheat. Non-invasive techniques such as these are good for researching the dynamics of nitrogen partitioning inside grains. Non-invasive phenotyping can be used to investigate N uptake and partitioning, which are impacted by N availability and interactions with the environment. When leaf-N contents of cultivars are compared under conditions of variable N availability, it may be feasible to obtain insight into their relative N response capabilities, i.e., germplasm that retains leaf-N content and growth when N supply is lowered or abolished.

In the field, sensors in HTP should be carried over the canopy. Ground-based gantry structures, unmanned aerial aircraft (UAVs), phenobuggies, and adapted agricultural vehicles are also examples of systems. Additionally, each of them has restrictions in terms of payload, price, and resolution. Ground-based devices have a higher spatial resolution but a lower throughput than UAVs operating near to the canopy. Unlike unmanned aerial vehicles, ground-based systems can accommodate larger sensors such as SWIR hyperspectral cameras.

For NUE, RGB and multispectral cameras can be used to analyse plant biomass, architecture, and chlorophyll-based indices such as the normalised difference vegetation index (NDVI). Wheat yield has a high correlation with spectral reflectance coefficients from multispectral cameras. Multispectral cameras may now be used to perform non-destructive studies of physiological parameters. These novel metrics, even though they do not directly monitor plant N status, may be relevant to NUE performance. N may also be determined directly in leaf tissue using hyperspectral reflectance, which is a field-applicable technique. Additionally, light detection and ranging (LIDAR) may be utilised to concurrently determine the biomass and nitrogen distribution of leaves and stems in a canopy. For example, recent field applications of high-throughput NUE phenotyping have included the categorising of sorghum growth in response to nitrogen fertilisation, which was used to aid in the selection of genomics-assisted sorghum breeding lines. When UAVs equipped with near-infrared green-blue (NIR-GB) cameras were used to evaluate canopy height, correlations between projected and actual canopy height were found to be 0.678 at high N and 0.842 at low N. Alternatively, in rice, ground-based high-resolution thermography (HTP) was used to determine canopy height, temperature, and reflectance ratios on a population of 1516 recombinant inbred lines (RILs), all of which were found to be highly correlated with biomass, leaf area index, flowering time, and nitrogen availability. In this large mapping population, our approaches successfully identified genomic regions associated with yield and yield-related traits. This research was made feasible by the use of high-throughput phenotyping, which enabled it to be done significantly faster and at a lower cost than conventional phenotyping. More importantly, it permitted the manual measurement of parameters that would have been difficult to measure manually, as well as the repeated measurement of such values without causing damage.

## 13.3 Phenotyping Based on Longitudinal Phenotypic Characteristics

Numerous universities across the globe now have access to fully automated and programmed high-throughput platform systems based on non-destructive phenotyping methods (Furbank and Tester 2011). It is feasible to speed the phenotyping process by utilising these platforms, which include imaging equipment, high-tech automated sensors, and computing capabilities (Furbank 2009). Three-dimensional imaging, far-infrared imaging, fluorescence imaging, hyperspectral imaging, near-infrared imaging, visible light scanning, magnetic resonance imaging and positron lemnatec technology, emission tomography, phenonet, phenomobile, phenocopter, and helium-filled aerostats are among the imaging techniques used on these platforms. There are several HTP platforms available. These systems blend plant mobility assessment with accurate phenotypic imaging, efficient imaging algorithms, and phenotypic data analytics to deliver a holistic view of plant mobility. These platforms for high-throughput plant phenotyping have evolved into very capable equipment paired with cutting-edge computer systems (Paproki et al. 2012).

The advantage of HTP is that it can track a crop's growth throughout its life cycle in order to determine how it responds to its environment. Longitudinal traits, infinite-dimensional traits (Kirkpatrick and Heckman 1989), or function-valued traits reflect physiological and biochemical responses to environmental stresses and developmental processes that occur repeatedly or continuously over the course of an individual's lifetime. Genetic pathways established from longitudinal features may enhance the capacity of complicated polygenic traits to be predicted in multivariate contexts and aid in the discovery of global or time-specific QTL. HTP systems offer a greater potential for detecting time-specific molecular events induced by yet-to-be-identified important genes in GWAS or for forecasting future phenotypes in longitudinal genomic prediction. Momen et al. (2019) use genetic data to predict longitudinal shoot development trajectories in rice using a random regression model. Momen et al. (2019) predicted shoot growth trajectories in well-watered and water-limited settings using B-splines and Legendre polynomials under varied longitudinal cross-validation conditions. The study revealed that the frequency of phenotypic screening might be decreased without sacrificing prediction accuracy. Moreira et al. (2013) emphasised the critical role of contemporary statistical genetic and modelling tools in crop improvement. Using data from wheat field-based HTP systems, it was identified that many temporal QTL correlations with modifying the genetic control of plant height. Despite recent improvements in quantitative genetics, genomics, and computer science, plant breeding's phenotyping remains a difficult task. To appreciate the links between genotype and phenotypic dynamics, significant advances in phenotyping, molecular technology, and statistical techniques are necessary.

### 13.3.1 3D Imaging (Three-Dimensional Imaging)

Researchers have proved that it is feasible to measure three-dimensional (3D) surface information from plants during the previous three decades. The ability to visualise the plant architecture enables tracking the geometric progress of the plant and parameterization of plant canopies, individual plants, and plant elements and components. Due to the non-destructive nature of 3D measurement, a monitoring system may be gradually implemented over time. It is critical to accomplish this in 3D to distinguish between plant movement and correct growth at the plant and organ level. Plant functional and structural models are developed to connect geometry and function, to separate movement from growth, to visualise and quantify diurnal patterns, and to monitor the effect of environmental stress on plant growth. All processes generate point clouds. Each point in the cloud is identified by three coordinates: $x$, $y$, and $z$. These coordinates enable the point to be located in three-dimensional space.

Depending on the measurement instrument, this coordinate may be complemented by information on the intensity or colour of the reflected light in the direction of recording. Due to the separation of 3D imaging into two components, one was based on dynamic lighting and the other on passive approaches. Active lighting sensors detect the canopy of a plant using an active light source, whereas passive lighting sensors detect the canopy of a plant using ambient light. Triangulation-based systems and time of flight measurements are examples of active measurement techniques. Laser triangulation (LT) and structured light (SL) technologies are based on triangulation, whereas terrestrial laser scanning (TLS) and time of flight (ToF) cameras are based on time of flight (ToF). Light field cameras (LF) and structure from motion (SfM) methods are examples of passive techniques. Detecting water scarcity for plants has grown more enticing during drought circumstances. Theoretically, either colour or morphological response may be used to identify water deficiency in plants (Revollon et al. 1998). Colour identification is simplified by ignoring the geometrical characteristics of the plants. However, it takes a little longer to detect a noticeable shift in the colour of the leaves. According to Zhao et al. (2021), the laser scanner is an excellent tool for observing wilting plant symptoms in three dimensions and II, the created leaf wilting index may be used to describe the morphological response of drought-stressed zucchini. To map QTLs governing important sections of the rice root architecture, semi-automated 3D imaging and digital phenotyping are used (Topp et al. 2013). It has been researched the yield characteristics, root architecture, imbibition, and germination rates of *Arabidopsis thaliana* under drought stress.

### 13.3.2 Near-Infrared Imaging

It is possible to examine the infrared radiation emitted by an item merely by looking at it using infrared thermal imaging. This imaging system obtains high-resolution pictures by using the inherent molecular movements of molecules that create infrared light (Kastberger and Stachl 2003). Infrared imaging makes use of two unique wavelength ranges of light: the near-infrared (NIR) range of 0.9–1.55 m and the far-infrared (Far-IR) range of 7.5–13.5 m. Near-infrared imaging is performed by utilising two unique wavelength ranges of light: one in the near-infrared (NIR) range of 0.9–1.55 m and another in the far infrared (NIR) range. NIR imaging is used to monitor soil and leaf water content, as well as mobility. In comparison, FIR cameras are used to determine the soil's temperature and moisture content (Weirman 2010). The soil NIR calculation is used to calculate the rate of water absorption by the roots. NIR imaging is used in seed kernels to determine the quantity of starch in the leaves and the amount of oil and protein in the kernels, according to the manufacturer (Jones et al. 2009). Near-infrared (NIR) technology was used to screen rice and soybean germplasm for drought tolerance (Seelig et al. 2008). When wheat and barley are exposed to salt stress, NIR researchers seek for osmotic resistance traits (Lobet et al. 2011). When used with high-throughput phenotyping, it can help in the standardisation and simplicity of phenotypic data collection as well as the creation of repository databases for QTL meta-analyses. Compared to a decade ago, our present capability for high-throughput molecular profiling far outpaces our skill for acquiring useful phenotypic data. The growth in the finding and profiling of single nucleotide polymorphisms (SNPs) across several crops is the most evident manifestation of this phenomenon. Nonetheless, there has been a rising recognition in recent years of the importance for more integrated, interdisciplinary, and field-based research to mitigate the negative implications of water shortage, particularly in poor nations.

High-throughput under pots, it is possible to maintain strict control over the water deficit imposed on various genotypes and the homogeneity of stress intensity, which is difficult to perform in field circumstances, particularly when the genotypes under test differ in phenology and/or biomass. Despite this, marijuana research is constrained by a number of intrinsic restrictions that must be carefully reviewed and addressed in order to provide useful findings applicable to real-world circumstances (Passioura 2006). When characteristics are investigated in binary form, such as photoperiod sensitivity, and when environmental factors have minimal effect on the target trait or are well defined (e.g., light vs. darkness), the procedure of phenotyping under controlled conditions is quite uncomplicated (e.g., light vs. darkness). However, when environmental conditions fluctuate during the day and, as is the case with development, when the goal quality is quantified (e.g., temperature, light intensity, soil water status, and so on), the issue quickly becomes more difficult (e.g., leaf elongation rate). Due to the dynamic nature of the phenotype in this context, it is best described as a sequence of reaction curves to environmental inputs. This laborious approach necessitates exact control of environmental conditions. As a result, it is necessary to quantify (1) the physical variable(s) that characterise the degree of water stress (e.g., pot weight; soil moisture; and so on); and (2) give a specific amount of water to each pot in the experiment. To facilitate and standardise the gathering of precise phenotypic data in glasshouse-grown plants, these procedures, which have already been used by a number of commercial enterprises and big government organisations, may now be automated through the use of high-throughput phenotyping systems. Continuous phenotypic evaluation of thousands of plants may be conducted in an automated and non-destructive way using cutting-edge technology such as imaging, robotics, and computer equipment, resulting in considerable cost savings. Regrettably, the installation and operation expenses of these platforms are relatively high in comparison. When used in conjunction with near-infrared spectroscopy and spectral reflectance techniques, as discussed below, digital photography and canopy quality evaluation may result in a more rapid gathering of high-throughput phenotypic data for particular traits.

### 13.3.3 Digital Imaging

Digital image analysis enables the precise evaluation of plant features that would otherwise require considerable time and effort. Digital images provide a number of advantages over more traditional methods of light interception computation, including the ability to examine photographs quickly on a computer. Video image analysis offers a non-destructive, dynamic assessment of canopy features and crop development.

Digital image analysis enables the precise evaluation of plant features that would otherwise require considerable time and effort. A critical example is the measurement of canopy properties. Digital images provide a number of advantages over more traditional methods of light interception computation, including the ability to examine photographs quickly on a computer. Video image analysis offers a non-destructive, dynamic assessment of canopy features and crop development. Digital photography is also advantageous for analysing root characteristics in trials, which are frequently hampered by a lack of suitable tools for continuous, non-destructive assessments that enable accurate examination at higher resolution scales, which is required to investigate the kinetics of root development systems. In this context, a non-invasive technique based on digital image sequence processing was employed to quantify highly resolved spatial-temporal activities inside the *Arabidopsis* root development zone.

### 13.3.4 Reflectance Spectroscopy and Near-Infrared Spectroscopy

Near-infrared spectroscopy is a fascinating component of high-throughput phenotyping systems for determining integrative features and spectral reflectance of plant canopies. They represent an interesting opportunity for the collection of integrative features with a high temporal resolution. Crop canopy spectral reflectance may be determined in the visible and near-infrared regions of the electromagnetic spectrum using sensors installed on tractors or digital cameras mounted on handheld devices. Remote sensing has improved in our understanding of how leaf reflectance and emittance change with species, leaf thickness, canopy structure, leafage, nutritional condition, and, most critically, water status. This data was utilised to develop multiple vegetative indices for crop canopies that may be used to quantify agronomic characteristics (e.g., leaf area, crop cover, biomass, yield, and so on). To extract useful information from the plot spectrum, calibration models for anticipating phenotypic values are necessary. The application of spectral reflectance in well-controlled experimental circumstances successfully monitors the pigment composition of photosynthetic plants, determines their water status, and early detects abiotic stress.

### 13.3.5 Modelling of Virtual Phenotypes

As we progress through the piecemeal (i.e., gene by gene) unravelling of gene functions and the attempt to understand how these functions ultimately affect the phenotype, the majority of models allow us to simulate virtual phenotypes derived from any combination of different factors—alleles, environmental variables, and so on. Modelling is a step toward a more complete approach to systems biology in a variety of ways. Plants are capable of responding to a broad variety of environmental conditions because their gene networks are coordinated. Controlled environmental factors like as temperature, transpiration, and soil water status must be employed to accurately and rapidly monitor the phenotypic of each accession for the target characteristic (e.g., leaf elongation). It is preferable to do this sort of research in a controlled setting. Maize QTL parameters were discovered for mapping population lines and then genetically analysed, allowing for the modelling of leaf development in new lines defined by their QTL alleles. As a result, quantitative trait loci (QTLs) for plant responses can be identified using this technique. Any "virtual genotype" composed of a particular combination of alleles may be expected to function well in any environment. While this method is now limited to a few critical features and genetic systems, it has significant potential.

Crop growth may be modelled using more sophisticated models that take environmental factors into account. As a result, they may be used to investigate the effect of certain features on the seasonal dynamics of crop water usage and carbon assimilation. However, because the algorithms are still crude, the effects of genes or QTLs cannot usually be simulated at the crop level, except for constitutive traits such as phenology, binary traits related to environmental triggers such as a flowering response to photoperiod, or when organ-level QTL models can be combined with crop level QTL models. They have historically been the major method for judging whether a certain characteristic will be useful throughout time. The effect of stay-green simulated and a trait hypothesised to promote drought tolerance, on 547 location-season combinations. However, when stress levels were high, this trait was beneficial; however, when stress levels were low, it was detrimental.

When a population of people is selected for the genes contributing to the network, incomprehensible hurdles are produced by non-linear effects connected with the genes functioning in networks. Despite the potential benefits of modelling, making exact predictions across genotypes remains difficult. Recently, close range hyperspectral imaging (HSI) has been investigated as a non-invasive approach for assessing plant traits in high-throughput plant phenotyping platforms (HTPP). Additionally, HSI has been used in the area of plant biology to investigate the reactions of plants to environmental stresses such as fungal infection or depletion. Numerous physiological and biochemical changes occur in plants when they develop under stress, altering the photosynthetic system, organs, water content, leaf surface area, and internal structures. These modifications may have an effect on the optical characteristics of the leaf. Vegetation indices (VIs) are a typical way for evaluating plant features using HSI. They are defined as ratios or linear combinations of reflectance at a few specified wavelengths. Utilising VIs mitigates the impacts of scale factors such as slope effects and variations in lighting conditions. Drought stress's complicated physiological consequences have an influence on a wide range of wavelengths. As a result, VIs may miss essential information, resulting in decreased discriminating accuracy.

Radiative transfer modelling (RTM) inversion is another widely used technique for deriving plant properties from reflectance data (RTM). In RTM inversion, model parameters including as chlorophyll concentration, water content, dry matter, and canopy structure are determined using look-up tables and optimisation processes. The unsuitability of these techniques is a frequent source of worry, as different combinations of vegetation variables may produce comparable spectra. Additionally, this strategy is inefficient at close range because physically-based leaf or canopy RTMs are difficult to adapt to the special challenges associated with near-range lighting. This study revealed that HSI may be a non-destructive and rapid method for monitoring drought stress responses in individual plants over time. The suggested approach utilises a data-driven strategy that includes clustering, band selection, and a spectral similarity measure to identify and recover from drought stress. The analytical approach was validated in a HTPP using maize plants that had been subjected to various types of drought stress throughout their vegetative growth. According to the trials, the technique discriminated clearly between plants suffering from water deficiency stress and healthy plants at an early stage of stress development. Additionally, the strategy demonstrated plant regeneration following a period of re-watering. This demonstrates the utility of HSI as a novel technique for high-throughput phenotyping investigations, which may help breeders gain a better understanding of the genetics of drought tolerance. Additionally, the recommended method is not restricted to drought stress; if there is an interest in monitoring plant process dynamics at the plant scale, it might be applied to a variety of other types of systemic stress.

## 13.4 Recent Uses of HTP for Water and Nitrogen Stress Utilising the Sensors and Platforms

### *13.4.1 Spectral Signature and Stress Caused by a Water Deficiency*

Typically, the reflectance spectra of a crop's plants are comparable throughout all of the crop's plants. On the other hand, water-deficit stress circumstances result in observable changes in reflectance spectra—the reflectance patterns of plants under various water-deficit stress conditions, i.e., a drop in relative water content—that can be studied in the laboratory. The water content of the combination fluctuates between 96.5 and 0.7%. In comparison to the dried plant, the fresh plant had a reduced reflectance (Fig. 13.1). As the RWC value decreases from the greatest to the lowest value, the SWIR band reflectance increases as well. Water absorption properties between 1400 and 1900 nm are deteriorating as reflectance increases. A similar trend of increasing reflectance with decreased water content was identified when the wavelength range was 350–700 nm. The blue and red portions of the spectrum (which correspond to the chlorophyll a and b absorption bands, respectively) exhibited a tendency toward higher reflectance when water content decreased due to chlorophyll loss. It has been observed that as leaves are dried, the wavelength range of 1400–1925 nm is shifted toward shorter wavelengths and the spectral reflectance increases. As the relative water content decreased, the absorption properties in the 1400–1550 and 1850–1900 nm wavelength ranges increased shallower. The decrease in absorption is produced by a weakening of the water absorption

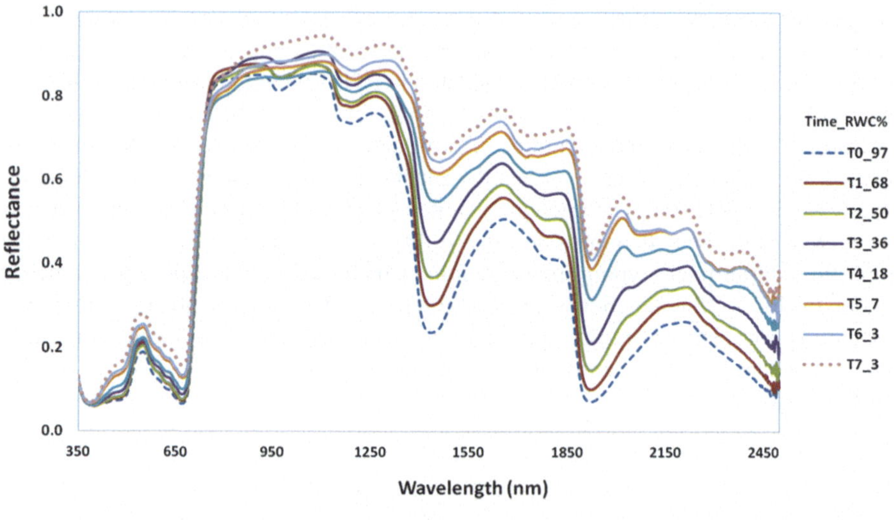


**Fig. 13.1** Representative mean spectral reflectance observations of the genotypes with decreasing RWC (%) in rice leaves, showing the percentage of RWC and corresponding spectra at different time intervals

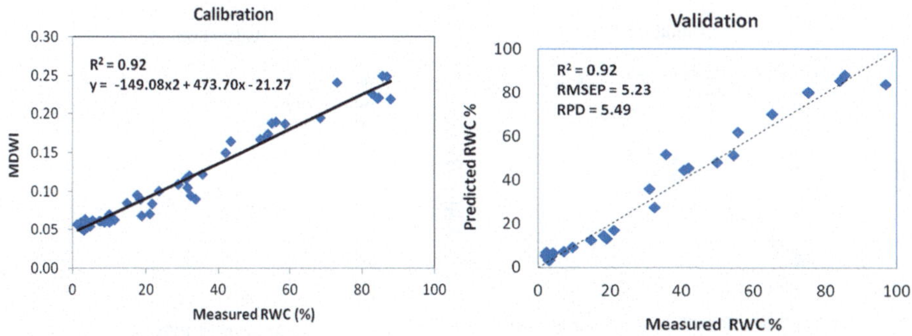


**Fig. 13.2** The Calibration model developed through the relationship between MDWI and Measured RWC (%) and its validation. (Calibration—$N = 55$ and validation—$N = 25$.) The solid black line is the regression line, and the dotted line is the 1:1 line

properties as a result of the decrease in water content. At wavelengths spanning from 810 to 1350 nm, the scattering in spongy mesophyll displayed a similar pattern of increasing reflectance with decreased water content as seen in the previous research. Additionally, absorption in the middle infrared region (1100–2500 nm) is a zone of intense absorption, primarily by water in a fresh leaf and, secondarily, by dry matter (e.g., protein, lignin, cellulose) when the leaf wilts, which becomes more visible with a decrease in RWC.

The Maximum Difference Water Index (MDWI) was shown to have the strongest correlation with an R2 of 0.92 for both the calibration and validation sets, indicating that it was the most accurate conventional water band indicator (Fig. 13.2). The MDWI is derived by averaging the maximum reflectance value between max1500–1750 nm and the lowest reflectance value between min 1500 and 1750 nm, which both fall within the atmospheric window between 1500 and 1750 nm. The MDWI worked effectively in the 1500–1750 nm wavelength range because it provides for the most optimum mix of numerator and denominator wavelengths. In both investigations. It was discovered that dynamic selection of superior absorption characteristics under varied plant water-deficit stress conditions resulted in improved outcomes.

By utilising the lambda versus lambda contour plotting technique, it is possible to effectively choose the optimal mix of wavebands for producing effective spectral indices. We identified hotspot positions with high correlation values between the two variables by using contour maps of R2 values obtained from linear regression between RWC and all possible combinations of RSI (ratio spectral index—ratio approach) and NDSI (normalised difference spectral index—normalised difference approach) (normalised difference spectral index—normalised difference approach). The hotspots found at 1233 and 1305 nm produced a single R2 value for both RSI and NDSI, the highest result ever recorded for each parameter. As a result, we chose the most refined RWC combinations with the highest R2 value, which included the ratio index (R1233, R1305) and the normalised difference ratio index (R1233, R1305) for RWC. Using the ratio index and the normalised difference ratio index,

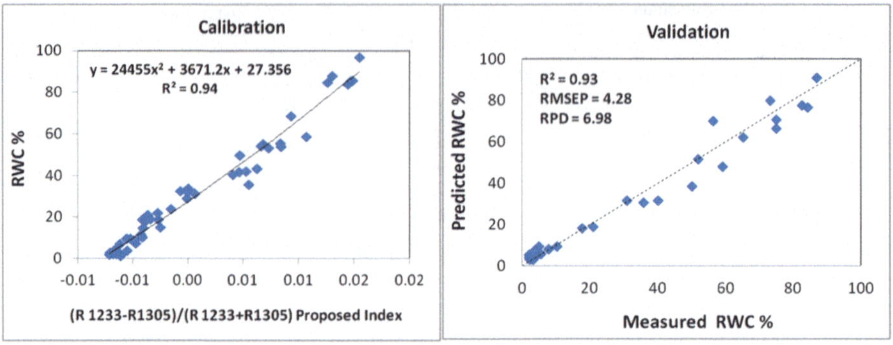


**Fig. 13.3** The proposed normalised difference ratio index $(R_{1233} - R_{1305})/(R_{1233} + R_{1305})$ for prediction of RWC. (Calibration—$N = 55$ and validation—$N = 25$.) The solid black line is the regression line, and the dotted line is the 1:1 line

the second-order polynomial equation was the most accurate in predicting RWC (R2Cal = 0.94 and RMSEP = 4.27, respectively). The second-order polynomial equation (R2Cal = 0.94 and RMSEP = 4.28, respectively) was shown to be the most accurate for predicting RWC when the ratio index and the normalised difference ratio index (R2Cal = 0.94 and RMSEP = 4.28, respectively) were used (Fig. 13.3). When both the RSI and NDSI R2 values are computed for calibration and validation, statistical significance is reached. A result is statistically significant when the *P*-value for the calibration and validation findings is less than 0.05.

### *13.4.2 Multivariate Techniques for RWC Estimation*

This study determined that the PLSR followed by MLR was the most effective strategy for developing RWC prediction models out of all the multivariate strategies that were assessed. The model equation produced by applying PLSR-MLR approaches can also be applied to the monitoring of water content in rice fields. The combination of PLSR and ANN was the second-best model produced. The support vector regression technique was also a valuable tool, with adequate results. The SVR determines the maximum-margin hyperplane, which helps to reduce the prediction error by a factor of 2. In addition, because the ANN is susceptible to outliers, its prediction was shown to be subpar when applied to the entire dataset. The random forest is an ensemble tree classifier that performs as well as a decision tree system in terms of accuracy. When compared to other classifiers, the RF performed as an intermediate classifier. In this investigation, it was discovered to be marginally superior to PLSR. Because every coefficient in the PLSR equation has a corresponding root mean square error (RMSE), the equation is more vulnerable to variation. As a result, when compared to the PLSR-MLR combination, the PLSR model generated using all of the *x* variables produced intermediate results. The following is the order of performance of the multivariate models in terms of R2 and RMSEP: PLSR-MLR > PLSR-ANN > SVR > RF > PLSR > ANN > SVR > RF > PLSR > ANN. This order of performance is also corroborated by the value of RPD for all models, which indicates that they perform in that order. There are numerous advantages to multivariate techniques when compared to conventional indices-based approaches, such as self-identification and removal of outliers, the use of principal components, the capability of coping with multi-collinearity and the use of a decision tree approach among others. A multivariate technique uses all of the water absorption-related bands, which results in a significant boost in the accuracy of the model by revealing improved sensitivity to changes in the RWC.

On the other hand, index-based techniques make use of only two or three notable water absorption bands. The MLR and ANN models were effectively applied to PLSR chosen optimal wavebands in this study, resulting in a significant increase in the model's accuracy as a result. The PLSR model was more efficient because it used optimal wavebands as input, eliminated the multi-collinearity problem in MLR (Arya et al. 2022), and gave outliers-free $x$ variables to the ANN, which helped to improve the efficiency of the PLSR model.

### 13.4.3 Differential Response of Rice Genotypes to Water Stress

The water loss behaviour of the different rice genotypes over time, starting with the first spectral measurement taken within 2 h of harvesting (T0) and ending with the last spectral measurement taken after 27 h (T11) (Fig. 13.1). Between the rice genotypes, Pusa44 and IR64 exhibited only the slightest difference between RWC and Vandana exhibited the most significant difference between RWC and Vandana. When the index difference between consecutive measures was determined, Vandana was found to have the smallest change, followed by IR64 and Pusa44; Nagina22 and Pusa Basmati6 were found to have the most change. This occurs because the least amount of change in the plant's water content or the least amount of difference in leaf reflectance occurs when the plant is maximally turgid, and the most water is lost. As a result, it was discovered that these genotypes were more resistant to water stress. Vandana was also identified as a water-deficit stress-tolerant variety grown in the field under drought circumstances.

### 13.4.4 Spectral Discrimination of Rice Genotypes

CART analysis was used to minimise the number of influential bands ($n = 1759$) selected by ANOVA analysis to a smaller number of bands capable of optimally separating the rice genotypes, as determined by the ANOVA analysis. In the case of rice, the spectral bands are as follows: 350; 532; 553; 665; 717; 730; 823, 887, 910, 960; 1440; 1960; 1973; 1979; 2009; 2296 and 2313; and 1440 nm. As a result of this investigation, we can conclude that the 17 wavelengths for rice can distinguish between genotypes. The separability index (J–M distance) between each genotype pair and the stress levels was calculated using the CART wavelengths chosen for the study.

### 13.4.5 Field Phenotyping with Drone Platform

RGB images and multispectral images from missense with red edge camera captured from 60 m height were pre-processed and mapped as true colour composite and Blue band based normalised difference vegetation index (BNDVI) (Fig. 13.4). Some of the fields were found fallow as some early genotypes were harvested in the control treatment with were recommended doses of irrigation and nitrogen. BNDVI was found the best for biomass mapping, having the highest $R^2$ of 0.59 (Fig. 13.4). It is the normalised difference ratio of energy received in the NIR and Blue bands of the sensor and expressed as BNDVI = (NIR − Blue)/(NIR + Blue). Based on the BNDVI value, the heat map of 182 genotypes was prepared, capturing their response to differential treatment of drought and nitrogen.

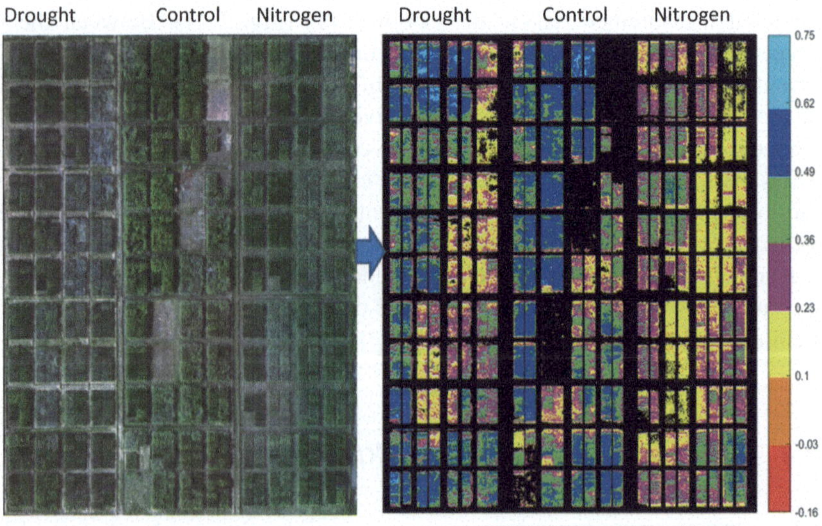


**Fig. 13.4** Multispectral (MicaSense) image of Rice Field Experimentation taken from the drone

## 13.5 Challenges and Opportunities

Foliar plant stress phenotyping is essential for plant phenotyping and precision farming in the agricultural industry. Various scales are involved, from the level of a single leaf to the level of larger plant canopies and entire fields. As a result, improvements in the accuracy and automation of stress severity evaluations have the potential to increase the rate of genetic gain across crops while also facilitating the creation of new management strategies. Plant stress evaluations evaluate the visible signs or symptoms of stress and the evolution of stress on individual plant units (e.g., leaves, stems, or roots) at the leaf, canopy, plot, and field levels and the effects of stress on the entire plant population. The use of these measures is required for evaluating a stress measuring system for both qualitatively and quantitatively expressed features, and it also allows for uniformity between scales. Compared to numerous qualitative qualities that are evident and understandable in their manifestation, quantitative stress-resistance traits present additional difficulties in the phenotyping process. When quantitative features are expressed, they are rarely found in defined infection groups and are more commonly seen in a continuous distribution, making stress measurement more challenging. There are many different types of rating scales that have been designed to deal with the complexity of quantitative rating qualities, including nominal, ordinal, interval, and ratio scales. Different scales express continuous or discrete properties, and they can be universal or particular to a single host-pathogen connection depending on the scale used. ML will bring value to agricultural technologies based on field stress phenotyping by enabling extensive data gathering facilitated by the unmanned aerial system (UAS) technology and ground robots in conjunction with machine learning (ML). This will increase farmers' decision-making capability and crop yields. Plant stress data gathering techniques, on the other hand, need to be improved and made more standardised. To further facilitate machine learning-based real-time plant water and nitrogen phenotyping on a broad scale, a seamless pipeline of data gathering, curation, and analytics should be built.

## 13.6 Conclusion

A greater knowledge of the genetic base and physiological mechanisms behind drought-adaptive characteristics is necessary to maximise the potential of germplasm resources and genomics technology to increase drought resistance. Precision and cost-effective phenotyping will be advantageous in this approach. Techniques/approaches that enable precise control of the water regime and a reduction in experimental noise, in combination with high-throughput platforms, would significantly simplify the selection of excellent phenotypic data and the cost-effectiveness of phenotyping. The elimination of the "statistical veil" that surrounds QTLs and impairs our ability to accurately assess their implications and predict the potential of novel QTL allele combinations would assist in dispelling the "statistical veil," at least in part. Regardless of how precise our phenotyping is, the vast majority of QTLs influencing the measured phenotype will go undiscovered by researchers. For instance, this phenomenon is referred to as the "iceberg effect" by me. Similarly, the majority of genetic factors influencing quantitative characteristics are likely to remain undiscovered because their effects are simply too small to be presented statistically significant, similar to an iceberg hidden beneath the surface of the water and thus invisible to the naked eye. As a result, implementing innovative crossing schemes and approaches will remain challenging. In comparison to MAS, genome-wide selection emphasises precision phenotyping while avoiding QTL finding. Drought-adaptive characteristics will become increasingly important as the cost of sequencing and genotyping continues to fall. Cost-effective phenotyping will become more important for identifying drought-adaptive features and customising cultivars for farming in drought-prone locations. It is hoped that as a result of this study, there would be more interest in phenotyping and a greater understanding and respect for the critical role it plays in medicine. Due to recent advancements in the use of non-invasive high-throughput sensors, the phenomics platform is now capable of collecting hundreds of phenotypic characteristics rather than just a few traits as was previously achievable using traditional methodologies. Performance evaluation studies have demonstrated that in some cases, a controlled environment and field phenomics may be a valuable supplemental tool. In some circumstances, such as biotic stress, resource efficiency, and positional cloning, phenomics may be used in place of time-consuming and labour-intensive field phenotyping. To guarantee that optimal growth conditions and therapies are implemented in controlled environment phenomics, treatment design in a controlled environment must take environmental phenotyping data from TPE into account. Significant progress has been achieved in extracting phenotypic characteristics from VIS, IR thermal, and near-infrared imaging sensors, as well as in correlating data to traits. Further advancements in hyperspectral imaging systems and the integration of non-invasive imaging sensors for physiological processes and metabolite phenotyping, such as NIR and FTIR spectroscopy, are necessary to characterise organisms at the molecular level. Despite the fact that the many sensors employed in phenomics create hundreds of distinct image properties, only a subset of them are utilised. Additionally, it is critical to apply machine learning patterns and feature extraction, as well as a multimodal method that incorporates phenomics data from several types of sensors, while phenotyping stress. It is anticipated that phenomics-based analytical breeding will aid in the breakdown of production barriers and increase crop sensitivity to adverse field conditions. To expedite gene discovery and analytical crop breeding, it would be advantageous to have a phenome bank that has curated and labelled phenome data, as well as functional connections to genome banks. To investigate natural variability in agricultural plants, forward genetic techniques such as QTL mapping and genome-wide association studies (GWAS) are often used. QTL mapping is a sort of linkage mapping in which the genetic links between two parent lines and their segregating offspring are mapped using the variations between them. HTP is ushering in a new age for breeders by leveraging functional genomics research, machine learning, and genome selection approaches to enable breeders to harness nature's dynamics and accelerate crop development at unprecedented rates.

## References


Arya S, Sandhu KS, Singh J, Kumar S (2022) Deep learning: as the new frontier in high-throughput plant phenotyping. Euphytica 218(4):1–22

Awika JM (2011) Major cereal grains production and use around the world. In: Advances in cereal science: implications to food processing and health promotion. American Chemical Society, Washington, DC, pp 1–13

Fiorani F, Schurr U (2013) Future scenarios for plant phenotyping. Annu Rev Plant Biol 64:267–291. https://doi.org/10.1146/annurev-arplant-050312-120137

Furbank RT (2009) Plant phenomics: from gene to form and function. Funct Plant Biol 36:10–11

Furbank RT, Tester M (2011) Phenomics--technologies to relieve the phenotyping bottleneck. Trends Plant Sci 16:635–644

James RA, Sirault XRR (2012) Infrared thermography in plant phenotyping for salinity tolerance. In: Shabala S, Cuin TA (eds) Plant salt tolerance: methods and protocols. Humana Press, Totowa, NJ, pp 173–189

Jones HG, Serraj R, Loveys BR, Xiong L, Wheaton A, Price AH (2009) Thermal infrared imaging of crop canopies for the remote diagnosis and quantification of plant responses to water stress in the field. Funct Plant Biol 36:978–989

Kastberger G, Stachl R (2003) Infrared imaging technology and biological applications. Behav Res Methods Instrum Comput 35:429–439

Kirkpatrick M, Heckman N (1989) A quantitative genetic model for growth, shape, reaction norms, and other infinite-dimensional characters. J Math Biol 27:429–450

Kumar S, Raju D, Sahoo RN, Viswanathan C (2016) Phenomics: unlocking the hidden genetic variation for breaking the barriers in yield and stress tolerance. Indian J Plant Physiol 21:409–419

Lobet G, Pages L, Draye X (2011) A novel image-analysis toolbox enabling quantitative analysis of root system architecture. Plant Physiol 157:29–39

Mishra AK, Singh VP (2010) A review of drought concepts. J Hydrol 391:202–216

Momen M, Campbell MT, Walia H, Morota G (2019) Predicting longitudinal traits derived from high-throughput phenomics in contrasting environments using genomic Legendre polynomials and B-splines. G3 (Bethesda) 9(10):3369–3380

Moreira FF, Oliveira HR, Volenec JJ, Rainey KM, Fiorani F, Schurr U (2013) Future scenarios for plant phenotyping. Annu Rev Plant Biol 64:267–291

Paproki A, Sirault X, Berry S, Furbank R, Fripp J (2012) A novel mesh processing-based technique for 3D plant analysis. BMC Plant Biol 12(1):1–13

Passioura J (2006) Increasing crop productivity when water is scarce—from breeding to field management. Agric Water Manag 80(1–3):176–196

Ray DK, Mueller ND, West PC, Foley JA (2013) Yield trends are insufficient to double global crop production by 2050. PLoS One 8:e66428

Revollon P, Chasseriaux G, Riviere LM, Gardet R (1998) The use of image processing for tracking the morphological modification of forsythia following an interruption of watering. In: Proceedings of the international conference on agricultural engineering. AgEng, Oslo

Seelig HD, Hoehn A, Stodieck LS (2008) The assessment of leaf water content using leaf refectance ratios in the visible, near-, and short-wave-infrared. Int J Remote Sens 29:3701–3713

Topp CN, Anjali SI, Anderson JT (2013) 3D phenotyping and quantitative trait locus mapping identify core regions of the rice genome controlling root architecture. Proc Natl Acad Sci U S A 110:1695–1704

Varshney RK, Terauchi R, McCouch SR (2014) Harvesting the promising fruits of genomics: applying genome sequencing technologies to crop breeding. PLoS Biol 13:528–530

Weirman A (2010) Plant phenomics teacher resource. http://www.plantphenomics.org.au/files/teacher/Final_Phenomics_for_word_with_images.doc. Accessed 9 Nov 2021

Yang W, Duan L, Chen G, Xiong L, Liu Q (2013) Plant phenomics and high-throughput phenotyping: accelerating rice functional genomics using multidisciplinary technologies. Curr Opin Plant Biol 16:180–187. https://doi.org/10.1016/j.pbi.2013.03.005

Zhao Y, Zheng B, Chapman SC, Laws K, George-Jaeggli B, Hammer GL, Jordan DR, Potgieter AB (2021) Detecting sorghum plant and head features from multispectral UAV imagery. Plant Phenomics 2021:9874650. https://doi.org/10.34133/2021/9874650

# Chapter 14
# Metabolomics as a Selection Tool for Abiotic Stress Tolerance in Crops

**M. D. Prathibha, S. G. Harsha, G. A. Geetha, A. N. Lokesha, and K. S. Shivashankara**

**Abstract** Abiotic stresses cause aberrations in plant primary metabolism which alters the cell homeostasis. Plants try to maintain the metabolic homeostasis during stress by modifying the metabolic networks leading to the altered production of metabolites. Metabolites are more closely related to plant phenotype than the transcripts or proteins alone. Metabolomics reflect the overall combination of protein interaction, gene expression, and other plant regulatory mechanisms. Among all—omics methods, metabolomics is the most versatile, since it may be used to directly link to a phenotype, hence it can be a species or genotype specific character. In this chapter, we discuss various abiotic stresses affecting the plants and their metabolic responses specifically associated with stress tolerance. Most of the studies have elucidated the important role played by primary (sugars, amino acids, Proline, Putrescine, Spermine, GABA, Krebs cycle intermediates, etc.) and secondary metabolites (anthocyanins and flavonoids) in stress tolerance. Relative accumulation or non-accumulation of certain metabolites, phenotypic response along with the corresponding data of gene expression provides a way for identification of metabolomic QTLs (mQTLs) and metabolomic-assisted genome wide association study (mGWAS). Rapidly emerging fields of metabolomics, with mGWAS and mQTLs help in better selection of breeding lines and crop improvement programs.

**Keywords** Abiotic stress · Physiological traits and responses · Primary and secondary metabolites · Crop improvement

M. D. Prathibha · G. A. Geetha · A. N. Lokesha · K. S. Shivashankara (✉)
Division of Basic Sciences, ICAR-Indian Institute of Horticultural Research, Bangalore, Karnataka, India
e-mail: shivashankara.ks@icar.gov.in

S. G. Harsha
Division of Basic Sciences, ICAR-Indian Institute of Horticultural Research, Bangalore, Karnataka, India

Department of plant Physiology, Indira Gandhi Krishi Vishwavidyalaya, Raipur, India



H. M. Mamrutha et al. (eds.), *Translating Physiological Tools to Augment Crop Breeding*, https://doi.org/10.1007/978-981-19-7498-4_14

## 14.1 Introduction

Increased food demand from rising population is causing an increased pressure on agricultural systems across the world. All main crop yields have reached a plateau, and chaotic urbanization (which reduces agricultural land) and extreme weather occurrences pose new dangers to food production ecosystems: floods, extended droughts, excessive heat, and late spring frosts are becoming frequent, need quick, and ongoing action (Voss-Fels et al. 2019). In these conditions, plant scientists must primarily concentrate on discovering traits that would ensure consistent higher yields. To identify suitable plant characters/traits toward crop improvement, huge advances are made in numerous "Omics" techniques, like phenomics, genomics, epigenomics, metabolomics, proteomics, and transcriptomics. The "Omics" techniques have improved the speed and precision of crop breeding programs toward direction of nutrient-rich and climate-smart varieties. Unlike proteomics, transcriptomics, genomics, the metabolomic approach denotes the final assessable responses of cell that cannot be deduced directly through any other omics approaches (Pyne et al. 2019). Hence, among various omics techniques metabolomics is considered as most predictive one, which provides the closest link between phenotype and genotype. The most complex of the omics techniques, metabolomics has been receiving attention in agricultural science for the last two decades, notably for trait mapping and trait specific parents choices (Parry and Hawkesford 2012).

Metabolism creates energy along with building blocks. These are necessary for plant's growth, maintenance, communication, reproduction, and protection (Barupal and Fiehn 2017). Metabolites are tiny biological compounds engaged in energy conversion to biosynthesis that perform critical activities like interaction of plant with environment. They are created when metabolism occurs. It is also utilized to perceive the molecular functioning and health of an individual organism (Volkova and Geras'kin 2018). Plants are well-known as factories that produce a large variety of natural substances, estimated to number over 200,000 which are important in growth and development. Metabolites that are structurally different with various biochemical properties and functions have vital part in plant active metabolism. Primary metabolites and secondary metabolites are continually generated throughout plant ontogenesis via complicated metabolic events (Pyne et al. 2019). Plants require primary metabolites for the making of lipids, amino acids, and also carbohydrates. They impact plants' normal growth and development by altering the tricarboxylic acid pathway, glycolysis, and photosynthesis. Variations in primary metabolite production can cause changes in photosynthetic rates and imbalance in osmotic adjustments in plants. Secondary metabolites like carotenoids, flavonoids, atropine, and phytic acid are produced by primary metabolism. These are generated due to various stress situations, including high temperature, salt, cold, drought, and insect-pest assault, and are not vital for plant survival. Plant metabolome contains many specific secondary metabolites for providing resistance for abiotic and biotic stresses, such as alkaloids, phenolics, and terpenoids. Many of these specific compounds have found as one-of-a-kind biomarkers for measuring plant performance under stressful environments, and are crucial components of many crop improvement activities.

Metabolomics words and concepts were first coined less than two decades ago, for gaining knowledge of biological networks by studying metabolism in great detail (Fiehn et al. 2008). The major goal of metabolomics/metabolite profiling is to know the overall changes in metabolites in living system. The chemical fingerprint of any phenotype can be determined by systematic identification and also quantification of biochemical response to a specific condition by an organism (Cambiaghi et al. 2017; Chaudhary et al. 2019). It is a strong tool that allows for a thorough examination of how metabolic network is controlled (Weckwerth and Fiehn 2002). There are various features to differentiate between metabolite profiling, metabolite fingerprinting, and metabolomics. Metabolite profiling entails measuring hundreds of metabolites in living organism. This requires an efficient and reliable protocol for extraction, separation. Metabolic fingerprinting is use of a wide analytical method to find noteworthy variances between two samples or genotypes. Study of all/complete metabolites in the system is known as metabolomics. No individual technique is selective, comprehensive, and sensitive to measure/identify all metabolites. It is still not technically possible and undoubtedly require an additional complementary platform or technologies to identify/discover/measure metabolites (Weckwerth 2003). Many techniques such as Gas chromatography mass spectrometry (GC-MS), Liquid chromatography mass spectrometry (LC-MS), one Ultra-high-performance liquid chromatography mass spectrometry (UHPLC-MS), Liquid chromatography quadrupole-mass spectrometry (LC-Q-MS), Triple quadrupole liquid chromatography coupled to mass spectrometry (LC-QQQ-MS), Liquid chromatography quadrupole-time of flight-mass spectrometry (LC-Q-TOF-MS), Gas chromatography coupled to time-of-flight mass spectrometry (GC-TOF-MS), Direct infusion tandem mass spectroscopy (DI-MS/MS), Nuclear Magnetic Resonance (NMR), Fourier-transform ion cyclotron resonance mass spectrometry (FT-ICR-MS), Electrospray ionization mass spectroscopy (ESI-MS), Electron impact ionization mass spectroscopy (EI-MS) are generally used in metabolomics. Major advantage and disadvantages of these techniques are enumerated (Table 14.1). The blend of LC and GC has always been the favored method since both methods characterize non-volatiles and volatile metabolites, respectively.

## 14.2 Approaches in Metabolomics

Many scientific literature studies are focused on one or few metabolites prevalent in plant metabolism. Investigating a single metabolite or path at once, in contrast, is not just time demanding, but also ignores the system's overall reaction to treatments. The value of this shortcoming in the former technique, HTP (high throughput) metabolomics has progressed to include two approaches, namely targeted

**Table 14.1** Analytical techniques used in metabolomics along with advantages and disadvantages

| Sl. no. | Technique or analytical platform | Advantages | Disadvantages |
|---|---|---|---|
| 1 | Gas chromatography coupled with mass spectrometry (GC-MS) | Effective method for separation, fewer matrix effects faster scanning rates, low cost high chromatographic resolution, high reproducibility, high repeatability | Needs chemical derivatization (nonvolatile metabolites to volatile derivatives), stereo- isomers, alpha-beta anomers are not differentiated, chemical standards required to absolute quantification |
| 2 | Liquid chromatography coupled with mass spectrometry (LC-MS) | Identifies/distinguishes specific thermolabile metabolites, polar metabolites and high molecular weight compound don't need of derivatization, high selectivity, unbiased identification | Lack of automated procedure, every study group identifies own library, doesn't differentiate stereo-isomers, alpha-beta anomers |
| 3 | Ultra-high-performance liquid chromatography coupled with mass spectrometry (UHPLC-MS) | Identifies complex samples, higher efficiency, greater sensitivity, selectivity, high resolution, faster separating, and lower solvent consumptions | Analysis time is long |
| 4 | Liquid chromatography quadrupole-mass spectrometry (LC-Q-MS) | Quantitative and qualitative analysis, qualitative analysis used for obtaining structural information | Restricted for ions present in an exclusive mass unit range |
| 5 | Triple quadrupole liquid chromatography coupled to mass spectrometry (LC-QQQ-MS) | Accurate, more sensitive, comprehensive ion detection, identifies primary ions in addition to secondary ions, high reproducibility | Approves the channel of a specific mass ion category |
| 6 | Liquid chromatography quadrupole-time of flight-mass spectrometry (LC-Q-TOF-MS) | Larger and accurate mass resolution, higher detection sensitivity | Difficulty to set up high mass spectrum libraries, difficult to identify low abundance ions |
| 7 | Gas chromatography coupled to time-of-flight mass spectrometry (GC-TOF-MS) | Higher accuracy to measure mass, fast operating time | High cost |
| 8 | Matrix-assisted laser-desorption and ionization time-of-flight mass spectroscopy (MALDI-TOF-MS) | Fast and economical in reagent usage, less time to sample processing, flexible system to analyze proteins | Difficulty to set up high mass spectrum libraries, many libraries |
| 9 | CE-MS | Minimal sample preparation, high sensitivity does not require the derivatization, very fast, efficient, low | Stereo-isomers, alpha-beta anomers not differentiated |

(continued)

**Table 14.1** (continued)

| Sl. no. | Technique or analytical platform | Advantages | Disadvantages |
|---|---|---|---|
| | | consumption of sample, and reagents | |
| 10 | Direct infusion tandem mass spectroscopy (DI-MS/MS) | Minimalistic sample preparation, high throughput, rapid sample analysis, highly sensitive, good reproducibility, simple data analysis than LC and GC MS | Higher instrumental cost, competitive ionization, lack of variation between isomers, matrix effects |
| 11 | Nuclear magnetic resonance (NMR) | Simultaneous identification and quantification of metabolites, ease in sample preparation, non-destructive, high reproducibility, chemical standards not required, non-invasive structural analysis, and detect isotopes | High cost, small dynamic range, lower sensitivity, lower resolution, and detects only fewer metabolites |
| 12 | Fourier-transform ion cyclotron resonance mass spectrometry (FT-ICR-MS) | High accuracy at sub ppm concentrations | Absence of method validation, low scanning rates, higher cost |
| 13 | Electrospray ionization mass spectroscopy (ESI-MS) | Softest ionization technique, no matrix interference. Easily interfaced with LC, practical mass range up to 70,000 Da, multiple charging for better mass accuracy | Difficulty in cleaning. Low tolerance or mixture. Multiple charging leads to confusing |
| 14 | Electron impact ionization mass spectroscopy (EI-MS) | Well established, many fragmentation libraries. No suppression. Easily interfaced to GC. Suits for insoluble and non-polar samples | Difficulty in parent identification. Need volatile samples. Need thermal stability. No interface to LC. Only for low mass compounds |

metabolomics and non-targeted metabolomics (Goodacre et al. 2004; Patti et al. 2012).

## 14.2.1 Targeted Metabolomics

This approach includes identifying already known metabolites that has well-defined chemical structure and quantifying them using genuine chemical standards (Albinsky et al. 2010; Dudley et al. 2010). Targeted metabolomics is ineffective to analyze the role of an unknown gene (over or under expressed) in the biological system under investigation. This is the result of non-availability of a complete information of all metabolic disturbances that are affecting gene expression. Hence here the analysis is biased toward metabolites for which authentic standards are available (Kueger et al. 2012). Hence, this method can only analyze a limited quantity of metabolites. For this, GC or LC, along with triple quadrupole-mass spectrometry is the analytical platforms of choice (Lisec et al. 2006).

### *14.2.2 Non-targeted Metabolomics*

By using metabolic fingerprinting or metabolic profile, non-targeted metabolomics broadens its investigation from known to unknown compounds (Allwood et al. 2010). Non-targeted metabolomics yields raw data that is examined using multivariate (MVA) statistical methods to extract useful data about metabolic markers whose concentrations fluctuate significantly between samples. For non-targeted metabolomics, GC or LC combined with time-of-flight (TOF) is a typical analytical platform.

## 14.3 Metabolomic Approach Applications (Crop/Plant Research)

### *14.3.1 Phytochemical Diversity and Phenotyping*

Metabolite profiling in large number of genotypes will enable to group them into metabolic phenotypes and thus extend the usefulness in functional genomics ways. Furthermore, it also gives a biochemical blueprint of an individual genotype. Metabolite profile is also used to identify genotypes by the zone of cultivation (Anastasiadi et al. 2009).

### *14.3.2 Functional Genomics*

Variations in gene expression can be more closely assessed by using the metabolite profiling than the assessment of mRNA or proteins. Transcripts and proteins are altered as a result of environmental control of gene regulatory systems, and their levels may not accurately represent their product. Metabolite levels are more closely interrelated to the phenotype than DNA, RNA, or protein. Gene-to-metabolite networks can be used to find genes that encode enzymes that control metabolite levels (Moreno-Risueno et al. 2010). Transcript changes with the corresponding changes in the metabolite levels indicate a direct gene effect.

### 14.3.3 Environmental/Abiotic Stresses

Stresses affect the metabolic pathways in plants which is an adaptive mechanism to thrive under adverse conditions. Metabolite profiling can be used to quantify changes in these substances (Bowen and Northen 2010). There are two types of stress reactions in plants: primary and secondary. Primary events are mainly those that occur immediately as a direct result of the stress and can be adaptive or protective. Secondary events are time-dependent and are involved in dysfunction-related symptoms.

### 14.3.4 Crop Improvement

Plant metabolomic technology can be used to identify the genotypes with the trait of interest and can be directly used for crop improvement programs (Carreno-Quintero et al. 2013). Many agronomic traits are multigenic and involve quantitative trait loci, leading to the production of a metabolite or a group of metabolites. Although the genes encoding these quantitative characteristics (QTLs) are found on distinct chromosomes, their products mainly metabolites, interact together to determine a phenotype. Metabolomics, on the other hand, could be a good way to correlate phenotypes to QTLs (Kliebenstein 2007; Keurentjes 2009). Metabolite QTLs are specific areas of the genome that co-localize with variations in metabolite levels (mQTLs).

## 14.4 Abiotic Stresses and Metabolomic Responses

Abiotic stresses are defined as any variation in a plant's natural environment (water, temperature, light, nutrients, salts, and metal ions) that disrupts the plant's normal growth, physiology, and metabolism. Plants use an array of regulatory methods to flourish in the adverse environmental changes to establish themselves successfully in the new habitat. Plants' response to stressful environments are dynamic, through the deviations in the metabolite types and levels that subsidize to physiological and morphological adaptations (Muscolo et al. 2015). Plants acquire adaptations such as osmotic potential alterations, cell wall flexibility modifications, enzymatic and non-enzymatic antioxidant system mediated reactive oxygen species (ROS) scavenging, and controlled energy dissipation via regulated pigment biosynthesis. All the metabolic responses are intended toward tolerance, sensitivity, or avoidance toward stressful conditions. Plants need a broad array of metabolites (primary and secondary) to achieve all of the aforementioned changes for effective stress adaption (Fig. 14.1). Plants' primary metabolites are those that are engaged directly in their normal growth, overall development, and reproduction. Among the primary

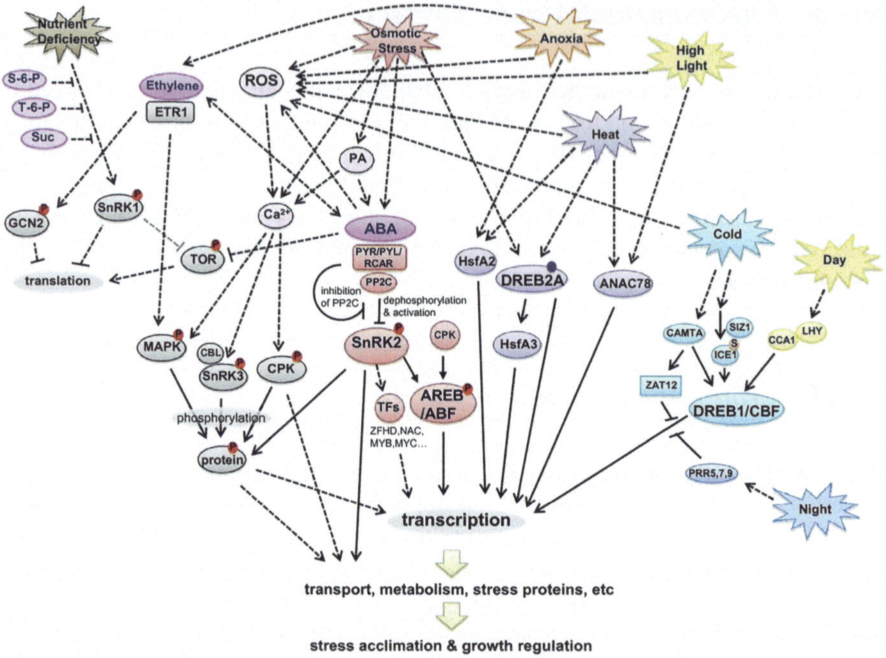


**Fig. 14.1** Basic signaling network model for plant responses to abiotic stresses (Cramer et al. 2011)

metabolites, sugars, amino acids, organic acids, fatty acids, and hormones are largely involved in stress response. Among the secondary metabolites, three types of secondary compounds are important: (1) compounds participating in acclimatization process, (2) molecules involved in signal transduction, (3) by-products from disrupted homeostasis due to stress. Small molecular weight organic osmoprotectants like amino acids (proline, serine, glutamine, and asparagine), amines (glycine betaine [GB] and polyamines), carbohydrates, and γ-aminobutyric acid (GABA) are among the metabolites conferring osmoprotection. GB reduces free energy level and protects the proteins (Fraire-Velázquez and Balderas-Hernández 2013). Non-enzymatic antioxidants (glutathione and ascorbate) and the carotenoid compounds involved in the dissipation of excitation energy are among the antioxidant metabolites. Changes among these metabolic products may be common to all stressors or unique to a single one (Muscolo et al. 2015). Metabolomics plays an important role in discovering new metabolites that help us in better understanding the mechanisms underpinning plant's regulative management of abiotic stresses, either by stress avoidance or stress tolerance strategies. The severity of stress injury is determined by the developmental period of plant during which it is exposed to stressful condition, besides the intensity and duration of the exposure. Agricultural sector is very badly affected by abiotic stressors, and the situation is worsening with progressive climate change. Abiotic stress tolerant varieties should be released to reduce the yield losses. Basic knowledge on essential metabolites and proteins required for growth and survival, their modification in response to diverse stressors, and the accurate interpretation of stress tolerance processes offer chances in creation of reliable biomarkers for plant abiotic stress tolerance. List of metabolites identified under various abiotic stresses are given in Table 14.2. The succeeding sections cover the unique modulations in numerous metabolites within plants for adaptation against various environmental stresses.

### *14.4.1 Drought Stress*

A major hindrance in achieving targeted food production over the decades is frequent occurrence of environmental stresses such as drought, heat, salinity, flooding, chilling, insects, pests, and so on (Verma and Deepti 2016). Drought stress dominates all of them with respect to frequency of occurrence, severity, and disturbance it creates in a plant's system. Insufficient rainfall and its erratic distribution causes water deficit and leads to drought stress (Chaves and Oliveira 2004; Bartels and Sunkar 2005). To survive, plants show a wide range of responses against drought and involve complex biological signals at cellular and molecular levels that ultimately are mediated by various metabolites (Wang et al. 2016). Any improvement program to overcome drought requires thorough knowledge on interactions between the metabolites synthesized and its linkage with the plant phenotype to mitigate harmful effects of drought.

#### 14.4.1.1 Osmolytes

Plant's capability to tolerate drought effects is governed by its capacity to maintain specific primary and secondary metabolites that regulate various tolerance responses. Ultra-performance liquid chromatography—high resolution mass spectrometry based untargeted profiling identifies, detects, quantifies, estimates, and analyzes metabolite changes under drought and discovered that amino acids such as tryptophan, proline, histidine, and isoleucine (Khan et al. 2019), organic acids, sugars—sucrose, trehalose, mannitol, and pinitol (Mibei et al. 2018), polyamines, nitrogenous compounds, polyphenols like caffeic, chlorogenic, *p*-coumaric, ellagic, gallic, and ferulic acids, flavonoids like rutin and apigenin (Kermani et al. 2019) accumulated in large quantities within root and shoot of drought-affected plants. These metabolites function in conferring osmoprotection, maintaining cell membrane integrity, and antioxidant activity, directly involved in lipid and carbohydrate biosynthesis, which induces the overall protection against drought.

Osmoadaptation is the initial response in plant system to confer early tolerance against drought, which involves rapid accumulation of compatible solutes accelerating cellular osmotic pressure and increasing water potential gradient thereby

**Table 14.2** Metabolites identified under different abiotic stress conditions

| Abiotic stress | Species name | Metabolites identified | Reference |
|---|---|---|---|
| Drought stress | *Zea mays* | Pyruvic acid, succinic acid, phenylalanine, adenine, and putrescine | Witt et al. (2012) |
| | *Triticum* spp. | Proline, sucrose, fructose, mannose, malic acid, glutamate, γ-aminobutyric acid | Guo et al. (2018) |
| | *Arabidopsis thaliana* | Acetic acid, jasmonic acid | Kim et al. (2017) |
| | *Oryza sativa* cv. *Nipponbare* | Glucose, erythritol/threitol, galactinol, raffinose, GABA, lactic acid dimer, salicylic acid, and 2-amino-adipic acid | Todaka et al. (2016a, b) |
| | *Triticeae* | Sucrose, mannitol, trehalose, maltose, glutamate, organic acids (succinate, citrate, aspartate, and gluconate), proline alanine, glycine, methionine, threonine, asparagines, phenylalanine, homocysteine, serine, valine, and tyrosine | Ullah et al. (2017) |
| | *Arabidopsis thaliana* | Phenylalanine, tryptophan, isoleucine, leucine, valine, galactose, glucose, maltose, mannose, sucrose nitrate and starch | Pires et al. (2016) |
| | *Arabidopsis thaliana* | Anthocyanins and flavonoids | Nakabayashi et al. (2014) |
| | *Thellungiella salsuginea* | Spermidine and proline | Lugan et al. (2010) |
| | *P. pinaster* | Ornithine, spermidine, galactinol, and proline | Gechev et al. (2013) |
| | *Zea mays* | Tyrosine, proline, myoinositol, and succinate | Alla et al. (2012) |
| Heat stress | *Eucalyptus globulus* | α-Glycerophosphate, shikimate, salicylic acid, jasmonic acid, mannitol, sorbitol, inositol succinate, malate, citrate, quinate, and glycerate | Correia et al. (2018) |
| | *Lycopersicon esculentum Mill.* cv. *Puhong 968* | Spermidine | Sang et al. (2017) |
| | *Zea mays* | Isoleucine, valine, threonine, 4-aminobutanoate, glycine, serine, and myoinositol | Obata et al. (2015) |
| | *Solanum lycopersicum* | Alanine, allantoil, arachidic acid, 2-ketoisocaproic acid, putrescine, fructose | Luengwilai et al. (2012) |
| | *Nicotiana tabacum* | Proline, putrescine, spermine | Cvikrová et al. (2012) |
| | *Lycopersicon esculentum* | Putrescine and spermine | Cheng et al. (2009) |
| | *Arabidopsis thaliana* | Phosphatidylethanolamine and monogalactosyldiacylglycerol | Degenkolbe et al. (2012) |
| | *Glycine max* | Proline, fructose, flavanoids, and tocopherols | Chebrolu et al. (2016) |

(continued)

**Table 14.2** (continued)

| Abiotic stress | Species name | Metabolites identified | Reference |
|---|---|---|---|
| | *Spinacia oleracea* | Zeaxanthin | Latowski et al. (2011) |
| | *Lactuca sativum* | Quercetin 3-(6″-malonyl-glucoside) | DuPont et al. (2000) |
| Salt stress | *Oryza sativa* | Proline, ornithine, aspartate, and methionine | Liu et al. (2013) |
| | *Arabidopsis thaliana* | γ-aminobutyric acid (GABA) | Renault et al. (2010) |
| | *Triticum turgidum* ssp. *Durum* | γ-aminobutyric acid (GABA) proline, threonine, leucine, 43 glutamic acid, glycine, mannose, and fructose | Borrelli et al. (2018) |
| | *Hordeum* spp. | Alanine, asparagine, 4-hydroxy-proline, arginine, citrulline, glutamine, phenylalanine, proline, succinate, 2-oxoglutarate, fumarate, maleate, and malate | Cao et al. (2017) |
| | *Triticum durum* cv. *Ofanto* | Glycine betaine, proline, GABA, asparagine, glutamate | Woodrow et al. (2016) |
| | *L. latifolium* | Cyclitols, β-AB, citrate, and aspartate | Gagneul et al. (2007) |
| | *Nitraria tangutorum* | Alanine, glutamate, acetamide, and malate | Ni et al. (2015) |
| | *Hardium vulgare* | Leucine and pyruvate, glutamate, leucine, myoinositol, organic acids, succinate, citrate, pyruvate, alanine | Wu et al. (2013), Chen et al. (2007) |
| | *Nicotiana tabacum* | Hypoxanthine, phytate, allantoin, GABA, and alanine | Zhang et al. (2011) |
| | *Porteresia coarctata* | D-Pinitol | Sengupta et al. (2008) |
| Flooding | *Cajanus cajan* | Starch | Pearson et al. (2013) |
| | *Pinus sylvestris* | Fructose | Pearson et al. (2013) |
| | *Prunus persica* | Spermidine | Tuo et al. (2015) |

facilitating continued water uptake via osmosis. GC-TOF-MS profiling of primary metabolites and LC-DAD profiling of secondary metabolites reveal that levels of kaempferol 3-*O*-diglycoside, quercetin 3-*O*-6′′-malonylglycoside, galactinol, quercetin, hydroxybenzoate, proline, and threonine increased several folds under drought. These compounds are involved in giving significant tolerance through maintaining osmotic adjustment, membrane stabilization, redox buffering, radical scavenging, signaling, protein hydrotroping, nitrogen, and carbon storage. These genotypes were found to maintain yields over drought conditions indicating these molecules like quercetin 3-*O*-6′′-malonylglycoside, proline, and galactinol, can be employed as promising biomarkers for yield performance during drought (Goufo et al. 2017).

Elevated $CO_2$ has a positive role in alleviating negative impacts of drought stress in cucumbers. It was observed that mechanism underlying high $CO_2$ induced stress tolerance involves alteration of metabolite composition of plants, which was confirmed by UHPLC-Q-TOF MS technique. Isoleucine, alanine, arginine, isoferulic acid, salicylic acid, *m*-coumaric acid, ethisterone capric acid, succinic acid semialdehyde, L-glutamate, L-gluconolacton, (*R*)-mevalonic acid 5-phosphate, pyrocatechol isomaltose, 1,2,3-trihydroxybenzene, trehalose, D-maltose, L-fucose, raffinose, and stachyose contents increased rapidly in response to elevated $CO_2$ concentrations and plants could sustain the drought stress for a longer duration (Li et al. 2018).

#### 14.4.1.2 Hormones

Several hormones mediate the tolerance response of plants under drought conditions and the hormone induced responses involve a complex cross-talk with plant primary and secondary metabolites resulting in their alteration ultimately protecting plants against drought. Jasmonic acid enhances expression of certain metabolites like phenolic phytoalexins, ascorbate peroxidase, alkaloids, taxanes, terpenoids, and coumarins (Van der Fits and Memelink 2000). Auxins increase flavonoid and aliphatic glucosinolate accumulation which help in stomatal closure and elevate drought responses by involving cross-talk with other hormones (Salehin et al. 2019). Cytokinins upregulate osmolytes like γ-aminobutyric acid, proline, glycine, alanine, sucrose, fructose, maltose, ribose, and organic acids which help in osmoprotection and elevating stress-responsive messengers ultimately protecting plants from stress damage under drought. Salicylic acid application ameliorates drought stress damages by positively regulating biosynthesis of terpenes, glyceric acid, succinate, leucine, disaccharide, saccharic acid, and phenolic compounds which helps in ameliorating drought stress (Jogawat et al. 2021).

#### 14.4.1.3 Gene-Metabolite Linkage

Transcriptome validation enables us to link the genes that encode different metabolites responsible for governing tolerance against drought. Flavonoids, amino acids, peroxidase, SOD, mevalonate derivatives, fatty acids, and glycans mediate drought tolerance response by increasing antioxidant activity, maintaining cell genome integrity during stressful conditions, and modulating hormonal profile. Genes such as α-galactosidase, β-galactosidase, galactinol synthase, peroxygenase 4, flavonoid glucosyltransferase, and flavonol synthase, ABC sub-family G, and NRT1/PTR family genes and many others were expressed several folds higher in drought-affected onion (Ghodke et al. 2020).

### *14.4.2 Heat Stress*

Flavonoids act as the most bioactive secondary metabolites in plants under stress and serve as reactive oxygen species (ROS) scavengers. Among them, anthocyanins are the key compounds having multiple roles as ultraviolet and high light irradiation absorbents, antioxidants that scavenge ROS, and agents acting against microorganisms in defense responses. TT18, UGT79B2, and UGT79B3 are the important genes that participate in proanthocyanin biosynthetic pathway. *Arabidopsis* plants overexpressing these genes show enhanced tolerance against drought and this response was mainly because of amplified anthocyanin levels, and mutants lacking these genes were unable to produce anthocyanin, thus susceptible to drought injury. This indicates the close association between anthocyanin accumulation and drought (Li et al. 2017).

Increasing global warming due to changing climate is a prime reason for elevated temperatures, and it is predicted that the ongoing levels of anthropogenic activities lead to acute rise in annual daily maximum temperature, which would be elevated by approximately 1–3 °C by 2050. Along with this, increasing depletion of ozone across the globe further raises temperature of the earth and in turn increases the amount of ultraviolet radiation reaching the earth's surface (Barnes et al. 2019). These changes all together increase micro-climate temperature and pose a detrimental environment for cropping, leading to heat injuries. Plants are constantly adapting to these changing environments involving extensive variation in their metabolic activities, to tolerate heat stress caused by excess temperatures.

#### 14.4.2.1 Pollen Metabolites

Metabolomics helps us to precisely identify, quantify, validate, and figure out the interactions within the plants at a metabolic level which directly or indirectly assist in plant protection under high temperatures. Heat stress is highly injurious during early stages of crop development as in germination and final stages of reproductive growth and seed development. It hinders pollen-stigma interactions and reduces fertilization and fruit set. To overcome these ill effects of high temperature, plants produce many metabolites, mostly flavonoids (flavonols and anthocyanins) that act as antioxidants, which increase the plasticity of the pollen wall and maintain pollen viability. Pollen tube analysis of tomatoes under high-temperature conditions shows improved buildup of flavanols and thus level of pollen flavanols can be considered as biomarkers for enhanced pollen thermotolerance and reproductive achievement in heat stress conditions (Rutley et al. 2021).

Gas chromatography mass spectrometry (GC-MS) based non-targeted metabolite profiling of heat stress affected tomato flowers revealed the function of HSP's (heat shock proteins) in conferring thermotolerance during another development and pollen tube growth by increasing the amount of putrescine, galactonic acid, succinic acid, and γ-amino butyric acid. Excess accumulation of phenylpropanoids also helps in acclimation of floral cells against heat stress (Fragkostefanakis et al. 2016).

#### 14.4.2.2 Membrane Protection

Heat stress disturbs typical functioning of plant by disrupting photosynthesis, respiration, hormone signaling, oxidative injury to membranes, and cause cell lysis in quick succession. This necessitate the plant to act as a whole to mitigate heat stress damage and can only be accomplished by brisk changes in metabolites that are part of important pathways, altering other associated plant mechanisms to prevent adverse effects. Metabolites are associated with maintaining ion homeostasis, vital metabolism, and antioxidant activities. Untargeted metabolic profiling with GCMS and LC-MS shows differential production of ribose, xylose, deoxyribose, gluconate, xylitol, alanine, lysine, isoleucine, methionine, daidzein, glycitin, daidzin, formononetin, syringic acid, genistein, and genistin in soybean during heat stress conditions, and regulate glycolysis, pentose phosphate pathway, TCA cycle, starch biosynthesis, and nitrogen metabolism helping plants to sustain excess heat (Das et al. 2017).

Non-targeted metabolome profiling has revealed that many metabolites specifically accumulate in plants under heat stress environments, such as polyols, myo-inositol, putrescine, glycerol, cysteine, serine, citrulline, glutamine, homocitrulline, alanine, threonine, ornithine, uridine diphosphate glucose, cytokinin, coumaryl alcohol, α L-rhamnose, root ginsenosides, and protopanaxatriols (Jochum et al. 2007; Wada et al. 2020; Wei et al. 2020). These metabolites mediate heat tolerance by enhancing membrane permeability, ROS scavenging enzymes synthesis, elevating hormonal signaling, production of other metabolites, and stabilizing photosynthetic pigments.

Higher night temperatures are harmful to germination, seedling growth and plant development and heat-tolerant plant species are capable of withstanding heat stress by changing the levels of specific metabolites. Quinic acid, shikimic acid, tyrosine, galactinol, and glucose-6-phosphate increased under high night temperatures and also showed a progressive increase as stress conditions prolonged, whereas sugars like threitol, xylitol, acetol, and trehalose content remained constant at stressed and non-stressed conditions, but the levels of lysine, myo-inositol, arbutin, xylose, citric acid, glyceric acid, and glycolic acid decreased as the night temperature increased. This reveals the role of various metabolites at distinct stages and different concentrations in providing heat tolerance (Dhatt et al. 2019).

#### 14.4.2.3 Gene-Metabolite Linkage

Transcriptomics based investigation of genes engaged in metabolite biosynthesis in heat stress conditions confirm the role of metabolites in alleviating heat stress and providing acclimation. Polyamines such as putrescine, spermine, spermidine, cadaverine, thermospermine, norspermine, and norspermidine are responsible in promoting photosynthesis, increasing the antioxidant capacity and osmotic adjustment in plants under high temperatures (Chen et al. 2019). Under high temperatures of 42 °C, genes regulating polyamine biosynthesis pathway, namely *SlARG1/2*, *SlAIH,* and *SlCPA* started increasing with stress duration and reached maximum level within 24 h in tomato leaves. These genes were also positively controlled by the production of HSPs. This reveals the possible interaction between hormones, proteins, transcription factors, and metabolites that act simultaneously in alleviating heat stress injuries (Upadhyay et al. 2020).

## *14.4.3 Salinity Stress*

Soil salinity is becoming a serious problem in recent years affecting crop growth in large areas. Presently, 23 million hectare of arable land is salt-affected, and is projected to increase to over 50% of the arable land leading to severe pressure on food production (Kumar and Sharma 2020; Mukhopadhyay et al. 2021). As the present efforts to reclaim the salt-affected soils is not enough, we need to develop salt tolerant and resistant varieties that can grow normally without being affected by excessive salts. Lack of complete knowledge about the plant metabolism dynamics under salinity conditions is affecting our approach in developing salt tolerant varieties. In this regard, metabolomics approach offers us a broad range of opportunities in understanding and unraveling the several stress-responsive changes in metabolites that confer resistance to salt stress.

### 14.4.3.1 Antioxidants

In this dominating era of omics techniques for high throughput studies of genes, transcripts, proteins, and metabolites, metabolomics helps in understanding metabolome pool which is the final step in gene expression. Thus, it helps in deciphering key tolerance mechanisms against salinity stress in plants. Several metabolites that serve as osmoprotectants were found to play a direct role in conferring resistance to salt (Shulaev et al. 2008). Recently, metabolomics studies in tomato plants irrigated with 200 mM NaCl showed that there was a rapid rise in phenolic and saponins content. These metabolites serve in elevating the hormone actions and thereby imparting an improved tolerance response against salinity (Abdel-Farid et al. 2020).

Targeted metabolomic analysis in leaves of salt tolerant and susceptible rice varieties has shown surprising results in terms of increased serotonin and gentisic acid content in salt tolerant varieties, which are modern-day hormone and signaling molecules respectively, mainly found in animals. Serotonin acts against ROS scavengers, delays senescence, and alters signaling responses with its derivative metabolite melatonin to combat stress and ill effects of NaCl. Gentisic acid induces defense proteins, elevates salicylic acid intermediates, and can act in a complementary way to salicylic acid under stressed conditions to overcome its ill effects. An elevated level of these two metabolites along with their relative components ferulic acid and vanillic acid govern resistance against salinity in tolerant rice varieties and thus can be considered as biomarkers for NaCl-associated stress (Gupta and De 2017).

Metabolome characterization of *Lonicera japonica Flos* (LJF), a perennial twining woody vine by Cai et al. (2020) revealed that relative content of caffeoylquinic acid (CQA) derivatives like chlorogenic acid, isochlorogenic acid A, neochlorogenic acid, 1,3-*O*-dicaffeoylquinic acid, 1,4-*O*-dicaffeoylquinic acid; flavonoids like quercetin, luteoloside, lonicerin, luteolin-5-*O*-β-D-glucopyranoside, flavoyadorinin-B, and iridoids such as loganin, swertiamarine, vogeloside, and secologanin increased in low salt-treated samples of LJF ($<$100 mM NaCl). At medium salt stress (200 mM NaCl), these compounds were further increased by 0.5–3 folds in supplement to the increase in other metabolites like hyperoside and genistein. But at higher salt concentration of 300 mM NaCl, there was a sharp decline in these compounds compared to control and in contrast iridoids, namely centauroside and secologanin increased profusely. These shikimic acid pathway compounds exhibit antibacterial, antiviral, anti-inflammatory, cytoprotective, immunomodulatory, and antihyperglycemic responses, and are responsible for scavenging the ROS and protecting cellular membrane's integrality under excess salts. These metabolites can be employed as markers of stress tolerance and would be helpful in efficient, effective, and rapid screening of plants for salinity tolerance.

Salt-induced phenotypic damages to plants at specific cell functions can be understood by targeted metabolomics (Pan et al. 2020). Flavonoids such as anthocyanins, flavonols, flavones, and flavanones, namely cyanidin 3-*O*-glucoside, cyanidin 3,5-*O*-diglucoside, and cyanidin 3-*O*-rutinoside, that are involved in color reaction in roots that indirectly cause an intrinsic response against salt stress were increased by 3–7 folds. And in contrast metabolites involved in biomass production such as carbohydrates, amino acids and their derivatives, nucleotides, and vitamins were decreased in foxtail millet at 150 mM NaCl salt concentration.

#### 14.4.3.2 Osmoprotectants

Halophytic plants that survive toxic salt concentrations in soil show increased accumulation of numerous metabolites which include organic acids, mainly citric acid, galacturonic acid, gluconic acid, and hexonic acid, sugars like glucose, glycerol, fructose, xylitol, D-mannitol, and turanose and amino acids phenylalanine and tyrosine. These compounds maintain cellular ionic homeostasis, act as precursors for other secondary metabolites, maintain energy requirement, and stabilize sugar transport and thus responsible for early tolerance response against salinity (Panda et al. 2021). It is interesting to understand the response of the biosynthetic pathways of these metabolites to salinity stress, so that we can develop clear strategies to link them to the QTLs.

#### 14.4.3.3 Gene-Metabolite Linkage

Sugar beet is known to have very high salinity tolerance and understanding its mechanism gives an insight into the cellular responses to salinity. Differentially expressed metabolites were studied in sugar beet to understand the metabolomic changes leading to the salinity acclimation. Amino acids L-glutamine and L-asparagine were significantly increased during the early phases of stress whereas, in later stages, organic acids, namely benzoic acid, cis-aconitate, L-malic acid, and alpha-ketoglutarate were increased. Majority of lipids also increased at the end of the stress period. Besides these many vitamins, sugars increased as the stress progressed. Along with these metabolites, transcriptome analyses also revealed the differential expression of many gene. Many of these DEGs were initially upregulated and then downregulated, whereas few showed opposite trend, indicating that transcriptomic responses for tolerance against salinity are highly time-specific. Among them, overexpressed genes during initial and later stages of the stress (first and seventh day) were mainly responsible for imparting salinity tolerance and they included genes for LEA proteins, expansin-like B, 2-hydroxyflavanone dehydratase, GDSL esterase/lipase (At5g55050), spermine synthase, ureide permease 2, polygalacturonase, auxin responsive protein IAA29, and BAG family molecular chaperone regulator-6 (Liu et al. 2020).

## 14.5 Metabolomics-Assisted Crop Improvement

Increased demand for food and fodder necessitates the continuous development of cultivars which can thrive environmental perturbations and give better yield and quality produce. Faster, high predictable, and cost-effective breeding methods are required in new agriculture to expand crop yields and quality (Khakimov et al. 2014). This necessitates the need of high-throughput analytical and cutting-edge techniques like metabolomics for crop breeding to expand crop yields during stress conditions, genetic modification, and biomarker development to evaluate the quality of final products.

Using metabolites as major plant phenotype has many rewards. Metabolomics is useful tool in analyzing linkages between phenotypes and QTLs, mQTLs, Whole Genome Association Studies (GWAS) using mQTL (mGWAS), to detect and determine the modifications in metabolic adaptation during various stresses (Beckles and Roessner 2012; Templer et al. 2017). Discovery of mQTLs/mGWAS through metabolite profiling have strengthened the status of metabolomics in metabolic marker-assisted plant breeding activities. Quite a few mQTLs/mGWAS have found association of genomic area and metabolic content in several crop species. To mention few such examples, carotenoids in tomato, kernel composition, vitamin-A content, starch content, lignocellulosic biomass quality, phytate in mustard, secondary metabolite composition in rice, and multi-traits in sugar beets. Usage of mQTL/mGWAS to detect fruit size, taste, color, and nutrient level in tomato has evidenced that the metabolomics approach is a useful technique/tool to know how plants cope with various environmental stresses. The results thereby assisted breeders in improving crop tolerance to various environmental stresses. Thus, metabolomic-assisted varietal development meaningfully enhances the precision and efficiency of plant breeding (Christ et al. 2018; Fraser et al. 2007; Harrigan et al. 2007; Hill et al. 2015; Matsuda et al. 2012; Schauer et al. 2006).

The complete examination of metabolites linked with plant growth and development, and the responsiveness to diverse stress circumstances, may lead to find metabolic biomarkers for specific environmental conditions or plant growth phases. This study is critical during introduction of chosen biological character to a crop plant to boost yield or sustainability of production (Salekdeh and Komatsu 2007). However, metabolic biomarker discovery necessitates widespread use of univariate in addition for multivariate data analysis of metabolic data. In rice, GC-MS-based non-targeted metabolic profiling has been done to distinguish a set of biomarkers to various developing periods with their role during pest interaction (Agarrwal et al. 2014; Tarpley et al. 2005). Tomato flesh and seeds exhibited modified metabolic composition during all its developmental stages, an observation that was helpful in stage-specific breeding activities. Salinity stress adaptive metabolites of cultivated and wild soybeans indicated the major biomarker sets comprised of fatty acids, sugars, amino acids, and organic acids (Li et al. 2019). Identifying stage specific and stress specific metabolite biomarkers will help for a targeted approach in crop breeding activities. Furthermore, the amalgamation of data from genomics and metabolomics yields fresh insights into gene annotation. The integrated omics approach has been beneficial in identifying the genomic regions regulating the quality and quantity of metabolites. (Abdelrahman et al. 2019; Fridman and Pichersky 2005).

## 14.6 Conclusion

Metabolome research using modern metabolomics methods has revealed as viable tool for studying plant metabolic adaptations to principal stress conditions. The metabolome reflects the mixture of genetic background and environmental influences, ensuing in a more precise description of the phenotype of a specific plant species. Plants orchestrate various responses to harmful abiotic stimuli aimed at stress avoidance, defense, or resistance, depending on stress tolerance. Among all metabolic reactions that occur in plants, changes in plant primary metabolism under different conditions are most visible and measured easily. The changes involve variations in sugar, amino acids, sugar alcohol levels, along with photosynthesis, respiration, and tricarboxylic acid cycle intermediates in response to abiotic stress. On contrary, variations in secondary metabolism, are more species-specific and extremely exclusive to the particular stress scenario.

The metabolomic method has become increasingly promising in categorizing the processes by which plants adapt to endure in stressful situations due to combination of diverse approaches. To comprehend the networking and harmonization of metabolic ways, metabolomics data must be integrated with data from other "omics" fields like genomes, transcriptomics, and proteomics. Plant biologists can benefit from the finding of metabolite biomarkers in two ways: (1) better study of regulatory factors disrupting plant metabolism under different stress conditions and (2) developing strategies based on metabolomics data to improve stress tolerance by metabolic engineering or breeding. Development of mQTLs for crop assortment for abiotic stress tolerance would aid in resolving the issues arising from varying environmental circumstances. Recently released plant genome information, along with newer and more sophisticated metabolite profiling methods, will be useful for faster crop improvement. Utilization of the growing area of metabolomics, along with mQTL/mGWAS will pave a way for better choice/selection of elite breeding lines. The finding of metabolic markers in plant stress response shows the metabolomics-assisted crop improvement program is the key to future food security and the developing of plants that can tolerate severe environments. Metabolomic profiling in agricultural plants are used to generate new crop improvement techniques in the near future. Establishing a metabolite pathway which specifically responds under a given stress will help in selection of suitable genotypes for stress tolerance breeding programs. Single metabolite usage as a marker is almost ruled out since the stress response is multi-trait controlled. Therefore, looking into the importance of a stress specific response like, osmolytes or ROS scavenging, membrane and protein protection, signal transduction, cell homeostasis, should be selected for identifying the tolerant genotypes and linking the metabolites with the phenotype.

## References

Abdel-Farid IB, Marghany MR, Rowezek MM, Sheded MG (2020) Effect of salinity stress on growth and metabolomic profiling of Cucumis sativus and Solanum lycopersicum. Plants (Basel) 9(11):1626

Abdelrahman M, Hirata S, Sawada Y, Hirai MY, Sato S, Hirakawa H, Mine Y, Tanaka K, Shigyo M (2019) Widely targeted metabolome and transcriptome landscapes of Allium fistulosum–A. cepa chromosome addition lines revealed a flavonoid hot spot on chromosome 5A. Sci Rep 9(1): 1–15

Agarrwal R, Bentur JS, Nair S (2014) Gas chromatography mass spectrometry based metabolic profiling reveals biomarkers involved in rice-gall midge interactions. J Integr Plant Biol 56(9): 837–848

Albinsky D, Kusano M, Higuchi M, Hayashi N, Kobayashi M, Fukushima A, Mori M, Ichikawa T, Matsui K, Kuroda H, Horii Y (2010) Metabolomic screening applied to rice FOX Arabidopsis lines leads to the identification of a gene-changing nitrogen metabolism. Mol Plant 3(1): 125–142

Alla MMN, Khedr AHA, Serag MM, Abu-Alnaga AZ, Nada RM (2012) Regulation of metabolomics in Atriplexhalimus growth under salt and drought stress. Plant Growth Regul 67(3):281–304

Allwood JW, Clarke A, Goodacre R, Mur LA (2010) Dual metabolomics: a novel approach to understanding plant–pathogen interactions. Phytochemistry 71(5–6):590–597

Anastasiadi M, Zira A, Magiatis P, Haroutounian SA, Skaltsounis AL, Mikros E (2009) 1H NMR-based metabonomics for the classification of Greek wines according to variety, region, and vintage. Comparison with HPLC data. J Agric Food Chem 57(23):11067–11074

Barnes PW, Williamson CE, Lucas RM, Robinson SA, Madronich S, Paul ND, Bornman JF, Bais AF, Sulzberger B, Wilson SR, Andrady AL (2019) Ozone depletion, ultraviolet radiation, climate change and prospects for a sustainable future. Nat Sustain 2(7):569–579

Bartels D, Sunkar R (2005) Drought and salt tolerance in plants. Crit Rev Plant Sci 24(1):23–58

Barupal DK, Fiehn O (2017) Chemical similarity enrichment analysis (ChemRICH) as alternative to biochemical pathway mapping for metabolomic datasets. Sci Rep 7(1):1–11

Beckles DM, Roessner U (2012) Plant metabolomics: applications and opportunities for agricultural biotechnology. In: Plant biotechnology and agriculture. Academic, Amsterdam, pp 67–81

Borrelli GM, Fragasso M, Nigro F, Platani C, Papa R, Beleggia R, Trono D (2018) Analysis of metabolic and mineral changes in response to salt stress in durum wheat (Triticum turgidum ssp. durum) genotypes, which differ in salinity tolerance. Plant Physiol Biochem 133:57–70

Bowen BP, Northen TR (2010) Dealing with the unknown: metabolomics and metabolite atlases. J Am Soc Mass Spectrom 21(9):1471–1476

Cai Z, Chen H, Chen J, Yang R, Zou L, Wang C, Chen J, Tan M, Mei Y, Wei L, Yin S (2020) Metabolomics characterizes the metabolic changes of Lonicerae Japonicae Flos under different salt stresses. PLoS One 15(12):e0243111

Cambiaghi A, Ferrario M, Masseroli M (2017) Analysis of metabolomic data: tools, current strategies and future challenges for omics data integration. Brief Bioinform 18(3):498–510

Cao D, Lutz A, Hill CB, Callahan DL, Roessner U (2017) A quantitative profiling method of phytohormones and other metabolites applied to barley roots subjected to salinity stress. Front Plant Sci 7:2070

Carreno-Quintero N, Bouwmeester HJ, Keurentjes JJ (2013) Genetic analysis of metabolome–phenotype interactions: from model to crop species. Trends Genet 29(1):41–50

Chaudhary J, Khatri P, Singla P, Kumawat S, Kumari A, Vikram A, Jindal SK, Kardile H, Kumar R, Sonah H, Deshmukh R (2019) Advances in omics approaches for abiotic stress tolerance in tomato. Biology 8(4):90

Chaves MM, Oliveira MM (2004) Mechanisms underlying plant resilience to water deficits: prospects for water-saving agriculture. J Exp Bot 55:2365–2384

Chebrolu KK, Fritschi FB, Ye S, Krishnan HB, Smith JR, Gillman JD (2016) Impact of heat stress during seed development on soybean seed metabolome. Metabolomics 12:1–14

Chen Z, Cuin TA, Zhou M, Twomey A, Naidu BP, Shabala S (2007) Compatible solute accumulation and stress-mitigating effects in barley genotypes contrasting in their salt tolerance. J Exp Bot 58(15–16):4245–4255

Chen D, Shao Q, Yin L, Younis A, Zheng B (2019) Polyamine function in plants: metabolism, regulation on development, and roles in abiotic stress responses. Front Plant Sci 9:1945

Cheng L, Zou Y, Ding S, Zhang J, Yu X, Cao J, Lu G (2009) Polyamine accumulation in transgenic tomato enhances the tolerance to high temperature stress. J Integr Plant Biol 51:489–499

Christ B, Pluskal T, Aubry S, Weng JK (2018) Contribution of untargeted metabolomics for future assessment of biotech crops. Trends Plant Sci 23(12):1047–1056

Correia B, Hancock RD, Amaral J, Gomez-Cadenas A, Valledor L, Pinto G (2018) Combined drought and heat activates protective responses in Eucalyptus globulus that are not activated when subjected to drought or heat stress alone. Front Plant Sci 9:819

Cramer GR, Urano K, Delrot S, Pezzotti M, Shinozaki K (2011) Effects of abiotic stress on plants: a systems biology perspective. BMC Plant Biol 11(1):1–14

Cvikrová M, Gemperlová L, Dobrá J, Martincová O, Prásil IT, Gubis J, Vanková R (2012) Effect of heat stress on polyamine metabolism in proline-over-producing tobacco plants. Plant Sci 182: 49–58

Das A, Rushton PJ, Rohila JS (2017) Metabolomic profiling of soybeans (*Glycine max L.*) reveals the importance of sugar and nitrogen metabolism under drought and heat stress. Plants 6(2):21

Degenkolbe T, Giavalisco P, Zuther E, Seiwert B, Hincha DK, Willmitzer L (2012) Differential remodelling of the lipidome during cold acclimation in natural accessions of *Arabidopsis thaliana*. Plant J 72:972–982

Dhatt BK, Abshire N, Paul P, Hasanthika K, Sandhu J, Zhang Q, Obata T, Walia H (2019) Metabolic dynamics of developing rice seeds under high night-time temperature stress. Front Plant Sci 10:1443

Dudley E, Yousef M, Wang Y, Griffiths WJ (2010) Targeted metabolomics and mass spectrometry. Adv Protein Chem Struct Biol 80:45–83

DuPont MS, Mondin Z, Williamson G, Price KR (2000) Effect of variety, processing, and storage on the flavonoid glycoside content and composition of lettuce and endive. J Agric Food Chem 48:3957–3964

Fiehn O, Wohlgemuth G, Scholz M, Kind T, Lee DY, Lu Y, Moon S, Nikolau B (2008) Quality control for plant metabolomics: reporting MSI-compliant studies. Plant J 53(4):691–704

Fragkostefanakis S, Mesihovic A, Simm S, Paupière MJ, Hu Y, Paul P, Mishra SK, Tschiersch B, Theres K, Bovy A, Schleiff E (2016) HsfA2 controls the activity of developmentally and stress-regulated heat stress protection mechanisms in tomato male reproductive tissues. Plant Physiol 170(4):2461–2477

Fraire-Velázquez S, Balderas-Hernández VE (2013) Abiotic stress in plants and metabolic responses. In: Abiotic stress—plant responses and applications in agriculture. InTechOpen, Rijeka, pp 25–48

Fraser PD, Enfissi EM, Goodfellow M, Eguchi T, Bramley PM (2007) Metabolite profiling of plant carotenoids using the matrix-assisted laser desorption ionization time-of-flight mass spectrometry. Plant J 49(3):552–564

Fridman E, Pichersky E (2005) Metabolomics, genomics, proteomics, and the identification of enzymes and their substrates and products. Curr Opin Plant Biol 8(3):242–248

Gagneul D, Aïnouche A, Duhazé C, Lugan R, Larher FR, Bouchereau A (2007) A reassessment of the function of the so-called compatible solutes in the halophytic *Plumbaginaceae Limonium latifolium*. Plant Physiol 144:1598–1611

Gechev TS, Benina M, Obata T, Tohge T, Sujeeth N, Minkov I (2013) Molecular mechanisms of desiccation tolerance in the resurrection glacial relic *Haberlea rhodopensis*. Cell Mol Life Sci 70:689–709

Ghodke P, Khandagale K, Thangasamy A, Kulkarni A, Narwade N, Shirsat D, Randive P, Roylawar P, Singh I, Gawande SJ, Mahajan V (2020) Comparative transcriptome analyses in contrasting onion (Allium cepa L.) genotypes for drought stress. PLoS One 15(8):e0237457

Goodacre R, Vaidyanathan S, Dunn WB, Harrigan GG, Kell DB (2004) Metabolomics by numbers: acquiring and understanding global metabolite data. Trends Biotechnol 22(5):245–252

Goufo P, Moutinho-Pereira JM, Jorge TF, Correia CM, Oliveira MR, Rosa EA, António C, Trindade H (2017) Cowpea (*Vigna unguiculata L. Walp.*) metabolomics: osmoprotection as a physiological strategy for drought stress resistance and improved yield. Front Plant Sci 8:586

Guo R, Shi L, Jiao Y, Li M, Zhong X, Gu F, Liu Q, Xia X, Li H (2018) Metabolic responses to drought stress in the tissues of drought-tolerant and drought-sensitive wheat genotype seedlings. AoB Plants 10(2):ply016

Gupta P, De B (2017) Metabolomics analysis of rice responses to salinity stress revealed elevation of serotonin, and gentisic acid levels in leaves of tolerant varieties. Plant Signal Behav 12(7): e1335845

Harrigan GG, Martino-Catt S, Glenn KC (2007) Metabolomics, metabolic diversity and genetic variation in crops. Metabolomics 3(3):259–272

Hill CB, Taylor JD, Edwards J, Mather D, Langridge P, Bacic A, Roessner U (2015) Detection of QTL for metabolic and agronomic traits in wheat with adjustments for variation at genetic loci that affect plant phenology. Plant Sci 233:143–154

Jochum GM, Mudge KW, Thomas RB (2007) Elevated temperatures increase leaf senescence and root secondary metabolite concentrations in the understory herb Panax quinquefolius (Araliaceae). Am J Bot 94(5):819–826

Jogawat A, Yadav B, Lakra N, Singh AK, Narayan OP (2021) Crosstalk between phytohormones and secondary metabolites in the drought stress tolerance of crop plants: a review. Physiol Plant 172(2):1106–1132

Kermani SG, Saeidi G, Sabzalian MR, Gianinetti A (2019) Drought stress influenced sesamin and sesamolin content and polyphenolic components in sesame (*Sesamum indicum L.*) populations with contrasting seed coat colors. Food Chem 289:360–368

Keurentjes JJ (2009) Genetical metabolomics: closing in on phenotypes. Curr Opin Plant Biol 12(2):223–230

Khakimov B, Bak S, Engelsen SB (2014) High-throughput cereal metabolomics: current analytical technologies, challenges and perspectives. J Cereal Sci 59(3):393–418

Khan N, Bano A, Rahman MA, Rathinasabapathi B, Babar MA (2019) UPLC-HRMS-based untargeted metabolic profiling reveals changes in chickpea (Cicer arietinum) metabolome following long-term drought stress. Plant Cell Environ 42(1):115–132

Kim JM, To TK, Matsui A, Tanoi K, Kobayashi NI, Matsuda F, Habu Y, Ogawa D, Sakamoto T, Matsunaga S, Bashir K (2017) Acetate-mediated novel survival strategy against drought in plants. Nat Plants 3(7):1–7

Kliebenstein DJ (2007) Metabolomics and plant quantitative trait locus analysis—the optimum genetical genomics platform? In: Concepts in plant metabolomics. Springer, Dordrecht, pp 29–44

Kueger S, Steinhauser D, Willmitzer L, Giavalisco P (2012) High-resolution plant metabolomics: from mass spectral features to metabolites and from whole-cell analysis to subcellular metabolite distributions. Plant J 70(1):39–50

Kumar P, Sharma PK (2020) Soil salinity and food security in India. Front Sustain Food Syst 4:174

Latowski D, Kuczyńska P, Strzałka K (2011) Xanthophyll cycle—a mechanism protecting plants against oxidative stress. Red Rep 16:78–90

Li P, Li YJ, Zhang FJ, Zhang GZ, Jiang XY, Yu HM, Hou BK (2017) The Arabidopsis UDP-glycosyltransferases UGT79B2 and UGT79B3, contribute to cold, salt and drought stress tolerance via modulating anthocyanin accumulation. Plant J 89(1):85–103

Li M, Li Y, Zhang W, Li S, Gao Y, Ai X, Zhang D, Liu B, Li Q (2018) Metabolomics analysis reveals that elevated atmospheric CO2 alleviates drought stress in cucumber seedling leaves. Anal Biochem 559:71–85

Li X, Peng T, Mu L, Hu X (2019) Phytotoxicity induced by engineered nanomaterials as explored by metabolomics: perspectives and challenges. Ecotoxicol Environ Saf 184:109602

Lisec J, Schauer N, Kopka J, Willmitzer L, Fernie AR (2006) Gas chromatography mass spectrometry–based metabolite profiling in plants. Nat Protoc 1(1):387–396

Liu D, Ford KL, Roessner U, Natera S, Cassin AM, Patterson JH, Bacic A (2013) Rice suspension cultured cells are evaluated as a model system to study salt responsive networks in plants using a combined proteomic and metabolomic profiling approach. Proteomics 13:2046–2062

Liu L, Wang B, Liu D, Zou C, Wu P, Wang Z, Wang Y, Li C (2020) Transcriptomic and metabolomic analyses reveal mechanisms of adaptation to salinity in which carbon and nitrogen metabolism is altered in sugar beet roots. BMC Plant Biol 20(1):1–21

Luengwilai K, Saltveit M, Beckles DM (2012) Metabolite content of harvested micro-tom tomato (*Solanum lycopersicum* L.) fruit is altered by chilling and protective heat-shock treatments as shown by GC–MS metabolic profiling. Postharvest Biol Technol 63:116–122

Lugan R, Niogret MF, Leport L, Guégan JP, Larher FR, Savouré A (2010) Metabolome and water homeostasis analysis of *Thellungiella salsuginea* suggests that dehydration tolerance is a key response to osmotic stress in this halophyte. Plant J 64:215–229

Matsuda F, Okazaki Y, Oikawa A, Kusano M, Nakabayashi R, Kikuchi J, Yonemaru JI, Ebana K, Yano M, Saito K (2012) Dissection of genotype–phenotype associations in rice grains using metabolome quantitative trait loci analysis. Plant J 70(4):624–636

Mibei EK, Owino WO, Ambuko J, Giovannoni JJ, Onyango AN (2018) Metabolomic analyses to evaluate the effect of drought stress on selected African Eggplant accessions. J Sci Food Agric 98(1):205–216

Moreno-Risueno MA, Busch W, Benfey PN (2010) Omics meet networks—using systems approaches to infer regulatory networks in plants. Curr Opin Plant Biol 13(2):126–131

Mukhopadhyay R, Sarkar B, Jat HS, Sharma PC, Bolan NS (2021) Soil salinity under climate change: challenges for sustainable agriculture and food security. J Environ Manag 280:111736

Muscolo A, Junker A, Klukas C, Weigelt-Fischer K, Riewe D, Altmann T (2015) Phenotypic and metabolic responses to drought and salinity of four contrasting lentil accessions. J Exp Bot 66(18):5467–5480

Nakabayashi R, Yonekura-Sakakibara K, Urano K, Suzuki M, Yamada Y, Nishizawa T, Matsuda F, Kojima M, Sakakibara H, Shinozaki K, Michael AJ (2014) Enhancement of oxidative and drought tolerance in Arabidopsis by overaccumulation of antioxidant flavonoids. Plant J 77(3): 367–379

Ni J, Yang X, Zhu J, Liu Z, Ni Y, Wu H (2015) Salinity-induced metabolic profile changes in *Nitraria tangutorum* Bobr. suspension cells. Plant Cell Tissue Organ Cult 122:239–248

Obata T, Witt S, Lisec J, Palacios-Rojas N, Florez-Sarasa I, Yousfi S, Araus JL, Cairns JE, Fernie AR (2015) Metabolite profiles of maize leaves in drought, heat, and combined stress field trials reveal the relationship between metabolism and grain yield. Plant Physiol 169(4):2665–2683

Pan J, Li Z, Dai S, Ding H, Wang Q, Li X, Ding G, Wang P, Guan Y, Liu W (2020) Integrative analyses of transcriptomics and metabolomics upon seed germination of foxtail millet in response to salinity. Sci Rep 10(1):1–16

Panda A, Rangani J, Parida AK (2021) Unraveling salt responsive metabolites and metabolic pathways using non-targeted metabolomics approach and elucidation of salt tolerance mechanisms in the xero-halophyte Haloxylon salicornicum. Plant Physiol Biochem 158:284–296

Parry MA, Hawkesford MJ (2012) An integrated approach to crop genetic improvement F. J Integr Plant Biol 54(4):250–259

Patti GJ, Yanes O, Siuzdak G (2012) Metabolomics: the apogee of the omics trilogy. Nat Rev Mol Cell Biol 13(4):263–269

Pearson M, Saarinen M, Nummelin L, Heiskanen J, Roitto M, Sarjala T, Laine J (2013) Tolerance of peat-grown Scots pine seedlings to waterlogging and drought: morphological, physiological, and metabolic responses to stress. For Ecol Manag 307:43–53

Pires MV, Pereira Júnior AA, Medeiros DB, Daloso DM, Pham PA, Barros KA, Engqvist MK, Florian A, Krahnert I, Maurino VG, Araújo WL (2016) The influence of alternative pathways of respiration that utilize branched-chain amino acids following water shortage in Arabidopsis. Plant Cell Environ 39(6):1304–1319

Pyne ME, Narcross L, Martin VJ (2019) Engineering plant secondary metabolism in microbial systems. Plant Physiol 179(3):844

Renault H, Roussel V, El Amrani A, Arzel M, Renault D, Bouchereau A, Deleu C (2010) The Arabidopsis pop2-1 mutant reveals the involvement of GABA transaminase in salt stress tolerance. BMC Plant Biol 10(1):1–16

Rutley N, Miller G, Wang F, Harper JF, Miller G, Lieberman-Lazarovich M (2021) Enhanced reproductive thermotolerance of the tomato high pigment 2 mutant is associated with increased accumulation of flavonols in pollen. Front Plant Sci 12:779

Salehin M, Li B, Tang M, Katz E, Song L, Ecker JR, Kliebenstein DJ, Estelle M (2019) Auxin-sensitive Aux/IAA proteins mediate drought tolerance in Arabidopsis by regulating glucosinolate levels. Nat Commun 10(1):1–9

Salekdeh GH, Komatsu S (2007) Crop proteomics: aim at sustainable agriculture of tomorrow. Proteomics 7(16):2976–2996

Sang Q, Shan X, An Y, Shu S, Sun J, Guo S (2017) Proteomic analysis reveals the positive effect of exogenous spermidine in tomato seedlings' response to high-temperature stress. Front Plant Sci 8:120

Schauer N, Semel Y, Roessner U, Gur A, Balbo I, Carrari F, Pleban T, Perez-Melis A, Bruedigam C, Kopka J, Willmitzer L (2006) Comprehensive metabolic profiling and phenotyping of interspecific introgression lines for tomato improvement. Nat Biotechnol 24(4):447–454

Sengupta S, Patra B, Ray S, Majumder AL (2008) Inositol methyl transferase from a halophytic wild rice, *Porteresia coarctata* Roxb. (Tateoka): regulation of pinitol synthesis under abiotic stress. Plant Cell Environ 31:1442–1459

Shulaev V, Cortes D, Miller G, Mittler R (2008) Metabolomics for plant stress response. Physiol Plant 132:199–208

Tarpley L, Duran AL, Kebrom TH, Sumner LW (2005) Biomarker metabolites capturing the metabolite variance present in a rice plant developmental period. BMC Plant Biol 5(1):1–12

Templer SE, Ammon A, Pscheidt D, Ciobotea O, Schuy C, McCollum C, Sonnewald U, Hanemann A, Förster J, Ordon F, von Korff M (2017) Metabolite profiling of barley flag leaves under drought and combined heat and drought stress reveals metabolic QTLs for metabolites associated with antioxidant defense. J Exp Bot 68(7):1697–1713

Todaka D, Zhao Y, Yoshida T, Kudo M, Kidokoro S, Mizoi J, Yamaguchi-Shinozaki K, Shinozaki K, Fernie AR, Sato M et al (2016a) Temporal and spatial changes in gene expression, metabolite accumulation and phytohormone content in rice seedlings grown under drought stress conditions. Plant J 90:61–78

Todaka D, Zhao Y, Yoshida T, Kudo M, Kidokoro S, Mizoi J, Kodaira KS, Takebayashi Y, Kojima M, Sakakibara H, Toyooka K (2016b) Temporal and spatial changes in gene expression, metabolite accumulation and phytohormone content in rice seedlings grown under drought stress conditions. Plant J 90(1):61–78

Tuo XQ, Li S, Wu QS, Zou YN (2015) Alleviation of waterlogged stress in peach seedlings inoculated with Funneliformis mosseae: changes in chlorophyll and proline metabolism. Sci Hortic 197:130–134

Ullah N, Yüce M, Gökçe ZNO, Budak H (2017) Comparative metabolite profiling of drought stress in roots and leaves of seven Triticeae species. BMC Genomics 18:969

Upadhyay RK, Fatima T, Handa AK, Mattoo AK (2020) Polyamines and their biosynthesis/catabolism genes are differentially modulated in response to heat versus cold stress in tomato leaves (*Solanum lycopersicum L.*). Cell 9(8):1749

van der Fits L, Memelink J (2000) ORCA3, a jasmonate-responsive transcriptional regulator of plant primary and secondary metabolism. Science 289(5477):295–297

Verma A, Deepti S (2016) Abiotic stress and crop improvement: current scenario. Adv Plants Agric Res 4:00149

Volkova PY, Geras'kin SA (2018) 'Omic'technologies as a helpful tool in radioecological research. J Environ Radioact 189:156–167

Voss-Fels KP, Stahl A, Wittkop B, Lichthardt C, Nagler S, Rose T, Chen TW, Zetzsche H, Seddig S, Majid Baig M, Ballvora A (2019) Breeding improves wheat productivity under contrasting agrochemical input levels. Nat Plants 5(7):706–714

Wada H, Hatakeyama Y, Nakashima T, Nonami H, Erra-Balsells R, Hakata M, Nakata K, Hiraoka K, Onda Y, Nakano H (2020) On-site single pollen metabolomics reveals varietal differences in phosphatidylinositol synthesis under heat stress conditions in rice. Sci Rep 10(1): 1–11

Wang X, Cai X, Xu C, Wang Q, Dai S (2016) Drought-responsive mechanisms in plant leaves revealed by proteomics. Int J Mol Sci 17(10):1706

Weckwerth W (2003) Metabolomics in systems biology. Annu Rev Plant Biol 54(1):669–689

Weckwerth W, Fiehn O (2002) Can we discover novel pathways using metabolomic analysis? Curr Opin Biotechnol 13(2):156–160

Wei S, Yang X, Huo G, Ge G, Liu H, Luo L, Hu J, Huang D, Long P (2020) Distinct metabolome changes during seed germination of lettuce (*Lactuca sativa L.*) in response to thermal stress as revealed by untargeted metabolomics analysis. Int J Mol Sci 21(4):1481

Witt S, Galicia L, Lisec J, Cairns J, Tiessen A, Araus JL, Palacios-Rojas N, Fernie AR (2012) Metabolic and phenotypic responses of greenhouse-grown maize hybrids to experimentally controlled drought stress. Mol Plant 5(2):401–417

Woodrow P, Ciarmiello LF, Annunziata MG, Pacifico S, Iannuzzi F, Mirto A, D'Amelia L, Dell'Aversana E, Piccolella S, Fuggi A, Carillo P (2016) Durum wheat seedling responses to simultaneous high light and salinity involve a fine reconfiguration of amino acids and carbohydrate metabolism. Physiol Plant 159(3):290–312

Zhang J, Zhang Y, Du Y, Chen S, Tang H (2011) Dynamic metabolomic responses of tobacco (*Nicotiana tabacum*) plants to salt stress. J Proteome Res 10:1904–1914

# Chapter 15
# Remote Sensing Algorithms and Their Applications in Plant Phenotyping

**Raju Bheemanahalli, B. Santhana Krishnan, Nuwan Kumara Wijewardane, Sathishkumar Samiappan, and K. Raja Reddy**

**Abstract** Assessing phenotypic traits associated with physiology, biochemistry, and plant health based on leaf spectral reflectance properties has become an important high throughput tool in agriculture research. Precise quantification and monitoring of plant responses to stresses (abiotic or biotic) help researchers' phenotype different genetic resources, map genetic loci, and choose donors for trait development. Studies have shown the potential use of leaf hyperspectral reflectance in assessing the plant phenotype under biotic and abiotic stress conditions. We compiled wavebands or reflectance strongly related to the lab and field-based measurements for pigments, leaf nitrogen, and leaf water content. This chapter also highlights the recent applications of hyperspectral reflectance in plant phenotyping, stress diagnosis, species classification, and robust statistical methods. Furthermore, we discuss the need for advanced analytical tools and their potential applications in plant phenotyping.

**Keywords** Electromagnetic spectrum · Diagnosis · Hyperspectral reflectance · Pigments · Leaf nitrogen · Phenotyping · Remote sensing · Sensors

---

R. Bheemanahalli (✉) · K. R. Reddy
Department of Plant and Soil Sciences, Mississippi State University, Mississippi State, MS, USA
e-mail: rajubr@pss.msstate.edu

B. S. Krishnan · S. Samiappan
Geosystems Research Institute, Mississippi State University, Starkville, MS, USA

N. K. Wijewardane
Department of Agricultural & Biological Engineering, Mississippi State University, Mississippi State, MS, USA

 
H. M. Mamrutha et al. (eds.), *Translating Physiological Tools to Augment Crop Breeding*, https://doi.org/10.1007/978-981-19-7498-4_15

## 15.1 Introduction

Based on evolutionary history, genetic background, and stress adaptation capacity, plants readily express differences in physiology and biochemistry, which can be detected using leaf or canopy spectral reflectance changes (Roitsch et al. 2019; Grzybowski et al. 2021). Along with the differences in morphology or ideotype at the plant level, plants have also evolved to synthesize different combinations of leaf pigments and other compounds to support growth and development under various environmental conditions. Improved detection and monitoring of plant health have been vital areas of research under a rapidly changing climate. Needless to say that high throughput non-invasive sensing-based phenotyping is faster, cheaper, reliable, unbiased, extendable, and robust.

Remote sensing techniques have been utilizing leaf optical properties (reflectance, absorbance, and transmittance) to understand leaf pigments (visible range), cell structure (near-infrared), and biochemical properties (shortwaves) using the leaf or canopy signatures. When a leaf or canopy is exposed to electromagnetic (EM) radiation, the *reflected radiation* can be measured and recorded as *reflectance spectrum.* Sensor modalities to study reflectance can be grouped into two categories (based on the signal type): optical and EM. Other choices include acoustic (ultrasound), magnetic (MRI), X-ray, thermal, electronic (electron-microscope observations), and mechanical sensors. To measure object (e.g., leaf or canopy) properties, most light/sound-based signal processing sensors/algorithms rely on the signal frequency, and the transduction depends on the following:

- Amplitude of the input/reflected/refracted signal
- Directionality of the input/reflected/refracted signal
- Time-of-flight measurement of the received signal

Let us recap a *(conveniently biased)* sampling of definitions from image processing. An image is the numerical representation of any real-world picture; it is a multi-dimensional array (or matrix) of numbers, each representing the amount of signal intensity sampled in a physical region, known as the *pixel.* There are a few choices to both describe the relative intensities of the different colors at each pixel (color space) and store images/parameters (Gonzalez and Woods 2018). Object-light interaction is dependent on both the incident signal characteristics (power level, frequency, bandwidth, etc.) and the structure and chemical constituents of the object. The signal's frequency is invariant as it travels through a medium. Thus, the velocity (and wavelength) of light traveling from vacuum through a non-vacuum medium will decrease. The ratio of light velocities between the medium and vacuum is the *medium's refractive index (RI).* Characterizing the object's structure based on the reflection/refraction signal can be found in more detail in text describing Snell's law or Fermat's least time principle.

Recall the wavelength ranges of the EM spectrum (Fig. 15.1). The EM spectrum is divided into four major wavelengths ($\lambda$) bands: visible (350–700 nm), near-infrared (NIR, 700–2500 nm), mid-infrared (MIR, 2500–25,000 nm), and

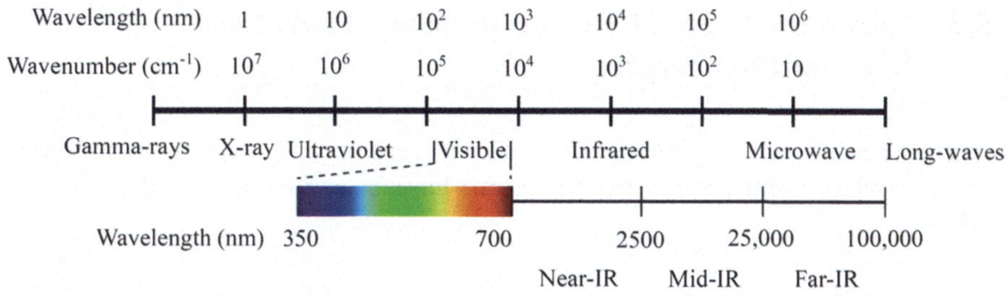


**Fig. 15.1** Electromagnetic spectrum

far-infrared (FIR, 25,000–100,000 nm) (Lin et al. 2004; Manley 2014). The EM wave (light/IR) penetrates the surface of the specimen, and its penetration depth is defined as the depth at which the incident light intensity is reduced by 99%. It has been observed (Lammertyn et al. 2000; Qi et al. 2010) that the penetration depth is inversely proportional to the wavelength, which implies that the IR signal can penetrate deeper into the cell structure.

When EM strikes on the matter or canopy, three interactions can occur: *absorb*, *reflect*, and *transmit*. In plants, the EM absorbed can be used for photosynthesis, converted to heat, and re-emitted as fluorescence (van Bezouw et al. 2019). The leaf surface can reflect part of the incident EM radiation, which can ultimately reach an observer to perceive as color. The remaining portion of the incident radiation can transmit through the leaf and leave from the other side. All these interactions are related to each other and can be denoted as in Eq. (15.1)

$$I_\lambda = A_\lambda + R_\lambda + T_\lambda, \tag{15.1}$$

where $I_\lambda$ is the incident radiation energy, $A_\lambda$ is the absorbed energy, $R_\lambda$ is the reflected energy, and $T_\lambda$ is the transmitted energy at wavelength $\lambda$.

As Eq. (15.1) indicates, all these phenomena are wavelength-dependent since the absorbance depends on the properties of the leaf, more specifically, different chemical compounds or bonds present in the leaf. Chemical bonds absorb EM photons at specific wavelengths causing vibrational and rotational changes in the molecules (Türker-Kaya and Huck 2017). These absorbances at different wavelengths, called spectral signatures or fingerprints, can effectively estimate the leaf's other compounds or related properties. For example, the leaf nitrogen (N) and chlorophyll molecules have spectral signatures at 650 and 940 nm, which are used to estimate leaf nitrogen and chlorophyll contents (Wood 1993; Blackmer et al. 1994; Daughtry et al. 2000). Most fundamental spectral signatures occur in the mid-infrared region (MIR). The overtones and combinational bands can extend to near-infrared or visible spectrum regions. A compilation of these spectral signatures and relevant properties can be found in Türker-Kaya and Huck (2017).

## 15.2 Common Types of Remote-Sensing Tools Used in Plant Phenotyping

Cameras/sensors with different illumination ranges and capture capabilities (red-green-blue, multispectral, hyperspectral) have been routinely used in diverse studies.

### *15.2.1 Visible Light Sensor*

This is a cheap and ubiquitously available sensor. The output (in its simplest form) is an 8 bits per pixel grayscale image. An RGB color image has 24 bits per pixel (one byte for each primary color, red, blue, and green). The range of human eye color perception is quite large, and unsurprisingly, the RGB space does not span the human color gamut. A filter attenuates (or entirely blocks) a subset of wavelengths and allows the complementary subset without attenuation. Many pre-processing options at the sensor and post-processing procedures on the image are available (Pereira et al. 2017; Ngugi et al. 2021).

### *15.2.2 Infrared Sensor*

The leaf typically reflects 40–60% of incident near-IR energy from its spongy mesophyll. In terms of data capture/storing data, the IR is identical to a standard camera (visible light). The difference is in the light–object interaction. For example, an IR spectrum can replace commonly used laboratory methods for moisture content measurements like using a gravimetric oven or performing Karl Fischer titration.

### *15.2.3 Hyperspectral Sensor*

While visible/infra-red imaging uses a wide range of frequencies (broadband) in the illuminating light source, the hyperspectral imaging sensor generates a series of images, each one representing the intensity distribution within a specific narrow-spectral band. For example, let us consider an object observed through the visible-IR wavelength range between 550 and 850 nm in steps of 5 nm. This set of 61 images, one for each frequency band, is arranged to form a three-dimensional (3D) hypercube used for further processing. This grouping of the frequency data in each pixel region using narrow bands (hyperspectral imaging) generates a number of independent measurements, thus strengthening any classification/detection procedures (Huang et al. 2014; Schmilovitch et al. 2014).

The most common types of remote-sensing cameras (RGB, multispectral, hyperspectral, light detection and ranging, LiDAR, and thermal) used in agricultural research are unmanned aerial vehicles (UAVs) based or handled spectroradiometers.

- RGB (red, green, and blue) cameras are cheap, most widely used across the platform, and have a high spatial resolution.
- Multispectral cameras: this platform consists of sensors with different lenses. Typically, there will be five sensors (red, green, blue, red-edge, and near-infrared), with each sensor sensitive in one spectral region.
- Hyperspectral cameras cover the 400–1000 nm spectral region with relatively narrow bands (<20 nm).
- Spectrometers cover the 350–2500 nm spectral region with high spectral resolution (1 nm).

The difference between *hyperspectral* and *multispectral* imaging is in their spectral resolution. A multispectral image is a collection of independent data from non-overlapping spectral bands. Usually, the number of bands is few (up to ten). A list of different sensors and their application has been complied in the Index Database (https://www.indexdatabase.de/; Roitsch et al. 2019). A spectrometer (or spectroradiometer) is used to measure both the wavelength and amplitude of the light reflected from the leaf's surface. Wavelength is detected by the position on the sensor that the signal hits. The detected light intensity for each frequency band is obtained using refraction (prism) or diffraction (grating). On the other hand, hand-held spectrometers (Fig. 15.2a) are relatively low cost, easy to collect data and process compared to UAVs.

The key advantage of using hand-held spectrometers is (a) free from confounding effects of illumination, water band, and leaf or camera angle, (b) data processing and analysis require less time than data collected using the UAVs, and (c) data can easily be collected from controlled and field experiments. With the growing interest in sensors application in agriculture, several studies demonstrated the use of hyperspectral data collected from the spectrometer. The spectral bands are often used to establish an empirical relationship between given leaf properties and spectral data. Genomic techniques have been significantly advanced; however, throughput phenotyping remains the major bottleneck to molecular breeding. To take advantage of genomic resources to dissect complex stress tolerance, hyperspectral spectrometers could be used for high throughput phenotyping (Grzybowski et al. 2021). The following sections will provide information on how we can use spectrometers to take advantage of hyperspectral reflectance in plant stress physiology or phenotyping.

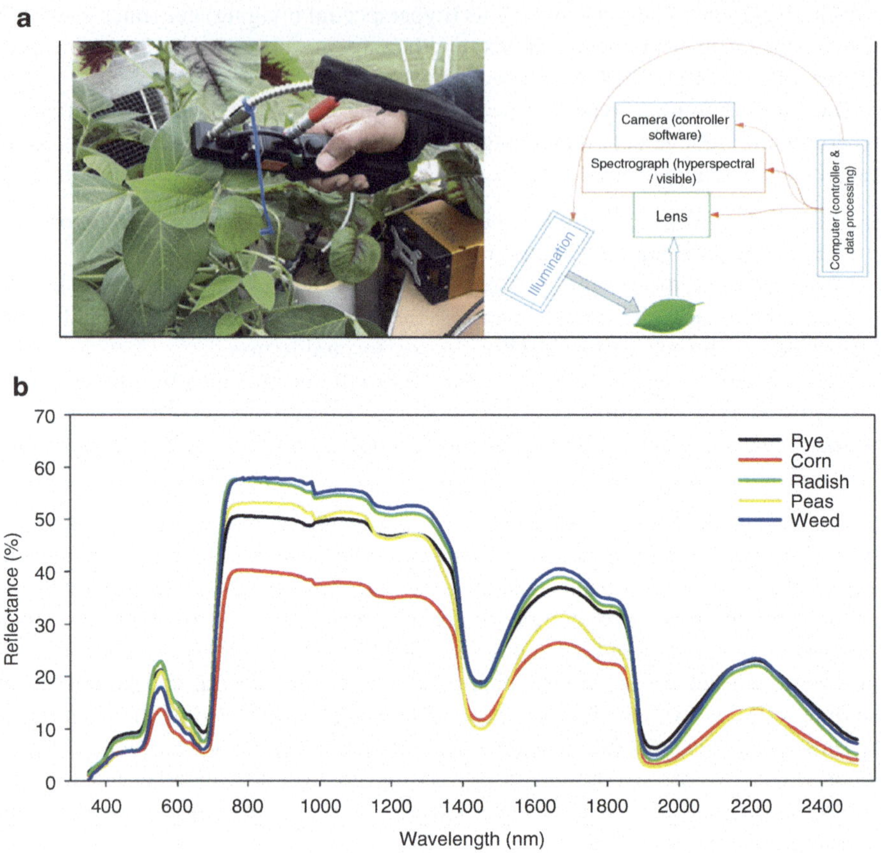


**Fig. 15.2** Handheld spectroradiometer (**a**) and hyperspectral reflectance data (**b**) of different crop species

## 15.3 Leaf Reflectance Relationship with Plant Phenotypes in Different Crops

A typical leaf hyperspectral reflectance (350–2500 nm) spectrum is shown in Fig. 15.2b. It has low reflectance in the visible region (350–700 nm). Most of the energy is absorbed by the pigments such as chlorophyll and carotenoids for photosynthesis (Zur et al. 2000). There is a peak in the green region (~550 nm, Fig. 15.2b), which has comparatively less absorbance than other colors in the spectrum, causing the leaf to be perceived as the color "green" to the human eye. With changes in leaf pigments due to senescence or species differences, these spectra can show different peaks in the visible region (Sims and Gamon 2002; Morley et al. 2020), showing different colors to the human eye. A comparison of RGB and visible light reflectance implies the possibility of using both to classify plant species (Fig. 15.2b). The first part of the NIR region (700–1200 nm) is dominated by the light scattering at air-cell wall interfaces and characterized by high reflectance and transmittance. An increase in leaf reflectance at the transition from red to NIR, producing a spectral feature called *red edge* (Gitelson et al. 1996). Researchers have observed correlations between the red-edge position and chlorophyll content, plant stress, plant phenological stages (Yendrek et al. 2017; Silva-Perez et al. 2018; Mir et al. 2019; Reynolds et al. 2020). The second part of the NIR region (1200–2500 nm) is predominantly the light absorption by water. Two leading water absorption bands are at 1450 and 1950 nm (Hoffer and Johannsen 1969), shown in Fig. 15.2b. Leaf absorbance (reflectance) of green light is far less (more) than red light when the plants are stressed.

Observed variations in spectral signatures among the species (Fig. 15.2b) depend on the genetic, physiological, and biochemical properties of the leaf such as temperature, ion-concentration, pigments, concentrations of organic acids, and so on. Chlorophyll pigments reflect strongly around the green band with a green/green-yellow color. Anthocyanin pigments, red in color, provide a high reflectance around the red band. Sugars have a higher reflectance at the NIR. An obvious indication of the temporal change in absorption/reflectance characteristics of visible light on a leaf is color change in response to species or stress or age and nutrient deficiency. Thus, biologists have been perfecting relationships between individual ions and plant health.

### *15.3.1 Nitrogen and Pigments Associated Leaf Reflectance Spectra*

Plants accumulate specific pigments during the courses of growth and development under different stresses. The pigments and their ratios determine the physiological characteristics and plant health status. Chlorophylls are the dominant pigment of green leaves and an expensive molecule because each ring contains four nitrogen atoms. Chlorophyll molecules absorb sunlight at different wavelengths (chlorophyll-a absorbs red-orange light and chlorophyll-b absorbs blue-purple light). As a result, chlorophyll index has been used as a proxy indicator of leaf or canopy nitrogen content and plant health. Traditional methods of determining pigments and tissue nitrogen are costly and time consuming.

Current advances in remote sensing have allowed timely data collection to assess plants' crop growth and nutrient (nitrogen) status. Studies highlighted in Table 15.1 used traditional and sensor-based approaches and identified spectral bands associated with pigments and nitrogen content. Studies reported in Table 15.1 used leaf spectral reflectance data collected using a portable ASD Field Spec FR spectroradiometer (Analytical Spectral Devices Inc., Boulder, CO, USA) compared with the wet lab measured observations (shoot or leaf nitrogen). These studies tested the relationship between the measured values and spectral formulations or indices (Table 15.1). Leaf nitrogen and chlorophyll concentrations of sorghum leaves were

**Table 15.1** Summary of selected research papers used leaf reflectance-based vegetative indices or bands in different crops. *R* represents reflectance at a specific wavelength. $R^2$ indicates the association between measured and vegetative indices or bands.

| Crop | Phenotype | Formula | $R^2$ | Reference |
|---|---|---|---|---|
| Bermuda grass | Forage nitrogen | *R915/R515* | 0.44 | Zhao et al. (2005b) |
| | | *R915/R705* | 0.51 | |
| Castor bean | Carotenoid | *R575/R675* | 0.80 | Reddy and Matcha (2010) |
| | | *R705/675* | 0.80 | |
| | Leaf chlorophyll | *R545* | 0.72 | |
| | | *R555/R675* | 0.91 | |
| | | *R555/R455* | 0.93 | |
| | | *R705/R655* | 0.92 | |
| Cotton | Leaf area index | *1.16*R800 – R670/(R800 + R670) + 0.16* | 0.75 | Zhao et al. (2007) |
| | | *(R900 – R680)/(R900 + R680)* | 0.73 | |
| | Leaf carotenoids | *R415/R685* | 0.79 | Tarpley et al. (2000) |
| | Leaf chlorophyll | *R415/R695* | 0.72 | |
| | Leaf nitrogen | *R415/R710* | 0.70 | |
| | | *R915/R515* | 0.65 | Zhao et al. (2005b) |
| | Leaf or shoot nitrogen | *R915/R705* | 0.78 | |
| | | *R715/R405* | 0.65 | Read et al. (2002) |
| | | *R795/R755* | 0.70 | |
| Maize | Carotenoids | *R712/R809* | 0.50 | Zhao et al. (2003) |
| | Chlorophyll a | *R712/R1088* | 0.55 | |
| | Chlorophyll b | *R712/R1097* | 0.58 | |
| | Total chlorophyll | *R712/R1088* | 0.59 | |
| | Leaf nitrogen | *R712/R1040* | 0.55 | |
| | | *R575/R526* | 0.69 | |
| Sorghum | Leaf chlorophyll | *R555/R465* | 0.65 | Zhao et al. (2005a) |
| | | *R555/685* | 0.62 | |
| | | *R715/R455* | 0.62 | |
| | | *R1075/R735* | 0.66 | |
| | Leaf nitrogen | *R405/R555* | 0.63 | |
| | | *R405/R715* | 0.68 | |
| | | *R555/425* | 0.62 | |
| | | *R715/R415* | 0.65 | |
| | | *R1075/R555* | 0.55 | |
| | | *R1075/R725* | 0.65 | |

significantly associated with the reflectance (*R*) ratios of R405/R715 ($R^2 = 0.68$) and R1075/R735 ($R^2 = 0.66$) (Zhao et al. 2005a).

Likewise, measured leaf chlorophyll or carotenoids showed a stronger relationship with the reflectance ratios of *R575/R526* ($R^2 = 0.69$) or *R712/R809* ($R^2 = 0.50$) in maize (*Zea mays*), respectively (Zhao et al. 2003). In cotton (*Gossypium hirsutum*), *R715/R405*, *R795/R755*, *R415/R710*, *R915/R515*, and *R915/R705* showed a stronger association with measured leaf or shoot nitrogen in cotton (Table 15.1). In general, spectral bands *R405*, *R415*, *R515*, *R705*, *R710*, *R755*, and *R915* are associated with leaf nitrogen content or concentration across cotton studies (Tarpley et al. 2000; Read et al. 2002; Zhao et al. 2005b). Further, reflectance ratios of *R555/R455* and *R705/R675* showed strong correlations with total chlorophyll ($R^2$ = 0.93) and carotenoids ($R^2 = 0.80$) in castor bean (*Ricinus communis* L.), respectively (Reddy and Matcha 2010). The weakest relation was noted in Bermuda grass between forage nitrogen and *R915/R515* ($R^2 = 0.44$). The nitrogen and pigment-specific bands (Table 15.1) could be used for non-destructive phenotyping. This information can be utilized to make appropriate nitrogen application decisions in the field.

### *15.3.2 Correlations of Mid-Day Leaf Water Potential with Spectral Indices*

Drought is one of the most significant abiotic stress factors limiting the production of crops. Most field crops are sensitive to drought stress or soil moisture deficit during growth and development (Galieni et al. 2021). High throughput detection or quantification of plant responses to drought stress is one of the most critical crop improvement programs. Here we discuss a study that utilized reflectance indices such as normalized difference vegetation index (NDVI), normalized water index (NWI) in response to evapotranspiration-based irrigation treatments (100, 80, 60, 40, and 20% ET) in soybean. Mid-day leaf water potential was measured 42–57 days after sowing (DAS) using a pressure chamber method (Wijewardana et al. 2019). On the same plants, leaf reflectance measurements were taken between 1100 and 1200 h using a portable ASD spectroradiometer. Soil moisture stress increased soybean canopy reflectance in the visible spectrum range (400–700 nm), especially for 60, 40, and 20% ET compared to the control and 80% ET (Fig. 15.3a). The lower reflectance in the NIR region further confirmed the soil moisture stress-induced reduction in transpiration and stomatal conductance. Beyond the NIR region (1300–2500 nm), which is a function of leaf-water content and leaf thickness, reported to increase spectral reflectance in soil moisture stressed plants. The midday leaf water potential was strongly correlated (Fig. 15.3a) with NDVI ($R^2 = 0.82$) and NWI16 ($R^2 = 0.72$). Soil moisture stress-induced reduction in chlorophyll and canopies size resulted in lower NDVI values. These findings suggest that the NIR region could be used as a proxy to describe the canopy mass under water-deficit conditions (Lobos et al. 2014; Elsayed et al. 2015).

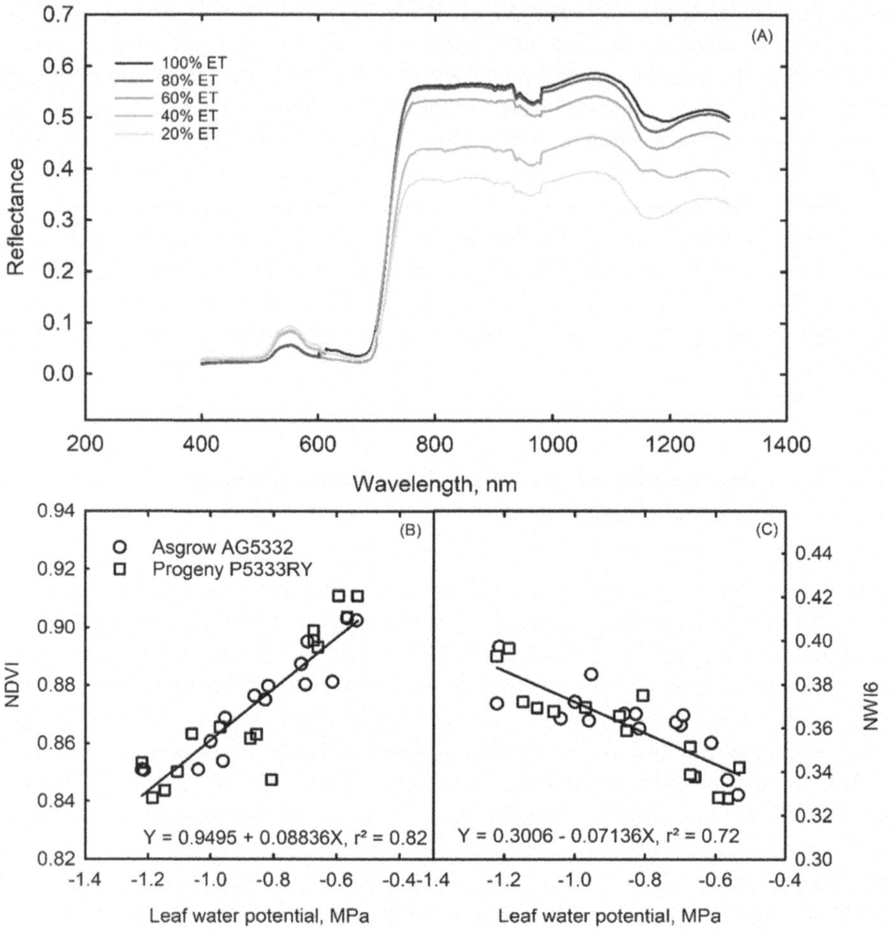


**Fig. 15.3** Soybean leaf hyperspectral reflectance of the five soil moisture stress treatments (**a**). The relationships between leaf water potential and leaf spectral indices (Normalized Difference Vegetation Index-NDVI, **b**; Normalized Water Index-NWI, **c**) for soybean subjected to five evapotranspiration-based irrigation treatments. For experimental details refer Wijewardana et al. (2019)

## 15.4 Advancement in the Analysis of Hyperspectral Reflectance Data

Conventionally, the spectral signatures are used to calculate simple indices, which are then used for testing the correlations among indices and plant phenotype (Table 15.1 and Fig. 15.3). However, this does not necessarily work for all the properties of interest or yield (Grzybowski et al. 2021). Estimating such leaf properties is not a simple regression task since overtones and combinational bands often contaminate these signatures, especially in the near-infrared (NIR) region. Hence, the leaf spectrum is a complex graph consisting of many overlapping peaks requiring complex multivariate modeling techniques for property estimations (Manley 2014).

The independent information from each band of the hyperspectral image increases the data dimension. While using the full spectral information in a classifier, we can only use a finite number of training samples to estimate many parameters necessary to describe the large data dimension. When the cardinality of the training samples is less than the number of parameters, the estimator performance naturally declines (Dalponte et al. 2009); this effect is called the *Hughes phenomenon*. An optimal pair of training samples, data dimensions exist in most practical circumstances. A dimensionality reduction step (Principal Component Analysis, PCA; Fisher's Linear Discriminant Analysis, LDA; and Stepwise LDA, SLDA) would help bring down redundant features to reduce the dimensions of the hyperspectral observations (Duda et al. 2021). In brief, PCA obtains the best projections of the data. Only a few of the projected components have significant variance (information), and only these fewer "principal" components are used. While PCA is unsupervised and not classification-centered, the LDA (can also be run in supervised mode) maximizes the class separability between the two classes. In SLDA, the discrimination model is refined at each step where a decision on whether to retain or discard a subset of features is taken.

The efficacy of the maximum likelihood (ML) statistical classifier in combination with PCA/LDA/SLDA has been used to classify hyperspectral signatures (Samiappan et al. 2021). These analyses have the capabilities to investigate temporal misalignments between training and testing conditions. These modeling techniques can range from classical partial least squares regression (PLSR) to more modern machine learning techniques such as artificial neural networks, convolutional neural networks, and deep learning (Grzybowski et al. 2021).

### 15.4.1 Leaf Reflectance for Early Season Disease Diagnosis

Disease diagnosis and accurate estimation of disease incidence are fundamental in agriculture production. For example, the root-knot nematode (RKN) (*Meloidogyne incognita*) is a significant threat to cotton production. The RKN, a soilborne roundworm, threatens cotton and other crops. The formation of RKN affects a series of physiological traits, root system architecture, and final yield. Early diagnosis of RKN is meaningful in reducing disease spread, financial loss and facilitating real-time management practice. Leaf hyperspectral reflectance (350–2500 nm) has been used to understand early-season RKN damages on cotton (Fig. 15.4). Cotton genotypes (Rk-Rn-1, nematode-resistant and M8, nematode susceptible) with contrasting responses to nematode infection at the early vegetative stage were inoculated with nematodes. Temporal hyperspectral reflectance scans were taken 10, 30, and 60 days after the dose of nematode infestation (DAI) using PSR+ 3500 Spectroradiometer

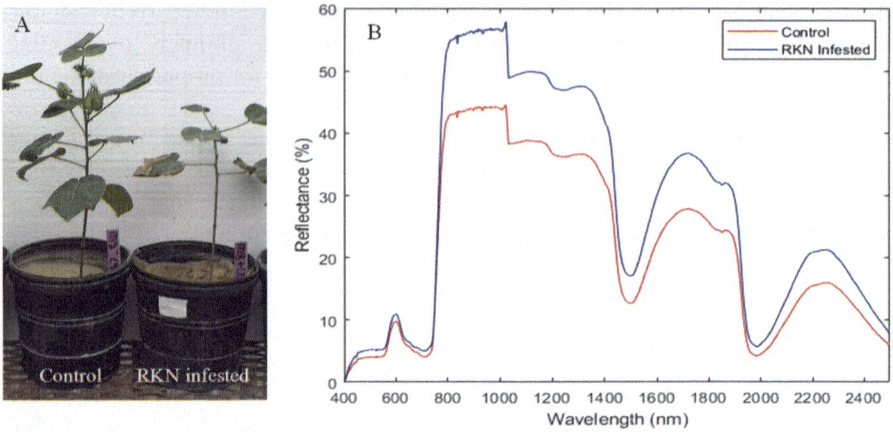


**Fig. 15.4** Cotton plant with and without root nematode infestation (**a**). The right graph (**b**) shows the reflectance of control and root-knot nematode-infested leaf of cotton

(Spectral Evolution, MD, USA) to discriminate the cotton plants with invisible RKN disease symptoms. We applied statistical supervised learning algorithms to classify RKN infested cotton from the control group using the spectral range is 350–2500 nm (2150 bands) with a spectral resolution of 2.8 nm at 700 nm, 8 nm at 1500 nm, and 6 nm at 2100 nm full width at half maximum (Fig. 15.4b). A maximum likelihood classifier with a dimensionality reducer revealed temporal hyperspectral changes in response to RKN before the plant's visual symptoms appear. The classifier performed with >90% accuracy using hyperspectral data while classifying healthy cotton plants from RKN infected plants (Fig. 15.4). These results show the possibility of using leaf reflectance data to diagnose RKN infestation in cotton, and the same techniques could be used for other diseases in other crops. The effect caused by the diseases can be non-invasively analyzed using hyperspectral data at the early growth stage.

### *15.4.2 Crop Species Discrimination Using Hyperspectral Data*

Leaf reflectance has been used for monitoring plant health, nutrient status, and stress tolerance in a range of crops (Yendrek et al. 2017; Silva-Perez et al. 2018; Mir et al. 2019; Reynolds et al. 2020; Grzybowski et al. 2021). Differences in leaf morph-physiological properties among different species lead to varying spectral signatures, which can be used for the classification of species (Fig. 15.2), abiotic (i.e., soil moisture stress treatments, Fig. 15.3) and biotic stresses (root-knot nematode, Fig. 15.4). For example, leaf spectral reflectance of showing a clear discrimination between corn and soybean in spectral-domain (Fig. 15.5a) and principal component

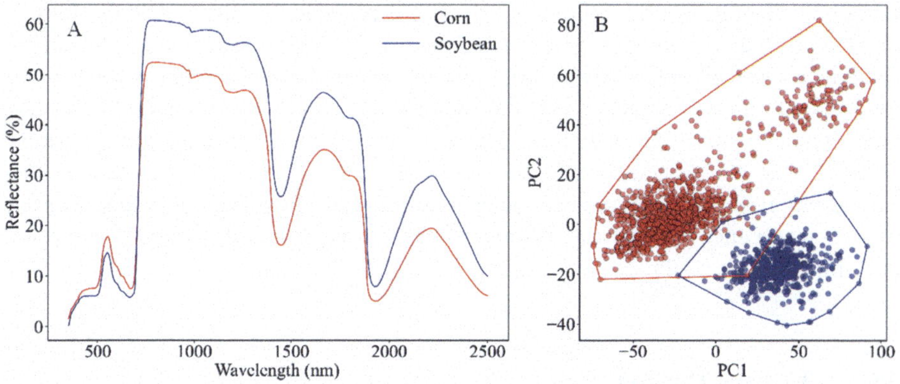


**Fig. 15.5** Average corn and soybean leave spectra (**a**) and their separation in principal components (PCs) space (**b**). Corn and soybean leaf hyperspectral signatures were collected from plants grown under controlled environment conditions during the reproductive stage (Bheemanahalli et al., Plant Stress Physiology Lab, Mississippi State University, unpublished). Leaf reflectance observations were collected using the PSR+ 3500 Spectroradiometer (Spectral Evolution, MD, USA)

domain (Fig. 15.5b). It is evident from the results that the two species: corn and soybean can be discriminated utilizing the leaf hyperspectral information (Fig. 15.5). However, this is not always guaranteed, especially when intraspecies spectral variability exceeds interspecies. Natural processes such as leaf senescence, nutrients, and stress can further complicate species separation (Reynolds et al. 2020).

### *15.4.3 Leaf Reflectance and Soil Nutrients*

Spectroscopy has been used in soil analysis for a few decades to estimate soil properties rapidly and cost-effectively. However, uncertainties are associated with estimated soil nutrient content as plant-available nutrients (Singh et al. 2019). The leaf level spectral measurements can bypass the soil nutrient analysis and show the direct impact of available nutrients on plants. Nutrients and minerals in plants in organic and inorganic fractions create spectral signatures used for estimations. For example, leaf N, mainly present in organic form in chlorophyll, absorbs light in the visible region due to the C–C and C=C bonds in the porphyrin rings and the magnesium ion (Sims and Gamon 2002). In addition, the mid-infrared (MIR) region has primary absorptions of many leaf properties such as cellulose, hemicellulose, lignin, xylan, cutin, carbohydrates, and proteins (Türker-Kaya and Huck 2017). Inorganic leaf nutrients such as Ca, K, and other micronutrients can also be derived from spectra due to their association with the organic compound functional groups or organic matrix (Yarce and Rojas 2012; Prananto et al. 2021). A compilation of past studies using near-infrared (NIR) to derive plant nutrients can be found in Prananto et al. (2020). In addition to NIR, the MIR region has also been utilized to derive numerous plant properties, including cell wall compounds (Jiang et al. 2009; Largo-Gosens et al. 2014), protein structure (Kumar et al. 2016), tissue and taxa differentiation (Huck-Pezzei et al. 2012), and water content (Ullah et al. 2012). Since the leaf level spectral measurements can better indicate the available nutrients in the soil; it can effectively be used in nutrient deficiency diagnosis or management. A study showed that leaf spectra could identify corn's N, P, K, and Ca deficiencies (Al-Abbas et al. 1972). Similarly, other nutrients and physiological processes such as leaf senescence (Ivanova and Singh 2003) can be detected using the leaf-level spectra.

## 15.5 Future Perspectives

Plant physiologists, geneticists, and breeders are interested in leaf biochemical and physiological properties, usually measured destructively. Destructive measurements limit the speed of phenotyping, where hundreds of samples are needed to characterize variations in nutrients, metabolites, and photosynthetic traits (Grzybowski et al. 2021). Genomic techniques have been significantly advanced; however, throughput phenotyping has remained the primary bottleneck. To take advantage of genomic resources to dissect complex stress tolerance, leaf hyperspectral combined with spectroscopy can serve as an effective and efficient tool for a comprehensive assessment of complex plant performance to stresses. In recent years considerable progress has been made in building spectral features for crop improvement and the technologies to deliver from hand-held instrumentation to drones-assisted platforms. Numerous studies have shown that leaf and canopy spectra can help derive proxies associated with physiological and biochemical traits, which is tremendously helpful in studying heritability, genetic loci discovery, and improving breeding efficiency. Nevertheless, spectral-assisted breeding and selection will gradually evolve as data analysis and tools become more accessible for breeders and physiologists. In the future, spectral-assisted breeding could be seen as a revolutionary approach to enhance genetic gain under various environments.

## References

Al-Abbas AH, Barr R, Hall JD, Crane FL, Baumgardner MF (1972) Spectral characteristics of normal and nutrient-deficient maize leaves. Information note 111472. Purdue University, West Lafayette

Blackmer TM, Schepers JS, Varvel GE (1994) Light reflectance compared with other nitrogen stress measurements in corn leaves. Agron J 86:934–938

Dalponte M, Bruzzone L, Vescovo L, Gianelle D (2009) The role of spectral resolution and classifier complexity in the analysis of hyperspectral images of forest areas. Remote Sens Environ 113:2345–2355

Daughtry CST, Walthall CL, Kim MS, de Colstoun EB, McMurtrey JE (2000) Estimating corn leaf chlorophyll concentration from leaf and canopy reflectance. Remote Sens Environ 74:229–239

Duda RO, Hart PE, Stork DG (2021) Pattern classification, 2nd edn. John Wiley & Sons, New York

Elsayed S, Rischbeck P, Schmidhalter U (2015) Comparing the performance of active and passive reflectance sensors to assess the normalized relative canopy temperature and grain yield of drought-stressed barley cultivars. Field Crop Res 177:148–160

Galieni A, D'Ascenzo N, Stagnari F, Pagnani G, Xie Q, Pisante M (2021) Past and future of plant stress detection: an overview from remote sensing to positron emission tomography. Front Plant Sci 11:1975

Gitelson AA, Merzlyak MN, Lichtenthaler HK (1996) Detection of red edge position and chlorophyll content by reflectance measurements near 700 nm. J Plant Physiol 148:501–508

Gonzalez RC, Woods RE (2018) Digital image processing, 4th edn. https://www.pearson.com/content/one-dot-com/one-dot-com/us/en/higher-education/program.html. Accessed 18 Dec 2021

Grzybowski M, Wijewardane NK, Atefi A, Ge Y, Schnable JC (2021) Hyperspectral reflectance-based phenotyping for quantitative genetics in crops: progress and challenges. Plant Commun 2:100209

Hoffer RM, Johannsen CJ (1969) Ecological potential in spectral signatures analysis. In: Johnson PO (ed) Remote sensing in ecology. University of Georgia Press, Athens, pp 1–16. Accessed 18 December 2021

Huang M, Wang Q, Zhang M, Zhu Q (2014) Prediction of color and moisture content for vegetable soybean during drying using hyperspectral imaging technology. J Food Eng 128:24–30

Huck-Pezzei VA, Pallua JD, Pezzei C, Bittner LK, Schönbichler SA, Abel G, Popp M, Bonn GK, Huck CW (2012) Fourier transform infrared imaging analysis in discrimination studies of St. John's wort (Hypericum perforatum). Anal Bioanal Chem 404:1771–1778

Ivanova DG, Singh BR (2003) Nondestructive FTIR monitoring of leaf senescence and elicitin-induced changes in plant leaves. Biopolymers 72:79–85

Jiang B, Shen Z, Shen J, Yu D, Sheng X, Lu H (2009) Germination and growth of sponge gourd (*Luffa cylindrica*) pollen tubes and FTIR analysis of the pollen tube wall. Sci Hortic 122:638–644

Kumar S, Lahlali R, Liu X, Karunakaran C (2016) Infrared spectroscopy combined with imaging: a new developing analytical tool in health and plant science. Appl Spectrosc Rev 51:466–483

Lammertyn J, Peirs A, De Baerdemaeker J, Nicolaı B (2000) Light penetration properties of NIR radiation in fruit with respect to non-destructive quality assessment. Postharvest Biol Technol 18:121–132

Largo-Gosens A, Hernández-Altamirano M, García-Calvo L, Alonso-Simón A, Álvarez J, Acebes JL (2014) Fourier transform mid infrared spectroscopy applications for monitoring the structural plasticity of plant cell walls. Front Plant Sci 5:303

Lin M, Al-Holy M, Al-Qadiri H, Kang D-H, Cavinato AG, Huang Y, Rasco BA (2004) Discrimination of intact and injured listeria monocytogenes by Fourier transform infrared spectroscopy and principal component analysis. J Agric Food Chem 52:5769–5772

Lobos GA, Matus I, Rodriguez A, Romero-Bravo S, Araus JL, del Pozo A (2014) Wheat genotypic variability in grain yield and carbon isotope discrimination under Mediterranean conditions assessed by spectral reflectance. J Integr Plant Biol 56:470–479

Manley M (2014) Near-infrared spectroscopy and hyperspectral imaging: non-destructive analysis of biological materials. Chem Soc Rev 43:8200–8214

Mir RR, Reynolds M, Pinto F, Khan MA, Bhat MA (2019) High-throughput phenotyping for crop improvement in the genomics era. Plant Sci 282:60–72

Morley PJ, Jump AS, West MD, Donoghue DNM (2020) Spectral response of chlorophyll content during leaf senescence in European beech trees. Environ Res Commun 2:071002

Ngugi LC, Abelwahab M, Abo-Zahhad M (2021) Recent advances in image processing techniques for automated leaf pest and disease recognition—a review. Inf Process Agric 8:27–51

Pereira CS, Morais R, Reis MJCS (2017) Recent advances in image processing techniques for automated harvesting purposes: a review. In: 2017 intelligent systems conference (IntelliSys), pp 566–575
Prananto J, Minasny B, Weaver T (2020) Near infrared (NIR) spectroscopy as a rapid and cost-effective method for nutrient analysis of plant leaf tissues. Adv Agron 164:1–49
Prananto JA, Minasny B, Weaver T (2021) Rapid and cost-effective nutrient content analysis of cotton leaves using near-infrared spectroscopy (NIRS). PeerJ 9:e11042
Qi Y, Heisler GM, Gao W, Vogelmann TC, Bai S (2010) Characteristics of UV-B radiation tolerance in broadleaf trees in southern USA. In: Gao W, Slusser JR, Schmoldt DL (eds) UV radiation in global climate change: measurements, modeling and effects on ecosystems. Springer, Berlin, pp 509–530
Read JJ, Tarpley L, McKinion JM, Reddy KR (2002) Narrow-waveband reflectance ratios for remote estimation of nitrogen status in cotton. J Environ Qual 31:1442–1452
Reddy KR, Matcha SK (2010) Remote sensing algorithms for castor bean nitrogen and pigment assessment for fertility management. Ind Crop Prod 32:411–419
Reynolds M, Chapman S, Crespo-Herrera L, Molero G, Mondal S, Pequeno DNL, Pinto F, Pinera-Chavez FJ, Poland J, Rivera-Amado C, Saint Pierre C, Sukumaran S (2020) Breeder friendly phenotyping. Plant Sci 295:110396
Roitsch T, Cabrera-Bosquet L, Fournier A, Ghamkhar K, Jiménez-Berni J, Pinto F, Ober ES (2019) Review: new sensors and data-driven approaches—a path to next generation phenomics. Plant Sci 282:2–10
Samiappan S, Bheemanahalli R, Zhou M, Brooks J, Wubben M (2021) Early detection of root-knot nematode (*Meloidogyne incognita*) infestation in cotton using hyperspectral data. In: 2021 IEEE international geoscience and remote sensing symposium IGARSS, pp 5849–5852
Schmilovitch Z, Ignat T, Alchanatis V, Gatker J, Ostrovsky V, Felföldi J (2014) Hyperspectral imaging of intact bell peppers. Biosyst Eng 117:83–93
Silva-Perez V, Molero G, Serbin SP, Condon AG, Reynolds MP, Furbank RT, Evans JR (2018) Hyperspectral reflectance as a tool to measure biochemical and physiological traits in wheat. J Exp Bot 69:483–496
Sims DA, Gamon JA (2002) Relationships between leaf pigment content and spectral reflectance across a wide range of species, leaf structures and developmental stages. Remote Sens Environ 81:337–354
Singh K, Majeed I, Panigrahi N, Vasava HB, Fidelis C, Karunaratne S, Bapiwai P, Yinil D, Sanderson T, Snoeck D, Das BS, Minasny B, Field DJ (2019) Near infrared diffuse reflectance spectroscopy for rapid and comprehensive soil condition assessment in smallholder cacao farming systems of Papua New Guinea. Catena 183:104185
Tarpley L, Reddy KR, Sassenrath-Cole GF (2000) Reflectance indices with precision and accuracy in predicting cotton leaf nitrogen concentration. Crop Sci 40:1814–1819
Türker-Kaya S, Huck CW (2017) A review of mid-infrared and near-infrared imaging: principles, concepts and applications in plant tissue analysis. Molecules 22:168
Ullah S, Skidmore AK, Naeem M, Schlerf M (2012) An accurate retrieval of leaf water content from mid to thermal infrared spectra using continuous wavelet analysis. Sci Total Environ 437:145–152
Van Bezouw RFHM, Keurentjes JJB, Harbinson J, Aarts MGM (2019) Converging phenomics and genomics to study natural variation in plant photosynthetic efficiency. Plant J 97:112–133
Wijewardana C, Alsajri FA, Irby JT, Krutz LJ, Golden B, Henry WB, Gao W, Reddy KR (2019) Physiological assessment of water deficit in soybean using midday leaf water potential and spectral features. J Plant Interact 14:533–543
Wood C (1993) Relationships between chlorophyll meter readings and leaf chlorophyll concentration, N status, and crop yield: a review. Proc Agron Soc NZ 23:1–9
Yarce CJ, Rojas G (2012) Near infrared spectroscopy for the analysis of macro and micronutrients in sugarcane leaves. Sugar Industry/Zuckerindustrie 137:707–710

Yendrek CR, Tomaz T, Montes CM, Cao Y, Morse AM, Brown PJ, McIntyre LM, Leakey ADB, Ainsworth EA (2017) High-throughput phenotyping of maize leaf physiological and biochemical traits using hyperspectral reflectance. Plant Physiol 173:614–626

Zhao D, Reddy KR, Kakani VG, Read JJ, Carter GA (2003) Corn (*Zea mays* L.) growth, leaf pigment concentration, photosynthesis and leaf hyperspectral reflectance properties as affected by nitrogen supply. Plant Soil 257:205–218

Zhao D, Reddy KR, Kakani VG, Reddy VR (2005a) Nitrogen deficiency effects on plant growth, leaf photosynthesis, and hyperspectral reflectance properties of sorghum. Eur J Agron 22:391–403

Zhao D, Reddy KR, Starks PJ (2005b) Assessment of N status of cotton and forage plant using remote sensing. In: Li CJ et al (eds) Plant nutrition for food security, human health and environmental protection. Tsinmingo University Press, Bejing, pp 1104–1105

Zhao D, Reddy KR, Kakani VG, Read JJ, Koti S (2007) Canopy reflectance in cotton for growth assessment and lint yield prediction. Eur J Agron 26:335–344

Zur Y, Gitelson A, Chivkunova O, Merzlyak M (2000) The spectral contribution of carotenoids to light absorption and reflectance in green leaves. Papers in natural resources 272. https://digitalcommons.unl.edu/natrespapers/272

# Chapter 16
# Endophyte-Mediated Crop Improvement: Manipulation of Abiotic Stress-Specific Traits

**K. H. Dhanyalakshmi, N. Pallavi, Rachana K. Pawar, and Karaba N. Nataraja**

**Abstract** Crop production has been extremely challenging in many parts of the world mainly due to the influence of the environmental factors on crop growth, development, and yield. In the wake of a changing climate scenario, the impact would be more severe, further affecting crop yields. There is an urgent need to develop climate-resilient crop varieties by incorporating appropriate stress adaptive traits through breeding or genetic manipulation approaches. Endophyte-mediated activation or manipulation of stress adaptive traits is now considered to be a promising alternative to these approaches, as it can build stress resilience in existing crop varieties, rather than breeding new varieties. But the success depends upon the choice of the endophyte, the nature of the interaction of the endophytes with the host, and the activation of specific target traits in the host. This chapter focuses on the potential of using endophytes as a tool to manipulate the major stress adaptive traits in crop plants under stressful conditions such as drought, salinity, and high temperature.

**Keywords** Endophyte · Abiotic stress · Trait manipulation · Crop improvement

K. H. Dhanyalakshmi
Plant Molecular Biology Laboratory, Department of Crop Physiology, University of Agricultural Sciences Bangalore, GKVK Campus, Bengaluru, India

Present Address: College of Agriculture, Padannakkad, Kerala Agricultural University, Kasaragod, Kerala, India

N. Pallavi · R. K. Pawar · K. N. Nataraja (✉)
Plant Molecular Biology Laboratory, Department of Crop Physiology, University of Agricultural Sciences Bangalore, GKVK Campus, Bengaluru, India
e-mail: nnkaraba@uasbangalore.edu.in



H. M. Mamrutha et al. (eds.), *Translating Physiological Tools to Augment Crop Breeding*, https://doi.org/10.1007/978-981-19-7498-4_16

## 16.1 Introduction

Agricultural production is primarily dependent on three important factors, i.e., crop genetics, natural resources, and agronomic management practices. Significant progress has been achieved in improving crop yields since the green revolution, through manipulation of crop genetics, better utilization of natural resources, and improved agronomic management practices. However, the dynamic biotic and abiotic environment that the sessile crops encounter during their growth and development has been a major bottleneck in achieving potential yields. Such challenging environments alter plant morphology, physiology, biochemistry, and molecular biology, ultimately limiting crop yields. Abiotic stresses such as drought, salinity, high temperature, and cold stresses have been more damaging due to their impact on a larger cropping area on the globe, affecting the yields of a greater number of crop species (Papagiannaki et al. 2014; Fahad et al. 2017; Zorb et al. 2019). Moreover, there has been an additional toll on crop yields due to the changing climate, which intensifies the existing scenario (Zhao et al. 2017; Wing et al. 2021). Hence, concerted efforts are essential to increase crop productivity under changing climatic conditions, and to meet the food demands of a growing population. There has been attempts to identify novel traits and stress adaptive pathways, in diverse crops including perennial plants (Dhanyalakshmi et al. 2016; Dhanyalakshmi and Nataraja 2021), and significant progress has been made in introgressing stress adaptive traits into modern crop varieties through conventional, molecular breeding, and transgenic approaches (Pruthvi et al. 2014; Karaba et al. 2007). But under the rapidly changing climatic scenario, there is a need for evolving eco-friendly and sustainable strategies to manipulate stress adaptive traits in crop plants, in a rapid way. Endophyte-mediated trait manipulation is now widely accepted as an important alternative approach to manipulate crop adaptation under stressful conditions.

## 16.2 Endophytes: An Ideal Tool for Manipulating Plant Traits

Plants have an inseparable association with diverse microorganisms, throughout the course of evolution. These microorganisms include archaea, protists, bacteria, and fungi, and are collectively called the plant microbiota. The early studies on plant microbiota, however, focused on those inhabiting the plant roots and on deciphering their interrelationships that shape the functional relevance of the association (Berg and Smalla 2009), subsequently looking into the phyllospheric microbiota, i.e., the microbiota associated with the aerial plant parts (Vorholt 2012). Furthermore, many of the microbes reside on plant surfaces as epiphytes while some penetrate the plant surfaces, and colonize various organs as endophytes (Hardoim et al. 2015). The endophytic population within the plants mostly comprises bacteria and fungi, and the plants and these microbial partners share an intricate relationship to perform diverse biological and ecologically important functions (Turner et al. 2013).

### 16.2.1 Defining Endophytes

deBary (1866) used the term endophyte for the first time to represent all the microorganisms that colonize the internal plant tissues. This definition was subsequently refined by several studies. As per Carroll (1986), endophytes are those microorganisms that colonize plants causing asymptomatic infections within the tissues. This excludes the pathogenic fungi and mutualists like mycorrhizal fungi (Carroll 1986). The definition of endophytes by Petrini (1991) accommodates all the microorganisms that inhabit plant organs and colonized the tissues for some time, causing no apparent harm to the host. This considered those endophytic organisms with a longer epiphytic nature and the latent pathogens that colonize a host for a period of time without causing symptoms (Petrini 1991). In simple terms, endophytes are microorganisms colonizing the inter- and/or intra-cellular spaces in various plant parts such as roots, leaves, berries, seeds, flowers, etc. and without causing visible symptoms on the host (Compant et al. 2011; Hardoim et al. 2015; Zhang et al. 2019).

### 16.2.2 Endophyte-Host Plant Association and Benefits

The endophytic lifestyle of microorganisms has many direct and indirect benefits to their host plants. One of the best-studied examples is the symbiotic relationship between the root nodule colonizing rhizobia and the legume plants is one of the best-studied examples wherein the endophyte helps the hosts through nitrogen fixation (Santoyo et al. 2016). Several studies have demonstrated that the endophyte host association confers numerous benefits to the host plants and also plays crucial roles in modifying plant growth and development, enhancing resilience against diverse biotic and abiotic stresses (Chitnis et al. 2020; Morelli et al. 2020).

The growth promotion can be a consequence of the direct or indirect action of the endophytes on the host. The direct effect is primarily by facilitating the acquisition of nutrients by the hosts and by producing plant growth-promoting molecules. This involves the fixing atmospheric nitrogen (Madhaiyan et al. 2015), solubilizing soil phosphorus (Varga et al. 2020), production of plant hormones like auxin (Suebrasri et al. 2020), gibberellic acid (Hamayun et al. 2017), synthesis of several other bioactive compounds (Gouda et al. 2016), resistance to diseases (Abdel-Motaal et al. 2020), tolerance to abiotic stress conditions such as drought (Jeong et al. 2021), salinity (Manasa et al. 2020; Hyde et al. 2019), high temperature (Ali et al.

**Table 16.1** Advantages of the endophyte association in host plants

| Sl. no | Crop | Endophyte species | Benefit(s) to the host plants | Reference |
|---|---|---|---|---|
| 1 | Sugarcane | *Klebsiella* sp. | Nitrogen fixation, improved growth | Lin et al. (2015) |
| 2 | Jatropha | *Methylo bacterium* sp. | Fix nitrogen and improved seed and biomass production | Madhaiyan et al. (2015) |
| 3 | *Phyllanthus amarus* | *Acinetobacter* sp. (ACMS25)<br>*Bacillus* sp. (PVMX4) | Solubilization of phosphorus (P), enhances antioxidant capacity, and phenolic content | Joe et al. (2016) |
| 4 | Soybean | *Porostereum spadiceum AGH786* | Improved growth under stress through the production of bioactive gibberellins, enhanced endogenous level of isoflavones like daidzen and genistein | Hamayun et al. (2017) |
| 5 | *Festuca rubra* | *Epichloe festucae* | Production of antifungal protein | Tian et al. (2017) |
| 6 | Cucumber | *CpE* | Improved tolerance to heat stress, enhanced accumulation of total sugars, soluble proteins, antioxidant enzymes and bioactive compounds like flavonoids and saponins | Ali et al. (2018) |
| 7 | Wheat | *Penicillium ruqueforti* Thom | Improved uptake of nutrients, tolerance to heavy metals | Ikram et al. (2018) |
| 8 | Jerusalem artichoke | *Macrophomina phaseolina BUP2/3*<br>*Diaporthe phaseolorum BUP3/1* | Phosphate solubilization, production of extracellular enzymes (amylase, protease, cellulase, and xylanase), growth promotion via biosynthesis of auxin | Suebrasri et al. (2020) |
| 9 | Tomato | *Aspergillus flavus* | Improved growth and secondary metabolite production, resistance to the pathogen (*A. phragmospora)* | Abdel-Motaal et al. (2020) |
| 10 | Rice | *Fusarium oxysporum (MH511104)* | Improved salt stress tolerance | Manasa et al. (2020) |
| 11 | Arabidopsis | *Piriformospora indica* | Freezing stress tolerance | Jiang et al. (2020) |
| 12 | Arabidopsis | *Kosakonia cowanii GG1* | Stimulate growth under drought through the production of exopolysaccharides | Jeong et al. (2021) |

2018), cold (Jiang et al. 2020), heavy metal (Ikram et al. 2018), etc. (Table 16.1). Apart from these, endophytes also produce many volatile organic compounds, antimicrobial compounds, enzymes, etc. which indirectly promote host plant growth and fitness (Santoyo et al. 2016; Tian et al. 2017).

### 16.2.3 Host Fitness by Holobiont Regulation

The endophyte-host association is as old as 60 million years (Sprent 2008). The influence of the endophytes on the host phenotype as reflected in the host fitness has been a desirable target for manipulating plant traits. Plants and their associated microorganisms called the microbiome are now considered to be a single functional entity called the holobiont (Zilber-Rosenberg and Rosenberg 2008). Here, the hologenome (combined genome of the host and the associated microbiome) function as a single genetic unit (Bordenstein and Theis 2015) to determine the function of the holobiont. The single genetic unit is further considered to be a single unit of selection, and hence the genomes of both the host and the microbiome are expected to be transmitted to next generation (Zilber-Rosenberg and Rosenberg 2008).

### 16.2.4 Transmission of Endophytes

Studies indicate that bacterial endophytes are transmitted vertically through seeds and pollen which is more likely, and horizontal transmission is possible through insects, soil, and atmosphere (Frank et al. 2017). The *Bacillus* spp. and *Microbacterium* spp. endophytes from switch grass were identified in the seeds of the previous generation crop, indicating a transgenerational transmission (Gagne-Bourgue et al. 2013). Another example is available in wheat grass (Ringelberg et al. 2012).Vertical (Hodgson et al. 2014) and horizontal transmission (Wiewiora et al. 2015) have been observed in fungal endophytes. However, it would be highly desirable if beneficial symbionts are vertically transmittable and are able to confer benefits to the next generation as well.

### 16.2.5 Endophytes as Potential Tools for Manipulating Stress Tolerance Traits

It is well known that plant–microbe associations are evolved to co-exist and influence the host phenotype. There are sufficient examples to show that these associations can improve the host's fitness, and can help the host plant in its adaptation to stressful conditions. Several studies indicate that endophytes are inevitable for plants to face the biotic and abiotic challenges in their habitat (Redman et al. 2002). Such habitat-adapted symbiotic relationships can be exploited for manipulating stress adaptive traits in non-host systems (Sangamesh et al. 2018; Hyde et al. 2019; Manasa et al. 2020). Moreover, it is promising that endophyte-host associations are also transmittable to subsequent crop generations. For these reasons, endophyte enrichment is increasingly considered as an ecologically sound and a potential tool to manipulate stress tolerance traits in economically important crop plants.

## 16.3 Endophytes Improve Stress Tolerance Traits in Crop Plants

Under the changing climate scenario, all crops are prone to stressful conditions in the field, either in singular or in combination. Survival under stressful conditions depends upon the type of the stress involved, its intensity, the tissue or organ affected, and on the activation of adaptation strategies (Cheng et al. 2013; Kissoudis et al. 2016). Hence, plants generally utilize a wide array of molecular, biochemical, and morpho-physiological strategies to respond to and acclimate to stressful conditions. Endophytes are known to confer or trigger many of these stress adaptive traits in their target hosts, through direct or indirect mechanisms. The various stress adaptive traits triggered by endophytes under major stress conditions of drought, salinity, and high temperature are discussed below.

### 16.3.1 *Activation of Drought Traits*

In many parts of the globe, water is already a limiting factor for agriculture. Plants generally rely on escape, avoidance, and tolerance strategies to face water deficit (Tuberosa 2012; Claeys and Inzé 2013), which involves various morpho-physiological and biochemical strategies conferring stress tolerance. Endophytes are known to activate several such traits in their hosts, which can be grouped into the traits associated with plant water relations (water conservation and water uptake) and cellular level tolerance (Dhanyalakshmi et al. 2019). This involves modifications of root development, root growth, and biomass as observed in *Chenopodium*, maize (González-Teuber et al. 2018; Li et al. 2019), improved water use efficiency, reduced stomatal conductance, and stomatal density in rice (Rho et al. 2018a), improved photosynthesis, stomatal conductance, and transpiration rate in sorghum (Zhang et al. 2017), improved abscisic acid (ABA) concentration in rice (Rho et al. 2018a), enhanced the accumulation of osmotically active solutes like glycine betaine, proline, soluble sugars, and organic acids and activation of reactive oxygen species (ROS) scavenging enzymes as observed in tall fescue and pinus (Nagabhyru et al. 2013; Zhou et al. 2021), enhanced accumulation of secondary metabolites, enhanced activities of the enzymes linked with secondary metabolism in sorghum (Zhang et al. 2017), etc. Moreover, endophytes trigger the expression of diverse stress-responsive genes in their host plants (Sherameti et al. 2008) that further prepares the crop to tolerate stress. In addition, there are several cases wherein endophytes directly confer the tolerance in plants through the secretion/ biosynthesis of osmolytes, hormones, and other bioactive compounds as noticed in maize and cucumber (Cohen et al. 2009; Waqas et al. 2012).

### 16.3.2 Activation of Salt Stress Tolerance Traits

Soil salinity is another important abiotic stress that limits crop productivity, globally. In the current trend of increasing soil salinization, by 2050, 50% of the cultivated land would be impacted by soil salinity (Hasanuzzaman et al. 2014; Hossain 2019). Accumulation of $Na^+$ and $Cl^-$ in soils interferes with growth and developmental processes in plants, through the induction of osmotic, ionic, and oxidative stresses (Tanveer and Shabala 2018). Salt exclusion, osmotic adjustment, and tissue tolerance are three important traits by which plants tolerate salt stress (Roy et al. 2014). This involves numerous strategies such as minimizing the uptake of $Na^+$ by the roots or by increasing the efflux of $Na^+$ back to the soil, $Na^+$ sequestration within the cell, and $Na^+$ exclusion, induction of osmotically active solutes, and oxidative stress tolerance (Zhu 2003; Munns et al. 2012; Shabala 2013). Endophytes activate many of these traits in plants under salt stress. Endophytes mitigate the toxic effect of $Na^+$ in salt-stressed plants by enhancing the uptake of $K^+$ to maintain the $Na^+$:$K^+$ ratio as observed in fescues (Sabzalian and Mirlohi 2010), likely by modulating the expression levels of $Na^+$:$K^+$ ion channels (Abdelaziz et al. 2017). In barley, endophyte-induced modification of root system architecture could regulate salt acquisition and translocation (Waller et al. 2005). Endophytes trigger the accumulation of several osmolytes like proline, polyamines, and organic acids in barley and chickpea (Egamberdieva et al. 2017; Chen et al. 2019), which can function to maintain the integrity of cell membranes, stabilize enzymes, and detoxify ROS. Endophyte-mediated activation of the antioxidative system to reduce the levels of ROS like hydrogen peroxide ($H_2O_2$) is demonstrated in barley and soybean (Egamberdieva et al. 2017; Asaf et al. 2018). Endophyte-induced accumulation of endogenous growth regulators like salicylic acid, and the accumulation of bioactive compounds like flavonoids (daidzein and genistein) minimize the stress effect on soybean plants (Khan et al. 2011). Another well known mechanism by which endophytes confer salt stress tolerance is by the activation of 1-aminocyclopropane-1-carboxylate (ACC) deaminase, that cleaves an ethylene precursor, to lower stress-related ethylene levels (Todorovic and Glick 2008; Win et al. 2018). Endophytes are also capable of altering the gene expression in plants under salt stress (Abdelaziz et al. 2017), which involves the activation of the salt stress-responsive genes, and that of the genes associated with the secondary metabolic pathways like phenylpropanoid and lignin pathways (Bajaj et al. 2018). The endophytic association helps the plants to improve the total biomass (Rho et al. 2018b), nutrient stoichiometry (Song et al. 2015), maintain ionic and hormonal balance and overall metabolism (Jan et al. 2019), thereby imparting tolerance to salt stress and better growth under salt stress conditions.

### 16.3.3 Activation of Heat Stress Tolerance Traits

Every plant species has an optimum temperature for growth and development, defined within the boundaries of its upper maximum and lower minimum temperature requirement for survival. A rise in temperature above the optimum and beyond a threshold level adversely affect growth, development and its productivity. Over the past decades, there was a consistent rise in global temperatures (Bita and Gerats 2013) and a 1.5 °C rise in global average surface temperatures is expected by 2100 (IPCC 2019) while every 1 °C change in temperature can cause considerable reduction in crop yields (Zhao et al. 2017). The major mechanisms evolved by plants to cope with the rising temperatures are the induction of diverse heat shock proteins (HSPs), accumulation of osmoprotectants, activation of ROS scavenging systems, stress signaling cascades, and phytohormones (Wahid et al. 2007). Endophytes are capable of enhancing its host's tolerance to high-temperature stress by improving the levels of osmolytes like proline, and other amino acids, sugars, and by reducing membrane injury (Ali et al. 2009). In cucumber, a significant increase in the accumulation of secondary metabolites like flavonoids and saponins under endophyte treatment has contributed for its tolerance to heat stress (Ali et al. 2018). There are several endophytes characterized to mitigate the heat stress effect by reducing lipid peroxidation and ROS levels as observed in sunflower and soybean (Ismail et al. 2020) and by enhancing the levels of antioxidant scavengers like superoxide dismutase, ascorbate reductase, and glutathione content in their host, soybean (Khan et al. 2020). Endophytes can activate the levels of HSPs under heat stress (Khan et al. 2020). Endophyte-mediated enhanced accumulation of phytohormones like ABA, indole butyric acid, and gibberellic acid, have led to heat stress tolerance in soybean (Park et al. 2017). Recent studies indicate that endophytes can induce transcriptome reprogramming via methylation of heat stress memory genes (Shekhawat et al. 2021).

## 16.4 Endophyte-Mediated Manipulation of Traits in Plants: Options and Challenges

The existing evidence from various host systems indicate that endophyte enrichment activates stress adaptive traits in host plants, besides improving their growth. There are several advantages to using endophyte-mediated activation of stress adaptive traits in plants when compared to the other approaches. Endophyte enrichment can be used as ideal approach to activate the diverse traits already encoded in the genomes of existing crop varieties. The study by Sampangi-Ramaiah et al. (2020) has demonstrated that the inoculation of salt stress-tolerant endophyte from Pokkali rice could activate salt stress tolerance traits in an otherwise salt-sensitive rice variety. There are several examples wherein a single endophyte can confer tolerance to multiple abiotic stresses (Hubbard et al. 2014). Co-inoculation of multiple trait-specific endophytes can be an important strategy to address diverse issues. For example, the enrichment of soybean using the endophytes (*Paecilomyces formosus* LHL10 and *Penicillium funiculosum* LHL06) producing the hormones gibberellins and indole-3-acetic acid, and are tolerant to multiple abiotic stresses (heavy metals, drought, and high temperature) promoted its growth and imparted tolerance to multiple abiotic stresses (Bilal et al. 2020). It is also important to identify specific endophytes that can confer specific traits. For example, endophytes that synthesize 1-aminocyclopropane-1-carboxylate (ACC deaminase) (*Pseudomonas* spp., *Sebacina vermifera*) specifically function to lower the stress induced ethylene levels in their hosts through the degradation of its precursor and thus conferring tolerance against salt stress (Barazani et al. 2007; Win et al. 2018). It is also a huge challenge to identify the endophytes which confer desired traits in a wide host range as in the case of *P. indica*.

## 16.5 Conclusions

Targeted trait manipulation in crops has resulted in significant improvements in crop yield and crop quality. Multiple traits have been pyramided in the modern crop varieties and for sustained food production under the conditions of climate change (Fig. 16.1), we need to think of alternative strategies for trait activation. The endophyte-mediated trait manipulation discussed in this chapter would be useful to build crop resilience against abiotic stresses. The endophyte enrichment approach would be eco-friendly, and will not have any regulatory issues as seen in the genetic manipulation technologies of crop improvement. However, there is a need to understand the endophyte–plant interaction to avoid deleterious effects under field conditions and also for exploiting the technology on a commercial scale.

**Acknowledgement** KNN would like to thank The World Bank and the Indian Council of Agricultural Research (ICAR), Government of India, New Delhi (ICAR-CAAST-Project, F.No./NAHEP/CAAST/ 2018-19) and ICAR-NICRA F.No.2-13(34)/2021-22 for the financial support.

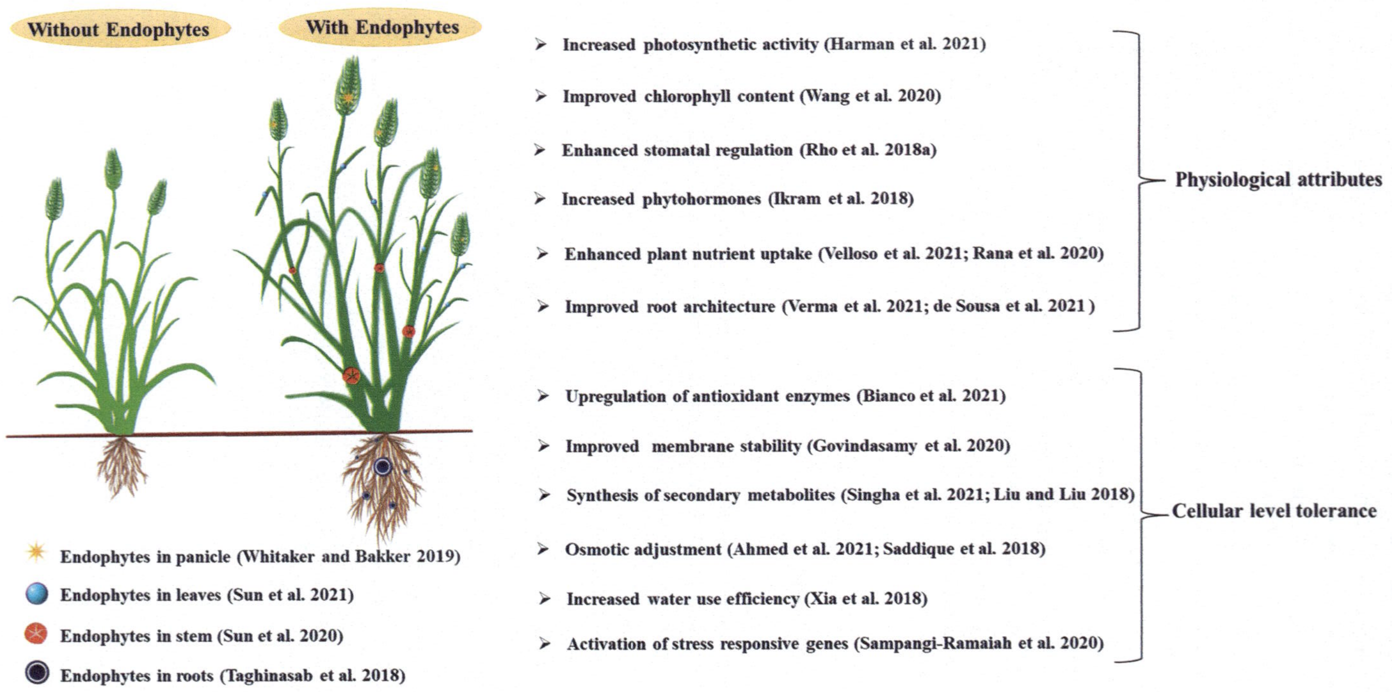


**Fig. 16.1** Schematic representation of benefits of endophyte enrichment and endophyte activated process/traits in plants

## References

Abdelaziz ME, Kim D, Ali S, Fedoroff NV, Al-Babili S (2017) The endophytic fungus *Piriformospora indica* enhances *Arabidopsis thaliana* growth and modulates Na /K homeostasis under salt stress conditions. Plant Sci 263:107–115

Abdel-Motaal F, Kamel N, El-Zayat S, Abou-Ellail M (2020) Early blight suppression and plant growth promotion potential of the endophyte *Aspergillus flavus* in tomato plant. Ann Agric Sci 65(2):117–123

Ahmed S, Heo TY, Roy Choudhury A, Walitang DI, Choi J, Sa T (2021) Accumulation of compatible solutes in rice (*Oryza sativa* L.) cultivars by inoculation of endophytic plant growth promoting bacteria to alleviate salt stress. Appl Biol Chem 64(1):1–4

Ali SZ, Sandhya V, Grover M, Kishore N, Rao LV, Venkateswarlu B (2009) *Pseudomonas* sp. strain AKM-P6 enhances tolerance of sorghum seedlings to elevated temperatures. Biol Fertil Soils 46(1):45–55

Ali AH, Abdelrahman M, Radwan U, El-Zayat S, El-Sayed MA (2018) Effect of *Thermomyces* fungal endophyte isolated from extreme hot desert-adapted plant on heat stress tolerance of cucumber. Appl Soil Ecol 124:155

Asaf S, Hamayun M, Khan AL, Waqas M, Khan MA, Jan R, Lee IJ, Hussain A (2018) Salt tolerance of *Glycine max*. L induced by endophytic fungus *Aspergillus flavus* CSH1, via regulating its endogenous hormones and antioxidative system. Plant Physiol Biochem 128:13–23

Bajaj R, Huang Y, Gebrechristos S, Mikolajczyk B, Brown H, Prasad R, Varma A, Bushley KE (2018) Transcriptional responses of soybean roots to colonization with the root endophytic fungus *Piriformospora indica* reveals altered phenylpropanoid and secondary metabolism. Sci Rep 8(1):1–8

Barazani O, von Dahl CC, Baldwin IT (2007) *Sebacina vermifera* promotes the growth and fitness of *Nicotiana attenuata* by inhibiting ethylene signaling. Plant Physiol 144(2):1223–1232

Berg G, Smalla K (2009) Plant species and soil type cooperatively shape the structure and function of microbial communities in the rhizosphere. FEMS Microbiol Ecol 68(1):1–13

Bianco C, Andreozzi A, Romano S, Fagorzi C, Cangioli L, Prieto P, Cisse F, Niangado O, Sidibé A, Pianezze S, Perini M (2021) Endophytes from African Rice (*Oryza glaberrima* L.) efficiently colonize Asian Rice (*Oryza sativa* L.) stimulating the activity of its antioxidant enzymes and increasing the content of nitrogen, carbon, and chlorophyll. Microorganisms 9(8):1714

Bilal S, Shahzad R, Imran M, Jan R, Kim KM, Lee IJ (2020) Synergistic association of endophytic fungi enhances *Glycine max* L. resilience to combined abiotic stresses: heavy metals, high temperature and drought stress. Ind Crops Prod 143:111931

Bita C, Gerats T (2013) Plant tolerance to high temperature in a changing environment: scientific fundamentals and production of heat stress-tolerant crops. Front Plant Sci 4:273

Bordenstein SR, Theis KR (2015) Host biology in light of the microbiome: ten principles of holobionts and hologenomes. PLoS Biol 13(8):e1002226

Carroll GC (1986) The biology of endophytism in plants with particular reference to woody perennials. In: Fokkema NJ, van den Heuvel J (eds) Microbiology of the phyllosphere. Cambridge University Press, Cambridge, pp 205–222

Chen T, Li C, White JF, Nan Z (2019) Effect of the fungal endophyte *Epichloë bromicola* on polyamines in wild barley (*Hordeum brevisubulatum*) under salt stress. Plant Soil 436(1):29–48

Cheng C, Gao X, Feng B, Sheen J, Shan L, He P (2013) Plant immune response to pathogens differs with changing temperatures. Nat Commun 4(1):1–9

Chitnis VR, Suryanarayanan TS, Nataraja KN, Prasad SR, Oelmüller R, Shaanker RU (2020) Fungal endophyte-mediated crop improvement: the way ahead. Front Plant Sci 11:1588

Claeys H, Inzé D (2013) The agony of choice: how plants balance growth and survival under water-limiting conditions. Plant Physiol 162(4):1768–1779

Cohen AC, Travaglia CN, Bottini R, Piccoli PN (2009) Participation of abscisic acid and gibberellins produced by endophytic *Azospirillum* in the alleviation of drought effects in maize. Botany 87(5):455–462

Compant S, Mitter B, Colli-Mull JG, Gangl H, Sessitsch A (2011) Endophytes of grapevine flowers, berries, and seeds: identification of cultivable bacteria, comparison with other plant parts, and visualization of niches of colonization. Microb Ecol 62(1):188–197

de Sousa SM, de Oliveira CA, Andrade DL, de Carvalho CG, Ribeiro VP, Pastina MM, Marriel IE, de Paula Lana UG, Gomes EA (2021) Tropical *Bacillus* strains inoculation enhances maize root surface area, dry weight, nutrient uptake and grain yield. J Plant Growth Regul 40(2):867–877

deBary A (1866) Morphologie und physiologie der pilze, flechten und myxomyceten. Engelmann

Dhanyalakshmi KH, Nataraja KN (2021) Universal stress protein-like gene from mulberry enhances abiotic stress tolerance in *Escherichia coli* and transgenic tobacco cells. Plant Biol 23(6):1190–1194

Dhanyalakshmi KH, Naika MB, Sajeevan RS, Mathew OK, Shafi KM, Sowdhamini R, Nataraja KN (2016) An approach to function annotation for proteins of unknown function (PUFs) in the transcriptome of Indian mulberry. PLoS One 11(3):e0151323

Dhanyalakshmi KH, Mounashree DC, Vidyashree DN, Earanna N, Nataraja KN (2019) Options and opportunities for manipulation of drought traits using endophytes in crops. Plant Physiol Rep 24(4):555–562

Egamberdieva D, Wirth SJ, Shurigin VV, Hashem A, Abd Allah EF (2017) Endophytic bacteria improve plant growth, symbiotic performance of chickpea (*Cicer arietinum* L.) and induce suppression of root rot caused by *Fusarium solani* under salt stress. Front Microbiol 8:1887

Fahad S, Bajwa AA, Nazir U, Anjum SA, Farooq A, Zohaib A, Sadia S, Nasim W, Adkins S, Saud S, Ihsan MZ (2017) Crop production under drought and heat stress: plant responses and management options. Front Plant Sci 8:1147

Frank AC, Saldierna Guzmán JP, Shay JE (2017) Transmission of bacterial endophytes. Microorganisms 5(4):70

Gagne-Bourgue F, Aliferis KA, Seguin P, Rani M, Samson R, Jabaji S (2013) Isolation and characterization of indigenous endophytic bacteria associated with leaves of switchgrass (*Panicum virgatum* L.) cultivars. J Appl Microbiol 114(3):836–853

González-Teuber M, Urzúa A, Plaza P, Bascuñán-Godoy L (2018) Effects of root endophytic fungi on response of *Chenopodium quinoa* to drought stress. Plant Ecol 219(3):231–240

Gouda S, Das G, Sen SK, Shin HS, Patra JK (2016) Endophytes: a treasure house of bioactive compounds of medicinal importance. Front Microbiol 7:1538

Govindasamy V, George P, Kumar M, Aher L, Raina SK, Rane J, Annapurna K, Minhas PS (2020) Multi-trait PGP rhizobacterial endophytes alleviate drought stress in a senescent genotype of sorghum [*Sorghum bicolor* (L.) Moench]. 3 Biotech 10(1):13

Hamayun M, Hussain A, Khan SA, Kim HY, Khan AL, Waqas M, Irshad M, Iqbal A, Rehman G, Jan S, Lee IJ (2017) Gibberellins producing endophytic fungus *Porostereum spadiceum* AGH786 rescues growth of affected soybean. Front Microbiol 8:686

Hardoim PR, Van Overbeek LS, Berg G, Pirttilä AM, Compant S, Campisano A, Döring M, Sessitsch A (2015) The hidden world within plants: ecological and evolutionary considerations for defining functioning of microbial endophytes. Microbiol Mol Biol Rev 79(3):293–320

Harman GE, Doni F, Khadka RB, Uphoff N (2021) Endophytic strains of *Trichoderma* increase plants' photosynthetic capability. J Appl Microbiol 130(2):529–546

Hasanuzzaman M, Nahar K, Alam M, Bhowmik PC, Hossain M, Rahman MM, Prasad MN, Ozturk M, Fujita M (2014) Potential use of halophytes to remediate saline soils. Biomed Res Int 2014:589341

Hodgson S, de Cates C, Hodgson J, Morley NJ, Sutton BC, Gange AC (2014) Vertical transmission of fungal endophytes is widespread in forbs. Ecol Evol 4(8):1199–1208

Hossain MS (2019) Present scenario of global salt-affected soils, its management, and importance of salinity research. Int Res J Biol Sci 1:1–3

Hubbard M, Germida JJ, Vujanovic V (2014) Fungal endophytes enhance wheat heat and drought tolerance in terms of grain yield and second-generation seed viability. J Appl Microbiol 116(1): 109–122

Hyde KD, Xu J, Rapior S et al (2019) The amazing potential of fungi: 50 ways we can exploit fungi industrially. Fungal Divers 97:1–136

Ikram M, Ali N, Jan G, Jan FG, Rahman IU, Iqbal A, Hamayun M (2018) IAA producing fungal endophyte*Penicillium roqueforti* Thom., enhances stress tolerance and nutrients uptake in wheat plants grown on heavy metal contaminated soils. PLoS One 13(11):e0208150

IPCC (2019) Technical summary [Pörtner H-O, Roberts DC, Masson-Delmotte V, Zhai P, Poloczanska E, Mintenbeck K, Tignor M, Alegría A, Nicolai M, Okem A, Petzold J, Rama B, Weyer NM (eds)]. In: Pörtner H-O, Roberts DC, Masson-Delmotte V, Zhai P, Tignor M, Poloczanska E, Mintenbeck K, Alegría A, Nicolai M, Okem A, Petzold J, Rama B, Weyer NM (eds) IPCC special report on the ocean and cryosphere in a changing climate

Ismail HM, Hussain A, Iqbal A, Khan SA, Lee IJ (2020) *Aspergillus niger* boosted heat stress tolerance in sunflower and soybean via regulating their metabolic and antioxidant system. J Plant Interact 15(1):223–232

Jan FG, Hamayun M, Hussain A, Jan G, Iqbal A, Khan A, Lee IJ (2019) An endophytic isolate of the fungus *Yarrowia lipolytica* produces metabolites that ameliorate the negative impact of salt stress on the physiology of maize. BMC Microbiol 19(1):1–10

Jeong S, Kim TM, Choi B, Kim Y, Kim E (2021) Invasive *Lactuca serriola* seeds contain endophytic bacteria that contribute to drought tolerance. Sci Rep 11(1):13307

Jiang W, Pan R, Wu C, Xu L, Abdelaziz ME, Oelmüller R, Zhang W (2020) *Piriformospora indica* enhances freezing tolerance and post-thaw recovery in *Arabidopsis* by stimulating the expression of CBF genes. Plant Signal Behav 15(4):1745472

Joe MM, Devaraj S, Benson A, Sa T (2016) Isolation of phosphate solubilizing endophytic bacteria from *Phyllanthus amarus* Schum&Thonn: evaluation of plant growth promotion and antioxidant activity under salt stress. J Appl Res Med Aromat Plants 3(2):71–77

Karaba A, Dixit S, Greco R, Aharoni A, Trijatmiko KR, Marsch-Martinez N, Krishnan A, Nataraja KN, Udayakumar M, Pereira A (2007) Improvement of water use efficiency in rice by expression of HARDY, an Arabidopsis drought and salt tolerance gene. Proc Natl Acad Sci U S A 104(39):15270–15275

Khan AL, Hamayun M, Ahmad N, Hussain J, Kang SM, Kim YH, Adnan M, Tang DS, Waqas M, Radhakrishnan R, Hwang YH (2011) Salinity stress resistance offered by endophytic fungal interaction between *Penicillium minioluteum* LHL09 and *Glycine max*. L J Microbiol Biotechnol 21(9):893–902

Khan MA, Asaf S, Khan AL, Jan R, Kang SM, Kim KM, Lee IJ (2020) Thermotolerance effect of plant growth-promoting *Bacillus cereus* SA1 on soybean during heat stress. BMC Microbiol 20(1):1–4

Kissoudis C, Sunarti S, Van De Wiel C, Visser RG, van der Linden CG, Bai Y (2016) Responses to combined abiotic and biotic stress in tomato are governed by stress intensity and resistance mechanism. J Exp Bot 67(17):5119–5132

Li X, He C, He X, Su F, Hou L, Ren Y, Hou Y (2019) Dark septate endophytes improve the growth of host and non-host plants under drought stress through altered root development. Plant Soil 439(1):259–272

Lin L, Wei C, Chen M, Wang H, Li Y, Li Y, Yang L, An Q (2015) Complete genome sequence of endophytic nitrogen-fixing *Klebsiella variicola* strain DX120E. Stand Genomic Sci 10(1):1–7

Liu J, Liu G (2018) Analysis of secondary metabolites from plant endophytic fungi. In: Plant pathogenic fungi and oomycetes 2018. Humana Press, New York, pp 25–38

Madhaiyan M, Alex TH, Ngoh ST, Prithiviraj B, Ji L (2015) Leaf-residing methylobacterium species fix nitrogen and promote biomass and seed production in *Jatropha curcas*. Biotechnol Biofuels 8(1):1–4

Manasa KM, Vasanthakumari MM, Nataraja KN, Shaanker RU (2020) Endophytic fungi of salt adapted *Ipomea pes-caprae* LR Br: their possible role in inducing salinity tolerance in paddy (*Oryza sativa* L.). Curr Sci 118(9):1448

Morelli M, Bahar O, Papadopoulou KK, Hopkins DL, Obradović A (2020) Role of endophytes in plant health and defense against pathogens. Front Plant Sci 11:1312

Munns R, James RA, Xu B, Athman A, Conn SJ, Jordans C, Byrt CS, Hare RA, Tyerman SD, Tester M, Plett D, Gilliham M (2012) Wheat grain yield on saline soils is improved by an ancestral $Na^+$ transporter gene. Nat Biotechnol 30(4):360–364

Nagabhyru P, Dinkins RD, Wood CL, Bacon CW, Schardl CL (2013) Tall fescue endophyte effects on tolerance to water-deficit stress. BMC Plant Biol 3(1):1–17

Papagiannaki K, Lagouvardos K, Kotroni V, Papagiannaki G (2014) Agricultural losses related to frost events: use of the 850 hPa level temperature as an explanatory variable of the damage cost. Nat Hazards Earth Syst Sci 14(9):2375–2386

Park YG, Mun BG, Kang SM, Hussain A, Shahzad R, Seo CW, Kim AY, Lee SU, Oh KY, Lee DY, Lee IJ (2017) *Bacillus aryabhattai* SRB02 tolerates oxidative and nitrosative stress and promotes the growth of soybean by modulating the production of phytohormones. PLoS One 12(3): e0173203

Petrini O (1991) Fungal endophytes of tree leaves. In: Microbial ecology of leaves. Springer, New York, pp 179–197

Pruthvi V, Narasimhan R, Nataraja KN (2014) Simultaneous expression of abiotic stress responsive transcription factors, AtDREB2A, AtHB7 and AtABF3 improves salinity and drought tolerance in peanut (*Arachis hypogaea* L.). PLoS One 9(12):e111152

Rana KL, Kour D, Kaur T, Sheikh I, Yadav AN, Kumar V, Suman A, Dhaliwal HS (2020) Endophytic microbes from diverse wheat genotypes and their potential biotechnological applications in plant growth promotion and nutrient uptake. Proc Natl Acad Sci India B Biol Sci 90(5):969–979

Redman RS, Sheehan KB, Stout RG, Rodriguez RJ, Henson JM (2002) Thermotolerance generated by plant/fungal symbiosis. Science 298(5598):1581

Rho H, Van Epps V, Wegley N, Doty SL, Kim SH (2018a) Salicaceae endophytes modulate stomatal behavior and increase water use efficiency in rice. Front Plant Sci 9:188

Rho H, Hsieh M, Kandel SL, Cantillo J, Doty SL, Kim SH (2018b) Do endophytes promote growth of host plants under stress? A meta-analysis on plant stress mitigation by endophytes. Microb Ecol 75(2):407–418

Ringelberg D, Foley K, Reynolds CM (2012) Bacterial endophyte communities of two wheatgrass varieties following propagation in different growing media. Can J Microbiol 58(1):67–80

Roy SJ, Negrão S, Tester M (2014) Salt resistant crop plants. Curr Opin Biotechnol 26:115–124

Sabzalian RM, Mirlohi A (2010) *Neotyphodium* endophytes trigger salt resistance in tall and meadow fescues. J Plant Nutr Soil Sci 173(6):952–957

Saddique MA, Ali Z, Khan AS, Rana IA, Shamsi IH (2018) Inoculation with the endophyte *Piriformospora indica* significantly affects mechanisms involved in osmotic stress in rice. Rice 11(1):34

Sampangi-Ramaiah MH, Dey P, Jambagi S, VasanthaKumari MM, Oelmüller R, Nataraja KN, Venkataramana Ravishankar K, Ravikanth G, Uma Shaanker R (2020) An endophyte from salt-adapted Pokkali rice confers salt-tolerance to a salt-sensitive rice variety and targets a unique pattern of genes in its new host. Sci Rep 1:1–4

Sangamesh MB, Jambagi S, Vasanthakumari MM, Shetty NJ, Kolte H, Ravikanth G, Nataraja KN, Uma Shaanker R (2018) Thermotolerance of fungal endophytes isolated from plants adapted to the Thar Desert, India. Symbiosis 75(2):135–147

Santoyo G, Moreno-Hagelsieb G, del Carmen O-MM, Glick BR (2016) Plant growth-promoting bacterial endophytes. Microbiol Res 183:92–99

Shabala S (2013) Learning from halophytes: physiological basis and strategies to improve abiotic stress tolerance in crops. Ann Bot 112(7):1209–1221

Shekhawat K, Saad MM, Sheikh A, Mariappan K, Al-Mahmoudi H, Abdulhakim F, Eida AA, Jalal R, Masmoudi K, Hirt H (2021) Root endophyte induced plant thermotolerance by constitutive chromatin modification at heat stress memory gene loci. EMBO Rep 22(3):e51049

Sherameti I, Tripathi S, Varma A, Oelmüller R (2008) The root-colonizing endophyte *Pirifomospora indica* confers drought tolerance in *Arabidopsis* by stimulating the expression of drought stress-related genes in leaves. Mol Plant-Microbe Interact 21(6):799–807

Singha KM, Singh B, Pandey P (2021) Host specific endophytic microbiome diversity and associated functions in three varieties of scented black rice are dependent on growth stage. Sci Rep 11(1):1–7

Song M, Chai Q, Li X, Yao X, Li C, Christensen MJ, Nan Z (2015) An asexual *Epichloë* endophyte modifies the nutrient stoichiometry of wild barley (*Hordeum brevisubulatum*) under salt stress. Plant Soil 387(1):153–165

Sprent JI (2008) 60Ma of legume nodulation. What's new? What's changing? J Exp Bot 59(5): 1081–1084

Suebrasri T, Harada H, Jogloy S, Ekprasert J, Boonlue S (2020) Auxin-producing fungal endophytes promote growth of sunchoke. Rhizosphere 16:100271

Sun X, Kosman E, Sharon A (2020) Stem endophytic mycobiota in wild and domesticated wheat: structural differences and hidden resources for wheat improvement. J Fungi 6(3):180

Sun ZB, Li HY, Lin ZA, Yuan L, Xu JK, Zhang SQ, Li YT, Zhao BQ, Wen YC (2021) Long-term fertilisation regimes influence the diversity and community of wheat leaf bacterial endophytes. Ann Appl Biol 179(2):176–184

Taghinasab M, Imani J, Steffens D, Glaeser SP, Kogel KH (2018) The root endophytes *Trametes versicolor* and *Piriformospora indica* increase grain yield and P content in wheat. Plant Soil 426(1):339–348

Tanveer M, Shabala S (2018) Targeting redox regulatory mechanisms for salinity stress tolerance in crops. In: Salinity responses and tolerance in plants, vol 1. Springer, Cham, pp 213–234

Tian Z, Wang R, Ambrose KV, Clarke BB, Belanger FC (2017) The *Epichloë festucae* antifungal protein has activity against the plant pathogen *Sclerotinia homoeocarpa*, the causal agent of dollar spot disease. Sci Rep 7(1):5

Todorovic B, Glick BR (2008) The interconversion of ACC deaminase and D-cysteine desulfhydrase by directed mutagenesis. Planta 229(1):193–205

Tuberosa R (2012) Phenotyping for drought tolerance of crops in the genomics era. Front Physiol 3: 347

Turner TR, James EK, Poole PS (2013) The plant microbiome. Genome Biol 14(6):1–10

Varga T, Hixson KK, Ahkami AH, Sher AW, Barnes ME, Chu RK, Battu AK, Nicora CD, Winkler TE, Reno LR, Fakra SC (2020) Endophyte-promoted phosphorus solubilization in *Populus*. Front Plant Sci 11:567918

Velloso CC, Ribeiro VP, Carvalho CG, Oliveira CA, Paula Lana UG, Marriel IE, Sousa SM, Gomes EA (2021) Tropical endophytic *Bacillus* species enhance plant growth and nutrient uptake in cereals. In: Endophytes: mineral nutrient management, vol 3. Springer, Cham, pp 157–180

Verma SK, Sahu PK, Kumar K, Pal G, Gond SK, Kharwar RN, White JF (2021) Endophyte roles in nutrient acquisition, root system architecture development and oxidative stress tolerance. J Appl Microbiol 131(5):2161–2177

Vorholt JA (2012) Microbial life in the phyllosphere. Nat Rev Microbiol 10(12):828–840

Wahid A, Gelani S, Ashraf M, Foolad MR (2007) Heat tolerance in plants: an overview. Environ Exp Bot 61(3):199–223

Waller F, Achatz B, Baltruschat H, Fodor J, Becker K, Fischer M, Heier T, Hückelhoven R, Neumann C, von Wettstein D, Franken P (2005) The endophytic fungus *Piriformospora indica* reprograms barley to salt-stress tolerance, disease resistance, and higher yield. Proc Natl Acad Sci U S A 102(38):13386–13391

Wang Z, Li C, White J (2020) Effects of *Epichloë* endophyte infection on growth, physiological properties and seed germination of wild barley under saline conditions. J Agron Crop Sci 206 (1):43–51

Waqas M, Khan AL, Kamran M, Hamayun M, Kang SM, Kim YH, Lee IJ (2012) Endophytic fungi produce gibberellins and indoleacetic acid and promotes host-plant growth during stress. Front Plant Sci 11:567918

Whitaker BK, Bakker MG (2019) Bacterial endophyte antagonism toward a fungal pathogen in vitro does not predict protection in live plant tissue. FEMS Microbiol Ecol 95(2):fiy237

Wiewiora B, Żurek G, Pańka D (2015) Is the vertical transmission of *Neotyphodium lolii* in perennial ryegrass the only possible way to the spread of endophytes? PLoS One 10(2):e011723

Win KT, Tanaka F, Okazaki K, Ohwaki Y (2018) The ACC deaminase expressing endophyte *Pseudomonas* spp. enhances NaCl stress tolerance by reducing stress-related ethylene production, resulting in improved growth, photosynthetic performance, and ionic balance in tomato plants. Plant Physiol Biochem 127:599–607

Wing IS, De Cian E, Mistry MN (2021) Global vulnerability of crop yields to climate change. J Environ Econ Manage 109:102462

Xia C, Christensen MJ, Zhang X, Nan Z (2018) Effect of *Epichloë gansuensis* endophyte and transgenerational effects on the water use efficiency, nutrient and biomass accumulation of *Achnatherum inebrians* under soil water deficit. Plant Soil 424(1):555–571

Zhang Q, Gong M, Yua J, Hou Y, Zhang H, Wang Y, Hou X (2017) Dark septate endophyte improves drought tolerance in sorghum. Int J Agric Biol 19(1)

Zhang Q, Acuña JJ, Inostroza NG, Mora ML, Radic S, Sadowsky MJ, Jorquera MA (2019) Endophytic bacterial communities associated with roots and leaves of plants growing in Chilean extreme environments. Sci Rep 9(1):1–2

Zhao C, Liu B, Piao S, Wang X, Lobell DB, Huang Y, Huang M, Yao Y, Bassu S, Ciais P, Durand JL (2017) Temperature increase reduces global yields of major crops in four independent estimates. Proc Natl Acad Sci U S A 114(35):9326–9331

Zhou XR, Dai L, Xu GF, Wang HS (2021) A strain of *Phoma* species improves drought tolerance of *Pinus tabulaeformis*. Sci Rep 11(1):1–11

Zhu JK (2003) Regulation of ion homeostasis under salt stress. Curr Opin Plant Biol 6(5):441–445

Zilber-Rosenberg I, Rosenberg E (2008) Role of microorganisms in the evolution of animals and plants: the hologenome theory of evolution. FEMS Microbiol Rev 32(5):723–735

Zorb C, Geilfus CM, Dietz KJ (2019) Salinity and crop yield. Plant Biol 21:31–38

# Chapter 17
# Influence of High Temperature Stress on Grain Crops

**P. V. V. Prasad, S. Gowsiga, and M. Djanaguiraman**

**Abstract** High temperature (HT) or heat stress is among the most important environmental stresses limiting productivity of major field grains. It is projected that in future climate most of the crop production regions will experience both increase in long-term average temperature and greater frequency of short periods (episodes) of HT stress. These will have negative impacts on yields. Here we summarized the impact of HT stress research done on major field grain crops (e.g., wheat, grain sorghum, millets, soybean, and groundnut). Above optimum daily mean temperatures decreased yield of crops. There are two most sensitive periods to HT stress for these crops: first during floral bud development (coinciding with sporogenesis) and the second at anthesis (coinciding with pollination and fertilization). Stress during these two stages decreases grain numbers. The decrease in grain numbers is a result of decreased gamete function (viability of pollen or ovules), fertilization (anther dehiscence, pollination; pollen tube growth and fertilization) or early embryo abortion. Similarly, HT during grain filling stage decreases seed size, mainly due to shorter grain filling duration. Combination of both fewer grain numbers, and grain size results in decreased yields. HT stress damaged chloroplast ultrastructure, membrane damage mediated through oxidative damage. This results in early leaf senescence leading to shorter green leaf area duration. HT stress increases respiration rates and decreases the net carbon available for growth. There are genetic differences in response to HT stress in these crops. However, the current breeding programs have not yet fully exploited the existing genetic variability. We must systematically evaluate germplasm including wild relative and exotic collections to identify diverse tolerant genotypes and use them in crop improvement programs. Use of high throughput phenotyping and integrating with genotypic

P. V. V. Prasad (✉)
Department of Agronomy, Kansas State University, Manhattan, KS, USA
e-mail: vara@ksu.edu

S. Gowsiga · M. Djanaguiraman
Department of Crop Physiology, Tamil Nadu Agricultural University, Coimbatore, India



H. M. Mamrutha et al. (eds.), *Translating Physiological Tools to Augment Crop Breeding*, https://doi.org/10.1007/978-981-19-7498-4_17

data will further help in identifying traits and genes associated with stress tolerance and breeding for HT tolerance in grain crops.

**Keywords** High temperature stress · Wheat · Sorghum · Millets · Soybean · Ground nut

## 17.1 Introduction

Wheat (*Triticum aestivum* L.), rice (*Oryza sativa* L.), maize (*Zea mays* L.), sorghum (*Sorghum bicolor* (L.) Moench), pearl millet (*Pennisetum glaucum* (L.) R. Br.), soybean (*Glycine max* (L.) Merrill), and peanut/groundnut (*Arachis hypogaea* L.) are some of the important food crops. These food crops are grown under diverse environmental conditions and experiences high temperatures (HT) at critical stages of crop development. Abiotic stresses like HT can cause changes in physiology, growth, and development, ultimately decreasing grain yield. Increase in mean temperature will also be associated with short episodes of high temperatures (IPCC 2013, 2021). Response of the crop to HT varies with phenological stage of the crop, crop species, and genotypes with crop species.

Temperature increase of 3–4 °C can decrease yields by 15–35% in many countries around the world (Ortiz et al. 2008). Meehl and Tebaldi (2004) and the Intergovernmental Panel on Climate Change (IPCC 2013, 2021) project more frequent and extreme events in crop production regions. Asseng et al. (2015) estimated that wheat productivity to fall by 6% per °C rise in temperatures. The simulations of Berg et al. (2011) indicates that in West Africa, sorghum and millet yield losses will be up to 41% due to HT, and if the temperature increase exceeds 2 °C, any rainfall change cannot overcome the negative impacts caused by HT. Hatfield et al. (2008) estimated that a 0.8–1 °C increase in temperature in Southeastern United States (US) will decrease soybean yield by 1.3–2.4%. Future climate change scenarios will decrease yield up to 5% per 1 °C increase in temperatures (Hatfield et al. 2011). Soybean yields are expected to decrease (Schlenker and Roberts 2009) under HT. Similarly, Tack et al. (2017) showed that the sorghum yields decreased by 17% per 2 °C increase in temperature. The uncertainty caused by temperature represents a more significant challenge to climate change impact for most crops in most regions (Lobell and Burke 2008).

Developing and using HT stress-tolerant cultivars is one of the efficient ways to produce stable yield. Breeding stress-tolerant cultivars requires improved understanding of mechanisms associated with tolerance and identification of tolerant genotypes using high throughput phenotyping and methods to introgress tolerant genes into modern cultivars. Each crop species has different response to HS. There are different cardinal temperatures (minimum, optimum, and maximum) for different developmental stages. Growth and development does not occur below the base temperature (minimum; $T_{min}$) and above ceiling temperature (maximum; $T_{max}$) and maximum growth and development occurs at optimum temperature ($T_{opt}$). Most crops have a higher $T_{opt}$ for vegetative phase when compared to reproductive phase,

**Table 17.1** Cardinal temperatures for different growth stages and yield of wheat, sorghum, pearl millet, peanut, and soybean

| Crop | Stage | $T_{min}$ (°C) | $T_{opt}$ (°C) | $T_{max}$ (°C) | Reference |
|---|---|---|---|---|---|
| Wheat | Sowing to emergence | 3.5 | 22.0 | 32.7 | Porter and Gawith (1999) |
| | Vernalization | −1.3 | 4.9 | 15.7 | |
| | Terminal spikelet | 1.5 | 10.6 | >20.0 | |
| | Anthesis | 9.5 | 21.0 | 31.0 | |
| | Grain filing | 9.2 | 20.7 | 35.4 | |
| Sorghum | Vegetative | 8 | 34 | 44 | Alagarswamy and Ritchie (1991) |
| | Reproductive | – | 32 | 40 | Prasad et al. (2006a) |
| | Yield | – | 23–25 | 35 | |
| | Vegetative | – | 34 | – | Hatfield et al. (2008) |
| | Reproductive | – | 26–34 | – | Hatfield et al. (2008) |
| Pearl millet | Germination | 8–13.5 | 34 | – | |
| | Vegetative stage | – | 32–35 | – | |
| | Reproductive stage | – | 22–35 | 40–45 | Djanaguiraman et al. (2018a) |
| | Grain or seed set | – | 22–25 | 38 | |
| | Grain or seed growth | – | 19–31 | – | |
| Peanut | Germination | 10–12 | – | – | |
| | Vegetative development | – | 29–33 | – | Bolhuis and De Groot (1959) |
| | Pollen viability and Seed set (%) | – | <31 | 39–40 | Prasad et al. (2003) |
| | Fruit set (%) | – | <33 | 43 | Prasad et al. (2001) |
| | Pod and seed yield, harvest index, and seed size | – | 23–24 | – | Prasad et al. (2003) |
| Soybean | Post-anthesis, single Seed growth rate, seed size, and yield | – | 23 | 39 | Hatfield et al. (2008) |

– Data not available

thus, the susceptibility to HT varies with the stage of plant development. Grain yield is a function of grain numbers and grain weight, both are sensitive to HT stress. Cardinal temperatures for key grain crops are indicated (Table 17.1).

The generally summary of HT influence on phenology, growth, physiological, reproductive and yield processes are summarized in Table 17.2. HT decreases the above (shoot) and below ground (root) growth and accelerates the leaf abscission and senescence process, which ultimately results in lower grain yield. Under HT stress plants show different cellular and molecular responses such as changes in cellular structure organization, increased membrane fluidity, decreased synthesis of

**Table 17.2** General effects (↑, increases; ↓, decreases; and =, no change) of high temperature stress on various growth and yield parameters in most crop species

| Trait | High temperature stress |
|---|---|
| Phenology | ↓↑ |
| Plant height | ↓ |
| Branching | ↓ |
| Leaf area | =↓ |
| Leaf thickness | ↓ |
| Dry matter | =↓ |
| Photosynthesis | ↓ |
| Photorespiration | = |
| Dark respiration | ↑ |
| Stomatal conductance | ↓ |
| Transpiration | ↑ |
| Water use efficiency | ↓ |
| Days to anthesis/maturity | ↓ |
| Percent fruit set | ↓ |
| Pollen viability | ↓ |
| Pollen production | ↓ |
| Ovule viability | ↓ |
| Number of grains | ↓ |
| Duration of grain filling | ↓ |
| Rate of grain filling | =↑ |
| Seed or fruit size | ↓ |
| Harvest index | ↓ |

normal proteins, increased synthesis of HSP (heat shock proteins), ROS (reactive oxygen species), phytohormones (e.g., abscisic acid, ABA; ethylene), and decreases antioxidants. HT shortens duration of crop cycle, duration of grain filling, increases respiration, decreases photosynthesis, and finally reduces yield (Prasad et al. 2008a, b; Barnabás et al. 2008). Damage of membranes and organ function could be associated with oxidative damage caused by increased ROS production (Prasad et al. 2017). The activity of the enzyme rubisco (ribulose-1,5-bisphophate carboxylase-oxigenase) is greatly altered under HT (Maestri et al. 2002). Inhibition of photosynthesis under HT stress is caused by damage to chloroplast membranes. Plant cellular membranes have long been proposed to be one of the prime targets of HT stress. Plants under HT stress adjusts their cellular membrane composition and saturation to maintain its stability and optimize photosynthesis and other cellular processes. Many plants show increased production of osmolytes which stabilizes the protein and membrane bilayer (Mirzaei et al. 2012). Secondary metabolites like flavonoids and anthocyanin are induced under HT as a mechanism of HT stress tolerance.

In general, HT stress at flowering stage causes sterility and loss of floret fertility (Prasad et al. 2017). However, the impacts vary with intensity, duration, and crop stage. HT at the flowering stage of the crop reduces the number of florets, floret fertility, and eventually crop yield (Prasad et al. 2008a, b, 2017). Major impacts of

HT are accelerated senescence, shortened vegetative and reproductive phases, decreased photosynthetic activity, and proteinaceous compounds (Al-Khatib and Paulsen 1984). In general, HT stress can reduce the grain yield of major cereals and legume (Prasad et al. 2017). Global wheat, rice, and soybean yields were shown to decrease by 6%, 3.2%, and 3.1%, respectively, for every 1 °C increase in global mean temperatures (Zhao et al. 2017).

## 17.2 Sensitive Stages to HT Stress in Grain Crops

Crop stages differ in their sensitivity to HT stress (Mitchell et al. 1993; Baker et al. 1989). Reproductive stages have relatively greater sensitivity compared to vegetative stages for most of the field grain crops. During reproductive phase, period of floral bud development (coinciding with micro- or mega-sporogenesis) and anthesis (coinciding with pollination and fertilization) are most sensitive to HT stress (Prasad et al. 1999a, 2008a, 2015; Prasad and Djanaguiraman 2014). In wheat, 8-6 days before anthesis and anthesis were extremely sensitive to HT stress (Prasad and Djanaguiraman 2014). Similarly, sorghum was most sensitive during 10 and 5 days before anthesis and at anthesis (Prasad et al. 2015). Data on pearl millet indicated that most sensitive periods were 10–12 days before anthesis and at anthesis (Djanaguiraman et al. 2018a). HT stress can also cause early embryo abortion in crops. Prasad et al. (1999a) reported that HTs at 6 days before and 15 days after anthesis decreased the numbers and weights of seeds. Key sensitive stages in different grain crops to HT stress are mentioned in Table 17.3.

**Table 17.3** Critical and sensitive stages to high temperature stress in selected crops

| Crops | Most sensitive stages | Key Reference |
|---|---|---|
| Wheat | 8–6 days before to anthesis, and pollination/ fertilization | Prasad and Djanaguiraman (2014) |
| Sorghum | 5–10 days before to anthesis and pollination/ fertilization | Prasad et al. (2015) |
| Pearl millet | 10–12 days before to anthesis and pollination/ fertilization | Djanaguiraman et al. (2018a) |
| Finger millet | 7–10 days before to anthesis and pollination/ fertilization | Opole et al. (2018) |
| Peanut | 4 days before to anthesis and anthesis/fertilization | Prasad et al. (2001) |
| Soybean | 3–5 days before to anthesis and anthesis/ fertilization | Prasad et al. (Unpublished) |

### 17.2.1 Impact of HT Stress on Wheat

Wheat crop is susceptible to HT during sporogenesis, flowering, and early grain filling phase (Prasad and Djanaguiraman 2014; Farooq et al. 2011). Barkley et al. (2014) reported wheat yield decreases of 21% with 1 °C increase in projected mean temperature. HT stress during anthesis and grain filling decreased chlorophyll content, photosynthesis as a result of damage to thylakoid membranes (Djanaguiraman et al. 2018b, 2020; Prasad et al. 2011). There was a negative relationship between membrane damage and leaf photosynthesis.

High temperature causes damage to photosystem II (PSII) due to increased production of ROS which can potentially affect the photosynthetic rate. The galactolipids such as mono galactosyl diacylglycerol (MGDG) and di galactosyl diacylglycerol (DGDG) are important chloroplast lipids. As lipids are susceptible to oxidative damage, a greater decrease of those lipid species was observed under HT (Djanaguiraman et al. 2018b). Damage to PSI and PSII complexes and proteins, and chlorophyll loss ultimately decreases photosynthetic rate (Mathur et al. 2014). Photochemical quenching (PQ) and electron transport rate (ETR) were negatively affected by HT stress (Brestic et al. 2012; Mathur et al. 2014). Greater non-PQ under HT shows inefficiencies in the production of ATP and $NADPH_2$. Antioxidants present in chloroplast are also downregulated due to excess ROS produced under HT, resulting in damaged membranes (Djanaguiraman et al. 2018b).

High temperature stress (35/25 °C) in wheat causes changes in lipid metabolism by generating ROS, thus decreasing the photosynthetic rate and ultimately reduces the grain yield (Djanaguiraman et al. 2018b). Decreased grain yield under HT in wheat was a result of negative impacts on spikelet fertility (48–56%), number of grains (56%), and grain weight (25%) (Prasad et al. 2011). Djanaguiraman et al. (2020) reported decreases in grain yield by 29 and 44% when HT was imposed during anthesis or grain filling stages due to lower percent seed set and smaller grain weight. Decreased seed set was due to abnormal structural changes in pollen and pistil caused by HT, leading to decreased fertilization and seed numbers (Saini et al. 1983; Farooq et al. 2011; Prasad and Djanaguiraman 2014). HT also causes acceleration and early leaf senescence leading to lower photosynthesis (Barlow et al. 2015). HT stress degenerates tapetal cells, affecting the nutrient translocation to pollen leading to sterility (Hess and Hesse 1994). Lower floret fertility under HT stress is a result of loss of gamete (pollen and ovule) viability, leading to poor fertilization, lower grain numbers, ultimately resulting in poor grain yield. Prasad et al. (2008b) reported that HT stress significantly decreased grain filling duration but not grain filling rate (Prasad et al. 2008b). Similar observations of decreased grain filling duration under HT was reported by Streck (2005) and Prasad and Djanaguiraman (2014).

High daytime and high night-time temperature may have differential impacts on physiology of reproductive function. High daytime temperature (35/15 °C) decreased the photosynthetic rate, antioxidant capacity, increased ROS, and plasma membrane damage. High daytime temperatures decreased seed set, grain numbers, and grain yield (Narayanan et al. 2015). High night-time temperature effects were like high daytime temperature for all the traits in wheat (Narayanan et al. 2015) as reported in soybean (Djanaguiraman et al. 2013a). The temperature average $T_{avg}$ rather than the difference in diurnal temperature range determines the impact of HT (Narayanan et al. 2015; Fischer 1985).

### 17.2.2 Impact of HT Stress on Sorghum

Sorghum is an important crop in arid and semiarid regions, where current mean air temperature are close to or slightly higher than the optimum for grain yield (Prasad et al. 2006a). Season long high temperature of 44/34 °C prohibited—emergence of panicle up to 110 days of sowing. Whereas partial panicle emergence was observed at 60 days after sowing at 40/30 °C, which was 20 days delay compared to OT (32/22 °C). Hence, it can be concluded that sorghum reproductive development and growth was hampered at ≥40/30 °C (Prasad et al. 2006a). Similarly, temperature increase decreased length and diameter of panicles, and number of branches per panicle (Prasad et al. 2006a).

HT stress can decrease PSII function, ETR, and photosynthesis. In addition, HT stress (a) decreased chlorophyll content by decreasing the chlorophyll *b* content, (b) causes premature leaf senescence by inducing the synthesis of ROS like superoxide radicals and hydrogen peroxide, (c) results in oxidative damage to membranes, (d) decreases net photosynthesis, and (e) increases rate of stomatal conductance and transpiration (Prasad et al. 2006a; Yan et al. 2011; Djanaguiraman et al. 2014, 2018c). The reduced photosynthetic rate might be due to the influence of HT stress on the PSII reaction center and the donor side of the oxygen-evolving complex than the acceptor side (Yan et al. 2011). Excessive energy released from electrons dissipated as heat increases the non-photochemical quenching under HT stress. However, lower non-photochemical quenching in OT indicated that plants could efficiently convert more light energy into chemical energy, possibly through managing the electron transport chain (Hideg et al. 2008). Djanaguiraman et al. (2014) showed that sorghum genotypes sensitive to HT showed decreased rubisco activase enzyme activity than tolerant genotypes under HT stress. However, HT stress increased the transcript levels of superoxide dismutase (SOD). The protein chaperones responsible for the activity of the rubisco activase enzyme are denatured under HT stress which deactivates the rubisco enzyme (Kim and Portis 2005). The increased SOD transcript level scavenges the superoxide radicle to $H_2O_2$; but the reduced levels of antioxidant enzymes such as ascorbate peroxidase (APX) and glutathione peroxidase (GPX) in chloroplast are insufficient to scavenge $H_2O_2$.

It is evident that HT during gamete development alters pollen, pistil, and ovary morphology and affects pollen, pistil, and ovule function (Djanaguiraman et al. 2018c). The morphological abnormalities at HT include irregular, deeply pitted, large micro-echniate ornamentation on pollen grains; and desiccated stigma and style (Djanaguiraman et al. 2018c). HT stress caused anatomical variation, with thickened tectum and foot layers, thinner columellae wall and absence of endexine and vacuolation in the intine. In OT, the vacuolation was not started and thin tectum and columellae wall (Djanaguiraman et al. 2018c). In addition, there were also significant damage to ovule and style anatomy and structure (Djanaguiraman et al. 2018c).

HT stress during the pro-gamic stage causes loss of pollen viability and stigma receptivity (Prasad et al. 2015, 2018). Studies also showed that HT decreased pollen germination, hexose transporters, and sugar conversion in anthers and pollen grains (Jain et al. 2011; Djanaguiraman et al. 2018c). HT induces the overproduction of superoxide radicle and hydrogen peroxide and decreased production of antioxidant enzymes in pollen and pistil. Overall, sorghum pollen grains were relatively more sensitive to HT than ovule or pistil (Djanaguiraman et al. 2018c).

High night-temperature alone also negatively impacts leaf physiology and reproductive function. HT stress causes carbon imbalance by impacting photosynthesis and respiration. High nigh temperature stress decreased chlorophyll content and increased thylakoid membrane damage (Prasad and Djanaguiraman 2011). Apart from this, the high night-time temperature decreased PSII, ETR, and photochemical quenching. In contrast, the non-photochemical quenching was increased by high night-time temperatures (Prasad and Djanaguiraman 2011). High night-time temperatures caused leaf membrane damage through increased ROS production (Prasad and Djanaguiraman 2011). The decrease in leaf photosynthetic rate might be due to the oxidative burst that happened in the thylakoid membrane due to high night-time temperatures. The higher non-photochemical quenching under high night-time temperature indicates absorbed light is safely dissipated as heat energy.

Decrease in pollen function at high night temperatures was associated with pollen ROS, pollen membrane damage, and pollen lipid molecular species (Prasad and Djanaguiraman 2011). High night-time temperatures decreased the polyunsaturated acyl lipid species and PA levels compared with optimum night-time temperature (Prasad and Djanaguiraman 2011). Overall high night-time temperature during the flowering stage reduced seed set and seed yield of sorghum.

### *17.2.3 Impact of HT Stress on Pearl Millet*

Pearl millet is produced in about 30 million ha in arid and semiarid regions. Pearl millet has been the emerging summer crop in the past two decades in regions of the northwestern states of India and other arid and semiarid regions where it is predominantly grown. It is also an important food security crops in several parts of West Africa. Though the crop is more tolerant to HT than any other major cereal grain crop, the yield of the crop can still be negatively impacted due to extremely hot summer temperatures.

As this crop is grown in hot and dry conditions, seedling establishment and emergence can be impacted by HT stress. The rate of germination decreased linearly with increasing temperature, at an HT of 46–48 °C the rate is almost zero (Khalifa and Ong 1990). Launders (1971) found that percentage and rate of germination and seedling growth increased until 33 °C and above which it decreases. Joshi et al. (1997) identified 30–35 °C as optimum for germination. At extreme HT stress (44/34 °C), the leaf photosynthetic rate decreased by 10% than OT, until that temperature, there is no decrease of photosynthetic rate in pearl millet (Djanaguiraman et al. 2018a).

Yield components (number of seeds and seed weight per panicle) were reduced to a greater extent under HT stress than optimum temperature (Djanaguiraman et al. 2018a). Decrease in the number of seeds/panicles may be due to a decrease in seed set percentage. Also, gametogenesis (10–12 days before anthesis) and anthesis were most sensitive to HT stress (Djanaguiraman et al. 2018a). Similar results were found in other legumes (Gross and Kigel 1994; Ahmed et al. 1992) and cereals (Prasad et al. 2008a; Prasad and Djanaguiraman 2014). Gupta et al. (2015) reported the booting stage as sensitive to HT stress, where active micro sporogenesis takes place; exposure of pearl millet to HT at this stage had greatly reduced the seed set percentage than under optimum temperature. Similar results were observed for drought stress (Mahalakshmi and Bidinger 1985) While, studies in rice indicate that flowering is more sensitive than booting stages (Farrell et al. 2006; Satake and Yoshida 1978).

HT stress during anthesis can reduce the floret fertility despite viable pollen (Prasad and Djanaguiraman 2014), which may be due to stigmatic dryness, poor stigma receptivity, improper pollen tube growth, poor pollen germination (Kakani et al. 2002; Prasad and Djanaguiraman 2014). In pearl millet, female reproductive organ, the pistil is highly sensitive to HT stress (Djanaguiraman et al. 2018a; Gupta et al. 2015). In *Arabidopsis*, HT caused ovule abortion (Whittle et al. 2009). In contrast, studies on wheat suggest that HT that caused pollen sterility does not have much effect on pistil (Saini and Aspinall 1982), as in chickpea (Devasirvatham et al. 2012) and in maize (Herrero and Johnson 1980). This indicated that tolerance/ susceptibility of gametes to HT is species-dependent (Herrero 2003).

High temperature increased the ROS production in both pollen and pistil. However, a higher increase was observed in the pistil. The ROS production rate is higher than its metabolism rate, causing oxidative damage (Apel and Hirt 2004). Higher ROS under HT in pistil could damage the female reproductive structure (membrane damage) and thus increase its sensitivity to HT than pollen (Djanaguiraman et al. 2014). Plants have a ROS scavenging mechanism to protect cells from ROS-dependent membrane damage. The major enzymes involved in this mechanism include CAT (catalase), POX (peroxidase), and SOD (superoxide dismutase). In pearl millet under HT, the activity of SOD that scavenges superoxide radicle and the activity of CAT and POX were decreased in both pollen and pistil. However, a higher decrease was observed in the pistil of pearl millet (Djanaguiraman et al. 2018a) indicating it to be more sensitive to HT than its pollen. The protogynous nature of pearl millet explains its pistil sensitivity to HT where the pistil is exposed to the outer environment, whereas in other crops like sorghum, soybean, rice, and wheat pollen and pistil are protected from the outer environment (Djanaguiraman et al. 2018a).

### 17.2.4 Impact of HT Stress in Finger Millet

Global finger millet production is about 4.5 million tons, and Africa produces ~2 million tons. The crop requires a minimum amount of moisture. It is grown mostly in arid and semiarid regions and/or under minimum precipitation. But in these regions, the temperature is ≥32/22 °C (Prasad et al. 2006a), which inhibits flowering (Directorate of Millets Development 2014). Also, the temperature in these regions is predicted to be further increased. This can decrease the yield potential of the crop to a greater extent.

Plant height, internodal length, tillers per plant were decreased when finger millet plants were exposed to season-long HT of 38/28 °C (Opole et al. 2018). High temperature stress decreased the chlorophyll index (Ashraf and Harris 2013). Opole et al. (2018) observed increased O/P ratio and decreased PSII quantum yield in finger millet under HT (38/28 °C), increased O/P ratio indicates the damage in the thylakoid membrane and its function. Decreased $F_v/F_m$ ratio may be due to alteration in energy balance. Similar observations were made in wheat (Prasad et al. 2008b) and sorghum (Djanaguiraman et al. 2014). Excess energy in PSII may lead to photo-oxidative damage by provoking ROS production (Roach and Krieger-Liszkay 2014).

The impact on leaf photosynthetic rate in finger millets under season long HT stress was not much evident (Opole et al. 2018) as observed in sorghum (Prasad et al. 2008a), pearl millet (Djanaguiraman et al. 2018a), and rice (Prasad et al. 2006b). Transpiration rate and stomatal conductance were increased under HT (Opole et al. 2018). The viscosity of water decreases under HT, thus more amount of water is supplied to the stomata, which helps in the stomatal opening (Von-Caemmerer and Evans 2015).

Opole et al. (2018) reported a decrease in grain numbers, grain size, and yield in finger millet when exposed to HT stress. The decrease in grain yield and seed number per panicle is because of decreased seed set associated with disfunction of pollen and pistil (Prasad et al. 2001; Young et al. 2004). Prasad et al. (2018) highlighted that the loss of pollen viability was associated with lower carbohydrate metabolism in pollen grains under HT. The other factors that cause malfunctioning of gametes may be alteration in lipid composition in cell walls, increased ROS production, decreased scavenging of ROS (Prasad et al. 2017).

HT stress (38/28 °C, 10 days) reduced seeds per panicle and seed yield per panicle in finger millet (Opole et al. 2018), as in other crops such as sorghum (Prasad et al. 2008a), wheat (Wollenweber et al. 2003) and pearl millet (Djanaguiraman et al. 2018a). There is a reduction in harvest index in plants exposed to HT (Opole et al. 2018), which may be due to a reduction in grain yield and above-ground biomass. Also, genotypic variation in finger millet was observed in response to HT for many traits including chlorophyll *a* fluorescence, panicle length, seed numbers, and weight (Opole et al. 2018).

### 17.2.5 Impact of HT Stress on Soybean

Soybean is widely grown all over the world. Soybean is rich in protein and oil content, and hence a major food crop. Abiotic stresses including HT limits soybean growth and productivity (Khan et al. 2007). Temperature in the range of 15–22 °C (emergence), 20–25 °C (flowering), and 15–22 °C (seed maturity) are optimum for soybean (Liu et al. 2008).

HT stress causes various morphological, physiological, and biochemical changes in plants, particularly in membranes that alter photosynthetic rate (Ashraf and Harris 2013). The principal targets of HT stress are thylakoid membranes and their accompanying PS II, causing damage to membranes of different cell organelles. These changes have resulted in premature leaf senescence (Djanaguiraman and Prasad 2010). The PSII photochemistry, leaf photosynthesis, stomatal conductance decreased under HT stress due to closure of PSII reaction center, eventually resulting in decreased electron transport (Bibi et al. 2008). HT stress led to decrease in stomatal conductance through increased ROS concentration and decreased antioxidant enzymes activity (Djanaguiraman et al. 2011b).

HT stress decreased the chlorophyll content (Djanaguiraman et al. 2011a, b, 2013a, 2019). There was a linear increase in chlorophyll loss under HT stress with the increasing duration compared to no change under optimum temperature (Djanaguiraman et al. 2011a, b). The study also showed that HT stress had increased thylakoid membrane damage (Djanaguiraman et al. 2011a). Lipid peroxidation of the thylakoid membrane may be a possible mechanism for chlorophyll degradation and decreased photosynthetic rate under HT stress, validated in soybean (Djanaguiraman and Prasad 2010; Djanaguiraman et al. 2011b).

HT stress increased the leaf thickness and specific leaf weight in soybean (Jumrani et al. 2017). Similarly, Djanaguiraman et al. (2011a) observed that HT stress increased thickness of palisade layers I and II, spongy, and epidermis tissues. In addition, stomatal number were affected by temperature. Reduced leaf area resulted in a significant increase in stomatal density on both surfaces of soybean leaves, indicating that this might be an adaptation strategy in soybean plants to cope with higher evaporative demand under HT conditions (Jumrani et al. 2017). Also, HT stress significantly decreased the individual stomatal diameter compared to optimum temperature (Djanaguiraman et al. 2011a).

Pollen formation, germination, and pollen tube expansion are all affected by HT, which may impact pollen function and, ultimately, pod and seed set. HT stress caused pollen morphological abnormalities (misshaped or oval), while under optimum temperature, the pollen grains were round and turgid (Djanaguiraman et al. 2013b). Salem et al. (2007) have shown that temperature of 38/30 °C at flowering caused shrivelling of pollen grains. In addition, there was tapetum degradation (Djanaguiraman et al. 2013b). The tapetum was vacuolated, and the cells showed autolysis than pollen grains of OT. The altered pollen morphologies have resulted in a decreased pod-set percentage in soybean. There was a significant genetic diversity among soybean lines for pod-set percentage under HT stress (Djanaguiraman et al.

2019). Seed yield components were reduced in soybean plants subjected to HT during the reproductive phase. These included number of pods, seeds, seed weight, and seed yield (Puteh et al. 2013). In addition, there were decrease in pod-set and seed yield under HT stress (Djanaguiraman and Prasad 2010; Djanaguiraman et al. 2011a).

High daytime temperature alone damaged thylakoids, decreased chlorophyll index and leaf photosynthesis (Djanaguiraman et al. 2019). Chlorophyll *a* fluorescence traits were significantly decreased under high daytime temperature stress (Djanaguiraman et al. 2013a). The pollen phospholipid species were significantly altered by high daytime temperature. High daytime temperature decreased the phosphatidic acid levels and increased phosphatidic choline levels (Djanaguiraman et al. 2013a). Decreased phosphatidic acid levels under high daytime temperature caused loss of polarity, changes in calcium gradient in pollen tube tips, and impacting tube growth. In vitro pollen germination and pod-set percentages were decreased under HT. The pollen morphology was altered due to high daytime temperature. The pollen was collapsed, flattened with no clear columella head under high daytime temperature. The tapetum cells were disintegrated under high daytime temperature, as revealed by rough exine ornamentation in the pollen grains resulting in lower fertilization, pod numbers, and seed yields (Djanaguiraman et al. 2019). High night-time temperature had negative impacts on leaf photosynthesis, reproductive biology like pollen tube growth, and the associated mechanism are like high daytime temperature (Djanaguiraman et al. 2013a).

### *17.2.6 Impact of HT Stress on Peanut/Groundnut*

Peanut or groundnut is a major oilseed crop grown in semiarid regions. In these regions, there is a relatively lower yields of 950 kg/ha in Africa, compared to 1770 kg/ha in Southeast Asia and >4200 kg/ha in North Americas (FAO 2020). The lower yield in many regions are due to higher air temperatures and dry conditions. For peanut, the optimum temperatures are 25 and 30 °C for reproductive and vegetative growth, respectively.

Temperatures from 32/22 to 44/34 °C did not impact leaf photosynthesis, whereas it increased conductance and transpiration (Prasad et al. 2003). However, increase in temperatures from 32/22 to 40/30 °C had positive impact on vegetative biomass, while further increase to 44/32 decreased vegetative dry matter (Prasad et al. 2003). In contrast, in the same temperature range, yield and harvest index of pods and seeds decreased with increasing temperature. Temperature above 33/23 °C decreased viability of pollen grains, resulting in lower seed set and seed numbers.

High temperature stress from floral bud formation to maturity have direct impact on the pod yield of peanut (Prasad et al. 1999a). There is a 50% reduction in the number of pegs at HT (38/22 °C) (Prasad et al. 1999a). Similarly, Ketring (1984) showed a 33% reduction at 35/22 °C. High temperature stress reduced the pod harvest index. In contrast, the flower numbers increased under HT. There was an increase in the rate of flower formation and not the duration of flowering (Prasad et al. 1999a). Peanut flowers were particularly sensitive to short periods (≤6 days) of HT (day/night 38/22 °C) from 6 days before to 15 days after flowering (Prasad et al. 1999b). Both high daytime (>34 °C) and high night-time (>28 °C) temperatures resulted in few pollen grains and poor viability (Prasad et al. 1999b). Furthermore, HTs during the first 6 h of the day had more adverse impacts (Prasad et al. 2000).

HT stress from flowering to podding increased flower numbers, but reduced fruit set, leading to fewer pods and lower pod yield. Reduced fruit set was due to negative impacts on microsporogenesis resulting in fewer pollen numbers and lower viability. Prasad et al. (1999b) showed that impacts of high air temperatures were greater at podding than at flowering, and pod numbers were reduced by 32% and 22%, respectively. Similarly, seed number and 100 seed weight were not affected by HT from flowering but were reduced by high air temperature from podding. There were significant reduction in total dry weights and pod yield when HT stress (38/22 °C) was imposed from the start of flowering or podding to maturity. Thus, pod yields were reduced by 18–26% due to HT (Prasad et al. 2000), and these results have been supported by Ketring (1984), Golombek and Johansen (1997), and Prasad et al. (1999a, b).

In peanut microsporogenesis occurs 3–6 days before the flowering stage, HT stress at this stage (6 days before flowering) greatly reduced peg number per plant (Prasad et al. 1999a, 2001), indicating microsporogenesis stage is also sensitive. Reduction in peg number at microsporogenesis stage is due to poor pollen viability and poor anther dehiscence. There was significant decrease in number of pegs and reproductive dry weight when HT stress occurred between 6 days before and 15 days after flowering. With maximum decrease occurring at 9 days after flowering, when first pegs appeared. The percentage of flowers that form pegs was reduced, whereas the number of pegs that set pods increased, which indicates that fertilization is most sensitive to HT (Prasad et al. 1999a, 2001).

High daytime temperature >38 °C decreases the number of pegs, number of pods, and ultimately pod yield (Ketring 1984). Pollen mortality occurs at temperatures >33 °C (De Beer 1963). Exposure of floral buds to temperature ≥39 °C for 1 day decreased fruit set compared to 28 °C (control) and effects differed with stage of flower buds (Prasad et al. 2001). There was decrease in pollen germination percentage and rate of pollen tube growth at higher temperatures which resulted in fewer pods. These results were similar with the findings of Talwar and Yanagihara (1999). An increase in daytime temperature per °C decreased the flower number by approximately 1.1 per plant, peg and pod numbers 0.9 per plant, and number of pollen grains by 390 per flower, and viability of pollen 1.9% (Prasad et al. 1999b).

Studies on high night temperature indicated the impact on total flower numbers. An increase in every 1 °C night temperature decreased the fruit set by 2.8%. Warm nights reduced pegs and pods from 7.7 to 5 per plant, and pollen production from 4398 to 2800 per flower, and pollen viability from 49 to 40% (Prasad et al. 1999b).

## 17.3 Conclusions and Future Prospects

High temperature stress impacts crop productivity by altering various molecular, biochemical, physiological, growth, development, and yield processes. Crop species have different cardinal temperatures for various developmental and growth processes. Above optimum temperatures will have negative impact on the yield depending upon the timing, intensity, duration of stress during the crop production. Reproductive stages are relatively more susceptible to HT stress. Within the reproductive period two stages (gametogenesis and flowering) are highly sensitive, decreasing viability of gametes and fertilization, loss of floret fertility and decreased seeds. While HT stress after seed set and during grain growth decreases duration of grain filling and individual grain weight. Both lower number and weight of grains decreases grain yield. It is important to develop strategies that minimizes negative impacts of HT stress on crops. An absolute resistance to HT cannot be achieved. Hence, the focus should be on tolerance, avoidance, and escaping HT stress in crops. Breeding for stress tolerance is a key component to developing tolerant genotypes or genotypes that can escape stress. Developing for HT stress tolerant genotypes requires (a) identification and use of key traits, (b) high throughput screening procedures, (c) diverse donor lines, and (d) understanding mechanisms of tolerance and their genetics and inheritance. The current crop breeding programs are using HT stress-tolerant traits like cell membrane stability, canopy temperature depression, extended green leaf area duration, early morning flowering, increased reproductive function, and improved carbon partitioning to grain in cultivated and wild relatives. Information on exact molecular and genetic mechanisms leading to HT stress-induced pollen and stigma sterility is limited and needs further studies. Current HT stress tolerance breeding programs exploit a narrow genetic pool which needs to be further enhanced by identifying tolerant lines in larger germplasm collections including wild relative and exotic collections. One of the limitations of screening large germplasm collection is availability of rapid and accurate high throughput phenotyping techniques. Advances in the phenotyping approaches using sensors (ground-based or aerial vehicles), physiological, molecular or biochemical markers can help in efficient and rapid screening. The whole-genome sequence, genetic and physical maps, and the available high throughput phenotyping and genotyping facilities have the potential to identify and enhance the breeding efficiency. Integration of phenotyping and genotyping data in simulation models can also help in determining the best genotypic traits and crop management practices specific to environments and social conditions to minimize negative impacts of abiotic stresses and improve crop yields.

## References


Ahmed FE, HalL AE, DeMason DA (1992) Heat injury during floral development in cowpea (*Vigna unguiculata*), Fabaceae. Am J Bot 79:784–791

Alagarswamy G, Ritchie JT (1991) Phasic development in CERES-sorghum model. In: Predicting crop phenology. CRC Press, Boca Raton, pp 143–152

Al-Khatib K, Paulsen GM (1984) Mode of high temperature injury to wheat during grain development. Physiol Plant 61(3):363–368

Apel K, Hirt H (2004) Reactive oxygen species: metabolism, oxidative stress, and signal transduction. Annu Rev Plant Biol 55:373–399

Ashraf M, Harris PJC (2013) Photosynthesis under stressful environments: an overview. Photosynthetica 51:163–190

Asseng S, Ewert F, Martre P, Rötter RP, Lobell DB, Cammarano D, Kimball BA, Ottman MJ, Wall GW, White JW, Reynolds MP (2015) Rising temperatures reduce global wheat production. Nat Clim Change 5:143–147

Baker JT, Allen LH Jr, Boote KJ, Jones P, Jones JW (1989) Response of soybean to air temperature and carbon dioxide concentration. Crop Sci 29(1):98–105

Barkley AP, Tack J, Nalley LL, Bergtold JS, Bowden RL, Fritz AK (2014) Weather, disease and wheat breeding effects on Kansas wheat varietal yields, 1985 to 2011. Agron J 106(1):227–235

Barlow KM, Christy BP, O'Leary GJ, Riffkin PA, Nuttall JG (2015) Simulating the impact of extreme heat and frost events on wheat crop production: a review. Field Crop Res 171:109–119

Barnabás B, Jäger K, Fehér A (2008) The effect of drought and heat stress on reproductive processes in cereals. Plant Cell Environ 31(1):11–38

Berg A, Sultan B, de Noblet-Ducoudré N (2011) Including tropical croplands in a terrestrial biosphere model: application to West Africa. Clim Change 104(3):755–782

Bibi AC, Oosterhuis DM, Gonias ED (2008) Photosynthesis, quantum yield of photosystem II and membrane leakage as affected by high temperatures in cotton genotypes. J Cotton Sci 12: 150159

Bolhuis GG, De Groot W (1959) Observations on the effect of varying temperatures on the flowering and fruit set in three varieties of groundnut. Netherlands J Agric Sci 7(4):317–326

Brestic M, Zivcak M, Kalaji HM, Carpentier R, Allakhverdiev SI (2012) Photosystem II thermostability in situ: environmentally induced acclimation and genotype-specific reactions in *Triticum aestivum* L. Plant Physiol Biochem 57:93–105

De Beer JF (1963) Influence of temperature on *Arachis hypogaea* L. with special reference to its pollen viability. Ph.D. thesis, State Agricultural University, Wageningen

Devasirvatham V, Gaur PM, Mallikarjuna N, Tokachichu RN, Trethowan RM, Tan DK (2012) Effect of high temperature on the reproductive development of chickpea genotypes under controlled environments. Funct Plant Biol 39(12):1009–1018

Directorate of Millets Development (2014) Status paper on coarse cereals (sorghum, pearl millet, finger millet, small millets, maize and barley). Department of Agriculture and Cooperation, Ministry of Agriculture, Government of India, Hyderabad, pp 87–99

Djanaguiraman M, Prasad PV (2010) Ethylene production under high temperature stress causes premature leaf senescence in soybean. Funct Plant Biol 37(11):1071–1084

Djanaguiraman M, Prasad PV, Al-Khatib K (2011a) Ethylene perception inhibitor 1-MCP decreases oxidative damage of leaves through enhanced antioxidant defense mechanisms in soybean plants grown under high temperature stress. Environ Exp Bot 71(2):215–223

Djanaguiraman M, Prasad PV, Boyle DL, Schapaugh WT (2011b) High-temperature stress and soybean leaves: leaf anatomy and photosynthesis. Crop Sci 51(5):2125–2131

Djanaguiraman M, Prasad PV, Schapaugh WT (2013a) High day-or nighttime temperature alters leaf assimilation, reproductive success, and phosphatidic acid of pollen grain in soybean [*Glycine max* (L.) Merr.]. Crop Sci 53(4):1594–1604

Djanaguiraman M, Prasad PV, Boyle DL, Schapaugh WT (2013b) Soybean pollen anatomy, viability and pod set under high temperature stress. J Agron Crop Sci 199(3):171–177

Djanaguiraman M, Prasad PV, Murugan M, Perumal R, Reddy UK (2014) Physiological differences among sorghum (*Sorghum bicolor* L. Moench) genotypes under high temperature stress. Environ Exp Bot 100:43–54

Djanaguiraman M, Perumal R, Ciampitti IA, Gupta SK, Prasad PV (2018a) Quantifying pearl millet response to high temperature stress: thresholds, sensitive stages, genetic variability and relative sensitivity of pollen and pistil. Plant Cell Environ 41(5):993–1007

Djanaguiraman M, Perumal R, Jagadish SV, Ciampitti IA, Welti R, Prasad PV (2018b) Sensitivity of sorghum pollen and pistil to high-temperature stress. Plant Cell Environ 41(5):1065–1082

Djanaguiraman M, Boyle DL, Welti R, Jagadish SV, Prasad PV (2018c) Decreased photosynthetic rate under high temperature in wheat is due to lipid desaturation, oxidation, acylation, and damage of organelles. BMC Plant Biol 18(1):1–7

Djanaguiraman M, Schapaugh W, Fritschi F, Nguyen H, Prasad PV (2019) Reproductive success of soybean (*Glycine max* L. Merril) cultivars and exotic lines under high daytime temperature. Plant Cell Environ 42(1):321–336

Djanaguiraman M, Narayanan S, Erdayani E, Prasad PV (2020) Effects of high temperature stress during anthesis and grain filling periods on photosynthesis, lipids and grain yield in wheat BMC. Plant Biol 20(1):1–2

FAO (2020) Agriculture statistics production year book (1997). FAO, Rome

Farooq M, Bramley H, Palta JA, Siddique KH (2011) Heat stress in wheat during reproductive and grain-filling phases. Crit Rev Plant Sci 30(6):491–507

Farrell TC, Fox KM, Williams RL, Fukai S (2006) Genotypic variation for cold tolerance during reproductive development in rice: screening with cold air and cold water. Field Crop Res 98(2–3):178–194

Fischer RA (1985) Number of kernels in wheat crops and the influence of solar radiation and temperature. J Agric Sci 105:447–461

Golombek SD, Johansen C (1997) Effect of soil temperature on vegetative and reproductive growth and development in three Spanish genotype of peanut [*Arachis hypogaea* (L.)] Peanut. Science 24(2):67–72

Gross Y, Kigel J (1994) Differential sensitivity to high temperature of stages in the reproductive development of common bean (*Phaseolus vulgaris* L.). Field Crop Res 36(3):201–212

Gupta SK, Rai KN, Singh P, Ameta VL, Gupta SK, Jayalekha AK, Verma YS (2015) Seed set variability under high temperatures during flowering period in pearl millet [*Pennisetum glaucum* (L.)]. Field Crop Res 171:41–53

Hatfield J, Boote KJ, Fay PA, Hahn GL, Izaurralde RC, Kimball BA, Mader TL, Morgan JA, Ort DR, Polley HW, Thomson AM, Wolfe DW (2008) Agriculture. In: The effects of climate change on agriculture, land resources, water resources, and biodiversity. A report by the U.S. Climate Change Science Program and the Subcommittee on Global Change Research, Washington, DC, p 362

Hatfield JL, Boote KJ, Kimball BA, Ziska LH, Izaurralde RC, Ort DR, Thomson AM, Wolfe D (2011) Climate impacts on agriculture: implications for crop production. Agron J 103:351–370

Herrero M (2003) Male and female synchrony and the regulation of mating in flowering plants. Philos Trans R Soc Lond B Biol Sci 358(1434):1019–1024

Herrero MP, Johnson RR (1980) High temperature stress and pollen viability of maize. Crop Sci 20: 796–800

Hess M, Hesse M (1994) Ultrastructural observations on anther tapetum development of freeze-fixed *Ledebouria socialis* Roth (Hyacinthaceae). Planta 192:421–430

Hideg E, Kos PB, Schreiber U (2008) Imaging of NPQ and ROS formation in tobacco leaves: heat inactivation of the water–water cycle prevents down-regulation of PSII. Plant Cell Physiol 49(12):1879–1886

IPCC Climate Change (2013) The physical science basis. In: Stocker TF, Qin D, Plattner GK, Tignor M, Allen SK, Boschung J, Nauels A, Xia Y, Bex V, Midgley PM (eds) Contribution of working group I to the fifth assessment report of the intergovernmental panel on climate change. Cambridge University Press, Cambridge

IPCC Climate Change (2021) The physical science basis. In: Masson-Delmotte B, Zhai P, Pirani A, Conners SL (eds) Contribution of working group I to the sixth assessment report of the intergovernmental panel on climate change. Cambridge University Press, Cambridge

Jain M, Prasad PVV, Boote KJ, Allen LH, Chourey PS (2011) Effect of season-long high temperature growth conditions on sugar-to-starch metabolism in developing microspores of grain sorghum [*Sorghumbicolor* (L.) Moench]. Planta 227:67–79

Joshi AK, Pandya JN, Mathukia RK, Pethani KV, Dave HR (1997) Seed germination in pearl millet hybrids and parents under extreme temperature conditions. Gujarat Agric Univ Res J 23(1): 77–83

Jumrani KVS, Bhatia VS, Pandey GP (2017) Impact of elevated temperatures on specific leaf weight, stomatal density, photosynthesis and chlorophyll fluorescence in soybean. Photosynth Res 131(3):333–350

Kakani VG, Prasad PVV, Craufurd PQ, Wheeler TR (2002) Response of in vitro pollen germination and pollen tube growth of groundnut [*Arachis hypogaea* (L.)] genotypes to temperature. Plant Cell Environ 25:1651–1661

Ketring DL (1984) Temperature effects on vegetative and reproductive development of peanut. Crop Sci 24:87

Khalifa FM, Ong CK (1990) Effect of supra-optimal temperatures on germination of pearl millet [*Pennisetum glaucum* (L.)] hybrids. Ann Arid Zone 29:279–288

Khan AZ, Khan H, Ghoneim A, Khan R, Ebid A (2007) Seed quality and vigor of soybean as influenced by planting date, density and cultivar under temperature environment. Int J Agric Res 2(4):368–376

Kim K, Portis AR (2005) Temperature dependence of photosynthesis in Arabidopsis plants with modifications in rubisco activase and membrane fluidity. Plant Cell Physiol 46(3):522–530

Launders TE (1971) The effect of early season soil temperature on emergence of summer crops on the north-western plains of New South wales. Aust J Exp Agric Anim Hus 11(48):39–44

Liu X, Jin J, Wang G, Herbert S (2008) Soybean yield physiology and development of high-yielding practices in Northeast China. Field Crop Res 10(3):157–171

Lobell D, Burke M (2008) Why are agricultural impacts of climate change so uncertain? The importance of temperature relative to precipitation. Environ Res Lett 3:034007. https://doi.org/10.1088/1748-9326/3/3/034007

Maestri E, Klueva N, Perrotta C, Gulli M, Nguyen HT, Marmiroli N (2002) Molecular genetics of heat tolerance and heat shock proteins in cereals. Plant Mol Biol 48:667–681

Mahalakshmi V, Bidinger FR (1985) Water stress and time of floral initiation in pearl millet. J Agric Sci 105(2):437–445

Mathur S, Agrawal D, Jajoo A (2014) Photosynthesis: response to high temperature stress. J Photochem Photobiol 137:116–126

Meehl GA, Tebaldi C (2004) More intense, more frequent, and longer lasting heat waves in the 21st century. Science 305:994–997

Mirzaei M, Pascovici D, Atwell BJ, Haynes PA (2012) Differential regulation of aquaporins, small GTPases and V-ATPases proteins in rice leaves subjected to drought stress and recovery. Proteomics 12(6):864–877. https://doi.org/10.1002/pmic.201100389

Mitchell RAC, Mitchell VJ, Driscoll SP, Franklin J, Lawlor DW (1993) Effects of increased $CO_2$ concentration and temperature on growth and yield of winter wheat at two levels of nitrogen application. Plant Cell Environ 16(5):521–529

Narayanan S, Prasad PVV, Fritz AK, Boyle DL, Gill BS (2015) Impact of high nighttime and high daytime temperature stress on winter wheat. J Agron Crop Sci 201:206–218

Opole RA, Prasad PVV, Djanaguiraman M, Vimala K, Kirkham MB, Upadhyaya HD (2018) Threshold sensitive stages and genetic variability of finger millet to high temperature stress. J Agron Crop Sci 204(5):477–492

Ortiz R, Sayre KD, Govaerts B, Gupta R, Subbarao GV, Ban T, Hodson D, Dixon JM, Ortiz-Monasterio JI, Reynolds M (2008) Climate change: can wheat beat the heat? Agric Ecosyst Environ 126:46–58

Porter JR, Gawith M (1999) Temperatures and the growth and development of wheat: a review. Eur J Agron 10:23–36
Prasad PVV, Djanaguiraman M (2011) High night temperature decreases leaf photosynthesis and pollen function in grain sorghum. Funct Plant Biol 38:993–1003
Prasad PVV, Djanaguiraman M (2014) Response of floret fertility and individual grain weight of wheat to high temperature stress: sensitive stages and thresholds for temperature and duration. Funct Plant Biol 41:1261–1269
Prasad PVV, Craufurd PQ, Summerfield RJ (1999a) Sensitivity of peanut to timing of heat stress during reproductive development. Crop Sci 39:1352–1357
Prasad PVV, Craufurd PQ, Summerfield RJ (1999b) Fruit number in relation to pollen production and viability in peanut exposed to short episodes of heat stress. Ann Bot 84:381–386
Prasad PVV, Craufurd PQ, Summerfield RJ, Wheeler TR (2000) Effects of short episodes of heat stress on flower production and fruit set of groundnut [*Arachis hypogaea* (L.)]. J Exp Bot 51(345):777–784
Prasad PVV, Craufurd PQ, Kakani VG, Boote KJ (2001) Influence of high temperature during pre and post-anthesis stages of floral development on fruit-set and pollen germination in peanut. Aust J Plant Physiol 28:233–240
Prasad PVV, Boote KJ, Allen LH, Thomas JMG (2003) Super-optimal temperatures are detrimental to peanut (*Arachis hypogaea* L.) reproductive processes and yield at both ambient and elevated carbon dioxide. Field Crop Res 95:398–411
Prasad PVV, Boote KJ, Allen LH (2006a) Adverse high temperature effects on pollen viability, seed-set, seed yield and harvest index of grain sorghum [*Sorghum bicolor* (L.) Moench] are more severe at elevated carbon dioxide due to higher tissue temperatures. Agric For Meteorol 139:237–251
Prasad PVV, Boote KJ, Allen LH, Sheehy JE, Thomas JMG (2006b) Species, ecotype and cultivar differences in spikelet fertility and harvest index of rice in response to high temperature stress. Field Crop Res 95:398–411
Prasad PVV, Pisipati SR, Mutava RN, Tuinstra MR (2008a) Sensitivity of grain sorghum to high temperature stress during reproductive development. Crop Sci 48:1911–1917
Prasad PVV, Pisipati SR, Ristic Z, Bukovnik U, Fritz AK (2008b) Impact of night time temperature on physiology and growth of spring wheat. Crop Sci 48:2372–2380
Prasad PVV, Pisipati SR, Momcilovic I, Ristic Z (2011) Independent and combined effects of high temperature and drought stress during grain filling on plant yield and chloroplast EF-Tu expression in spring wheat. J Agron Crop Sci 97:430–441
Prasad PVV, Djanaguiraman M, Perumal R, Ciampitti I (2015) Impact of high temperature stress on floret fertility and individual grain weight of grain sorghum: sensitive stages and thresholds for temperature and duration. Front Plant Sci 6:820
Prasad PVV, Bheemanahalli R, Jagadish SVK (2017) Field crops and the fear of heat stress—opportunities, challenges and future directions. Field Crop Res 200:114–121
Prasad PVV, Djanaguiraman M, Jagadish SVK, Ciampitti IA (2018) Drought and high temperature stress and traits associated with tolerance. In: Sorghum: a state of the art and future perspectives. Wiley, New York, pp 241–265
Puteh AB, ThuZar M, Mondal MMS, Abdullah NAPB, Halim MRA (2013) Soybean [*Glycine max* (L.) Merrill] seed yield response to high temperature stress during reproductive growth stages. Aust J Crop Sci 7(10):1472–1479
Roach T, Krieger-Liszkay A (2014) Regulation of photosynthetic electron transport and photoinhibition. Curr Protein Pept Sci 15:351–362
Saini HS, Aspinall D (1982) Abnormal sporogenesis in wheat [*Triticum aestivum* (L.)] induced by short periods of high temperature. Ann Bot 49:835–846
Saini HS, Sedgley M, Aspinall D (1983) Effect of heat stress during floral development on pollen tube growth and ovary anatomy in wheat [*Triticum aestivum* (L.)]. Aust J Plant Physiol 10:137–144

Salem MA, Kakani VG, Koti S, Reddy KR (2007) Pollen-based screening of soybean genotypes for high temperatures. Crop Sci 47:219–231
Satake T, Yoshida S (1978) High temperature induced sterility in indica rices at flowering. Jpn J Crop Sci 47(1):6–17
Schlenker W, Roberts MJ (2009) Nonlinear temperature effects indicate severe damages to U.S. crop yields under climate change. Proc Natl Acad Sci 106(37):15594–15598
Streck NA (2005) Climate change and agroecosystems: the effect of elevated atmospheric $CO_2$ and temperature on crop growth, development and yield. Cienc Rural 35:730–740
Tack J, Lingenfelser J, Jagadish SVK (2017) Disaggregating sorghum yield reductions under warming scenarios exposes narrow genetic diversity in US breeding programs. Proc Natl Acad Sci 114(35):9296–9301
Talwar HS, Yanagihara S (1999) Physiological basis for heat tolerance during flowering and pod setting stage in groundnut [*Arachis hypogaea* (L.)]. JIRCAS working report, vol 14, pp 47–65
Von-Caemmerer S, Evans JR (2015) Temperature responses of mesophyll conductance differ greatly between species. Plant Cell Environ 38:629–637
Whittle CA, Otto SP, Johnston MO, Krochko JE (2009) Adaptive epigenetic memory of ancestral temperature regime in Arabidopsis thaliana. Botany 87(6):650–657
Wollenweber B, Porter JR, Schellberg J (2003) Lack of interaction between extreme high temperature events at vegetative and reproductive growth stages in wheat. J Agron Crop Sci 189:142–150
Yan K, Chen P, Shao HB, Zhang L, Xu G (2011) Effects of short-term high temperature on photosynthesis and photosystem II performance in sorghum. J Agron Crop Sci 197:400–408
Zhao C, Liu B, Piao S, Wang X, Lobell DB, Huang Y, Huang M (2017) Temperature increase reduces global yields of major crops in four independent estimates. Proc Natl Acad Sci 114(35): 9326–9331

# Chapter 18
# Morpho-physiological Basis of Finger Millet to Withstand Climatic Extremes: A Special Reference to Drought

**Y. A. Nanja Reddy and Y. N. Priya Reddy**

**Abstract** Drought is the major abiotic stress in semi-arid regions that limits the yield of rainfed crops. In the scenario of changing climate, finger millet is better suited for frequent drought episodes when compared to the traditional food crops like rice, wheat, and maize. Finger millet has climate resilience, superior grain quality, and fodder quality. There is a need to improve the productivity, since the productivity has stagnated in the last decade. Further improvement is possible through involving the physiological traits in breeding programs for drought adaptation. In this respect, this chapter provides updated information about the importance of finger millet, drought adaptation traits/mechanisms, and ways to enhance the grain yield of finger millet under moisture stress/rainfed conditions, especially at the whole plant level. Besides, this chapter will be a reference for finger millet research on drought adaptation.

**Keywords** Finger millet · Drought · Stress adaptive traits · Millets · Physiological breeding

## 18.1 Introduction

The world's population is increasing, and a severe food shortage is projected by 2050 (Gupta et al. 2017). Cereal grains are the major contributors of food, and yield of such major cereals, wheat, rice, and corn was improved over the years (Fischer and Edmeades 2010) but they are water-loving crops. With the global climate change, the total rainfall is projected to increase (Jalihal et al. 2019), but water is becoming a scarce input for agriculture as the frequency of drought episodes are increasing due to irregular rainfall pattern (Dash et al. 2009; Anonymous 2019). Among cereals, the

Y. A. N. Reddy (✉)
College of Horticulture, Munirabad, University of Horticultural Sciences, Bagalakote, Karnataka, India

Y. N. P. Reddy
Faculty of Biological Sciences, Friedrich Schiller University, Jena, Germany

 
H. M. Mamrutha et al. (eds.), *Translating Physiological Tools to Augment Crop Breeding*, https://doi.org/10.1007/978-981-19-7498-4_18

**Table 18.1** Effect of drought stress on grain yield of finger millet

| Growth stage | Stress period (days) | % Reduction in grain yield | Reference |
|---|---|---|---|
| 28 DAS to harvest | >80 | 109.8 | Maqsood and Azam Ali (2007) |
| Pot culture | 35–42 | 36.1 | Mohanabharathi et al. (2019) |
| Grand growth to ear-emergence | 25 | 18.0 | Nanja Reddy et al. (2020) |
| Ear emergence to grain filling | 36 | 36.6 | Krishna et al. (2021) |

finger millet is grown as a rainfed crop (Davis et al. 2019) combined with pulses or oilseeds and; as a sole crop (Dayakar et al. 2017) with less water requirement and high water use efficiency, and therefore finger millet would be an alternate crop to cultivate under semi-arid regions especially in the non-traditional regions (Dwivedi et al. 2012). However, productivity will be decreased by drought stress (DS) combined with moderately high temperatures (Maqsood and Azam Ali 2007; Nanja Reddy et al. 2020; Krishna et al. 2021).

Finger millet is well-known for its nutritional quality and health benefits when compared to other staple cereals. It is called "super grain" or functional food because, the grain is rich in fiber (18.9%), protein (6–8%), fat (1.3%), calcium (296 mg), carbohydrate (70–76%), lysine (2.86%), tryptophan (1.39%), methionine (2.86%), threonine and iron content (6.5 mg/100 g) (Ravindran 1991; Chethan and Malleshi 2007; Chandrashekhar 2010; Devi et al. 2014; Chandra et al. 2016; Kumar et al. 2016; Chauhan and Sarita 2018; Nanja Reddy et al. 2019a). At global level, about 3.5 billion people are suffering from calcium deficiency (Kumssa et al. 2015), for such people, the finger millet grain could be a better option (Puranik et al. 2017). Besides, the finger millet productivity in India is 1747 kg $ha^{-1}$ when compared to major rainfed crops, the sorghum (989 kg $ha^{-1}$) and pearl millet (1374 kg $ha^{-1}$; www.indiastatagri.com). Therefore, finger millet is a better option to meet the nutritional security in developing countries of Asia and Africa (Puranik et al. 2017), regional food security, and is the best cattle feed due to its 61% digestible nutrients (Baath et al. 2018).

In India, during 2019–2020, finger millet was cultivated in 1.004 m ha with a production of 1.755 mt (www.indiaagristat.com). More than 90% of finger millet area under cultivation is in rainfed conditions (Davis et al. 2019), where the occurrence of moisture stress for >25 to 30 days at one or the other crop growth stage is a common phenomenon. Although finger millet is known for its drought adaptation, the grain yield is reported to decrease by >100% depending upon the duration of drought stress (DS) and the crop growth stage (Table 18.1). Productivity has reached nearly stagnation in the last decade (Swetha 2011; Megha 2022) and is required to increase to 4.5 t/ha by 2025 (Borlaug 2002) to contribute to the national food production, and for regional food security in India and Africa. Therefore, approaches for increasing the yield of finger millet under rainfed conditions are pertinent in India and Africa.

Another important abiotic stress that combines the drought stress in semi-arid regions is the moderate high-temperature, where finger millet is mostly cultivated. Of the temperature extremes, relatively low temperature is not a constraint for finger millet production as the crop has been originated from the cool climates of Africa. Presently the crop is cultivated under rainfed situations where micro-climate temperature will be higher by 4–5 °C and decreases the grain yield (Yogeesh et al. 2016). For instance, the expected increase in temperature is by 4 °C, decreases the productivity of crops (Webb and Braun 1994). However, under such extreme temperatures, the finger millet is not generally cultivated, but identification of high-temperature tolerant varieties could influence the expansion of finger millet cultivation to non-traditional areas.

Research on finger millet started in 1920 in Karnataka, India by Coleman (1920), and improved by Lakshmanaiah (1967). Recent plant breeding efforts lead to the development of >120 varieties for rainfed cultivation (Gowda et al. 2014). However, the yield stagnation in recent years is reported (Swetha 2011; Megha 2022), and the present grain yield on farmer's field is only 50% of the research plots. Therefore, more research efforts are needed on finger millet because it meets the regional staple food demand and nutritional security for rural livelihood in addition to fodder security in semi-arid regions.

## 18.2 Physiological Basis of Drought Stress Adaptation

Drought is one of the important abiotic stresses that limit the finger millet productivity as it is cultivated as a rainfed crop in semi-arid regions. The crop experiences drought stress (DS) at any of the following growth stages like, seedling phase, early vegetative phase, grand growth phase, flowering, and grain filling phase. Under DS conditions, plants undergo modifications in morphological, physiological, cellular, and molecular traits to achieve better plant growth, development, and productivity. The general drought adaptation mechanisms developed by the plants are drought escape, drought avoidance, and drought tolerance (Levitt 1980).

### *18.2.1 Morpho-physiological Traits of Drought Escape*

Drought escape refers to the crops (plants) that complete their life cycle before the onset of severe drought stress and produce reasonable yield (Nanja Reddy et al. 2021a). Drought escape is associated high metabolic rates, cell division, cell expansion, high seedling vigor, rapid growth rates, early flowering; and completes their life cycle without altering the grain filling period of 35–40 days. It is reported that the short-duration genotypes are low-yielding due to lower biomass production, and the yield levels of the long-duration varieties are 22% higher than the early and medium duration genotypes under adequate rainfall (>695 mm) conditions (Udaya Kumar et al. 1986). However, the new varieties like GPU-28, GPU-48, GPU-26, Indaf-9, with short to medium duration (105–110 days) yields high when compared to the long-duration genotypes (Nanja Reddy et al. 2019b, 2021b). The short life cycle is a built-in mechanism of drought escape with a rapid growth rate, where a large leaf area need to be achieved through a higher rate of leaf appearance, high radiation use efficiency, and early flowering.

#### 18.2.1.1 Seed Germination

Seed germination and the crop establishment are the crucial stages for adequate plant population and productivity under DS conditions. The seed germination is mainly dependent upon soil moisture availability for the imbibition of water by the seed. Maintenance of soil moisture near to field capacity continuously with a 5 mm daily evaporation would help the seed to germinate within 2–3 days (Nanja Reddy et al. 2021a). However, providing field capacity once in 3 days takes 6 days to germinate, because top soil loose water by 5mm per day, and seeds will not be in contact with sufficient soil moisture. Therefore, 30 mm irrigation after sowing gives 100% germination and the seed must be sown in the top 2 cm surface. The rapid imbibition leads to increased alpha-amylase activity and thus seed germination (Poornima 2017). The speed of imbibition depends both on the soil moisture availability, and seed composition; the higher seed protein content is expected to have a higher rate of imbibition (Zhao et al. 2018). Therefore screening genotypes for high seed protein content could be a better avenue for rapid imbibition. In general, the short-duration genotypes are expected to germinate early than the long duration genotypes. However, the bold seed germinates early but the leaf production will be relatively faster in the case of short duration varieties (Lokesh et al. 2000; Unpublished; observation with cv. GPU-48 a small-seeded short duration variety in comparison to MR-6 a bold seeded long-duration variety). The other ways to increase the seed germination, seedling length, and vigor index are, seed hardening with water or nutrients or hormones (Sastry et al. 1982; Karivaratharaju and Ramakrishnan 1985). The mannitol stress (600 mM) was found to decrease the shoot length by sixfold in contrast the root length was increased by threefold as compared to the WW conditions (Mukami et al. 2019). The seed germination percentage was found zero in the 25% PEG-6000 medium (Gupta et al. 2014; Bangari 2020). Suitable genotypes for rainfed conditions can be identified using PEG at seedling level (Bangari 2020), and; the gravimetric approach and field conditions by with-holding the irrigation. Identification of traits for rapid seed germination and its incorporation into popular varieties could be pertinent for drought escape.

#### 18.2.1.2 Tillering

The direct-sown seed emerges out of the soil by 6–7 days after sowing and takes 10–12 days to produce 3–4 leaves, the start of tillering, and seedling establishment. The variety with high vigorous growth capacity will have early tillering, and the seedling stage extend up to the completion of tillering phase (up to 30–35 days after sowing, DAS). If the dry spell continues from 10 to 20 DAS, the seedling shows leaf wilting and higher evaporation from the soil surface, which further increases the leaf wilting, but decreases the tiller-growth and biomass accumulation. Under such DS situation, the genotype possessing mid-day leaf folding, thick stem with lesser leaf chlorosis, and lesser leaf wilting could be preferred. Importantly, the rapid recovery growth after re-watering is crucial although the plant turns to turgid condition within 24 h after irrigation (Krishna and Nanja Reddy 2021). Rapid canopy recovery covers the soil surface and reduces the evaporation. In transplanted finger millet under rainfed conditions, the variety should possess rapid recovery from transplanting shock with re-growth of roots. The basal tillering at the seedling stage is important for uniform and synchronous flowering and; the late formed culm branching leads to asynchronous flowering and unproductive tiller formation. In general, finger millet takes 1 week to recover from transplanting shock and start of tillering, therefore 12-day old seedlings could be better for transplanting especially in the case of short duration varieties. The early DS found to decrease the basal tillering and growth by 30% (Krishna and Nanja Reddy 2021). If the tillering stage is exposed to DS, the death of early formed tillers and survival of later formed tillers within 25–30 DAS is a requisite as the recovery growth of early formed tillers is difficult. Therefore, genetic resources need to be exploited for rapid recovery growth rates as the effect of DS is inevitable.

#### 18.2.1.3 Root Traits

Finger millet being a cereal crop has fibrous root system formed by the primary roots that originate from the embryo, and thick crown roots which arise from the base of the stem. The DS at the seedling stage (from 15 to 30 days) increases the root length (26.6%) but, the root volume and root dry weight were decreased by 37.0% and 41.8%, respectively (Krishna and Nanja Reddy 2021). The root system provides anchorage to the plant and helps in enhanced uptake of water and minerals, interacts with microbes to improve drought tolerance (Comas et al. 2013). QTLs (qLRDW.1, qLRDW.2, qHSDW.1, and qHRL.1) associated with root dry weight, shoot dry weight, and root length has been identified in finger millet (Ramakrishnan et al. 2017), and suggests the possibilities for improvement in the root system. Therefore, a deeper root system at the early stages is important for plant survival and growth.

#### 18.2.1.4 Flowering

In finger millet, the grain filling period remains unaltered much (35–40 days) and therefore, the crop duration depends on the duration of the vegetative phase. The pre-ponement in flowering is a key event to accomplish the drought escape phenomenon. The DS through the vegetative phase delays the flowering or inhibits completely (Unpublished). Increased days to flowering under DS conditions showed a negative relationship with grain yield (Suma 2014). Hence, the growth rate during the early vegetative phase should be high with high water use efficiency in short duration genotypes to produce high biomass and grain yield. The suitable short-duration varieties are GPU-48, GPU-45, and GPU-26 with 100–105 days duration, and these genotypes have shown yield advantage over the long duration genotypes especially under terminal DS conditions, because the short duration genotypes could complete their life cycle before the severe DS conditions occur. Therefore, the selection of short duration genotypes will offer drought escape in finger millet, in particular with moderate and effective tillering habit.

### *18.2.2 Morpho-physiological Traits of Drought Avoidance*

Drought avoidance is the adaptation of the crop plants to maintain higher plant water status for normal functioning of the plant without experiencing the stress. This can be achieved by increased water uptake for maintaining higher tissue water status and or by water conservation through minimizing water loss through various plant modifications.

#### 18.2.2.1 Root Traits

Finger millet has fibrous root system by combination of seminal, nodal, and lateral roots. The root traits those impart the drought avoidance are high root surface area, high root volume, appropriate root system architecture (RSA), deep root length, higher root diameter and root hairs, higher root/shoot ratio, higher aquaporin activity, resumption of rapid root growth after DS alleviation with lesser root senescence, and high root secretions (Nanja Reddy et al. 2021a).

A better root system under DS is essential for higher water and nutrient uptake. A moderate DS increases the root growth by compensating with the shoot growth (Naik et al. 2020; Krishna and Nanja Reddy 2021). The DS decreases the shoot system, probably due to the higher diversion of photosynthates towards root growth, which acts as a sink. However, the root length is also affected by severe DS at the seedling stage, 15–30 DAS (Krishna and Nanja Reddy 2021; Naik et al. 2020), while at later stages (45 DAS) the root biomass is more affected (Krishna and Nanja Reddy 2021). In finger millet, the root grows in length-wise quickly up to tillering with a subsequent slow growth till flowering or after. In contrast, the root volume and root dry weight increases slowly up to tillering with a speedy increase from tillering to ear emergence/ flowering with not much alterations after flowering (Krishna and Nanja Reddy 2021). The drought-tolerant variety, PR-202 accumulates higher root weight and deeper root system under DS conditions (Nanja Reddy et al. 2010, 2021a). Hence, it could be apt to screen genotypes for deep root length at tillering stage and; root biomass at the active vegetative stage for drought avoidance.

The root traits are manipulated through multiple gene actions and each gene contributes to a small effect for a QTL controlling a root trait. The root traits can be constitutive or adaptive. The semi-dwarf plants possess a shallow root system when compared to tall plants and the dwarf plant will have higher tiller number. The dry-land species like millets are likely to develop deeper root systems when compared to the wet-land species that can be ascribed to higher root respiration under aerobic situations. The coarse roots with higher root length density (RLD) in deeper soil depth are desirable under DS adaptation conditions (Comas et al. 2013). Therefore, the selection of varieties that produce deeper roots with the start of stress conditions and rapid root spread after stress alleviation could be highly relevant.

#### 18.2.2.2 Leaf Characters

In cereal crops, the leaf lamina is the major photosynthesizing organ in addition to leaf sheath. In finger millet, the green ear-head also contributes to productivity to the tune of 41% (Tieszen and Imbamba 1978; Sashidhar et al. 1984). Many reports with a few exceptions show a positive influence of leaf area index (LAI) at the time of flowering to the grain yield (Nanja Reddy et al. 2019b). Drought stress for 36 days from flag leaf through ear-emergence to grain filling period decreased the LAI by 16.5% from 6.63 in WW to 5.34 in DS condition for 20 days (Krishna et al. 2021). Similarly, a reduction in LAI with different DS durations and the crop growth stages has been reported (Maqsood and Azam Ali 2007; Aparna 2016). Under DS condition, the decrease in LAI could be due to decreased cell turgor and consequent decrease in cell cycle, cell size, cell number, and decreased leaf expansion rates. However, the green leaf lamina area at crop harvest has no influence on the grain yield (Assefa et al. 2013). Probably, as the crop reaches maturity; more the green leaf area could act as a sink rather than remobilization of assimilates. Therefore, leaf area at the flowering time could be important. The DS at tillering stage, decreased the leaf number by a greater extent 44% while, it was to a lesser extent when DS was imposed after tillering (21% reduction), and the recovery rates were 101.3% and 31.1%, respectively (Krishna and Nanja Reddy 2021).

Photosynthesis is a major physiological process that determines crop growth and grain yield. Finger millet as a $C_4$ species, the photosynthetic rate (Pn) will be high (Sastry et al. 1982; Ueno et al. 2006; Nanja Reddy 2020; Nanja Reddy et al. 2020), and is reported as high as 31.5 u Mol. $m^{-2}$ $s^{-1}$ under WW conditions. Under the DS at active growth period, finger millet takes 18–20 days for initiation of wilting symptoms at given water potential of 5.0 MPa (Suma 2014). However, the moderate DS with soil evaporation of 4.2 mm/day for 18–20 days decreases the Pn by 16.6% but the grain yield is less affected (9.5%; Mohanabharathi et al. 2019; Nanja Reddy et al. 2020). Therefore, there is a need for gas exchange trait improvement. In this regard, a drought adaptive variety, cv. PR-202 showed a meagre reduction in Pn by 4.4% due to DS conditions and it was attributed to higher stomatal conductance and transpiration rate, which might be due to higher water relations (Nanja Reddy et al. 2010, 2020). The DS found to reduce the stomatal conductance (20.1%), transpiration (25.0%), and the photosynthetic rate (16.6%; Nanja Reddy et al. 2020). The relationship between Pn and transpiration rate will be high under DS when compared to the WW conditions (Nanja Reddy et al. 2020) and hence, finger millet could be better adapted to rainfed situations. However, genotypes with lower stomatal frequency will adapt better to the rainfed situations (Udaya Kumar et al. 1986). The tolerant genotypes would maintain lower transpiration rates due to stomatal limitations (Kholova et al. 2010). Hence, it can be considered that a reduction in transpiration with lesser leaf area could be a significant plant adaptation to dehydration post-ponement under severe DS where soil water is limited at the deeper layers. But under mild-stress conditions, the deeper root system could be apt to draw more water for higher transpiration rates and crop productivity.

In drought-tolerant genotypes, the DS adapted plants will have smaller and thicker leaves, with reduced epidermal cell size and stomatal size and increased stomatal frequency (Bhutta et al. 2006). The reduction in leaf area due to stress will be attributed to the sensitivity of leaf expansion rates. Pre-treatment of flag leaf tissues with water stress during the early stage of development resulted in an irreversible reduction in leaf area with an increased leaf thickness (Zagdanska and Kozdoj 1994). The leaf thickness, an indirect measure of specific leaf weight (SLW) and have shown a direct positive relationship with photosynthetic rate in the finger millet genotypes (Suma 2014). SLW have shown a positive relationship with photosynthetic rate and biomass production (Subramanyam 2000). The drought-adapted varieties maintain a higher leaf thickness, SLW, and photosynthetic rate (Craufurd et al. 1999) with dense chlorophyll concentration. Therefore, it could be suitable if the leaf surface area is reduced with higher SLW to enhance the WUE under DS conditions (Sastry et al. 1982; Suma 2014). The increased leaf thickness could be a survival mechanism, where intercellular spaces might expand to protect the mesophyll cells.

Another parameter of drought avoidance is the leaf hairiness that reflects the PAR, thus reduce the heat load on the plant under DS conditions but with a reduction in photosynthetic rates. Besides, the epicuticular wax accumulation plays an important role in the conserving the water through reduced cuticular transpiration (Parvathi and Nataraja 2017), and therefore, epicuticular wax is a desirable trait of drought avoidance (Bondada et al. 1996). A drought-tolerant genotype PR-202 showed an increased wax accumulation when compared to a susceptible variety under DS condition (Anonymous 2006; Nanja Reddy et al. 2021a). Mid-day leaf folding could be a drought adaptation mechanism, with a fully opened leaf in the morning and evening times (Campbell 1990). Genotypic differences were observed in leaf folding and stay greenness of finger millet, which can be exploited for drought avoidance.

Besides, leaf temperature (LT) has been recognized as an indicator of plant water status and easy to assess the drought adaptation in finger millet under DS conditions (Ramya and Nanja Reddy 2018). The LT can be measured by CCATD (Anke Gowda et al. 1999) or by a digital infrared thermometer at a single plant level (Suma 2014). The LT have shown negative relationship with grain yield under DS conditions (Suma 2014; Ramya and Nanja Reddy 2018; Krishna et al. 2021). The optimum LT could be 28 °C for higher grain yield (Ramya and Nanja Reddy 2018; Mohanabharathi et al. 2019). The increase in leaf temperature could be due to decrease in leaf water potential, decreased stomatal conductance, and decreased transpiration rates (Mohanabharathi et al. 2019). Besides, the higher LT might disrupt PS-II and decrease the photosynthetic rate leading to a reduction in grain yield (Krishna et al. 2021). One of the examples is the cv. PR-202 maintained lower LT and also had higher recovery after drought stress alleviation. The contribution of LT towards grain yield under DS was to a tune of 46% among the physiological traits (Krishna et al. 2021). Therefore, LT could be easy and appropriate selection criteria for drought response.

## *18.2.3 Morpho-physiological Traits of Drought Tolerance*

Drought tolerance is the ability of the plant to maintain balanced metabolic activities at lower tissue water potential through osmotic adjustment and to produce relative grain yields. Osmotic adjustment is a decrease in the osmotic potential of the cell by accumulating compatible solutes, such as proline, organic acids, soluble sugars, and glycine betaine (Tadele 2016). At the cellular level, the drought-tolerant genotype is expected to possess higher osmo-regulation through higher osmolytes accumulation, better chloroplast integrity, higher membrane stability, and synthesis of antioxidant enzymes. This leads to higher crop growth rates, leaf expansion rates, and partitioning of dry matter to economic parts after the DS alleviation. Therefore, genotypes that can tolerate drought stress and with higher recovery growth rates after DS alleviation could be appropriate.

### 18.2.3.1 Water Relations

Maintenance of higher leaf water potential (LWP) is a key physiological trait of drought tolerance under DS conditions. Drought stress limit the water supply and decreases the LWP leading to decrease the turgor potential, stomatal conductance, photosynthesis, and eventually decreased grain yield. The DS results in shrinkage of cells, loss in turgor, disintegration of the cell membrane, and reduced metabolic processes (Mahajan and Tuteja 2006). The LWP was similar between tolerant and susceptible genotypes under well-watered (WW) conditions but, under DS conditions, the tolerant genotypes showed a lesser reduction (6.9%) as against a higher reduction 14.4% in susceptible genotypes of finger millet (Suma 2014).

Maintenance of higher relative water content (RWC) is an additional drought tolerance mechanism and a measure (Barrs and Weatherley 1962; Almeselmani et al. 2011; Suma 2014). RWC is the existing water status of the leaf compared to its water content, which can hold at full turgidity of leaf (Barrs and Weatherley 1962). The higher RWC reflects the capacity of a plant to absorb water under DS conditions. The RWC maintains higher at the flowering stage and decreases towards crop maturity (Kumar and Singh 1998). The genotypes collected from moisture-rich regions show a greater reduction in RWC under stress conditions (Assefa et al. 2013). The RWC ranges from 85 to 92% under rainfed cultivation and the genotypes which accumulated higher RWC produced a higher grain yield, suggesting that the higher RWC is a useful trait in achieving higher grain yield under rainfed conditions (Aparna and Bhargavi 2017). Naik et al. (2020) reported that the RWC ranges from 57.9 to 80.7 under WW conditions in susceptible and tolerant varieties respectively, while under DS it ranged from 52.93 to 68.44%, respectively. The drought-tolerant varieties maintained higher RWC indicating that the RWC is a good index for drought tolerance (Aparna et al. 2017). A reduction in RWC (by 2.58%) resulted in reduced stomatal conductance (21.7%), transpiration rate (25.3%), and photosynthetic rate (16.8%) (Nanja Reddy et al. 2010). The reduced RWC under DS found to speed up flag leaf senescence with a reduced grain filling period and decreased yield (Almeselmani et al. 2011). Therefore, higher RWC under DS can be used effectively in identification of drought-tolerant genotypes.

#### 18.2.3.2 Photosynthesis Pigments and Gas Exchange Traits

Chlorophyll is the foremost pigment system that decides the photosynthetic rate and grain yield of finger millet (Suma 2014; Pallavi et al. 2016). Drought stress hastens the chlorophyllase, and peroxidase activity to result in decreased chlorophyll content and enhancing the senescence process (Maqsood and Azam Ali 2007; Aparna 2016; Manjula 2016). The reduction of chlorophyll content was higher when the duration of DS period was increased (Tiwari et al. 2020). Extraction of chlorophyll in large number of genotypes using organic solvents is a tedious process (Nanja Reddy et al. 1990), and hence, SPAD chlorophyll meter reading (SCMR), a non-destructive, simple and portable diagnostic device has been used in used in estimation of chlorophyll index (Suma 2014). Maintenance of higher SCMR can improve the light absorption, $CO_2$ fixation, and the grain yield (Almeselmani et al. 2013). The high-yielding finger millet variety (Cv. BR-36) maintained higher chlorophyll index with a SPAD value of 31.08 (Aparna and Ansari 2017). The DS from ear emergence to grain filling decreased the SPAD value by 20% among 16 popular varieties of finger millet (Krishna and Nanja Reddy 2021), and by 15.5% amongst 181 finger millet genotypes (Ramya 2022). Suma (2014) reported that the drought-tolerant finger millet genotypes maintain higher SPAD values under DS conditions. Susceptible varieties showed a higher decrease in the chlorophyll content (by 44.8%) under

DS condition when compared to the tolerant varieties (Kotapati et al. 2014). The DS decreases the SCMR by 10 units when DS was given for 20 days (Assefa et al. 2013; Anitha et al. 2019; Mohanabharathi et al. 2019). However, the SCMR differences in value of less than 10 have no significant effect on grain yield (Aparna 2016). Therefore, the value of SCMR >50 (in comparison with cv. GPU-28, with SCMR of 40 units) at the ear-emergence stage could be considered in achieving higher grain yields of finger millet under rainfed conditions. The SCMR is also related to leaf thickness positively under DS conditions and has relevance to productivity (Krishna et al. 2021).

Besides, the chlorophyll stability index (CSI) under DS conditions is an indicator of the stress tolerance capacity in plants and was higher in tolerant genotypes when compared to susceptible genotypes (Bhatt et al. 2011). Therefore, maintenance of higher CSI under DS conditions with lower leaf senescence rates (stay-green character of top 3–5 leaves) could be used as trait of drought tolerance in finger millet. Maintenance of stay greenness of leaf is crucial for current photosynthesis especially during the grain filling phase (Van Oosterom et al. 1996). Stay greenness would maintain reserve carbohydrates in stem and; transport to the grain and; therefore, had a positive relationship to grain yield (Borrell and Douglas 1996). Stay green also helps in reducing lodging (Rosenow 1984). Maintenance of stay greenness (delayed leaf senescence) was associated with the higher cytokinin synthesis, and reduced ACC oxidase activity and ethylene synthesis (John et al. 1995). Therefore, reduced leaf senescence rates especially the top 3–5 leaves (not entire foliage) could be a better selection criterion for drought tolerance. Water use efficiency (WUE) is another important trait that plays a key role in drought adaptation. Under limited water conditions, the transpiration efficiency (TE) will be high as compared to the WW conditions, and higher TE is required to produce higher biomass and grain yield (Talwar et al. 2020).

#### 18.2.3.3 Osmolyte Accumulation and Antioxidant Enzymes

Abiotic stresses including the drought, generate reactive oxygen species (ROS) like superoxide ($O_2^-$), hydrogen peroxide, and hydroxyl radicals that cause membrane damage, degradation of chlorophyll, protein carbonylation, and inactivation of –SH group-containing enzymes. The ROS acts as signaling molecules at low concentrations, but at higher concentrations, cause damage the cells (Tadele 2016). To combat the harmful effects of ROS, cells generates non-enzymatic antioxidants like tocopherols, anthocyanins, flavonoids, carotenoids, osmolytes, and; the ROS scavenging enzymes like superoxide dismutase (SOD), catalase, peroxidase, ascorbate peroxidase, glutathione reductase (Bhatt et al. 2011).

Proline is one of the important osmo-protectants, known to involve in lessening the cytosolic acidosis. Accumulation of proline has a positive relationship with stress tolerance. An increased level of proline content helps the plant to survive against drought stress by increasing the osmotic strength of cell sap (Bhatt et al. 2011; Tiwari et al. 2020). The leaf proline content will be higher at the reproductive stage when compared to the vegetative stage and increases with the duration of stress (Tiwari et al. 2020). Genotypic variation in proline accumulation was observed under rainfed cultivation, which ranged from 0.39 to 0.77 ug $g^{-1}$ (Aparna and Bhargavi 2017). A three-fold increase in proline content was observed under DS when compared to the WW conditions (Kotapati et al. 2014). Drought tolerant variety, PR-202 found to accumulate a 10-fold increase in proline content against 7.5-folds in susceptible variety PES-400 under DS conditions (Bhatt et al. 2011), suggesting that the proline content accumulation confers the ability to lower the osmotic potential and protects the DNA, enzymes and cellular membranes from oxidative damage caused by DS. Therefore, the proline content could be used as a trait for drought tolerance measurement.

The peroxidation of membrane lipids causes loss in membrane leads to electrolyte leakage and MDA accumulation under drought stress. The MDA content does not differ between genotypes under control conditions, while under DS conditions, the MDA, phenol content, flavonoid content, and catalase activity were increased with the duration of the stress period and the tolerant genotypes maintained lower MDA in comparison to the susceptible varieties (Tiwari et al. 2020). The damage could be higher in sensitive variety (50%) as against 35% in tolerant variety (Bhatt et al. 2011). The peroxide levels were low in tolerant variety (70%) when compared to 147% in susceptible varieties under DS conditions (Bhatt et al. 2011). Furthermore, glutathione reductase (GR) was higher in the case of tolerant varieties as compared to susceptible varieties, to maintain higher reduced glutathione levels in cells in case of tolerant varieties, as the reduced glutathione is a non-enzymatic antioxidant in the cells.

Among the antioxidant enzymes, the SOD plays an important role in defense mechanisms under DS conditions. SOD helps in binding the highly reactive superoxide radical to hydrogen ion, and forms another reactive compound called hydrogen peroxide. Subsequently, oxidation of hydrogen peroxide by the catalase or ascorbate peroxidase forms the water molecules and liberates oxygen molecules. The SOD activity was found to increase under DS conditions especially in tolerant genotypes than in the susceptible variety. A higher SOD produces higher $H_2O_2$, therefore APX is important to convert the excess peroxide into water and oxygen. Hence the ratio of APX/SOD is important and it was 2.4 in tolerant genotypes against 1.77 in susceptible varieties (Bhatt et al. 2011). To fight against the oxidative damage, an increased activity of SOD, CAT, APX, GR, and LOX were activated under DS conditions in finger millet (Bhatt et al. 2011; Kotapati et al. 2014). Finger millet stressed for 6 days (45 to 51 DAS; to 45% of field capacity) showed an increase in anti-oxidant (proline) content. Besides, the MDA (malondialdehyde), $H_2O_2$ production, electrolyte leakage were higher under DS conditions. The relative drought adaptive variety, cv. PR-202 showed a higher SOD, GR, CAT, and APX when compared to the sensitive variety cv. PES-400 (Bartwal and Arora 2017). This tolerance of cv. PR-202 could be development of the variety was from hot region of Peddapuram, Andhra Pradesh, as compared to cv. PES-400, which was developed at low temperature high-altitude region (Bhatt et al. 2011). Therefore, screening of germplasm for higher antioxidant enzymes and or osmolytes accumulation could aid in development of donor lines for drought tolerance.

Cell membrane is the primary site of DS injury, thus, evaluation for cellular membrane integrity would be an appropriate criterion for stress tolerance (Sullivan and Ross 1979). Measurement of electrolyte leakage is the typical method for estimating the membrane damage in response to DS conditions. Leakage will vary with the membrane's ability to uptake and retain the solutes, which are genotype dependent (Agarie et al. 1995). Under DS conditions, the tolerant genotypes maintain a lower electrolyte leakage (high membrane integrity; Bhatt et al. 2011). Therefore, the membrane stability could be another important selectable trait for drought-tolerance.

#### 18.2.3.4 Partitioning of Photo-assimilates to the Grain

The harvest index (HI) is partitioning of biomass towards the grain during the reproductive phase. The yield improvement in several crops including finger millet was achieved through an increased HI (Fischer and Edmeades 2010; Swetha 2011; Megha 2022). The ear-emergence stage is most sensitive to DS, and during which the assimilate transport from leaf to the grain will be inhibited, induces sterility and decreases the grain yield. Under DS conditions, the remobilization of stem reserves will contribute to the grain yield by 75 to 100% (Van Herwaarden et al. 1998). During the grain-filling phase, the green leaves (for their respiratory demands) and the grains compete as sinks for the hydrolyzed carbohydrates in the stem. Hence, it would be apt to have rapid senescence rather than stay green at the physiological maturity stage (Ex. Genotype, GE-4683 with bold grain, Unpublished). The DS during grain filling phase would increase the translocation of carbohydrates to grain, could be an evolutionary mechanism (Unpublished). Therefore, selection should aim at high partitioning and high biomass production under DS conditions.

## 18.3 Approaches for Improving Yield Under Stress Conditions

The semi-arid regions are characterized by unpredicted rainfall with frequent dry spells for which finger millet could be an ideal crop. Several approaches are being employed for improvement of drought adaptation in crop species.

Utilization of genetic variability is the primary step in the breeding program for identification and selection of desirable drought adaptive traits. The National and international institutions, like ICRISAT, AICRP on Small Millets, Kenya, Ethiopia, Uganda, USA, and other countries made efforts to collect and maintain the germplasm accessions (Ceasar et al. 2018). India is maintaining 67% of the total finger millet accessions (33,650 numbers; Ceasar et al. 2018). The exploitation of genetic variability through an empirical approach (field screening) is a major selection mode even these days. The drought susceptibility index (DSI), under field conditions is one of the measures that provide the capacity of variety. In the selection of drought adaptive varieties, the varieties should possess similar grain yield under WW conditions, and having a lesser reduction in grain yield under DS conditions. Varieties with a low drought susceptibility index (DSI) values are called tolerant and vice-versa. Specific traits would be appropriate in selecting drought adaptive varieties in finger millet. Such traits will be useful to develop the MAGIC population, using the deeper root system to extract water or leaf wax accumulation for radiation reflectance (Nanja Reddy et al. 2010).

In the present day of molecular biology era, understanding of the molecular mechanisms of drought adaptation is important in identification of the genes associated it. The genes and QTLs that confer the drought tolerance have been identified. In this respect, the leaf rolling and wilting under DS conditions have shown a decreased solute potential with over-expression of many drought-responsive genes (Parvathi et al. 2013). Transgenic finger millet has been developed for over-expression of drought-responsive genes like metallothionein (Parvathi et al. 2013), mltD for osmotic tolerance (Hema et al. 2014), Ecdehydrin7 (Singh et al. 2015), EcCIPK31-like (Nagarjuna et al. 2016), EcGBF3 (Ramegowda et al. 2017), EcTAF6 (Parvathi and Nataraja 2017). Under DS condition, the stress-responsive genes like CIPK, GRX, LIP HV90, SRP, and zinc finger were over-expressed, and can be useful in marker-assisted selection (Suma 2014).

## 18.4 Conclusions and Future Prospects

Finger millet has an important role in meeting regional food security with rich health benefits and as a climate-resilient crop. The major health disorder in the present society is diabetes due to changes in food habits. In India, in pre-independence era, the irrigation projects were less, and there is no proper data about the diabetes population. However, after the irrigation projects in post-independent era, rice became an important cereal crop to meet the food demands, and in addition, the diabetic population is also in increasing trend. Because of the resistant starch, high fiber, and slow digestibility; the finger millet could be a better cereal food especially for the diabetic population. In addition, the finger millet has high water use efficiency as compared to the water-spending cereal crops like rice, wheat, and maize. The finger millet productivity will be highest at 75% field capacity rather than 100% field capacity (Anonymous 2011); hence, the finger millet has drought adaptation and is suitable to rainfed conditions. However, severe DS decreases the grain yields; therefore, there is a continuous necessity to develop drought adaptive variety with higher grain yields under rainfed cultivation in semi-arid regions. It is difficult to demarcate the one with another drought adaptation mechanism; hence, combined drought adaptation mechanisms could be better especially the physiological traits in addition to grain yield attributing traits. Furthermore, the development of multi-parent advanced generation inter-cross (MAGIC) population using the said traits could be of great use.

## References

Agarie S, Hanaoka N, Ueno O, Miyazaki A, Kubota F, Agata W, Kaufman PB (1995) Effects of silicon on tolerance to water deficit and heat stress in rice plants (*Oryza sativa* L.) monitored by electrolyte leakage. Plant Prot Sci 1(2):96–103. https://doi.org/10.1626/pps.1.96

Almeselmani M, Abdullah F, Hareri F, Naaesan M, Ammar MA, Kanbar OZ, Saud A (2011) Effect of drought on different physiological characters and yield component in different Syrian durum wheat varieties. J Agric Sci 3:127–133

Almeselmani M, Saud AR, Zubi HA, Zubi KA, Hareri F, Naaesan M, Ammar MA, Kanbar OZ, Naseef HA, Swed WA, Masalma M (2013) Selectable physiological traits for yield improvement in durum wheat grown under water deficit condition. Stand Sci Res Essays 1(3):41–51

Anitha K, Sritharan N, Ravikesavan R, Djanaguirama M, Senthil A (2019) Melatonin alters photosynthesis-related traits in finger millet (*Eleusine coracana* L) under drought conditions. Int J Chem Stud 7(3):2750–2754

Anke Gowda SJ, Sashidhar VR, Devendra R (1999) Genotypic variation in crop canopy air temperature differences and its association with growth and yield attributes in finger millet. Indian J Plant Physiol 4(4):262–265

Anonymous (2006) Annual report. All India Coordinated Small Millets Improvement Project, ICAR, GKVK, Bangalore

Anonymous (2011) Annual report. All India Coordinated Small Millets Improvement Project, ICAR, GKVK, Bangalore

Anonymous (2019) uasbanagalore.edu.in/index.php/research-en/agromateorology-en

Aparna K (2016) Identification of ragi genotypes for drought tolerance, high yield and seed quality parameters under kharif rainfed conditions. M.Sc. (Agri.) Thesis. Acharya N.G. Ranga Agricultural University, Guntur, Andhra Pradesh, India

Aparna K, Ansari ZG (2017) Evaluation of ragi genotypes on growth parameters and physiological attributes under kharif rainfed conditions. Int J Chem Stud 5(6):1899–1901

Aparna K, Bhargavi H (2017) Identification of drought-tolerant and high yielding genotypes in ragi under rainfed conditions. Agric Update 12(8):2142–2145

Aparna K, Ansari ZG, Sudhakar P (2017) Screening for water use efficiency traits and other drought-tolerant traits in finger millet (*Eleusine coracana* (L.) Gaertn.) genotypes under kharif rainfed conditions. Eur J Biotechnol Bio-sci 5(6):37–40

Assefa A, Fetene M, Tesfaye K (2013) Agro-morphological, physiological and yield related performances of finger millet [*Eleusine coracana* (l.) Gaertn.]. accessions evaluated for drought resistance under field condition. Asian J Agric Rural Dev 3(10):709–720

Baath GS, Northup BK, Gowda PH, Rocateli AC, Turner KE (2018) Adaptability and forage characterization of finger millet accessions in U.S. Southern Great Plains. Agronomy 8:2–9

Bangari PSM (2020) Field performance of finger millet genotypes differing in drought adaptive traits: physiological and biochemical analysis. M.Sc. (Agri.) Thesis, Department of Crop Physiology, University of Agricultural Sciences, Bangalore

Barrs HD, Weatherley PE (1962) A re-examination of the relative turgidity technique for estimating water deficits in leaves. Aust J Biol Sci 15:413–428

Bartwal A, Arora S (2017) Drought stress-induced enzyme activity and MDAR and APX gene expression in tolerant and susceptible genotypes of Eleusine coracana (L.). In Vitro Cell Dev Biol Plant 53:41–49

Bhatt D, Sharma P, Arora S, Negi M, Saxena SC, Dobriyal AK (2011) Responses to drought induced oxidative stress in five finger millet varieties differing in their geographical distribution. Physiol Mol Biol Plants 17(4):347–353

Bhutta WM, Ibrahim M, Tahira (2006) Comparison of water relations and drought-related flag leaf traits in hexaploid spring wheat (*Triticum aestivum* L.). Plant Soil Environ 52(5):234–238

Bondada B, Oosterhuis DM, Murphy JB, Kim KS (1996) Effect of water stress on the epicuticular wax composition and ultra-structure of cotton (*Gossypium hirsutum* L.) leaf, bract, and boll. Environ Exp Bot 36:61–67

Borlaug NE (2002) Feeding a world of 10 billion people: the miracle ahead. In Vitro Cell Dev Biol Plant 38:221–228

Borrell AK, Douglas ACL (1996) Maintaining green leaf area in grain sorghum increases yield in water limited environment. In: Forale MA, Henzel RG, Kneipp JF (eds) Proceedings of the third australian sorghum conference. Melbourne, Australian Institute of Agricultural Sciences

Campbell (1990) Derivation of an angle density function for canopies with ellipsoidal leaf angle distributions. Agric For Meteorol 49(3):173–176

Ceasar SA, Maharajan T, Ajeesh Krishna TP, Ramakrishnan M, Victor Roch G, Satish L, Ignacimuthu S (2018) Finger millet [*Eleusine coracana* (L.) Gaertn.] improvement: Current status and future interventions of whole genome sequence. Front Plant Sci 9(1054):1–16

Chandra D, Chandra S, Pallavi, Sharma AK (2016) Review of finger millet (*Eleusine coracana* (L) Gaertn): a power house of health benefiting nutrients. Food Sci Human Wellness 5:149–155

Chandrashekhar A (2010) Finger millet (*Eleusine coracana*). Adv Food Nutr Res 59:215–262

Chauhan ES, Sarita (2018) Effects of processing (germination and popping) on the nutritional and anti-nutritional properties of finger millet (*Eleusine coracana*). Curr Res Nutr Food Sci 6(2): 566–572

Chethan S, Malleshi NG (2007) Finger millet polyphenols: characterization and their nutraceutical potential. Am J Food Technol 2:582–592

Coleman LC (1920) The cultivation of ragi in Mysore. Bull. Dept. Agric., Mysore, Gen. Ser 11

Comas LH, Becker SR, Cruz VM, Byrne PF, Dierig DA (2013) Root traits contributing to plant productivity under drought. Front Plant Sci 4:442. http://www.frontiersin.org/Plant_Science/editorialboard

Craufurd PC, Wheelee TR, Ellis RH, Summer Field RJ, Williams JH (1999) Effect of temperature and water deficit on water use efficiency, carbon isotope discrimination and specific leaf weight in peanut. Crop Sci 39:136–142

Dash SK, Kulkarni MA, Mohanty UC, Prasad K (2009) Changes in the characteristics of rain events in India. J Geophys Res Atmos 11:D10. https://doi.org/10.1029/2008JD010572

Davis KF, Chhtre A, Rao ND, Singh D, De Fries R (2019) Sensitivity of grain yields to historical climate variability in India. Environ Res Lett 14:064013. https://doi.org/10.1088/1748-9326/ab22db

Dayakar RB, Bhaskarachary K, Arlene Christina GD, Sudha DG, Vilas AT (2017) Nutritional and health benefits of millets. ICAR Indian Institute of Millets Research (IIMR), Rajendranagar

Devi PB, Vijayabharathi R, Sathyabama S, Malleshi NG, Priyadarisini VB (2014) Health benefits of finger millet (*Eleusine coracana* L.) polyphenols and dietary fiber: a review. J Food Sci Technol 51:1021–1040

Dwivedi S, Upadhyaya H, Senthilvel S, Hash C, Fukunaga K, Diao X, Santra D, Baltensperger D, Prasad M (2012) Millets: genetic and genomic resources. In: Plant breeding reviews. Hoboken, Wiley-Blackwell, pp 247–375

Fischer RA, Edmeades GO (2010) Breeding and cereal yield progress. Crop Sci 50:85–97

Gowda MVC, Nanja Reddy YA, Pushpalatha N, Deepika M, Pramila CK, Jadhav SS (2014) Released varieties of small millets. ICAR—All India Coordinated Small Millets Improvement Project. University of Agricultural Sciences GKVK, Bangalore

Gupta S, Rashmi G, Kumar S, Sengar RS (2014) Assessment of finger millet (*Eleusine coracana* L.) genotypes for drought tolerance at early seedling growth stages. VEGETOS 27(3):227–230

Gupta SM, Arora S, Mirza N, Pande A, Lata C, Puranik S (2017) Finger millet: a "certain" crop for an "uncertain" future and a solution to food insecurity and hidden hunger under stressful environments. Front Plant Sci 8:643

Hema R, Vemanna RS, Sreeramulu S, Reddy CP, Senthil Kumar M, Udayakumar M (2014) Stable expression of mtlD gene imparts multiple stress tolerance in finger millet. PLoS One 9:e99110. https://doi.org/10.1371/journal.pone.0099110

Jalihal C, Srinivasan J, Chakraborty A (2019) Modulation of Indian monsoon by water vapor and cloud feedback over the past 22,000 years. Nat Commun. https://doi.org/10.1038/s41467-019-13754-6

John I, Drake R, Farrell A, Cooper W, Lee P, Horton P, Grierson D (1995) Delayed leaf senescence in ethylene deficient ACC oxidase anti-sense tomato plants: molecular and physiological analysis. Plant J 7:483–490

Karivaratharaju TV, Ramakrishnan V (1985) Seed hardening studies in two varieties of ragi (*Eueusine coracana*). Indian J Plant Physiol 28:243–248

Kholova J, Tom Hash C, Kakkera A, Kocova M, Vadez V (2010) Constitutive water-conserving mechanisms are correlated with the terminal drought tolerance of pearl millet [*Pennisetum glaucum* (L.) R. Br.]. J Exp Bot 61(2):369–377

Kotapati KV, Bhagathkumar P, Anithamma K, Reddy RP, Dinakararao A (2014) Response of antioxidative enzymes and lipoxygenase to drought stress in finger millet leaves (*Eleusine coracana* (L.) Gaertn.). Int J Plant Anim Environ Sci 4(3):644–653

Krishna SS, Nanja Reddy YA (2021) Root growth pattern in finger millet under well-watered and drought stress condition: its relation to shoot mass production. Plant Physiol Rep. https://doi.org/10.1007/s40502-021-00592-2

Krishna SS, Nanja Reddy YA, Ravi Kumar RL (2021) Assessment of traits for grain yield under drought in ûnger millet. Plant Physiol Rep 26(1):84–94

Kumar A, Singh DP (1998) Use of physiological indices as a screening technique for drought tolerance in oilseed Brassica species. Ann Bot 81:413–420

Kumar A, Metwal M, Kaur S, Gupta AK, Puranik S, Singh S, Singh M, Gupta S, Babu BK, Sood S, Yadav R (2016) Nutraceutical value of finger millet [*Eleusine coracana* (L.) Gaertn.] and their improvement using omics approaches. Front Plant Sci 7:934

Kumssa DB, Joy EJM, Ander EL, Watts MJ, Young SD, Walker S, Broadley MR (2015) Dietary calcium and zinc deficiency risks are decreasing but remain prevalent. Sci Rep 5:10974

Lakshmanaiah CH (1967) High yielding varieties of irrigated ragi. In: Proceedings of state level symposium on high yielding varieties. Administrative Training Institute, Mysore

Levitt J (1980) Responses of plants to environmental stresses. Physiological Ecology Series Academic, Ann Arbor

Lokesh K, Reddy N, Krishnappa K, Chandrakanthappa Reddy DMV, Swamy SN (2000) Effect of seed size on storability in finger millet. Curr Ther Res 29:122–126

Mahajan S, Tuteja N (2006) Cold, salinity, and drought stresses. Arch Biochem Biophys 444:139–158

Manjula DC (2016) Assessment of the physiological basis of drought tolerance in finger millet (*Eleusine coracana* (L.) Gaertn.) accessions differing in growth and yield under stress conditions. M.Sc. (Agri.) Thesis, Department of Crop Physiology, Univ. Agric. Sci., Bangalore

Maqsood M, Azam Ali AN (2007) Effects of drought on growth, development, radiation use efficiency and yield of finger millet (*Eleusine coracana*). Pak J Bot 39(1):123–134

Megha KC (2022) Comparative analysis of physiological, nutritional and yield attributing traits in finger millet varieties released over years. M.Sc. (Agri.) Thesis, Department of Crop Physiology, UAS, Bangalore

Mohanabharathi M, Sritharan N, Senthil A, Ravikesavan R (2019) Physiological studies for yield enhancement in finger millet under drought condition. J Pharmacogn Phytochem 8(3):3308–3312

Mukami A, Ngetich A, Mweu C (2019) Differential characterization of physiological and biochemical responses during drought stress in finger millet varieties. Physiol Mol Biol Plants 25: 837–846

Nagarjuna KN, Parvathi MS, Sajeevan RS, Pruthvi V, Mamrutha HM, Nataraja KN (2016) Full-length cloning and characterization of abiotic stress responsive CIPK31-like gene from finger millet, a drought-tolerant crop. Curr Sci 111:890–894

Naik LM, Mondam M, Vijayalakshmi G, Sivakumar Pinjari OB, Nanja Reddy YA, Khan PSSV (2020) Morpho-physiological and biochemical changes in finger millet [*Eleusine coracana* (L.) Gaertn.] under drought stress. Physiol Mol Biol Plants. https://doi.org/10.1007/s12298-020-00909-9

Nanja Reddy YA (2020) Studies on photosynthetic rate, anatomical characters, and grain yield in finger millet genotypes. Curr J Appl Sci Technol 39(23):31–39

Nanja Reddy YA, Dhurjati C, Krishnakumar AK (1990) A comparison of dimethyl sulphoxide (DMSO) and acetone extracts for the determination of chlorophyll in Hevea leaf tissue. Indian J Nat Rubber Res 3(2):131–134

Nanja Reddy YA, Ashok EG, Dhananjaya BN, Anjaneya Reddy B, Jayaramegowda Krishne Gowda KT, Channabyre Gowda MV (2010) Evaluation of finger millet genotypes for relative drought resistance. In: National symposium on climate change and rainfed agriculture. CRIDA, Hyderabad, pp 207–211

Nanja Reddy YA, Lavanyabai T, Prabhakar Ramamurthy V, Chame Gowda TC, Shankar AG, Gowda MVC (2019a) Bench mark values for grain iron content in finger millet (*Eleusine coracana* (L.) Gaertn.). Int J Curr Microbiol App Sci 8(6):502–506

Nanja Reddy YA, Gowda J, Ashok EG, Krishne Gowda KT, Gowda MVC (2019b) Higher leaf area improves the productivity of finger millet (*Eleusine coracana* (L.) Gaertn.) under rainfed conditions. Int J Curr Microbiol App Sci 8(5):1369–1377

Nanja Reddy YA, Jayarame Gowda Ashok EG, Krishne Gowda KT (2020) Effect of moderate drought stress on photosynthetic rate and grain yield in finger millet genotypes. Int J Curr Microbiol App Sci 9(05):2951–2959

Nanja Reddy YA, Priya Reddy YN, Ramya V, Suma LS, Narayana Reddy AB, Sanjeev Krishna S (2021a) Drought adaptation: approaches for crop improvement. In: Singh M, Sood S (eds) Millets and pseudo cereals: genetic resources and breeding advancements. Woodhead Publishing, Sawston

Nanja Reddy YA, Gowda J, Gowda KTK (2021b) Approaches for enhancing grain yield of finger millet (Eleusine coracana). Plant Genet Resour 19:229–237. https://doi.org/10.1017/S1479262121000265

Pallavi C, Joseph B, Aariff KMA, Hemalatha S (2016) Physiological parameters, leaf nitrogen content and grain yield of finger millet as affected by different sources of organic manures under INM in comparison with RDF. Int J Curr Res Biosci Plant Biol 3(8):123–130

Parvathi MS, Nataraja KN (2017) Discovery of stress responsive TATAbox binding protein associated Factor6 (TAF6) from finger millet (*Eleusine coracana* (L.) Gaertn). J Plant Biol 60:335–342. https://doi.org/10.1007/s12374-016-0574-6

Parvathi MS, Nataraja KN, Yashoda BK, Ramegowda HV, Mamrutha HM, Rama N (2013) Expression analysis of stress responsive pathway genes linked to drought hardiness in an adapted crop, finger millet (*Eleusine coracana*). J Plant Biochem Biotechnol 22:192–201

Poornima A (2017) Physiological traits for yield potential under optimal production environment in finger millet (*Eleusine coracana* (L) Gaertn.). M.Sc. (Agri.) thesis, Department of Crop Physiology, University of Agricultural Sciences, Bangalore

Puranik S, Kam J, Sahu PP, Yadav R, Srivastava RK, Ojulong H, Yadav R (2017) Harnessing finger millet to combat calcium deficiency in humans: challenges and prospects. Front Plant Sci 8:1311

Ramakrishnan M, Ceasar SA, Vinod KK, Duraipandiyan V, Ajeesh Krishna TP, Upadhyaya HD, Al-Dhabi NA, Ignacimuthu S (2017) Identification of putative QTLs for seedling stage phosphorus starvation response in finger millet (*Eleusine coracana* L. Gaertn.) by association mapping and cross species synteny analysis. PLoS ONE. https://doi.org/10.1371/journal.pone.0183261

Ramegowda HV, Gill US, Sivalingam PN, Gupta A, Gupta C, Govind G (2017) GBF3 transcription factor imparts drought tolerance in *Arabidopsis thaliana*. Sci Rep 7:9148. https://doi.org/10.1038/s41598-017-09542-1

Ramya V (2022) Identification and characterization of donor lines for drought tolerance in finger millet [*Eleusine coracana* (L.) gaertn]. Ph.D Thesis, Department of Crop Physiology, UAS, Bangalore

Ramya V, Nanja Reddy YA (2018) Leaf temperature: a screening trait for drought tolerance in finger millet. Mysore J Agric Sci 53(3):529–535

Ravindran G (1991) Studies on millets: proximate composition, mineral composition, and phytate and oxalate contents. Food Chem 39(1):99–107

Rosenow (1984) Breeding for resistance to root and stalk rots in Texas. In: Mughogho LK, Rosenberg G (eds) Sorghum root and stalk rots: a critical review. ICRISAT, Patancheru, pp 209–217

Sashidhar VR, Prasad TG, Seetharam A, Udaykumar M, Sastry KSK (1984) Ear photosynthesis in long glumed and normal glumed genotypes of finger millet (*Eleusine coracana* Gaertn.). Indian J Plant Physiol 27(1):104–107

Sastry KSK, Udayakumar M, Vishwanath HR (1982) Desirable plant characteristics in genotypes of finger millet (*Eleusine coracana* (L.) Gaertn.) for rainfed conditions. Indian Natl Sci Acad 48: 264–270

Singh RK, Singh VK, Raghavendrarao S, Phanindra MLV, Raman KV, Solanke AU (2015) Expression of finger millet EcDehydrin 7 in transgenic tobacco confers tolerance to drought stress. Appl Biochem Biotechnol 177:207–216

Subramanyam D (2000) Genotypic variability in photosynthetic characters in finger millet. Photosynthetica 38(1):105–109

Sullivan CY, Ross WM (1979) Selecting for drought and heat resistance in grain sorghum. In: Mussell H, Staples RC (eds) Stress physiology in crop plants. Wiley, New York, pp 263–281

Suma LS (2014) Characterization of selected germplasm accessions for drought tolerance in finger millet (*Eleusine coracana*). M.Sc. (Agri.) Thesis, Univ. Agric. Sci., Bangalore

Swetha TN (2011) Assessment of the contribution of physiological traits to grain yield during crop improvement of finger millet (*Eleusine coracana*). M.Sc. (Agri.) thesis, Department of Crop Physiology, University of Agriculture Sciences Bangalore

Tadele Z (2016) Drought adaptation in millets. In: Abiotic and biotic stress in plants: Recent advances and future perspectives. IntechOpen, London, pp 640–662

Talwar HS, Kumar S, Madhusudhana R, Ganapathy KN, Ronanki S, Tonapi VA (2020) Variations in drought tolerance components and their association with yield components in finger millet (*Eleusine coracana*). Funct Plant Biol 47:659–674

Tieszen LL, Imbamba SK (1978) Gas exchange of finger millet inflorescences. Crop Sci 18(3): 495–498

Tiwari A, Rastogi A, Singh V, Arunachalam A (2020) Effect of water stress on oxidative damage and antioxidant enzyme activity in finger millet and barnyard millet. Indian J Hill Farm 33(1): 36–45

Udaya Kumar M, Sashidhar VR, Prasad TG (1986) Physiological approaches for improving productivity of finger millet under rainfed conditions. In: Seetharam A, Riley KW, Harinarayana G (eds) Small millets in global agriculture: proceedings of the first international small millets workshop. Oxford & IBH Publishing Co. Pvt. Ltd, New Delhi, pp 179–208

Ueno O, Kawano Y, Wakayama M, Takeda T (2006) Leaf vascular systems in $C_3$ and $C_4$ Grasses: a two-dimensional analysis. Ann Bot 97(4):611–621

Van Herwaarden A, Angus JF, Richards RA, Farquhar GD (1998) 'Haying-off', the negative grain yield response of dryland wheat to nitrogen fertilizer. II. Carbohydrate and protein dynamics. Aust J Agric Res 49:1083–1093

Van Oosterom EJ, Jayachandran R, Bidinger FR (1996) Diallel analysis of the stay-green trait and its components in sorghum. Crop Sci 36:549–555

Webb P, Braun VJ (1994) Famine and food security in Ethiopia. Wiley, New York

Yogeesh LN, Naryana Reddy AB, Nanja Reddy YA, Channabyre Gowda MV (2016) High temperature tolerant genotypes of finger millet (*Eleusine coracana* L.). Nat Environ Pol Tech 15(4):1293–1296

Zagdanska B, Kozdoj J (1994) Water stress induced changes in morphology and anatomy of flag leaf of spring wheat. Acta Soc Bot Pol 63(1):61–66

Zhao M, Zhang H, Yan H, Qiu L, Baskin CC (2018) Mobilization and role of starch, protein, and fat reserves during seed germination of six wild grassland species. Front Plant Sci 9:234. https://doi.org/10.3389/fpls.2018.00234

# Chapter 19
# Comprehending the Physiological Efficiency of Millets Under Abiotic Stress

**Karnam Venkatesh, Elangovan Maruthamuthu, S. Srividhya, and Seva Nayak**

**Abstract** Millets are coarse cereals that include sorghum, pearl millet and small millets. Global concern over limited resources, climate change, food and nutritional insecurity has necessitated to improve agricultural productivity and overall sustainability. Owing to their $C_4$ anatomy ($C_4$ photosynthetic trait) and climate resilient adaptability traits, millets are model crops for sustainable agriculture solutions and climate change. Regarded as Climate Smart Crops, they have lesser requirement of water, inputs and crop maintenance so as to obtain the optimum yield and widely adaptable to semi-arid and harsh ecologies. Key features like short life cycle (12–14 weeks from seed to seed), short stature, rapid plant establishment, leaf venation features, deeper root systems and hardiness to stress make them climate resilient. Apart from the above said qualities, millets are regarded as nutri-grains due to their extremely superior nutritional benefits (especially high micronutrients), antioxidants, essential amino acids, dietary fibres, vitamins and minerals. However, attempts for exploiting the climate resiliency features for crop improvement in millets are still at infancy compared to other cereals. Also, there is a huge gap in bringing millet as staples as they were long neglected after major cereals took over the food systems. Reports also depict a major decline in their cultivation area, production and productivity around the globe. In this chapter, the special physiological features and adaption mechanisms which make millets climate resilient has been discussed in detail. The chapter also focuses on the status of available millet genetic resources in India and their efficient use in crop improvement which may further enhance climate resilience and yield potential in millets.

**Keywords** Millets · Sorghum · Minor millets · Climate resilience · Climate smart crops · Abiotic stress

---

K. Venkatesh (✉) · E. Maruthamuthu · S. Srividhya · S. Nayak
ICAR-Indian Institute of Millets Research, Hyderabad, India
e-mail: Karnam.Venkatesh@icar.gov.in



H. M. Mamrutha et al. (eds.), *Translating Physiological Tools to Augment Crop Breeding*, https://doi.org/10.1007/978-981-19-7498-4_19

## 19.1 Introduction

The small seeded cereals are commonly called as millets and grown all over the globe for food, fodder and fuel. Millets can be broadly classified as major and minor millets. On the one hand, Pearl millet (*Pennisetum glaucum*) and sorghum (*Sorghum bicolour*) comes under major millets while on the other hand, *Panicum sumatrense* (little millet), Echinochloa crusgalli (barnyard millet), *Paspalum scrobiculatum* (kodo millet), *Panicum miliaceum* (proso millet), *Eleusine coracana* (finger millet), *Setaria italica* (foxtail millet), Eragrostis tef (teff) and Urochloa ramosa (browntop millet) comes under minor millets (Dwivedi et al. 2012). Millets comes under $C_4$ crops and are a member of Poaceae family and Panicoideae and Chloridoideae subfamilies. Millets being commonly grown in the resource poor marginal environments are known for having excellent climate resiliency (Vetriventhan et al. 2020).

Millets are the 6th most significantly grown crops globally satisfying the energy and protein demands of huge populations living in India and Africa, and particularly for individuals living in dry and semi-arid countries of globe (Amadoubr and Le 2013). India, Nigeria, Niger, China, Mali and Burkina Faso are the major producers of millets globally with minor contribution from other countries (FAOSTAT 2020). Asia (52%) and Africa (45%) are the major producers of millets compared to Europe (2%) and United States of America (1%) (Fig. 19.1)

Rice and wheat are the chief crops fulfilling the energy and protein requirement of today's world population (Sharma et al. 2015). These major staples are totally reliant on constant supply of inorganic fertilizers, weeds and insect control measures for sustaining current production levels. The realization of climate change events such as excess heat and water scarcity (drought) in the major food producing regions of the world has caused yield losses in rice and wheat crops up to 70–80% (Leng and Hall 2019). Rice is extremely sensitive to germination under inadequate water supply conditions and wheat is highly sensitive to increased temperature for normal

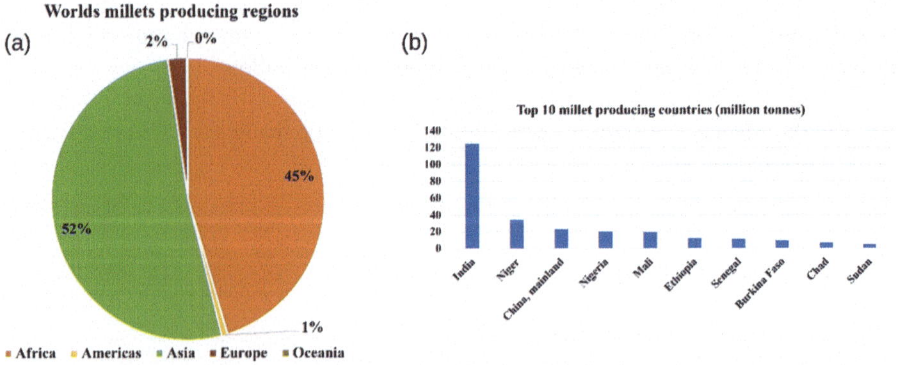


**Fig. 19.1** (**a**) Millet producing countries of the world and their share in global production. (**b**) Top ten millet production countries. (*Production units are in Metric Tons (MT)). (Data: FAOSTAT 2020, accessed from http://www.fao.org/faostat/en/#data/QC/visualize)

crop establishment and yield attainment. A projected increase of 4–5 °C of the average global temperature by the year 2100 may result in disappearance of wheat crop (Tollefson 2020). Millets, being climate resilient crops due to presence of unique traits inherent in them can serve as an excellent alternative to major cereals for sustaining the food production under harsher environmental conditions. This chapter briefly presents the physiological mechanisms which make millets resilient to various abiotic stresses.

## 19.2 Understanding the Physiology of Millets

Crops are vulnerable to climate change and variability, affecting India's food security. Several studies display the alarming trend of decline in crop production and productivity. Various climate change factors (decrease in overall rainfall, increased frequency of drought coupled with high temperature) has forced us to rethink on sustainable agriculture policies for building climate resilience. The semi-dry (arid) locations of India have extended by 10% in the past couple of decades, posing significant climate change and agrarian distress.

One of the effective strategies or solutions to avoid these vulnerabilities is to adapt climate smart crops. In the present agriculture scenario, Millets (known as super grains) are the promising crops that can cope with changing climate, sustain productivity and be resilient to abiotic stresses. Millets include Sorghum, Pearl millet and small or minor millets (Finger millet, barnyard millet, Proso millet, Foxtail millet, Kodo millet, little millet, Brown top millet, also recently Teff and Fonio). Of all the crops grown by humans, millets are most likely the oldest. Understanding the dire need to deliver the environmental, ecological and nutritional benefits of millets to customers, makers, and leaders, Realizing the urgent need to encourage the nutritional and ecological benefits of millets to stakeholders, UN announced the year 2023 as the International Year of Millets. The food and nutritional insecurity coupled with changing climate necessitates to exploit the millets crop for future climate.

Millets are considered as nutri-cereals as they hold abundant nutritional benefits, food and fodder value, ecological benefits, reduced reliance on chemical fertilizers and less sensitive to yield declines with lower inputs (Kole et al. 2015). Many of the millets are notable for their climate resiliency and have the ability to endure prolonged drier spells, higher temperatures nevertheless produce grains and feed and therefore considered as climate smart crops.

In spite of their many advantages, millets were overlooked and focused mainly on production and consumption of rice and wheat. Notwithstanding their many advantages, millets were disregarded and instead emphasis was given to rice and wheat. Presently, there is lack of diversity in our food regarding balance of nutrients, post the green revolution. Building climate resilience through millets for future climate scenarios necessitates understanding the physiology of millet crops. Given the above situation, this article highlights the importance for extensive research on millet crops to dissect their physiological, agronomical and decipher stress tolerance traits.

## 19.3 Millets as Model Crops for Stress Tolerance, their Traits Contributing for Climate Resilience

Millets owing to their climate resilient characteristics have now become model crops for cultivation in dry, semi-arid regions of the India. Coarse grains (like sorghum, pearl millet and small millets) with $C_4$ photosynthetic pathways which have evolved in the dry areas near the equatorial regions, are in general more drought tolerant, can withstand hot climates, and are more resilient to marginal soil conditions with less nutrient supply clubbed with higher $CO_2$ levels in atmosphere (Myers et al. 2014).

Millets contain many physiological and morphological traits which make them climate resilient and stress tolerant compared to rice, wheat and maize. Short life cycle (12–14 weeks from seed to seed), short stature, smaller leaf area, leaf venation features and deeper root systems, hardiness to stress conditions force us study them in detail (Li and Brutnell 2011).

Rominger (1959) considered millets as perfect example (model) crop for studying $C_4$ mechanism of photosynthesis. Foxtail millet (Setaria) being fastest growing grass has been well studied for functional and molecular characterization of $C_4$ photosynthesis at genomic level. According to Leegood (2002), there are three essential decarboxylase enzymes that play key role in C4 photosynthesis, nicotinamide adenine dinucleotide phosphate dependent malic enzyme (NADP-ME), nicotinamide adenine dinucleotide dependent malic enzyme (NAD-ME) and phosphoenolpyruvate carboxylase (PCK) (Fig. 19.2). For example, Proso millet and Amaranthus employs NAD-ME for decarboxylation; Pearl millet and Guinea grass (Panicum maximum) employs PCK; and Kodomillet, Setaria, Sorghum, Sugarcane and Maize employs NADP-ME.

The entire genome of foxtail millet was sequenced, and gene families from 15 different plant genomes were compared, revealing that more than thousand genes were unique in foxtail millet (Zhang et al. 2012). 586 genes were labelled as "reaction to water", suggesting that they may have key functions in drought and dehydration stress tolerance. The photosynthetic enzymes namely, carbonic anhydrase (CAH), malate dehydrogenase (MDH), malic enzyme (ME), phosphoenolpyruvate carboxylase (PEPC), phosphoenolpyruvate carboxylase kinase (PPCK) and pyruvate orthophosphate dikinase (PPDK) are reported and found similar to major cereals such as sorghum, maize and rice. According to scientific evidences, foxtail crop has maximum number of MDH and PPDK photosynthetic genes that can be transferred to study and improve other millet species. Thus, it is highly probable to discover regulation mechanism involved in cell-specific expression of $C_4$ photosynthesis genes by comparison of regulatory sequences of decarboxylases from these 3 lineages. This information can be used to improve

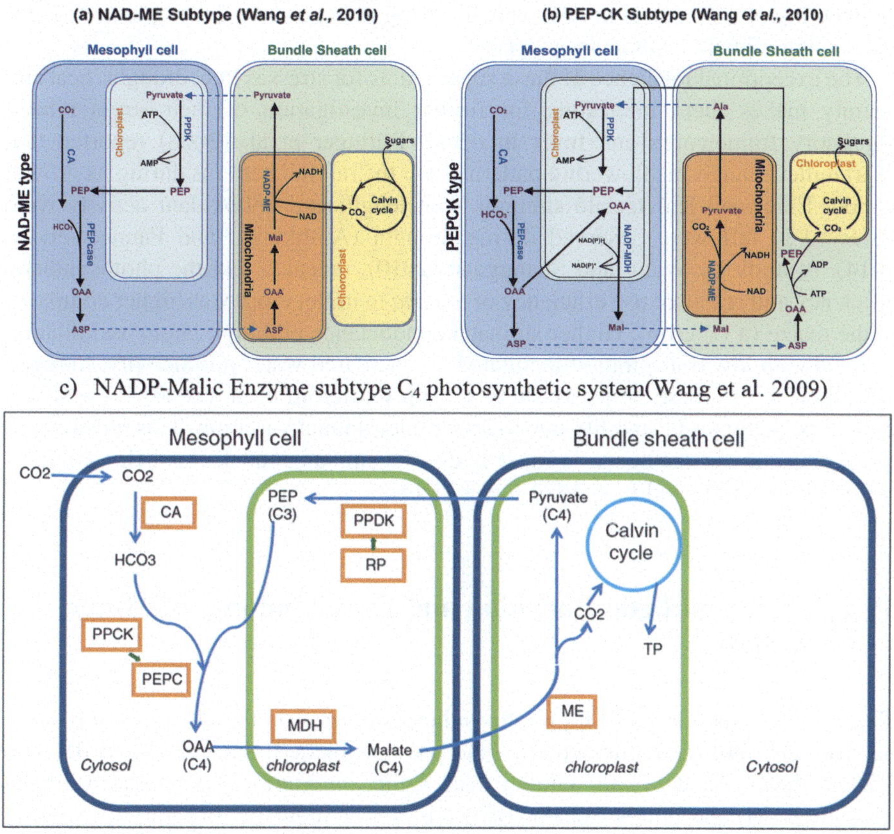


**Fig. 19.2** (a–c) Three main decarboxylase enzymes that are employed in $C_4$ photosynthesis

millets in particular and other interrelated grasses at large. The widespread similarity among genomes of cereals at gene level would work help in quick and efficient transfer of key genes/QTLs controlling stress tolerance.

Furthermore, millets benefit greatly from the $C_4$ photosynthetic trait. $C_4$ mechanism (*Kranz anatomy*) helps in maintaining higher $CO_2$ levels in bundle sheath cells, reduces photo-respirative loss (around 80%) and additionally, the carboxylase activity of RuBisCO is also enhanced (Sage and Zhu 2011). RuBisCO enzymes in $C_4$ plants can function normally under increased $CO_2$ concentrations with higher rate of photosynthesis at hotter temperature conditions. This confers additional benefits compared to $C_3$ plants, like higher water and nitrogen use efficiency by many folds (Sage and Zhu 2011). For example, Li and Brutnell (2011) reported that 1 g of dry biomass production by foxtail millet required just about 257–260 g of water compared to wheat and maize, respectively, requiring 510 and 470 g.

Improved overall plant growth, higher biomass allocation and partitioning are some of the additional benefits of $C_4$ photosynthesis in addition to WUE and NUE (Sage and Zhu 2011). Realizing the stress tolerance potential of millets and engineering of $C_4$ features in other cereals, using millet as a model crop is critical for developing climate resilient crop species.

The exceptional resilience of these super grains for stresses like, drought, heat and salinity makes them interesting for further investigation of their stress related regulatory frameworks and traits in detail. Bidinger et al. (2008) reported that pearl millet changes its flowering pattern as per the rainfall patterns during its growth period. Enhanced leaf tensile strength, root length and antioxidant activity were observed in teff were observed during drought (Ajithkumar and Panneerselvam 2014). A study by Wang and Shangguan (2010), revealed that the photosynthetic efficiency and resource use efficiency of *Setaria* (a millet crop) was higher compared to the maize (a $C_4$ crop). Higher stomatal conductance and high photo assimilation coupled with low transpiration facilitated the increased water productivity, nitrogen use efficiency in these crops. A study by van Oosterom et al. (2006) showed that millet had better yield stability due to lesser inter-annual variation. This unwavering quality of climate resilience is most likely the fundamental factor behind millets being widely cultivated in spite of lower yields.

## 19.4 Photosynthetic Variation and Leaf Anatomy of Minor Millets

The efficient functioning of $C_4$ photosynthesis depends on the quantitative balance between the photosynthetic subtypes and photosynthetic tissues of M and BS cells (Hatch 1987). As it is well known that, *Kranz anatomy* close proximity of the mesophyll (M) and bundle sheath (BS) cells is a prerequisite for rapid diffusion of photosynthetic metabolites and millet crops encompass this trait. In addition to this, millets have also been reported to include high vein density with denser networks of small longitudinal and transverse veins, while keeping a constant density of large longitudinal veins (Dengler 1994; Dengler and Nelson 1999; Govinda et al. 2012). Leaves of C4 grasses have primarily an efficient water distribution cascade compared to C3 grasses by having denser network of the small longitudinal and transverse veins. The leaf veins anatomy depicts a hierarchical order and have different structures. As reported by (Altus and Canny 1982):

- The enormous longitudinal veins run from the leaf blade into the sheath. The huge longitudinal veins assist essentially in longitudinal carriage of photosynthate outside the leaf blade.
- The little longitudinal veins serve fundamentally in gathering photosynthate from adjacent photosynthetic cells.
- The cross over (transverse) veins and little longitudinal veins assume an essential part in sideways carriage of photosynthate from the little to the enormous longitudinal veins (Fig. 19.3).

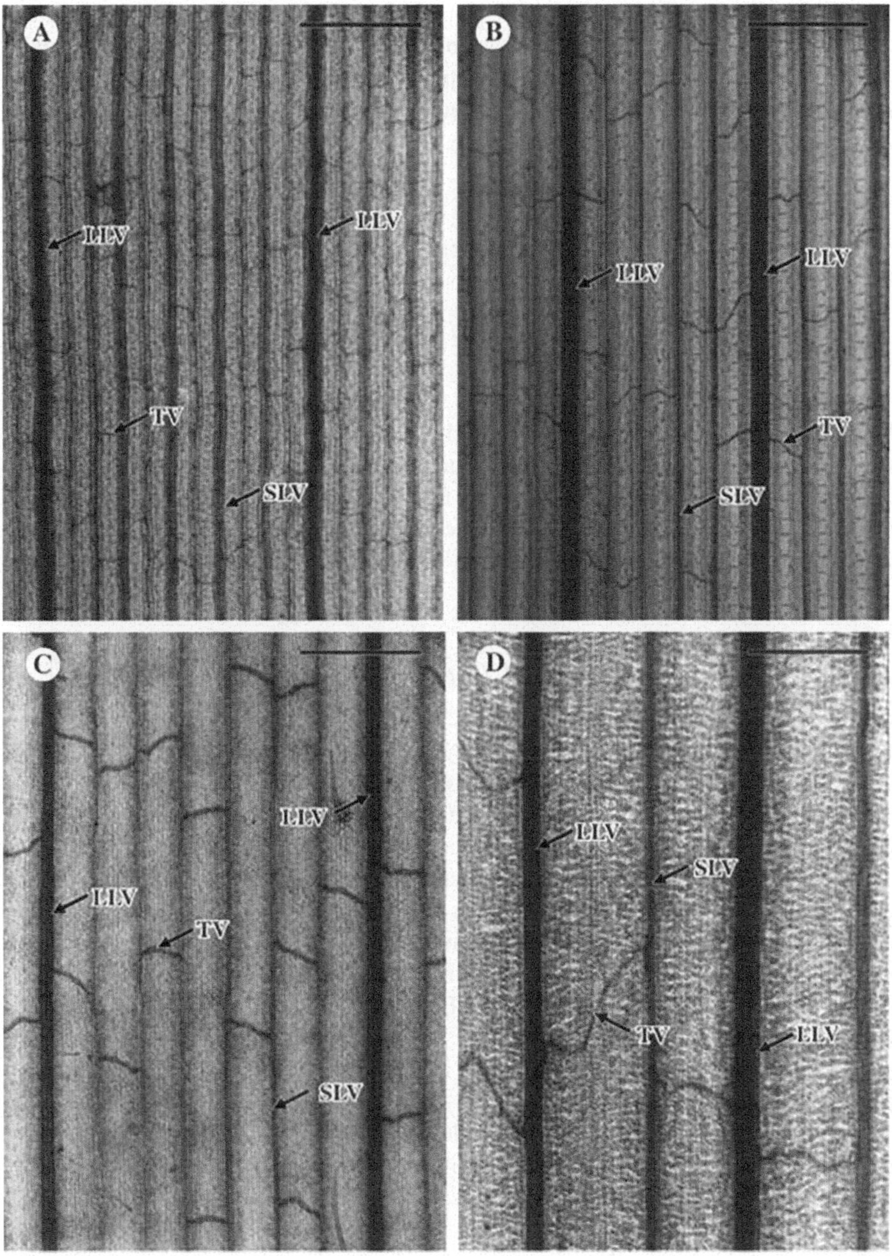


**Fig. 19.3** Paradermal view of cleared leaf blades of $C_3$ and $C_4$ grasses as reported by Ueno et al. (2006). (**a**) Digitaria sanguinalis, an NADP-ME $C_4$ species; (**b**) Eleusine indica, an NAD-ME $C_4$ species; (**c**) Panicum bisulcatum, a panicoid $C_3$ species; (**d**) Briza minor, a festucoid $C_3$ species

Despite these benefits, millets have been orphaned and there is a dearth of studies to unveil mechanisms that confer stress tolerance. To completely exploit the fundamental genetic potential of millets application of high-throughput phenotypic screening, conventional/molecular breeding procedures and transgenic technologies are needed.

## 19.5 Adaptation Mechanisms Underlying Climate Resilience of Millets

### *19.5.1 Escape Mechanism*

Plants completing their life cycle by reaching maturity before the onset of climatic stresses is referred to as escape. Quick establishment, rapid growth, early flowering and higher photosynthetic efficiency are few of the traits related to escape mechanism (Kooyers 2015). Earlier studies have shown that pearl millet escapes stress by adjusting its life cycle to rainfall pattern (Sivakumar 1992).

### *19.5.2 Avoidance by Way of Phenotypic Adjustments*

A greater amount of phenotypic plasticity is exhibited by crop plants to withstand any type of abiotic stresses as adaptation measures. Plants exhibit increased growth of certain plant organs such as pubescence on leaves and on leaf sheaths or increased accumulation of certain secretions such as wax on different plant parts exposed to atmosphere to withstand the adverse effects of stresses (Banowetz et al. 2008; Maes et al. 2001). Certain other alterations such as shortened life cycle, altered orientation of leaf, etc. which favours plants in saving most crucial resources(required for maintaining plant health) from escaping into atmosphere in the form of evapotranspiration under severe stress conditions (Banowetz et al. 2008; Maes et al. 2001). Blum et al. (2001) reported that whenever a plant senses any kind of abiotic stress, it reduces its maturity duration to complete its life cycle before onset of stress thereby avoiding adverse stress effects.

In few of the cereals especially millets leaf rolling is a common adaptation mechanism observed whenever stresses like heat and drought occurs (Sarieva et al. 2010). Few of the cereals try to escape the most sensitive developmental stages such as flowering and grain filling from stresses as observed in case of wheat and pearl millet (Hall 1992, 1993). Wax deposition on plant surfaces is also a commonly seen mechanism of stress adaptation in plants. Wax protects plants from excess radiation resulting in reduced evapotranspiration water loss under excess heat and drought stresses (Venkatesh et al. 2020). Additionally, wax deposition acts as a defensive shield protecting the vital organs of plants such as membranes and leaf structural components (Venkatesh et al. 2022).

Srivastava et al. (2012) reported that plants develop larger xylem vessels and larger, deeper roots to enhance water uptake under drought and heat conditions to compensate the water lost due to evapotranspiration. In other cases, lower canopy temperature as a resultant of increased evapotranspiration is also a survival mechanism under severe stress conditions.

### *19.5.3 Tolerance*

Capacity of the plant to accomplish normal growth and development to produce standard monetary yield under stress situation is called as tolerance. The plants have developed different tolerance mechanisms, for example, modifying ion transport frameworks, synthesis of late embryogenesis abundant (LEA) proteins, aggregation of osmoprotectant molecules, free-radical scavengers and controlling frameworks including factors like ubiquitin and dehydrin through signalling pathways and transcriptional control (Rodríguez et al. 2005; Wang et al. 2004). Regulation of stomatal closure leading efficient maintenance of plant water levels to support the vital plant functions essentially dependent on water under stress situations (Woodward et al. 2002).

Plants adapt to stress alternatively by increasing root growth (volume as well as length) to reach water available at deeper soil layers in order to take up more water also has been reported (Lehman and Engelke 1993). Broadened grain filling duration was likewise seen as a tolerance adaption and positive relationship of grain filling duration with better yields under stress conditions has been earlier reported (Yang et al. 2002). Changes in the photosynthetic pathway components and the enzymes required by the plants have been viewed as another tolerance mechanism adapted by plants. Enhanced proclivity of Rubisco the primary enzyme involved in fixation of carbon to $CO_2$ under stress conditions such as heat has been reported for in certain plants like *Limonium gibertii* (Parry et al. 2011). Increased photosynthesis activity at extremely higher temperatures above normal (Allakhverdiev et al. 2008; Ristic et al. 2007) and allocation of increased carbon and nitrogen uptake as stress tolerance mechanisms were also seen in plants (Xu et al. 2006).

### *19.5.4 Stress Recovery*

In this case the plants recover from severe stress situation once the normal climate returns to produce reasonable yield or biomass. This type of mechanism has been recorded in rabi sorghum which strives on the conserved soil moisture.

## 19.6 Millet Germplasm Resources Available for Abiotic Stress Tolerance Research

ICAR-IIMR, Hyderabad conserves a collection of 78,429 germplasm accessions received from various National and International centres. The collection includes, Sorghum (52,757 acc.), Finger millet (10,704), Foxtail millet (5096), Pearl millet (4973), Proso millet (1666), Barnyard millet (1661 acc.) etc being conserved in the medium term storage. Additionally, through fresh exploration trips, 2273 accessions were collected by ICAR-IIMR between 2000 to 2021 and 423 acc. were collected by other individuals from 15 millets growing states of the country. These available millet germplasm resources are being distributed to researchers worldwide for use in crop improvement programmes.

A total of 48 final products were with climate resilience developed through the use of ICAR-IIMR germplasm resources by the AICRP on Sorghum trials during 2007–2021. Maximum of 22 rabi sorghum varieties followed by six sweet sorghum varieties, seven kharif sorghum varieties, one kharif sorghum hybrid, five single-cut forage varieties, one each of sweet sorghum variety, sweet sorghum hybrid and dual-purpose variety, three specialty sorghums were contributed to the trials. In addition, a total of 27 millets genetic stocks registered with ICAR-NBPGR during 2002–2021 with unique trait superiority including the abiotic stress tolerance as one of the traits. The maximum of 20 acc. of sorghum genetic stocks registered followed by finger millet (5 acc.), little millet and barnyard millet (1 acc. each).

## 19.7 Conclusion

Millets hold enormous potential for food and nutrition security. Millets are nutrient-dense, have impressive health advantages, require relatively lower inputs and are tolerant or resistant to most biotic and abiotic stress factors. These characteristics highlight millets as a crop of choice for the global population, especially under the threat of climate change. Given the advantages of millet crops, switching from rice, wheat and maize to millets will be difficult. An alternative way should be substituting or promoting cultivation of climate resilient crops such as small millets in the regions prone to abiotic stresses would help minimize the impact of damage on food production. Additionally, socioeconomic policies and market demand creation can bring larger impact on crop selection by the farmers. A better option would be to incentivize small and marginal farmers to increase their crop diversity to reduce the sensitivity of water consuming crops to rainfall variability.

## References


Ajithkumar IP, Panneerselvam R (2014) ROS scavenging system, osmotic maintenance, pigment and growth status of Panicum sumatrenseroth. under drought stress. Cell Biochem Biophys 68: 587–595

Allakhverdiev SI, Kreslavski VD, Klimov VV, Los DA, Carpentier R, Mohanty P (2008) Heat stress: an overview of molecular responses in photosynthesis. Photosynth Res 98:541

Altus D, Canny M (1982) Loading of assimilates in wheat leaves. I. The specialization of vein types for separate activities. Funct Plant Biol 9:571. https://doi.org/10.1071/PP9820571

Amadoubr I, Le M (2013) Millets: nutritional composition, some health benefits and processing – a review. Emir J Food Agric 25:501. https://doi.org/10.9755/ejfa.v25i7.12045

Banowetz GM, Azevedo MD, Stout R (2008) Morphological adaptations of hot springs panic grass (Dichantheliumlanigunosum var sericeum (Schmoll)) to thermal stress. J Therm Biol 33:106–116. https://doi.org/10.1016/j.jtherbio.2007.08.006

Bidinger F, Sharma M, Yadav O (2008) Performance of landraces and hybrids of pearl millet [*Pennisetumglaucum (L.)* R. Br.] under good management in the arid zone. Indian J Genet Plant Breed 68:146–148

Blum A, Klueva N, Nguyen HT (2001) Wheat cellular thermotolerance is related to yield under heat stress. Euphytica 117:117–123. https://doi.org/10.1023/A:1004083305905

Dengler N (1994) Quantitative leaf anatomy of C3 and C4 grasses (Poaceae): bundle sheath and mesophyll surface area relationships. Ann Bot 73:241–255. https://doi.org/10.1006/anbo.1994.1029

Dengler NG, Nelson T (1999) Leaf structure and development in C4 plants. In: C4 plant biology. Elsevier, Amsterdam, pp 133–172. https://doi.org/10.1016/B978-012614440-6/50006-9

Dwivedi SL, Upadhyaya HD, Senthilvel S, Hash CT, Fukunaga K, Diao X, Santra D, Baltensperge D, Prasad M (2012) Millets: genetic and genomic resources. In: Plant breeding reviews. Wiley-Blackwell, Hoboken

FAOSTAT (2020) Data on millet producing countries. Food & Agriculture, Rome

Govinda R, Shanta K, Vivek T, Jolly C, Coe R, Samart W, Quick W et al (2012) Towards a C4 rice. Asian J Cell Biol 7:13–31

Hall AE (1992) Breeding for heat tolerance. Plant Breed Rev 10:129–168

Hall AE (1993) Physiology and breeding for heat tolerance in cowpea, and comparison with other crops. In: Adaptation of food crops to temperature and water stress. Food & Agriculture, Rome, pp 271–284

Hatch MD (1987) C4 photosynthesis: a unique elend of modified biochemistry, anatomy and ultrastructure. Biochim Biophys Acta 895:81–106. https://doi.org/10.1016/S0304-4173(87)80009-5

Kole C, Muthamilarasan M, Henry R, Edwards D, Sharma R, Abberton M, Batley J, Bentley A, Blakeney M, Bryant J, Cai H (2015) Application of genomics-assisted breeding for generation of climate resilient crops: progress and prospects. Front Plant Sci 6:563

Kooyers NJ (2015) The evolution of drought escape and avoidance in natural herbaceous populations. Plant Sci 234:155–162

Leegood RC (2002) C4 photosynthesis: principles of CO2 concentration and prospects for its introduction into C3 plants. J Exp Bot 53:581–590. https://doi.org/10.1093/jexbot/53.369.581

Lehman V, Engelke MC (1993) Heritability of creeping bentgrass shoot water content under soil dehydration and elevated temperatures. Crop Sci 33:1061–1066

Leng G, Hall J (2019) Crop yield sensitivity of global major agricultural countries to droughts and the projected changes in the future. Sci Total Environ 654:811–821. https://doi.org/10.1016/j.scitotenv.2018.10.434

Li P, Brutnell TP (2011) Setariaviridis and Setariaitalica, model genetic systems for the Panicoid grasses. J Exp Bot 62:3031–3037. https://doi.org/10.1093/jxb/err096

Maes B, Trethowan RM, Reynolds MP, Ginkel M, van Skovmand B (2001) The influence of glume pubescence on spikelet temperature of wheat under freezing conditions. Funct Plant Biol 28: 141. https://doi.org/10.1071/PP00049

Myers SS, Zanobetti A, Kloog I, Huybers P, Leakey AD, Bloom AJ, Carlisle E, Dietterich LH, Fitzgerald G, Hasegawa T, Holbrook NM (2014) Increasing CO2 threatens human nutrition. Nature 510(7503):139–142

Parry MA, Reynolds M, Salvucci ME, Raines C, Andralojc PJ, Zhu XG, Price GD, Condon AG, Furbank RT (2011) Raising yield potential of wheat. II. Increasing photosynthetic capacity and efficiency. J Exp Bot 62(2):453–467

Ristic Z, Bukovnik U, Prasad PV (2007) Correlation between heat stability of thylakoid membranes and loss of chlorophyll in winter wheat under heat stress. Crop Sci 47:2067–2073

Rodríguez M, Canales E, Borrás-Hidalgo O (2005) Molecular aspects of abiotic stress in plants. Biotecnol Apl 22:1–10

Rominger JM (1959) Taxonomy of Setaria (Gramineae) in North America. University of Illinois at Urbana-Champaign, Urbana-Champaign

Sage RF, Zhu XG (2011) Exploiting the engine of C4 photosynthesis. J Exp Bot 62:2989–3000. https://doi.org/10.1093/jxb/err179

Sarieva G, Kenzhebaeva S, Lichtenthaler H (2010) Adaptation potential of photosynthesis in wheat cultivars with a capability of leaf rolling under high temperature conditions. Russ J Plant Physiol 57:28–36

Sharma I, Tyagi B, Singh G, Venkatesh K, Gupta O (2015) Enhancing wheat production-A global perspective. Indian J Agric Sci 85:3–13

Sivakumar M (1992) Empirical analysis of dry spells for agricultural applications in West Africa. J Clim 5:532–539

Srivastava S, Pathak AD, Gupta PS, Shrivastava AK, Srivastava AK (2012) Hydrogen peroxide-scavenging enzymes impart tolerance to high temperature induced oxidative stress in sugarcane. J Environ Biol 33:657

Tollefson J (2020) How hot will Earth get by 2100? Nature 580:443–446

Ueno O, Kawano Y, Wakayama M, Takeda T (2006) Leaf vascular systems in C3 and C4 grasses: a two-dimensional analysis. Ann Bot 97:611–621. https://doi.org/10.1093/aob/mcl010

van Oosterom EJ, Weltzien E, Yadav OP, Bidinger FR (2006) Grain yield components of pearl millet under optimum conditions can be used to identify germplasm with adaptation to arid zones. Field Crop Res 96:407–421. https://doi.org/10.1016/j.fcr.2005.08.008

Venkatesh K, Gupta V, Singh G, Pratap Singh G (2020) Mitigating heat stress in wheat: integrating omics tools with plant breeding. In: Jha C, Nayyar H, Gupta S (eds) Heat stress in food grain crops: plant breeding and omics research. Bentham Science Publishers, Sharjah, pp 1–27. https://doi.org/10.2174/9789811473982120010003

Venkatesh K, Senthilkumar KM, Mamrutha HM, Singh G, Singh GP (2022) High-temperature stress in wheat under climate change scenario, effects and mitigation strategies. In: Climate change and crop stress. Elsevier, Amsterdam, pp 209–229. https://doi.org/10.1016/B978-0-12-816091-6.00014-6

Vetriventhan M, Azevedo VC, Upadhyaya HD, Nirmalakumari A, Kane-Potaka J, Anitha S, Ceasar SA, Muthamilarasan M, Bhat BV, Hariprasanna K, Bellundagi A (2020) Genetic and genomic resources, and breeding for accelerating improvement of small millets: current status and future interventions. Nucleus 63(3):217–239

Wang K, Shangguan Z (2010) Photosynthetic characteristics and resource utilization efficiency of maize (*Zea mays* L.) and millet (*Setariaitalica* L.) in a semi-arid hilly loess region in China. N Z J Crop Hortic Sci 38:247–254. https://doi.org/10.1080/01140671.2010.503987

Wang W, Vinocur B, Shoseyov O, Altman A (2004) Role of plant heat-shock proteins and molecular chaperones in the abiotic stress response. Trends Plant Sci 9:244–252

Wang Y, Zhi H, Li W, Li H, Wang Y, Huang Z, Diao X (2009) A novel genome of C and the first autotetraploid species in the Setaria genus identified by genomic in situ hybridization. Genet Resour Crop Evol 56(6):843–850

Woodward FI, Lake JA, Quick WP (2002) Stomatal development and CO2: ecological consequences. New Phytol 1:477–484
Xu S, Li J, Zhang X, Wei H, Cui L (2006) Effects of heat acclimation pretreatment on changes of membrane lipid peroxidation, antioxidant metabolites, and ultrastructure of chloroplasts in two cool-season turfgrass species under heat stress. Environ Exp Bot 56(3):274–285
Yang J, Sears RG, Gill BS, Paulsen GM (2002) Genotypic differences in utilization of assimilate sources during maturation of wheat under chronic heat and heat shock stresses. Euphytica 125(2):179–188
Zhang G, Liu X, Quan Z, Cheng S, Xu X, Pan S, Xie M, Zeng P, Yue Z, Wang W, Tao Y (2012) Genome sequence of foxtail millet (Setaria italica) provides insights into grass evolution and biofuel potential. Nat Biotechnol 30(6):549–554

# Chapter 20
# Role of Next Generation Sequencing in Trait Identification, Genetic Mapping, and Crop Improvement

**Sidak Minocha, Sakshi Vats, Amit Kumar Bhatt, Vijaya Chitnis, Ramesha Thimmappa, and Manoj Kumar**

**Abstract** As the world population is constantly growing, crop productivity must need to be increased to provide nutritious food to everyone, however, the challenge is climate change which needs to be addressed. Recent developments in technologies like NGS have revolutionized the understanding of mechanism of complex genetic trait inheritance. Availability of variety of reference genomes as well as associated NGS has made it possible to develop genome-wide high-density linkage maps in variety of plants including crops, enabling discovery of mutations as precise as SNP in cost-effective and time efficient manner. However, one of the biggest limitations for plant biologists has been to understand how to exploit the power of the data science due to lack of computational knowledge and efficiency. Recently many computational tools such as SHOREmap, MutMap, NGM, etc. have been developed and successfully implemented for mutation mapping using NGS data. These tools have been developed aiming to provide user-friendly and self-explanatory user interface to perform the data analysis as well as data interpretation. Some of the tools like SNP track integrates powerful statistical models which calculates precise allele distribution and also addresses sequencing artefacts leading to mis-scoring of mutations. Overall, mapping-by-sequencing approach using NGS data integrates genetic mapping with reference genome sequences to identify mutations leading to cloning of the genes for further characterisation.

**Keywords** NGS · Mapping · Molecular marker · SNP · Breeding

---

Sidak Minocha and Sakshi Vats contributed equally.

---

S. Minocha · S. Vats · A. K. Bhatt · R. Thimmappa · M. Kumar (✉)
Amity Institute of Genome Engineering (AIGE), Amity University, Noida, Uttar Pradesh, India
e-mail: mkumar18@amity.edu

V. Chitnis
Amrut Parnavati, Mudhol, Karnataka, India

 
H. M. Mamrutha et al. (eds.), *Translating Physiological Tools to Augment Crop Breeding*, https://doi.org/10.1007/978-981-19-7498-4_20

## 20.1 Introduction

Molecular sequencing technology has advanced at a rapid pace over the past decades, leading to accelerated research in fundamental as well as applied sciences. These advances have enabled faster and affordable sequencing of diverse group of organisms that facilitated the emergence of the genomics era in the late twentieth century. Considering the impact of human genome project, there was a realization of importance of plant genome research to meet the challenges of agriculture and environment globally. First step towards such plant genome research was "multinational coordinated *Arabidopsis thaliana* (*A. thaliana*) genome research project" (Kaul et al. 2000). The project aimed at identifying all *A. Thaliana* genes by functional genomics approach followed by genome sequencing by the end of the twentieth century.

Genome sequencing of *A. Thaliana* was a big milestone in the history of plant sciences, which paved the path to genomics era in plant sciences (Provart et al. 2021). Developments in plant genome had huge implications not only in identification and cloning of mutants but also providing novel insights on the shaping of plant genomes. The onset of Next Generation Sequencing (NGS) provided a more robust and cheaper alternatives to describe the sequence diversity across the *A. Thaliana* genome (James et al. 2013). NGS based data led to additional availability and information of SNPs, copy number, and structural variants with a significantly reduced false discovery rate (Trick et al. 2012; Yang et al. 2012; Bhat et al. 2016). Owing to its favourable genetics, it was paramount to study *A. Thaliana* genotypic and phenotypic variations. In this direction the "1001 genome project" was launched in 2008 to understand sequence variations among 1001 accessions of Arabidopsis at whole genome level (Sharbel et al. 2000). The 1001 genome project helped in detailed analysis of 1135 genomes including genome assembly of diverse collection of *A. thaliana,* transcriptome, and epigenome information (Ossowski et al. 2008; Alonso-Blanco et al. 2016). Additionally, NGS platform based studies allowed large-scale polymorphism studies in crops like rice, soybean, and maize with high-quality genetic maps (Lam et al. 2010; Jarquín et al. 2014).

The phenotypic and genotypic variations in rice and wheat crops are often influenced by the environmental factors, hence inheritance of these traits are complex and quantitative in nature (Ramkumar et al. 2010; Trick et al. 2012; Spindel et al. 2015; Zheng et al. 2016). The analysis of such complex links between genetic information and phenotypic characteristics is known as Quantitative Trait Loci (QTL). To understand this association between the phenotype of interest and DNA sequence variants present in an individual genome, the Genome-Wide Association Studies (GWAS) are used to map QTLs. The development of NGS has fast-tracked the acquisition of data required for the efficient GWAS mapping (Korte and Farlow 2013). The development replaced the earlier high-cost microarray-based tools used for developing first-generation haplotype maps in *A. thaliana* and rice that were utilized in GWA studies (Bhat et al. 2021). NGS applications can be extended to the identification of novel genes and characterizations of their functionality, gene expression studies, interactomics, and epigenomics (Varshney et al. 2014; Nguyen et al. 2019).

Along with whole genome DNA sequencing and RNA sequencing, another important application of NGS is high-throughput chromosome conformation capture (Hi-C) which determine the interactions of chromatin in the nucleus (Lieberman-Aiden et al. 2009). In conjugation with NGS, Hi-C enables the detection of chromatin interactions efficiently. Considering plants are not mobile, the study of folding of chromosomes can help in understanding the complex state of gene activity under different development stages as well as climate conditions (Grob et al. 2014; Dong et al. 2020; Montgomery et al. 2020; Naish et al. 2021; Yadav et al. 2021).

## 20.2 Principle of Genetic Mapping

Genetic mapping principally identifies the localization of a gene relevant to phenotypic characteristics based on genetic information thus, identification of a phenotype-governing mutation involves mapping it to a chromosomal region through genetic linkage analysis and determine the causal variant using functional studies (Beyer et al. 2007). For mapping analysis, a genetic population is generated with a specific phenotype, possessing well-defined markers. The fundamental principle of mapping suggests that closer the genetic marker to the allele of interest there will be co-segregation whereas, unlinked marker alleles will segregate randomly. Thus, the artificial selection of mutant recombinant individuals results in unique distribution of the alleles among the closely linked markers. Genotyping of phenotypically mutant recombinants help in obtaining mapping intervals and subsequently causal mutation is identified. Such conventional methodologies are time-consuming and laborious, as they require intensive analysis of each gene that falls under the narrowed region by functional approaches using hundreds or thousands of progenies in each generation. Therefore, new NGS based mapping methods have been developed.

### *20.2.1 Mapping Then and Now*

Before the advent of the next generation sequencing technologies and even now, a range of DNA and protein-based markers were used for assessing the genetic diversity among the populations for crop breeding. Further, morphological traits such as, size, shape, colour, and textures etc. were also used along with DNA, protein or biochemical markers. Though the information related to these markers are readily available and have simple setup, they had their own drawbacks of limited numbers and subject to environmental cues (Jonas and De Koning 2013; Crossa et al. 2017). Subsequently molecular markers-based polymorphisms were utilized, like, the variation in single base pair, translocation or inversions, variations in the tandem repeats and insertions or deletions. Further, markers based upon DNA length polymorphisms (RFLP), nucleotide sequence repeats polymorphisms (VNTRs, STRs) are also utilized to perform mutation mapping, as these cover the genomic landscape to a greater extent and are more reliable compared to phenotypic markers (Ramesh et al. 2020). Nevertheless, most of these markers are cumbersome to genotype, in general, however, with the help of the reference genomes, it has now become possible to utilize SNP markers and obtain finer genetic mapping intervals by genotyping recombinant mutants (Schneeberger and Weigel 2011). High-density SNP maps for different model organisms were subsequently published and made available, making them one of the ideal markers for mapping to identify causal gene mutation and cloning of the gene for further characterization (Jaganathan et al. 2020).

The genotyping of SNP markers was initially done using PCR combined with sanger's sequencing or restriction digestion. Development of the "Bulked Segregant Assay (BSA)" in plant systems enabled robust analysis of pooled DNA samples from recombinants, minimizing the time and labour to analyse each recombinant individually (Magwene et al. 2011; Klein et al. 2018). This was in parallel to the advent of genome-wide high-throughput genotyping methods such as hybridization microarrays, which in combination to BSA, were used to map a vast number of monogenic and complex traits in diverse organisms (Hazen et al. 2005; Ossowski et al. 2008; Smith et al. 2016). Despite these advances, trait mapping remained a two-step process, requiring large number of mutant recombinants along with rigorous analysis of the mapping interval to identify the causal gene.

The emergence of NGS technologies involving short and long read sequencing has revolutionized the approach of trait mapping. Application of NGS allows estimation of mapping interval and identification of candidate genes in a single integrated step known as mapping by sequencing (Mascher et al. 2014). The concept of mapping-by-sequencing was first implemented in *A. thaliana* in 2009 where EMS (Ethyl Methane Sulphonate) induced mutants were used to generate an F2 mapping population by outcrossing with polymorphic Ler-1 parent strain using Illumina sequencing technology with the help of sequenced reference genome (Schneeberger et al. 2009).

Mapping-by-sequencing was rapidly adopted in *Saccharomyces cerevisiae*, *Caenorhabditis elegans,* zebrafish and a wide variety of non-model organisms owing to its cost-effective and fast-paced nature (Wicks et al. 2001; Doitsidou et al. 2010; Wenger et al. 2010; Magwene et al. 2011; Voz et al. 2012). The key advantage of NGS based mapping is the ability to simultaneously probe the polymorphic markers, at a genome-wide level, to map and identify the causal mutation instead of analysing each marker individually which greatly reduced the time and laborious effort required. Several advancements have been made to the core technique of mapping-by-sequencing over the years to address the general challenges involved with it (Provart et al. 2021).

## 20.3 Application of NGS in Plant Research

Plant breeding has been the major driving force for the improvement of crops, feed, and fodder. Traditionally the conventional and marker-assisted breeding (MAB) approaches have been frequently employed in plant breeding. The conventional breeding makes use of genetic diversity among different species and their subsequent hybridisation and selection over number of generations to develop an improved version of the crop. However, this approach has few limitations such as the requirement of a long time-period (generally 5–12 years) to develop new crop variety, due to poor effectiveness for traits that have low heritability, presence of high environmental noise as well as its dependence on phenotypic selection. On the other hand, MAB relies on the use of molecular markers and aids in indirect selection of traits of interest.

Next Generation Sequencing has played a very imminent role in the identification, selection, and development of molecular markers. The data generated using NGS has led to the quantum leap in the development of large-scale plant genomic resources, including sequence and transcript data as well as physical maps for many crop varieties. Briefly, NGS technologies have facilitated in making the mapping of traits and marker-assisted breeding more feasible.

The collaborative efforts in large-scale sequencing projects led to a tremendous increase in our knowledge about plant reference genomes. This not only allowed scientists to gain the insights of plant genome structure and function but also guided them in the genome assembly of species that are closely related. Also, the reference genome sequences helped in mining of huge number of molecular markers and candidate genes.

Several molecular markers involved in identifying traits like enhanced yield, quality, and tolerance to environmental stresses have been discovered for crop plants like rice and wheat (Table 20.1). Newer methods such as genotyping by sequencing (GBS), proved as a robust and rapid approach towards genome-wide SNP discovery

**Table 20.1** Selected studies involving the use of NGS applications in mapping candidate genes and loci in different crop species

| Species | Reference(s) |
|---|---|
| Rice | Ramkumar et al. (2010), Abe et al. (2012), Yang et al. (2012), Takagi et al. (2013a), Spindel et al. (2015), Zheng et al. (2016), Zegeye et al. (2018) |
| Wheat | Periyannan et al. (2011), Trick et al. (2012), Periyannan et al. (2014), Rutkoski et al. (2014), Crossa et al. (2016) |
| Millet | Gimode et al. (2016), Ni et al. (2017), Antony Ceasar et al. (2018), Puranik et al. (2020), Wambi et al. (2020), Xu et al. (2021), Fukunaga et al. (2022) |
| Maize | Su et al. (2017), Ladejobi et al. (2019), Tran et al. (2020), Wang et al. (2020), Chen et al. (2021), Peng et al. (2021) |
| Soyabean | Qi et al. (2014), Lu et al. (2017), Kong et al. (2018), Li et al. (2019a, b), Prince et al. (2020), Bhat and Yu (2021) |
| Tomato | Celik et al. (2017), Razali et al. (2018), Wang et al. (2018), Liu et al. (2019), Topcu et al. (2021) |

and genotyping in a single step. Thus, marker-assisted selection could be widely used for identifying genes regulating key traits, and for GWAS studies.

## 20.4 Case Studies

### *20.4.1 Rice*

Rice (*Oryza sativa* L.) is a staple food crop for people all around the world and currently the major goal has been to increase its grain yield. The traditional method of breeding has proved to be inefficient for this ever-increasing demand for rice grains. Moreover, many factors such as extreme cold, soil salinity, pathogens, pests, and weeds, etc., greatly affect rice cultivation. Major crop losses are caused by biotic stresses such as bacterial blight and blast disease. Recent developments in NGS have boosted genome sequencing of thousands of rice varieties, thus opening avenues for the discovery of maker genes associated with the quality traits which were eventually used for marker-assisted selection breeding (Abe et al. 2012; Takagi et al. 2013a; Tribhuvan et al. 2018). One such example is the discovery of Pi65(t), a novel resistance gene conferring resistance to fungus *Magnaporthe oryzae* and the subsequent generation of a new rice variety Chuangxin1 in the Liaoning province of China (Zheng et al. 2016). The gene confering this resistance was mapped to a locus on chromosome 11 (30.42–30.85 M) using bulked segregant analysis in combination with NGS.

#### 20.4.1.1 WGS, RNA-Seq, Rad-Seq, Etc.

In recent times, NGS has been used as a primary approach to identify genetic variations and gene fusions in various species. With that, many user-friendly computational tools have been developed to detect these genetic variations/mutations using data generated from whole genome sequencing (WGS) and RNA sequencing. The identification and characterization of mutants from its complete genome has always been a great matter of interest. The arrival of the NGS approach has made whole genome sequencing (WGS) a very feasible and cost-effective method for genetic screening. It has revolutionized the ability to detect mutation/variation by underlying the phenotypic features. By mapping the data of interest with a reference genome it can provide an array of variant sequences, which has enriched the approach of "fast-forward" genetics. Whole genome sequencing has proven to be a beneficial approach for geneticists who perform both classical and forward based genetic mapping (includes, random based mutagenesis and phenotypic characterization) to model species. This made WGS to be used in routine for forward based genetic screening. However, still, WGS is an expensive method for large genomes, but it may not be if the analysis majorly focussed on mutation in the genome. Other alternatives such as whole exome sequencing (WES) (Kaur and Gaikwad 2017) and other DNA enriched mapping intervals can establish a powerful impact. In fact, enriched DNA pools enable the continuous mapping by sequencing, whereas low depth WGS could allow mutant pool for mapping almost the maximum region which facilitates a specific enrichment assay that could target the deep sequencing effort to the specific region.

Unlike WGS or WES, the restriction site-associated sequencing or RAD-sequencing does not need any initial information about sequence (Clugston et al. 2019). Before the NGS discovery, the polymorphic RAD genome-wide analysis was already in use for bulk segregating based-mutation mapping and even later it conveniently adapted the NGS approach. RAD sequencing is not enriched by genes and does not include any actual mutation and requires more efforts in target based fine-mapping. The RAD sequencing method provides ease in analysis as it does not require a reference genome for detecting mutation and allows to perform forward genetics over non-model species. But such mapping analysis which depends on the reference genome is still more informative as it could make use of linkage that is provided by order of sequence aligned to the reference genome.

RNA sequencing is the best alternative to WGS because it provides less complexity, as the transcriptome size is smaller than the genome and hence, is less expensive than WGS. It also simultaneously facilitates transcript expression analysis. In RNA sequencing, there is a limitation that genetic markers can be only genotyped if they are expressed. It even allows us to identify the effects of splice-site mutations.

## 20.5 Tools for Mapping

### *20.5.1 SHOREmap*

SHOREmap (http://bioinfo.mpipz.mpg.de/shoremap/) is a NGS-based analysis tool for genetic mapping (Schneeberger et al. 2009; Sun and Schneeberger 2015). In SHOREmap, SHORE stands for SHOrt Read as it uses the short reads, generated from high-throughput sequencing for reference-based genome assembly. SHOREmap comes with a mapping by sequencing approach which depends on markers that could be polymorphic between two alleles. It can manage different types of segregation populations generated by outcrossing as well as backcrossing the mutants. SHOREmap is implemented in the C/C++ programming language and it does not rely on any resequencing tool. The major function of SHOREmap is to use resequencing results that are stored in VCF files and preprocessed for the further analysis. SHOREmap includes inbuilt SNP markers and facilitates a function "creates" for creating an own marker selection list. The tool also provides "outcross" and "backcross" functions to calculate AFs in the mutant genomes. The "outcross" function makes prediction of the mapping interval, whereas the "backcross" function detects causal mutation candidates. The SHOREmap includes an "annotate" feature which provides predictive analysis, to know the functional impact of novel mutation over the gene.

### 20.5.2 NGM

The next generation sequencing technologies provided the advantage of directly mapping the mutation which causes phenotypic changes via direct sequencing of mutations. But several mapping approaches developed require some genetic analysis which could be very time-consuming. The Next Generation Mapping (NGM) (http://bar.utoronto.ca/NGM) is a user-friendly web tool which overcome such obstacles and provides robust mapping based on the de-novo method (Austin et al. 2011). It maps the location of EMS mutation from a small pool of F2 population sequencing. NGM uses statistical significance to determine contribution of the parental mutants and the mapping lines of each SNP in the F2 population. The information is then used for identifying candidate mutations. The NGM was successfully demonstrated for detecting the three genes involved in *Arabidopsis thaliana* cell-wall biology, in which mapping performed between ten F2 lines and illumina genome analyser dataset. The tool enables the mapping of any model organism and also could be beneficial to utilize in crops improvement.

### 20.5.3 MutMap

The MutMap is a popular mutation mapping tool which works based on bulk-segregant analysis of mutation progenies obtained from the F2 population (Fig. 20.1) (Abe et al. 2012; Fekih et al. 2013). It was developed for mapping monogenic based recessive gene-controlled traits. The tool was demonstrated by mapping mutation loci of rice which was responsible for colour changing in leaf from dark green to pale green. The MutMap is considered to be an alternative of SHOREmap and NGM (next generation mapping). SHOREmap and NGM perform distant related mapping between parent and bulk which obtain large populations of SNPs, whereas MutMap depends on crosses between mutant and its wild-type which directly detects the causal SNPs which are obtained during mutagenesis. Nowadays, MutMap is widely used because it focussed majorly on causal variants/SNPs which is considered to be a strong source for causing change in phenotypic behaviour. MutMap improvised versions were lately introduced, i.e., MutMap+, MutMap Gap, and QTL-seq which has expanded the analysis and provide a detailed prospect towards MutMap application (Fekih et al. 2013; Takagi et al. 2013b; Zegeye et al. 2018; Sugihara et al. 2022).

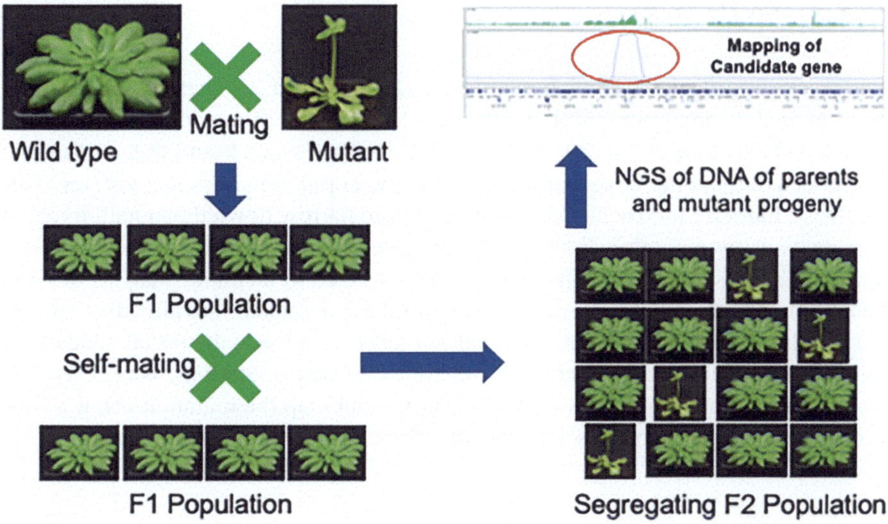


**Fig. 20.1** Schematic diagram showing identification of mutation through SNP mapping using NGS approach

## *20.5.4 Easymap*

Easymap (http://atlas.umh.es/genetics) is a GUI based application which uses mapping population to analyse the high-throughput data which is derived through various experiments such as EMS induced mutation or insertional mutation (Lup et al. 2021). The Easymap enables mapping-by-sequencing approach which provides ease to researchers for conducting experiments even without any bioinformatics experience. The tool provides features like web-based graphical interface, understandable installation script, mapping analysis, and produces user-friendly mapping reports. The tool comes with two pipelines for map point and insertional mutations. This UNIX operating system-based tool gives high performance and automates the data analysis process such as, reference-based alignment, variant calling, filtering and other parameters. The Easymap consists of a controller layer, a workflow layer that includes linkage analysis mapping and tagged sequence mapping and other common pipeline for mapping, and a tasks layer (third party programmes). The tool exhibits a simple API (Application Programming Interface) through which the command line and web interface could interact with others. This enables EasyMap to install either local or remote platforms with an advantage of access to command line and graphical user interface.

### 20.5.5 SNPTrack

SNPTrack (http://genetics.bwh.harvard.edu/snptrack/) is a fast method for mapping using NGS data combining SNP discovery, identification of mutation and sequence variation (Leshchiner et al. 2012). SNPTrack facilitates quick and precise through cost-effective mapping of causal mutations in the population of model species. This involves a hidden Markov model which enables to narrow down the mutation region in mutants using recombination breakpoints of chromosomes. It is hosted online at Harvard University server with user-friendly interface to facilitate analysis of NGS data. There are several reference genomes of model organisms are already available and developers offer to include any other reference genome to the portal. One of the biggest advantages of this method is its ability to map mutations using very few samples from mapping population that means it can map the mutation using as low as only five samples which is very useful for the traits which is hard to map.

### 20.5.6 SIMPLE

SIMPLE (https://github.com/wacguy/Simple) is a simple and user-friendly mapping pipeline which generates relevant candidate genes and their genomic locations (Wachsman et al. 2017). This tool requires the fastq reads files of wild-type and mutants DNA pools produced from NGS approach. It is easily accessed and installed as it does not require any prior bioinformatics experience but some initial preparation for executing the programme. After execution, users can access the relevant candidate genes and their figure files including the location of these candidate genes. SIMPLE is a comprehensive programme apart from that initial preparation steps, i.e., generating and uploading the fastq files and inputting species name, the user just needs to run the programme over the terminal. The programme allows both single-end and paired-end fastq files and includes various segregating populations like, M1 and M2 population. Thus, the tool does not need a back- cross approach.

### 20.5.7 MMAPPR

MMAPPR (http://yost.genetics.utah.edu/software.php) is a robust tool that enables mutation mapping analysis pipeline for pooled RNA seq(Hill et al. 2013). This automated application does not require any knowledge of parental strain or any pre-existing information about the SNP map data of the organism. The differential recombination frequencies of the genome are well adapted by this tool. It easily accommodates the noise which is generated due to RNA seq data. The tool estimates the allele frequency using the Euclidean distance method which is carried out by Loess regression analysis. It finds out the exact location where mutation occurs and generates a list of evident mutations in that particular genomic region. The MMAPPR tool can make productive use of RNA-seq data that is acquired from tissue or whole organisms which are used for transcriptome analysis and gene expression analysis in novel mutants.

## 20.6 Future Prospects

NGS technology provides massive data generated through parallel sequencing which offers ultra-high throughput leading to enormous scalability potential in short time scale (Slatko et al. 2018). NGS enabled us to sequence entire genomes or targeted regions of DNA or RNA which has transformed the biological sciences including crop science. Considering the complex nature of gene functions at DNA, RNA, and protein level it was never easy to identify the functions of unknown gene to characterize at organism level, however, with the help of versatility of NGS many uncovered trait mechanisms are being identified which can potentially be utilized to develop better crop in terms of yield as well as sustainability. With the development of newer cost-effective NGS technologies along with fourth generation sequencing era we can expect to have deeper informations to address complex trait heritability mechanisms which can enable us to have better crop varieties (Dumschott et al. 2020; Suzuki 2020).

## References

Abe A, Kosugi S, Yoshida K, Natsume S, Takagi H, Kanzaki H, Matsumura H, Yoshida K, Mitsuoka C, Tamiru M et al (2012) Genome sequencing reveals agronomically important loci in rice using MutMap. Nat Biotechnol 30(2):174–178

Alonso-Blanco C, Andrade J, Becker C, Bemm F, Bergelson J, Borgwardt KMM, Cao J, Chae E, Dezwaan TMM, Ding W et al (2016) 1,135 Genomes reveal the global pattern of polymorphism in Arabidopsis thaliana. Cell 166(2):481–491

Antony Ceasar S, Maharajan T, Ajeesh Krishna TP, Ramakrishnan M, Victor Roch G, Satish L, Ignacimuthu S (2018) Finger millet [Eleusine coracana (L.) Gaertn.] improvement: current status and future interventions of whole genome sequence. Front Plant Sci 9:1054

Austin RS, Vidaurre D, Stamatiou G, Breit R, Provart NJ, Bonetta D, Zhang J, Fung P, Gong Y, Wang PW et al (2011) Next-generation mapping of Arabidopsis genes. Plant J 67:715–725

Beyer A, Bandyopadhyay S, Ideker T (2007) Integrating physical and genetic maps: from genomes to interaction networks. Nat Rev Genet 89(8):699–710

Bhat JA, Yu D (2021) High-throughput NGS-based genotyping and phenotyping: role in genomics-assisted breeding for soybean improvement. Legum Sci 3:e81

Bhat JA, Ali S, Salgotra RK, Mir ZA, Dutta S, Jadon V, Tyagi A, Mushtaq M, Jain N, Singh PK et al (2016) Genomic selection in the era of next generation sequencing for complex traits in plant breeding. Front Genet 7:221

Bhat JA, Yu D, Bohra A, Ganie SA, Varshney RK (2021) Features and applications of haplotypes in crop breeding. Commun Biol 41(4):1–12

Celik I, Gurbuz N, Uncu AT, Frary A, Doganlar S (2017) Genome-wide SNP discovery and QTL mapping for fruit quality traits in inbred backcross lines (IBLs) of solanum pimpinellifolium using genotyping by sequencing. BMC Genomics 18(1):1–10. https://doi.org/10.1186/S12864-016-3406-7

Chen Z, Tang D, Hu K, Zhang L, Yin Y, Ni J, Li P, Wang L, Rong T, Liu J (2021) Combining QTL-seq and linkage mapping to uncover the genetic basis of single vs. paired spikelets in the advanced populations of two-ranked maize×teosinte. BMC Plant Biol 21(1):1–15

Clugston JAR, Kenicer GJ, Milne R, Overcast I, Wilson TC, Nagalingum NS (2019) RADseq as a valuable tool for plants with large genomes-A case study in cycads. Mol Ecol Resour 19:1610–1622

Crossa J, Jarquín D, Franco J, Pérez-Rodríguez P, Burgueño J, Saint-Pierre C, Vikram P, Sansaloni C, Petroli C, Akdemir D et al (2016) Genomic prediction of gene bank wheat landraces. Genetics 6:1819–1834

Crossa J, Pérez-Rodríguez P, Cuevas J, Montesinos-López O, Jarquín D, de Campos G, Burgueño J, González-Camacho JM, Pérez-Elizalde S, Beyene Y et al (2017) Genomic selection in plant breeding: methods, models, and perspectives. Trends Plant Sci 22:961–975

Doitsidou M, Poole RJ, Sarin S, Bigelow H, Hobert O (2010) C. elegans mutant identification with a one-step whole-genome-sequencing and SNP mapping strategy. PLoS One 5:e15435

Dong P, Tu X, Li H, Zhang J, Grierson D, Li P, Zhong S (2020) Tissue-specific Hi-C analyses of rice, foxtail millet and maize suggest non-canonical function of plant chromatin domains. J Integr Plant Biol 62:201–217

Dumschott K, Schmidt MHW, Chawla HS, Snowdon R, Usadel B (2020) Oxford Nanopore sequencing: new opportunities for plant genomics? J Exp Bot 71:5313–5322

Fekih R, Takagi H, Tamiru M, Abe A, Natsume S, Yaegashi H, Sharma S, Sharma S, Kanzaki H, Matsumura H et al (2013) MutMap+: genetic mapping and mutant identification without crossing in rice. PLoS One 8:e68529

Fukunaga K, Abe A, Mukainari Y, Komori K, Tanaka K, Fujihara A, Yaegashi H, Kobayashi M, Ito K, Ohsako T et al (2022) Recombinant inbred lines and next-generation sequencing enable rapid identification of candidate genes involved in morphological and agronomic traits in foxtail millet. Sci Rep 12(1):1–12

Gimode D, Odeny DA, De Villiers EP, Wanyonyi S, Dida MM, Mneney EE, Muchugi A, Machuka J, De Villiers SM (2016) Identification of SNP and SSR markers in finger millet using next generation sequencing technologies. PLoS One 11:e0159437

Grob S, Schmid MW, Grossniklaus U (2014) Hi-C analysis in Arabidopsis identifies the KNOT, a structure with similarities to the flamenco locus of Drosophila. Mol Cell 55:678–693

Hazen SP, Borevitz JO, Harmon FG, Pruneda-Paz JL, Schultz TF, Yanovsky MJ, Liljegren SJ, Ecker JR, Kay SA (2005) Rapid array mapping of circadian clock and developmental mutations in arabidopsis. Plant Physiol 138:990

Hill JT, Demarest BL, Bisgrove BW, Gorsi B, Su YC, Yost HJ (2013) MMAPPR: mutation mapping analysis pipeline for pooled RNA-seq. Genome Res 23:687–697

Jaganathan D, Bohra A, Thudi M, Varshney RK (2020) Fine mapping and gene cloning in the post-NGS era: advances and prospects. Theor Appl Genet 133:1791–1810

James GV, Patel V, Nordström KJV, Klasen JR, Salomé PA, Weigel D, Schneeberger K (2013) User guide for mapping-by-sequencing in Arabidopsis. Genome Biol 14:1–13

Jarquín D, Kocak K, Posadas L, Hyma K, Jedlicka J, Graef G, Lorenz A (2014) Genotyping by sequencing for genomic prediction in a soybean breeding population. BMC Genomics 15(1): 1–10. https://doi.org/10.1186/1471-2164-15-740

Jonas E, De Koning DJ (2013) Does genomic selection have a future in plant breeding? Trends Biotechnol 31:497–504

Kaul S, Koo HL, Jenkins J, Rizzo M, Rooney T, Tallon LJ, Feldblyum T, Nierman W, Benito MI, Lin X et al (2000) Analysis of the genome sequence of the flowering plant Arabidopsis thaliana. Nature 408(6814):796–815

Kaur P, Gaikwad K (2017) From genomes to gene-omes: exome sequencing concept and applications in crop improvement. Front Plant Sci 8:2164
Klein H, Xiao Y, Conklin PA, Govindarajulu R, Kelly JA, Scanlon MJ, Whipple CJ, Bartlett M (2018) Bulked-segregant analysis coupled to whole genome sequencing (BSA-Seq) for rapid gene cloning in maize. Genetics 8:3583–3592
Kong L, Lu S, Wang Y, Fang C, Wang F, Nan H, Su T, Li S, Zhang F, Li X et al (2018) Quantitative trait locus mapping of flowering time and maturity in soybean using next-generation sequencing-based analysis. Front Plant Sci 9:995
Korte A, Farlow A (2013) The advantages and limitations of trait analysis with GWAS: a review. Plant Methods 9:1–9
Ladejobi O, Mackay IJ, Poland J, Praud S, Hibberd JM, Bentley AR (2019) Reference genome anchoring of high-density markers for association mapping and genomic prediction in European winter wheat. Front Plant Sci 10:1278
Lam HM, Xu X, Liu X, Chen W, Yang G, Wong FL, Li MW, He W, Qin N, Wang B et al (2010) Resequencing of 31 wild and cultivated soybean genomes identifies patterns of genetic diversity and selection. Nat Genet 42:1053–1059
Leshchiner I, Alexa K, Kelsey P, Adzhubei I, Austin-Tse CA, Cooney JD, Anderson H, King MJ, Stottmann RW, Garnaas MK et al (2012) Mutation mapping and identification by whole-genome sequencing. Genome Res 22:1541–1548
Li D, Zhao X, Han Y, Li W, Xie F (2019a) Genome-wide association mapping for seed protein and oil contents using a large panel of soybean accessions. Genomics 111:90–95
Li S, Jia S, Hou L, Nguyen H, Sato S, Holding D, Cahoon E, Zhang C, Clemente T, Yu B (2019b) Mapping of transgenic alleles in soybean using a nanopore-based sequencing strategy. J Exp Bot 70:3825–3833
Lieberman-Aiden E, Van Berkum NL, Williams L, Imakaev M, Ragoczy T, Telling A, Amit I, Lajoie BR, Sabo PJ, Dorschner MO et al (2009) Comprehensive mapping of long range interactions reveals folding principles of the human genome. Science 326:289
Liu G, Zhao T, You X, Jiang J, Li J, Xu X (2019) Molecular mapping of the Cf-10 gene by combining SNP/InDel-index and linkage analysis in tomato (Solanum lycopersicum). BMC Plant Biol 19:1–11
Lu X, Xiong Q, Cheng T, Li QT, Liu XL, Bi YD, Li W, Zhang WK, Ma B, Lai YC et al (2017) A PP2C-1 allele underlying a quantitative trait locus enhances soybean 100-seed weight. Mol Plant 10:670–684
Lup SD, Wilson-Sánchez D, Andreu-Sánchez S, Micol JL (2021) Easymap: a user-friendly software package for rapid mapping-by-sequencing of point mutations and large insertions. Front Plant Sci 12:601
Magwene PM, Willis JH, Kelly JK (2011) The statistics of bulk segregant analysis using next generation sequencing. PLoS Comput Biol 7:e1002255
Mascher M, Jost M, Kuon JE, Himmelbach A, Abfalg A, Beier S, Scholz U, Graner A, Stein N (2014) Mapping-by-sequencing accelerates forward genetics in barley. Genome Biol 15:1–15
Montgomery SA, Tanizawa Y, Galik B, Wang N, Ito T, Mochizuki T, Akimcheva S, Bowman JL, Cognat V, Maréchal-Drouard L et al (2020) Chromatin organization in early land plants reveals an ancestral association between H3K27me3, transposons, and constitutive heterochromatin. Curr Biol 30:573–588.e7
Naish M, Alonge M, Wlodzimierz P, Tock AJ, Abramson BW, Schmücker A, Mandáková T, Jamge B, Lambing C, Kuo P et al (2021) The genetic and epigenetic landscape of the Arabidopsis centromeres. Science 374(6569):7489. https://doi.org/10.1126/SCIENCE.ABI7489
Nguyen K, Grondin A, Courtois B, Gantet P (2019) Next-generation sequencing accelerates crop gene discovery. Trends Plant Sci 24:263–274
Ni X, Xia Q, Zhang H, Cheng S, Li H, Fan G, Guo T, Huang P, Xiang H, Chen Q et al (2017) Updated foxtail millet genome assembly and gene mapping of nine key agronomic traits by resequencing a RIL population. Gigascience 6:1

Ossowski S, Schneeberger K, Clark RM, Lanz C, Warthmann N, Weigel D (2008) Sequencing of natural strains of *Arabidopsis thaliana* with short reads. Genome Res 18:2024–2033

Peng C, Mei Y, Ding L, Wang X, Chen X, Wang J, Xu J (2021) Using combined methods of genetic mapping and nanopore-based sequencing technology to analyze the insertion positions of G10evo-EPSPS and Cry1Ab/Cry2Aj transgenes in maize. Front Plant Sci 12:1519

Periyannan SK, Bansal UK, Bariana HS, Pumphrey M, Lagudah ES (2011) A robust molecular marker for the detection of shortened introgressed segment carrying the stem rust resistance gene Sr22 in common wheat. Theor Appl Genet 122:1–7

Periyannan S, Bansal U, Bariana H, Deal K, Luo MC, Dvorak J, Lagudah E (2014) Identification of a robust molecular marker for the detection of the stem rust resistance gene Sr45 in common wheat. Theor Appl Genet 127:947–955

Prince SJ, Vuong TD, Wu X, Bai Y, Lu F, Kumpatla SP, Valliyodan B, Shannon JG, Nguyen HT (2020) Mapping quantitative trait loci for soybean seedling shoot and root architecture traits in an inter-specific genetic population. Front Plant Sci 11:1284

Provart NJ, Brady SM, Parry G, Schmitz RJ, Queitsch C, Bonetta D, Waese J, Schneeberger K, Loraine AE (2021) Anno genominis XX: 20 years of Arabidopsis genomics. Plant Cell 33:832–845

Puranik S, Sahu PP, Beynon S, Srivastava RK, Sehgal D, Ojulong H, Yadav R (2020) Genome-wide association mapping and comparative genomics identifies genomic regions governing grain nutritional traits in finger millet *(Eleusine coracana L. Gaertn.)*. Plants People Planet 2: 649–662

Qi Z, Huang L, Zhu R, Xin D, Liu C, Han X, Jiang H, Hong W, Hu G, Zheng H et al (2014) A high-density genetic map for soybean based on specific length amplified fragment sequencing. PLoS One 9:863

Ramesh P, Mallikarjuna G, Sameena S, Kumar A, Gurulakshmi K, Reddy BV, Reddy PCO, Sekhar AC (2020) Advancements in molecular marker technologies and their applications in diversity studies. J Biosci 45(1):1–5. https://doi.org/10.1007/s12038-020-00089-4

Ramkumar G, Sivaranjani AKP, Pandey MK, Sakthivel K, Shobha Rani N, Sudarshan I, Prasad GSV, Neeraja CN, Sundaram RM, Viraktamath BC et al (2010) Development of a PCR-based SNP marker system for effective selection of kernel length and kernel elongation in rice. Mol Breed 4:735–740

Razali R, Bougouffa S, Morton MJL, Lightfoot DJ, Alam I, Essack M, Arold ST, Kamau AA, Schmöckel SM, Pailles Y et al (2018) The genome sequence of the wild tomato solanum pimpinellifolium provides insights into salinity tolerance. Front Plant Sci 9:1402

Rutkoski J, Poland JE, Singh RP, Huerta-Espino J, Bhavani S, Barbier H, Rouse MN, Jannink JL, Sorrells ME (2014) Genomic selection for quantitative adult plant stem rust resistance in wheat. Plant Genome. https://doi.org/10.3835/PLANTGENOME2014.02.0006

Schneeberger K, Weigel D (2011) Fast-forward genetics enabled by new sequencing technologies. Trends Plant Sci 16:282–288

Schneeberger K, Ossowski S, Lanz C, Juul T, Petersen AH, Nielsen KL, Jørgensen JE, Weigel D, Andersen SU (2009) SHOREmap: simultaneous mapping and mutation identification by deep sequencing. Nat Methods 68(6):550–551

Sharbel TF, Haubold B, Mitchell-Olds T (2000) Genetic isolation by distance in Arabidopsis thaliana: biogeography and postglacial colonization of Europe. Mol Ecol 9:2109–2118

Slatko BE, Gardner AF, Ausubel FM (2018) Overview of next generation sequencing technologies. Curr Protoc Mol Biol 122:e59

Smith HE, Fabritius AS, Jaramillo-Lambert A, Golden A (2016) Mapping challenging mutations by whole-genome sequencing. Genetics 6:1297–1304

Spindel J, Begum H, Akdemir D, Virk P, Collard B, Redoña E, Atlin G, Jannink JL, McCouch SR (2015) Genomic selection and association mapping in rice (Oryza sativa): effect of trait genetic architecture, training population composition, marker number and statistical model on accuracy of rice genomic selection in elite, tropical rice breeding lines. PLoS Genet 11:1–25

Su C, Wang W, Gong S, Zuo J, Li S, Xu S (2017) High density linkage map construction and mapping of yield trait qtls in maize (*Zea mays*) using the genotyping-by-sequencing (GBS) technology. Front Plant Sci 8:706

Sugihara Y, Young L, Yaegashi H, Natsume S, Shea DJ, Takagi H, Booker H, Innan H, Terauchi R, Abe A (2022) High-performance pipeline for MutMap and QTL-seq. PeerJ 18:10–13

Sun H, Schneeberger K (2015) SHOREmap v3.0: fast and accurate identification of causal mutations from forward genetic screens. Methods Mol Biol 1284:381–395

Suzuki Y (2020) Advent of a new sequencing era: long-read and on-site sequencing. J Hum Genet 65(1):1

Takagi H, Abe A, Yoshida K, Kosugi S, Natsume S, Mitsuoka C, Uemura A, Utsushi H, Tamiru M, Takuno S et al (2013a) QTL-seq: rapid mapping of quantitative trait loci in rice by whole genome resequencing of DNA from two bulked populations. Plant J 74:174–183

Takagi H, Uemura A, Yaegashi H, Tamiru M, Abe A, Mitsuoka C, Utsushi H, Natsume S, Kanzaki H, Matsumura H et al (2013b) MutMap-Gap: whole-genome resequencing of mutant F2 progeny bulk combined with de novo assembly of gap regions identifies the rice blast resistance gene. New Phytol 200:276–283

Topcu Y, Sapkota M, Illa-Berenguer E, Nambeesan SU, van der Knaap E (2021) Identification of blossom-end rot loci using joint QTL-seq and linkage-based QTL mapping in tomato. Theor Appl Genet 134:2931–2945

Tran QH, Bui NH, Kappel C, Dau NTN, Nguyen LT, Tran TT, Khanh TD, Trung KH, Lenhard M, Vi SL (2020) Mapping-by-sequencing via MutMap identifies a mutation in ZmCLE7 underlying fasciation in a newly developed EMS mutant population in an elite tropical maize inbred. Gene 11(3):281. https://doi.org/10.3390/GENES11030281

Tribhuvan KU, Sandhya KK, Sevanthi AM, Gaikwad K (2018) MutMap: a versatile tool for identification of mutant loci and mapping of genes. Indian J Plant Physiol 23(4):612–621

Trick M, Adamski NM, Mugford SG, Jiang CC, Febrer M, Uauy C (2012) Combining SNP discovery from next-generation sequencing data with bulked segregant analysis (BSA) to fine-map genes in polyploid wheat. BMC Plant Biol 12:1–17

Varshney RK, Terauchi R, McCouch SR (2014) Harvesting the promising fruits of genomics: applying genome sequencing technologies to crop breeding. PLoS Biol 12:e1001883

Voz ML, Coppieters W, Manfroid I, Baudhuin A, von Berg V, Charlier C, Meyer D, Driever W, Martial JA, Peers B (2012) Fast homozygosity mapping and identification of a zebrafish ENU-induced mutation by whole-genome sequencing. PLoS One 7:e34671

Wachsman G, Modliszewski JL, Valdes M, Benfey PN (2017) A SIMPLE pipeline for mapping point mutations. Plant Physiol 174:1307–1313

Wambi W, Otienno G, Tumwesigye W, Mulumba J (2020) Genetic and genomic resources for finger millet improvement: opportunities for advancing climate-smart agriculture. J Crop Improv 35(2):204–233

Wang Y, Jiang J, Zhao L, Zhou R, Yu W, Zhao T (2018) Application of whole genome resequencing in mapping of a tomato yellow leaf curl virus resistance gene. Sci Rep 8(1):1–11

Wang N, Yuan Y, Wang H, Yu D, Liu Y, Zhang A, Gowda M, Nair SK, Hao Z, Lu Y et al (2020) Applications of genotyping-by-sequencing (GBS) in maize genetics and breeding. Sci Rep 10(1):1–12

Wenger JW, Schwartz K, Sherlock G (2010) Bulk segregant analysis by high-throughput sequencing reveals a novel xylose utilization gene from *saccharomyces cerevisiae*. PLoS Genet 6: e1000942

Wicks SR, Yeh RT, Gish WR, Waterston RH, Plasterk RHA (2001) Rapid gene mapping in Caenorhabditis elegans using a high density polymorphism map. Nat Genet 28(2):160–164

Xu W, Liang M, Yang X, Wang H, Luo M (2021) Genomic resources of broomcorn millet: demonstration and application of a high-throughput BAC mapping pipeline. BMC Genomic Data 22:1–9

Yadav VK, Santos-González J, Köhler C (2021) INT-Hi-C reveals distinct chromatin architecture in endosperm and leaf tissues of *Arabidopsis*. Nucleic Acids Res 49:4371–4385

Yang H, Tao Y, Zheng Z, Li C, Sweetingham MW, Howieson JG (2012) Application of next-generation sequencing for rapid marker development in molecular plant breeding: a case study on anthracnose disease resistance in *Lupinus angustifolius* L. BMC Genomics 13(1):1–2. https://doi.org/10.1186/1471-2164-13-318

Zegeye WA, Zhang Y, Cao L, Cheng S (2018) Whole genome resequencing from bulked populations as a rapid QTL and gene identification method in rice. Int J Mol Sci 19(12):4000

Zheng W, Wang Y, Wang L, Ma Z, Zhao J, Wang P, Zhang L, Liu Z, Lu X (2016) Genetic mapping and molecular marker development for Pi65(t), a novel broad-spectrum resistance gene to rice blast using next-generation sequencing. Theor Appl Genet 129:1035–1044

# Chapter 21
# Application of Artificial Intelligence and Machine Learning in Agriculture

**Sudeep Marwaha, Chandan Kumar Deb, Md. Ashraful Haque, Sanchita Naha, and Arpan Kumar Maji**

**Abstract** Artificial intelligence (AI) is the branch of science that deals with the development of machines to mimic human intelligence. Machine learning (ML) is a subdomain of AI where the machine can learn automatically from data without being explicitly programmed. Agriculture is constantly pressed upon to produce more with less resource. AI and ML techniques have the capacity to optimize resource utilization by analysing agricultural data. It has changed the present-day face of farming by predicting various input parameters and forecasting post-harvest life of a crop. This chapter discusses the different AI and ML techniques available and how they have been used in different phases of the agriculture life cycle. This chapter includes vast range areas in agriculture that requires AI and ML. It includes soil, irrigation, and disease managements. Importance of AI in the field of plant phenomics also included in this chapter. The probable use of geographic information system (GIS) and remote sensing coupled with AI are discussed in this chapter.

**Keywords** Artificial intelligence (AI) · Machine learning (ML) · Agriculture · Recommender system · Phenomics · Geographic information system (GIS) · Remote sensing

## 21.1 Introduction

Artificial intelligence (AI) is the study of tools and technologies which are used to solve tasks that require human intelligence. Tasks such as natural language understanding, processing, generation, visual perception, decision making, and many more. Machine learning and deep learning are the two most widely used AI approaches. With breakthrough technologies, AI has transformed every aspect of life, including agriculture. With more than 50% workforce employed in agriculture, low expert to farmer ratio requires necessary AI interventions like automatic

---

S. Marwaha (✉) · C. K. Deb · M. A. Haque · S. Naha · A. K. Maji
ICAR-Indian Agricultural Statistics Research Institute, New Delhi, India
e-mail: sudeep@icar.gov.in

 
H. M. Mamrutha et al. (eds.), *Translating Physiological Tools to Augment Crop Breeding*, https://doi.org/10.1007/978-981-19-7498-4_21

diagnosis and recommendation of proper advisories. The major hurdles in agricultural production are decision making related to crop production, disease pest infestation, weather forecasting, yield prediction, advisory systems for enhanced crop productivity, etc. Agricultural productivity is mostly influenced by temperature, soil fertility, water availability, water quality, etc. For predicting these parameters accurately, improved AI techniques are being applied. While the technological explosion has made farming little easier, small, and marginal farmers still face many obstacles. Unlike other technologies AI has the potential to reach out to individual farmer much more easily and improve the life of farmer. The consideration of two life cycle, namely agriculture and farmers have a gigantic scope to intervene and enormously improve the same.

Agriculture life cycle starts from land preparation for the crop followed by seed sowing, irrigation, weeding, fertilizer application, pest and disease management, harvesting, post-harvest processing, storage, and marketing. Various AI techniques have the potential to affect and improve all the phases of the life cycle, some of which are already available and some still need to be worked on. In an ideal smart ecosystem, a farmer would be guided by an artificially intelligent assistant that would suggest the most appropriate date and method to prepare the land based on the GIS and remote sensing data of that region. Using a block chain and recommender system enabled supply chain, farmer would collect quality seeds to sow after land preparation. Scheduled weeding would be handled by low-cost smart weeding and fertigation (fertilization and irrigation) systems. The identification of pest and disease with their suitable management practices may be handled by AI enabled mobile applications. The yield prediction may be done through drone based smart application and the predicted yield will help in selecting the appropriate market and buyer (Fig. 21.1).

## 21.2 Artificial Intelligence, Machine Learning, and Deep Learning

The artificial intelligence is a very old field of study and has a rich history. Modern AI was formalized by John McCarthy, considered as father of AI. It was established as a branch of computer science around early 1950s. Primarily, the term artificial intelligence (AI) refers to a group of techniques that enable a computer or a machine to mimic the behaviour of humans in problem solving tasks. Formally, AI is described as 'the study of how to make the computers do things at which, at the moment, people are better' (Rich and Knight 1991; Rich et al. 2009). The main aim of AI is to programme the computer for performing certain tasks in humanly manner such as knowledgebase, reasoning, learning, planning, problem solving, etc. The machine learning (ML) techniques are the subset of AI which makes the computers/machines/programmes capable of learning and performing tasks without being explicitly programmed. The ML techniques are not just the way of mimicking

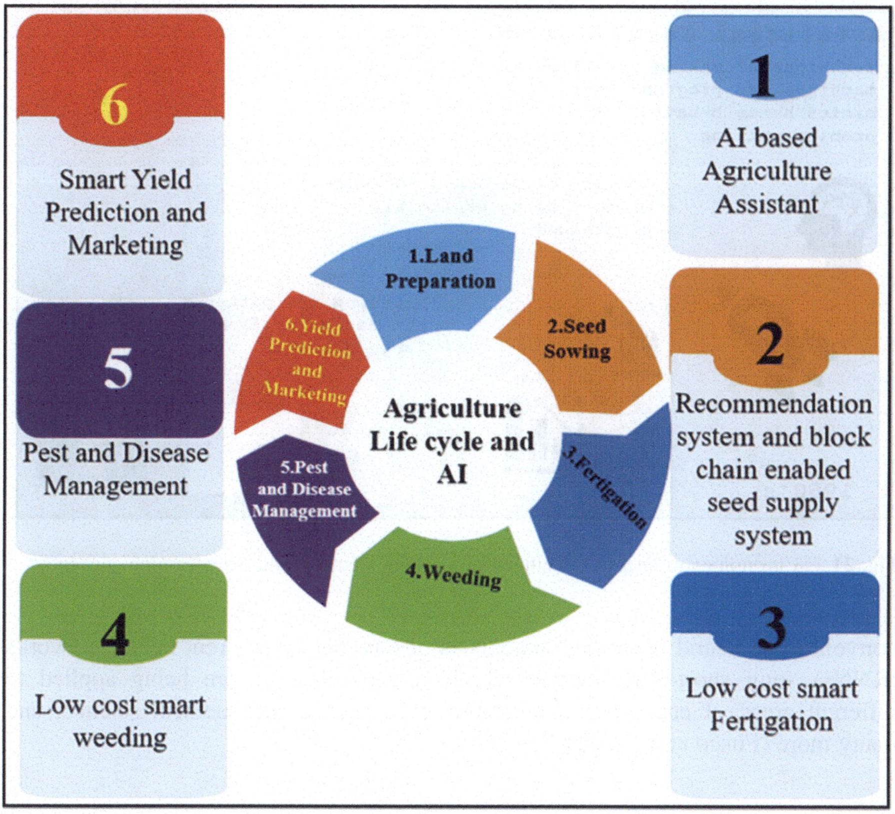


**Fig. 21.1** An ideal smart ecosystem for farmer

human behaviour but the way of mimicking how humans learn things. The main characteristics of machine learning are 'learning from experience' for solving any kind of problem. The methods of learning can be categorized into three types: (a) supervised learning algorithm is given with labelled data and the desired output, whereas (b) unsupervised learning algorithm is given with unlabelled data and identifies the patterns from the input data, and (c) reinforcement learning algorithm allows the ML techniques to capture the learnable things on the basis of rewards or reinforcement. Nowadays, deep learning (DL) techniques are the advanced version of machine learning algorithms gained huge popularity in the area of artificial intelligence based applications. The artificial neural networks (ANNs) clubbed with representation learning are the backbone of the deep learning concepts. These techniques allow a machine to learn patterns in the dataset with multiple levels of abstractions. The DL models are composed of a series of non-linear layers where each of the layer has the capability of transforming the low-level representations into higher-level representations, i.e. into a more abstract representations (LeCun et al. 2015). There are several DL algorithms available now-a-days such as deep

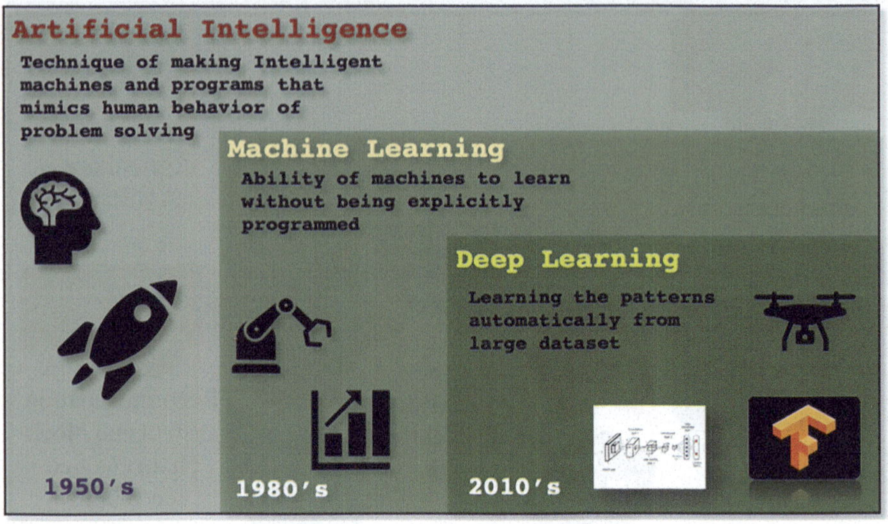


**Fig. 21.2** Chronology of artificial intelligence, machine learning, and deep learning concepts

convolutional neural networks (aka CNNs or convnets), recurrent neural networks (RNNs), long short-term memory (LSTM) networks that are being applied to different areas of engineering, bioinformatics, agriculture, medical science, and many more (Fusco et al. 2021) (Fig. 21.2).

## 21.3 Major Applications of AI and ML Techniques in Agriculture

In the present scenario, AI and ML techniques are being exponentially applied in the various areas of the agricultural domain. These areas can be categorized into the following groups: soil and water management, crop health management, crop phenotyping, recommender-based systems for crops, semantic web and ontology driven expert systems for crops and Geo-AI. The applications of AI, ML, and DL based techniques on these areas are discussed in the following sections.

### *21.3.1 Soil and Irrigation Management*

Soil and irrigation are the most viable components of agriculture. The soil and irrigation are the determinant factors for the optimum crop yield. In order to obtain enhanced crop yield and to maintain the soil properties, there is a requirement of appropriate knowledge about the soil resources. Irrigation scheduling becomes crucial when water resources are scarce. Therefore, the soil and irrigation related issues should be managed properly and cautiously to ensure a potential yield in crops. In this regard, AI and ML based techniques have shown potential ability to resolve soil and irrigation related issues in crops. A range of machine learning models such as regression-based models, support vector machines (or regressors), artificial neural networks, and random forest algorithm are being used. Many researchers have used remote-sensed data with the machine learning techniques for determining soil health parameters. Few significant studies in this area have been highlighted in the following sections.

#### 21.3.1.1 Soil Management

Besalatpour et al. (2012), Aitkenhead et al. (2012), and Sirsat et al. (2017) used different machine learning techniques such as multiple-linear regression (MLR), support vector regressors (SVR), random forests regressors (RFR) for the prediction of the physical and chemical properties of soil. Rivera and Bonilla (2020) and Azizi et al. (2020) worked on estimation and classification of aggregate stability of soil using conventional machine learning techniques as well as deep learning techniques from the publicly available soil properties datasets. Jha and Ahmad (2018) worked on prediction of microbial dynamics in soils using regression-based techniques which is very significant in improving the soil fertility. Patil and Deka (2016) and Taghizadeh-Mehrjardi et al. (2016) worked on predicting the evapotranspiration rate in crops using several machine learning techniques. Researchers worked on mapping the soil properties digitally using machine learning techniques as well as deep CNN techniques (Taghizadeh-Mehrjardi et al. 2016; Kalambukattu et al. 2018; Padarian et al. 2019; Taghizadeh-Mehrjardi et al. 2020). In these soil property mapping works, researchers have used several remotely sensed data and historical weather-based data. More details of these works are provided in Table 21.1.

#### 21.3.1.2 Irrigation Management

Zema et al. (2018) applied data envelopment analysis (DEA) with multiple linear regression (MLR) based techniques to improve the irrigation performance of the Water Users Associations. Ramya et al. (2020) and Glória et al. (2021) worked on IoT-based smart irrigation systems in which machine learning techniques were used to analyse the data. Agastya et al. (2021) and Zhang et al. (2018) applied deep CNN-based models for detection of irrigations using remote sensing (RS) data. Jimenez et al. (2021) worked on estimating the irrigation schedules based on the soil matric potential using a LSTM-based neural networks. More details are provided in Table 21.2

**Table 21.1** Soil management using AI and ML techniques

| Area of work | Techniques used | Remarks | Citation |
|---|---|---|---|
| Prediction of physical properties of soil | Support vector machines with simulated annealing | SVM obtained $R^2$ of 0.98 and MSE of 0.06 | Besalatpour et al. (2012) |
| Prediction of chemical properties of soil | Artificial neural network (ANN) model | Highest $r^2$ value obtained for clay property of 0.89 | Aitkenhead et al. (2012) |
| Classification of soil data | Random forest | More than 90% accuracy obtained | Sirsat et al. (2017) |
| Prediction of aggregate stability property of soil | ANN and GLM | ANN achieved highest performance with $R^2$ of 0.82 on test dataset | Rivera and Bonilla (2020) |
| Classification of soil aggregate stability | Deep learning models | ResNet50 Model obtained accuracy of 98.72% | Azizi et al. (2020) |
| Estimation of crop evapotranspiration | Extreme learning machine (ELM) algorithm | ELM performed better than the ANN and Hargreaves model | Patil and Deka (2016) |
| Estimation of mean reference evapotranspiration on monthly basis | SVM, multivariate adaptive regression splines (MARS), algorithms | MARS and SVM-RBF algorithms performed better | Taghizadeh-Mehrjardi et al. (2016) |
| Modelling the soil salinity variation in Arid Region | Genetic algorithm-based machine learning techniques | Obtained $R^2$ was 0.87 | Taghizadeh-Mehrjardi et al. (2016) |
| Digital mapping of soil nutrients in the hilly watersheds of Indian Himalayan region | ANN algorithms | For SOC<br>$R^2$: 0.83<br>and MSE: 0.05<br>For available nitrogen<br>$R^2$: 0.62<br>and MSE: 0.0006 | Kalambukattu et al. (2018) |
| Mapping of soil properties digitally | Deep learning techniques | 30% of error reduction observed | Padarian et al. (2019) |
| Digital soil mapping for predicting the particle size fractions | CNN and random forest algorithm | CNN model out-performed the random forest algorithm | Taghizadeh-Mehrjardi et al. (2020) |
| Microbial dynamics prediction | Regression-based techniques | $R^2$ of 0.99 | Jha and Ahmad (2018) |

### 21.3.2 *Crop Health Management*

Every year a significant amount of yield is damaged due to the attack of disease-causing pathogens and insect-pest infestation. In order to manage the spread of the

**Table 21.2** Irrigation management using AI and ML techniques

| Area of work | Techniques used | Remarks | Citation |
|---|---|---|---|
| Improved irrigation scheduling | Data envelopment analysis (DEA) was integrated with multiple linear regression (MLR) analysis | This approach potentially helpful for the irrigation decisionmakers | Zema et al. (2018) |
| IoT based system for smart irrigation in fields | SVR and bagging | Ensemble based techniques achieved 90% accuracy | Ramya et al. (2020) |
| IoT-based adaptive irrigation management system | Random forest, ANN, DT, and SVM | Random forest obtained best results with accuracy of 84.6% | Glória et al. (2021) |
| Irrigation detection using satellite images | CNN models with IoT based systems | Proposed CNN model was nine times better than traditional supervised models | Agastya et al. (2021) |
| Image-based recognition of center pivot irrigation systems | CNN-based model | Precision: 95.85% and recall: 93.33% | Zhang et al. (2018) |
| Estimate of irrigation schedules | LSTM network | $R^2$ ranges from 0.82 to 0.98 | Jimenez et al. (2021) |

diseases and insect pests, proper management practices should be applied at the earliest. Therefore, there is requirement of automatic diseases, pest identification system. In this regard, image-based diagnosis of diseases and pests have become de facto standard of automatic stress identification. This kind of automated detection methodology use sophisticated deep learning-based AI techniques that reduces the intervention of the human experts. There have been several attempts to diagnose the diseases as well as insects-pests in crops using deep learning techniques. In this section, some of the significant works in this field have been discussed briefly.

#### 21.3.2.1 Disease Identification

Mohanty et al. (2016) worked on disease diagnosis problem using deep CNN models. They used an open-source dataset named Plant Village (Hughes and Salathe 2016) containing 54,306 colour images of 26 diseases from 14 crops. Ferentinos (2018) worked on developing deep CNN-based models for recognizing 56 diseases from different crops. Barbedo (2019) applied transfer learning approach for diagnosis of diseases of 12 different crops. Too et al. (2019) applied pre-trained deep CNN models for the identification of diseases of 18 crops using the Plant Village data. Chen et al. (2020) applied a pretrained VGG Net network for classifying the diseases of Rice and Maize crop. Chen et al. (2020) and Rahman et al. (2020) worked on identifying the major diseases of Rice crop using deep learning approach. Lu et al. (2017), Johannes et al. (2017), Picon et al. (2019), and Nigam et al. (2021) applied

**Table 21.3** Disease identification using deep learning techniques

| Crop | Number of diseases | Techniques used | Remarks | Citation |
|---|---|---|---|---|
| 14 crops | 26 (diseased and healthy) | Pre-trained GoogleNet and AlexNet model | GoogleNet obtained 99.35% testing accuracy | Mohanty et al. (2016) |
| 25 crops | 58 (diseased and healthy) | AlexNet, VGGNet, GoogleNet | Highest accuracy: 99.53% | Ferentinos (2018) |
| 14 crops | 79 diseases | Transfer learning on GoogleNet | Average accuracy around 82% | Barbedo (2019) |
| 18 crops | 26 diseases | VGG 16, Inception V4, ResNet and DenseNet | DenseNet achieved the highest accuracy of 99.75% | Too et al. (2019) |
| Rice and maize | Rice: 4 Maize: 4 | Pre-trained VGGNet | Accuracy: 91.83% | Chen et al. (2020) |
| Rice | 5 diseases | Modified DenseNet | Accuracy: 98.63% | Chen et al. (2020) |
| | 5 diseases | Custom two-stage small CNN architecture | Accuracy of 93.3% | Rahman et al. (2020) |
| Wheat | 7 (diseased and Healthy) | Modified VGGNet model with localisation | Recognition accuracy: 97.95% | Lu et al. (2017) |
| | 3 diseases | Image processing techniques with deep learning model | AuC metrics higher than 0.80 | Johannes et al. (2017) |
| | 3 diseases | Deep CNN model | Accuracy more than 96% | Picon et al. (2019) |
| | 3 diseases | Custom CNN based model | Accuracy: more than 90% | Nigam et al. (2021) |
| Maize | 4 (diseases and healthy) | Modified LeNet model | 97.89 % accuracy on test dataset | Ahila Priyadharshini et al. (2019) |
| | 4 (diseases and healthy) | Custom CNN model | 92.85% accuracy | Sibiya and Sumbwanyambe (2019) |
| | 2 (MLB and healthy) | Base-line training or Inception model | Classification accuracy: 99.14% | Haque et al. (2021) |

deep CNN models for recognizing the most important diseases of wheat crop. Ahila Priyadharshini et al. (2019), Sibiya and Sumbwanyambe (2019), and Haque et al. (2021) used deep learning-based models for identifying images of diseases of maize crop. More details about the disease diagnosis work are discussed in Table 21.3

#### 21.3.2.2 Pest Identification

Pest identification problem is inherently different from disease identification problem. As compared to disease detection comparatively lesser studies have been reported in pest identification. Most of the works on insect-pest detection are based on localization and object detection concept. Some significant research works on the pest identification have been discussed in this section.

Cheeti et al. (2021) proposed insect-pest detection approach using advanced deep learning techniques for four pests of crops. They used YOLO (You Only Look Once) algorithm for detection and localisation of the pests in the image and AlexNet model as classifier to classify the images. Chen et al. (2021) employed three object detection models such Faster R-CNNs, SSDs, and Yolo-v4 for detecting and localizing the scale pest from the images. Their proposed approach achieved more than 89% accuracy for detecting and classifying the pest images. Fuentes et al. (2017) worked on combining the object detection based meta-architectures with deep CNN models for detecting the diseases crop pests of tomato crop. In this work, their proposed Faster R-CNN with VGG-16 model outperformed all the other models with mean average precision (mAP) of 83%. Li et al. (2020) proposed a deep learning-based disease and pest detection approach of rice crop using video dataset. They employed different state-of-the-art deep learning models such VGG16, ResNet-50, ResNet-101 with YOLO-v4 for the object detection purpose. Liu and Wang (2020) applied theYOLO-v3 object detector model for detecting and classifying the images of disease and pests of tomato crop.

### *21.3.3 Plant Phenotyping*

Non-destructive phenotypic measurement with high throughput imaging technique becoming extremely popular. High throughput imaging system produces a large number of images. Deduction of the phenotypic characteristics through image analysis is quick and accurate. A wide range of phenotypic study can be done using phenomics analysis. High throughput imaging system coupled with sophisticated AI technology like deep learning make this field more efficient and accurate. Phenomics has been used for studying several phenotypic characters like spike detection and counting, yield forecasting, quantification of the senescence in the plant, leaf weight and count, plant volume, convex hull, water stress, and many more as presented in Table 21.4.

**Table 21.4** Plant phenotyping using AL and ML techniques

| Area of working | Technology used | Country | Remarks | Citation |
|---|---|---|---|---|
| Spike detection | Neural network-based method using laws texture energy | Australia | Can identify spikes with an accuracy of over 80% | Qiongyan et al. (2014) |
| Forecasting of yield | Auxiliary information | India | Evaluation of different machine learning models | Elavarasan et al. (2018) |
| In field spike detection | Faster R-CNN | Australia | Average detection accuracy ranging from 88 to 94% | Hasan et al. (2018) |
| Spikelet counting | FCN | Norwich | Mean absolute error (MAE) and mean square error (MSE) of 53.0, 71.2 respectively, in counting | Alkhudaydi and Zhou (2019) |
| In-field counting of spikes | Context-augmented local regression networks | China | 91.01% counting accuracy | Xiong et al. (2019) |
| Spike detection and counting | U-Nets | India | Precision detection—99.93%, counting—99% | Misra et al. (2020) |
| Plant senescence | Colour thresholding | Australia | Quantifying the onset and progression of senescence | Cai et al. (2016) |
| Water-deficit stress | Spectral imaging | India | PLSR was the best model for prediction of RWC | Das et al. (2017) |
| Water status in plant | Hyperspectral reflectance | India | The models based on water band index (WBI), MSI, NDWI 1640, and NMDI become best | Ranjan et al. (2017) |
| NDVI | RGB image processing | USA | Simpler to use and more cost efficient than traditional dual-image NDVI or hyper-spectral imaging | Beisel et al. (2018) |
| Leaf fresh weight | ANN | India | Enhanced the fresh biomass prediction as compared to the conventional regression technique | Misra et al. (2020) |
| Water deficit stress | Thermal imaging and hyperspectral remote sensing | India | Optimal wavebands related to water deficit stress were evoked | Krishna et al. (2021) |

## *21.3.4 Recommender Systems*

Recommender systems (RS) help online users in decision making regarding products among a pile of alternatives. In general, these systems are software solutions which predict liking of a user for unseen items. RSs have been mainly designed to help users in decision making for areas where one is lacking enough personal experience to evaluate the overwhelming number of alternative items that a website has to offer (Resnick and Varian 1997). Recommender systems have proved its worth in many different applications such as e-commerce, e-library, e-tourism, e-learning, e-business, e-resource services, etc. by suggesting suitable products to users (Lu et al. 2015). RSs are used to introduce new/unseen items to users, to increase user satisfaction, etc. Recommendations are generated by processing large amount of historical data on the users and the products to be suggested. Most popular way of gathering users liking on a particular product is in terms of rating either in numerical scale (1–5) or ordinal scale (strongly agree, agree, neutral, disagree, strongly disagree). Other techniques of more knowledge-based recommendation are the use of Ontologies (Middleton et al. 2002) of user profiles or item descriptions, etc. The core task of a recommendation system is to predict the usefulness of an item to an individual user based on the earlier history of that item or by evaluating the earlier choices of the user. Collaborative way of user modelling (Konstan et al. 1997) is where ratings are predicted for <*user*, *item*> pair, $\overline{R}<u, i>$ based on a large number of ratings previously gathered by the system on individual <*user*, *item*> pairs. Another way of recommendation is to suggest items that are similar to the ones previously liked by the user, called Content based filtering (Wang et al. 2018; Smyth 2007). In a hybrid method of prediction, limitations by the earlier mentioned processes are tackled in various ways.

Agriculture has used recommender systems since 2015 and continues to do so. RSs have been explored to develop crop recommendation strategies based on soil and weather parameters, crop rotation practices, water management, suggestions on suitable varieties, recommendations for management practices, etc. It is absolutely essential for the farmers to receive recommendations on the best crop for cultivation. Kamatchi and Parvathi (2019) proposed a hybrid RS in combination with collaborative filtering, case-based reasoning, and artificial neural networks (ANNs) to predict future climatic conditions and recommendation of crops based on the predicted climate. Crop recommendations have been developed based on season and productivity (Vaishnavi et al. 2021) area and soil type (Pande et al. 2021) by using several machine learning algorithms such as support vector machine (SVM), random forest (RF), multivariate linear regression (MLR), K-nearest neighbour (KNN), ANN, etc. Ensemble techniques have been used to develop a collaborative system of crop rotation, crop yield prediction, forecasting, and fertilizer recommendation (Archana and Saranya 2020) to classify soil types into recommended crop types Kharif or Rabi based on specific physical and chemical characteristics, average rainfall and surface temperature (Kulkarni et al. 2018). Naha and Marwaha (2020) presented an Ontology driven context aware RS that can recommend land preparation methods, sowing time, seed rate, fertilizer management, irrigation scheduling and harvesting methods to Maize cultivators. Application of RSs has also penetrated in the e-agriculture domain by suggesting parts of agricultural machineries in online ordering (Ballesteros et al. 2021).

### 21.3.5 Semantic Web, Knowledge Base, and Natural Language Processing

The agricultural industry is a vast source of information. But most of it is stored in an unstructured way. That unstructured knowledge is merely understandable for machine. It is also having low accessibility for human too. The main objectives of the semantic web and knowledge base system is to make unstructured data into structured one. Semantic web and the knowledge base mainly facilitated by the ontology in the back end. Ontology is a formal, explicit specification of a shared conceptualization (Gruber 1991). Making of Ontology that facilitated the semantic web and knowledge base can be made across the agricultural domain to make the unstructured data into structured one. Many ontologies have already been developed in accordance with the Bedi and Marwaha (2004) in the agricultural domain. Saha (2011) developed an ontology on dynamic maize variety selection in different climatic conditions, Sahiram (2012) developed an ontology on rapeseed and mustard for identification of the variety in multiple languages; Das et al. (2017) developed an ontology for USDA soil taxonomy and another ontology was extended by Deb et al. (2015). Biswas et al. (2013) developed an ontology on microbial taxonomy and was extended by Karn (2014).

### 21.3.6 GIS and Remote Sensing Coupled with AI

GIS and remote sensing are helping agricultural community since long. The land use planning, land cover analysis, forest distribution, water distribution, water use pattern, crop rotation, and crop calendar analysis can be done by GIS and remote sensing. But when AI and machine learning are coupled with this technology it becomes more powerful. Machine learning and AI efficiently used for correct land classification and phonological change detection. From digital soil mapping to yield forecasting, from phenology detection to leaf area index a vast range of the area in agriculture can be handled by GIS and Remote sensing.

## 21.4 Framework for Phenology Study Using Artificial Intelligence

Based on the critical analysis of the reviews on the topic, one framework (as shown Fig. 21.3) has been deducted which depicts the mode of working of the methodologies of AI and Machine learning models used in agriculture. The whole framework can be subdivided into three distinct layers. The description of the layers are as follows:

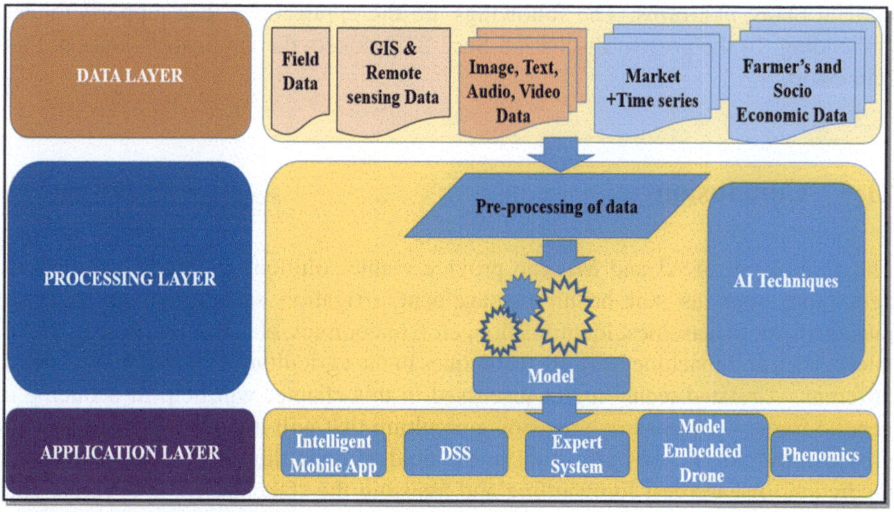


**Fig. 21.3** Framework depicting mode of working of the methodologies of AI and machine learning models used in predictive modelling of phenology

### *21.4.1 Data Layer*

This layer of the framework deals with the relevant data. As we know agriculture is a vast source of data with a wide range type. The remote sensing data with a longer period of time also has an importance along with the climatic data. With improvement of the technology, the image data coupled with time series data from Unmanned Aerial Vehicle (UAV) has been used in many studies. This layer is also used for any kind of pre-processing and making the data ready to use for the study.

### *21.4.2 Processing Layer*

This layer of the framework deals with the core part AI. It takes the input, i.e. the ready to use data from the data layer. It also takes care of the model development part of the framework. The ultimate product of the layer is a model that has promising result in the training and testing framework of the machine learning.

### *21.4.3 Application Layer*

This layer has a countless number of uses for the end users. Model which is developed in the previous layer has a great potential to be embedded and can be used in many platforms. The followings are the applications that supports model embedding. In the present discussion Sect. 21.3 covers the application layer of this framework.

## 21.5 Conclusion

The application of AI and ML can provide viable solutions to major problems in agriculture such as soil health management, irrigation scheduling, crop health management, disease/pest identification, crop phenomics, etc. The utility of artificial intelligence and machine learning techniques in the agriculture domain and survey of different AI related technologies discussed in this chapter will help in deducing a generic framework towards precision agriculture that will improve the overall crop productivity. AI is a powerful tool in the field of agriculture for accurate weather prediction, disease/pest forewarning and assisting the stakeholders in accurate and real-time prediction of various related parameters to obtain maximum yield at minimum cost. AI tools will transform the agriculture industry with better agricultural practices which in turn will benefit the farmers and aid in improving the economy of the country.

## References

Agastya C, Ghebremusse S, Anderson I, Vahabi H, Todeschini A (2021) Self-supervised contrastive learning for irrigation detection in satellite imagery. arXiv preprint arXiv:2108.05484

Ahila Priyadharshini R, Arivazhagan S, Arun M, Mirnalini A (2019) Maize leaf disease classification using deep convolutional neural networks. Neural Comput & Applic 31(12):8887–8895

Aitkenhead MJ, Coull MC, Towers W, Hudson G, Black HIJ (2012) Predicting soil chemical composition and other soil parameters from field observations using a neural network. Comput Electron Agric 82:108–116

Alkhudaydi T, Zhou J (2019) Spikeletfcn: counting spikelets from infield wheat crop images using fully convolutional networks. In: International Conference on Artificial Intelligence and Soft Computing, pp 3–13

Archana K, Saranya KG (2020) Crop yield prediction, forecasting and fertilizer recommendation using voting based ensemble classifier. Int J Comput Sci Eng 7(5):1–4

Azizi A, Gilandeh YA, Mesri-Gundoshmian T, Saleh-Bigdeli AA, Moghaddam HA (2020) Classification of soil aggregates: a novel approach based on deep learning. Soil Tillage Res 199: 104586

Ballesteros JM, Cartujano AR, Evaldez D, Macutay J (2021) Online ordering and recommender system of combine harvester parts and equipment with 3D modelling and augmented reality brochure for BLAZE equifarm and general merchandise. In: 11th international workshop on computer science and engineering (WCSE 2021), pp 174–179

Barbedo JGA (2019) Plant disease identification from individual lesions and spots using deep learning. Biosyst Eng 180:96–107

Bedi P, Marwaha S (2004) Designing ontologies from traditional taxonomies. In: Proceedings of International Conference on Cognitive Science, Allahabad, pp 324–329

Beisel NS, Callaham JB, Sng NJ, Taylor DJ, Paul A, Ferl RJ (2018) Utilization of single-image normalized difference vegetation index (SI-NDVI) for early plant stress detection. Appl Plant Sci 6(10):e01186

Besalatpour A, Hajabbasi MA, Ayoubi S, Gharipour A, Jazi AY (2012) Prediction of soil physical properties by optimized support vector machines. Int Agrophys 26:2

Biswas S, Marwaha S, Malhotra PK, Wahi SD, Dhar DW, Singh R (2013) Building and querying microbial ontology. Proc Technol 10:13–19

Cai J, Okamoto M, Atieno J, Sutton T, Li Y, Miklavcic SJ (2016) Quantifying the onset and progression of plant senescence by color image analysis for high throughput applications. PLoS One 11(6):e0157102

Cheeti S, Kumar GS, Priyanka JS, Firdous G, Ranjeeva PR (2021) Pest detection and classification using YOLO AND CNN. Ann Roman Soc Cell Biol 2021:15295–15300

Chen J, Zhang D, Nanehkaran YA, Li D (2020) Detection of rice plant diseases based on deep transfer learning. J Sci Food Agric 100(7):3246–3256

Chen JW, Lin WJ, Cheng HJ, Hung CL, Lin CY, Chen SP (2021) A smartphone-based application for scale pest detection using multiple-object detection methods. Electronics 10(4):372

Das B, Sahoo RN, Pargal S, Krishna G, Verma R, Chinnusamy V, Sehgal VK, Gupta VK (2017) Comparison of different uni-and multi-variate techniques for monitoring leaf water status as an indicator of water-deficit stress in wheat through spectroscopy. Biosyst Eng 160:69–83

Deb CK, Marwaha S, Malhotra PK, Wahi SD, Pandey RN (2015) Strengthening soil taxonomy ontology software for description and classification of USDA soil taxonomy up to soil series. In: 2015 2nd international conference on computing for sustainable global development (INDIACom), pp 1180–1184

Elavarasan D, Vincent DR, Sharma V, Zomaya AY, Srinivasan K (2018) Forecasting yield by integrating agrarian factors and machine learning models: a survey. Comput Electron Agric 155: 257–282. https://doi.org/10.1016/j.compag.2018.10.024

Ferentinos KP (2018) Deep learning models for plant disease detection and diagnosis. Comput Electron Agric 145:311–318

Fuentes A, Yoon S, Kim SC, Park DS (2017) A robust deep-learning-based detector for real-time tomato plant diseases and pests recognition. Sensors 17(9):2022

Fusco R, Grassi R, Granata V, Setola SV, Grassi F, Cozzi D, Petrillo A (2021) Artificial intelligence and COVID-19 using chest CT scan and chest X-ray images: machine learning and deep learning approaches for diagnosis and treatment. J Personal Med 11(10):993

Glória A, Cardoso J, Sebastião P (2021) Sustainable irrigation system for farming supported by machine learning and real-time sensor data. Sensors 21:3079

Gruber TR (1991) The role of common ontology in achieving sharable, reusable knowledge bases. Kr 91:601–602

Haque MA, Marwaha S, Arora A, Paul RK, Hooda KS, Sharma A, Grover M (2021) Image-based identification of maydis leaf blight disease of maize (Zea mays) using deep learning. Indian J Agric Sci 91(9):1362–1367. https://doi.org/10.56093/ijas.v91i9.116089

Hasan MM, Chopin JP, Laga H, Miklavcic SJ (2018) Detection and analysis of wheat spikes using convolutional neural networks. Plant Methods 14(1):1–13

Jha SK, Ahmad Z (2018) Soil microbial dynamics prediction using machine learning regression methods. Comput Electron Agric 147:158–165

Jimenez AF, Ortiz BV, Bondesan L, Morata G, Damianidis D (2021) Long short-term memory neural network for irrigation management: a case study from southern Alabama, USA. Precis Agric 22:475–492

Johannes A, Picon A, Alvarez-Gila A, Echazarra J, Rodriguez-Vaamonde S, Navajas AD, Ortiz-Barredo A (2017) Automatic plant disease diagnosis using mobile capture devices, applied on a wheat use case. Comput Electron Agric 138:200–209

Karn SK (2014) Strengthening and enhancing microbial taxonomy ontology. M.Sc. Thesis Dissertation. ICAR-IARI, New Delhi

Kalambukattu JG, Kumar S, Raj RA (2018) Digital soil mapping in a Himalayan watershed using remote sensing and terrain parameters employing artificial neural network model. Environ Earth Sci 77(5):1–14

Kamatchi SB, Parvathi R (2019) Improvement of crop production using recommender system by weather forecasts. Proc Comput Sci 165:724–732

Konstan JA, Miller BN, Maltz D, Herlocker JL, Gordon LR, Riedl J (1997) Grouplens: applying collaborative filtering to usenet news. Commun ACM 40(3):77–87

Krishna G, Sahoo RN, Singh P, Patra H, Bajpai V, Das B, Kumar S, Dhandapani R, Vishwakarma C, Pal M, Chinnusamy V (2021) Application of thermal imaging and hyperspectral remote sensing for crop water deficit stress monitoring. Geocarto Int 36(5): 481–498

Kulkarni NH, Srinivasan GN, Sagar BM, Cauvery NK (2018) Improving crop productivity through a crop recommendation system using ensembling technique. In: 2018 3rd international conference on computational systems and information technology for sustainable solutions (CSITSS), pp 114–119

LeCun Y, Bengio Y, Hinton G (2015) Deep learning. Nature 521(7553):436–444

Li D, Wang R, Xie C, Liu L, Zhang J, Li R, Liu W (2020) A recognition method for rice plant diseases and pests video detection based on deep convolutional neural network. Sensors 20(3): 578

Liu J, Wang X (2020) Tomato diseases and pests detection based on improved Yolo V3 convolutional neural network. Front Plant Sci 11:898

Lu J, Wu D, Mao M, Wang W, Zhang G (2015) Recommender system application developments: a survey. Decis Support Syst 74:12–32

Lu J, Hu J, Zhao G, Mei F, Zhang C (2017) An in-field automatic wheat disease diagnosis system. Comput Electron Agric 142:369–379

Middleton SE, Alani H, De Roure DC (2002) Exploiting synergy between ontologies and recommender systems. arXiv preprint. cs/0204012

Misra T, Arora A, Marwaha S, Chinnusamy V, Rao AR, Jain R et al (2020) SpikeSegNet-a deep learning approach utilizing encoder-decoder network with hourglass for spike segmentation and counting in wheat plant from visual imaging. Plant Methods 16(1):1–20

Mohanty SP, Hughes DP, Salathé M (2016) Using deep learning for image-based plant disease detection. Front Plant Sci 7:1419

Naha S, Marwaha S (2020) Context-aware recommender system for maize cultivation. J Commun Mobiliz Sustainable Dev 15(2):485–490

Nigam S, Jain R, Marwaha S, Arora A (2021) Wheat rust disease identification using deep learning. De Gruyter, Berlin

Padarian J, Minasny B, McBratney AB (2019) Using deep learning for digital soil mapping. Soil 5(1):79–89

Pande SM, Ramesh PK, Anmol A, Aishwarya BR, Rohilla K, Shaurya K (2021) Crop recommender system using machine learning approach. In: 2021 5th international conference on computing methodologies and communication (ICCMC), pp 1066–1071

Patil AP, Deka PC (2016) An extreme learning machine approach for modelling evapotranspiration using extrinsic inputs. Comput Electron Agric 121:385–392

Picon A, Alvarez-Gila A, Seitz M, Ortiz-Barredo A, Echazarra J, Johannes A (2019) Deep convolutional neural networks for mobile capture device-based crop disease classification in the wild. Comput Electron Agric 161:280–290

Qiongyan L, Cai J, Berger B, Miklavcic S (2014) Study on spike detection of cereal plants. In: 13 International Conference on Control Automation Robotics & Vision (ICARCV). IEEE, pp 228–233

Rahman CR, Arko PS, Ali ME, Khan MAI, Apon SH, Nowrin F, Wasif A (2020) Identification and recognition of rice diseases and pests using convolutional neural networks. Biosyst Eng 194: 112–120

Ramya S, Swetha AM, Doraipandian M (2020) IoT framework for smart irrigation using machine learning technique. J Comput Sci 16:355–363
Ranjan R, Sahoo RN, Chopra UK, Pramanik M, Singh AK, Pradhan S (2017) Assessment of water status in wheat (Triticum aestivum L.) using ground based hyperspectral reflectance. Proc Natl Acad Sci, India B Biol Sci 87(2):377–388
Resnick P, Varian HR (1997) Recommender systems. Commun ACM 40(3):56–58
Rich E, Knight K (1991) Artificial intelligence, 2nd edn. McGraw-Hill, New York
Rich E, Knight K, Nair SB (2009) Artificial intelligence, 3rd edn. Tata McGraw-Hill, New Delhi
Rivera JI, Bonilla CA (2020) Predicting soil aggregate stability using readily available soil properties and machine learning techniques. Catena 187:104408
Saha A (2011) Ontologies-based expert system for maize. M.Sc. Thesis Dissertation. ICAR-IARI, New Delhi
Sahiram (2012) Ontology based expert system for rapeseed-mustard crop. M.Sc. Thesis dissertation. ICAR-IARI, New Delhi
Sibiya M, Sumbwanyambe M (2019) A computational procedure for the recognition and classification of maize leaf diseases out of healthy leaves using convolutional neural networks. AgriEngineering 1(1):119–131
Sirsat MS, Cernadas E, Fernández-Delgado M, Khan R (2017) Classification of agricultural soil parameters in India. Comput Electron Agric 135:269–279
Smyth B (2007) Case-based recommendation. In: The adaptive web. Springer, Berlin, pp 342–376
Taghizadeh-Mehrjardi R, Ayoubi S, Namazi Z, Malone BP, Zolfaghari AA, Sadrabadi FR (2016) Prediction of soil surface salinity in arid region of central Iran using auxiliary variables and genetic programming. Arid Land Res Manag 30(1):49–64
Taghizadeh-Mehrjardi R, Mahdianpari M, Mohammadimanesh F, Behrens T, Toomanian N, Scholten T, Schmidt K (2020) Multi-task convolutional neural networks outperformed random forest for mapping soil particle size fractions in central Iran. Geoderma 376:114552
Too EC, Yujian L, Njuki S, Yingchun L (2019) A comparative study of fine-tuning deep learning models for plant disease identification. Comput Electron Agric 161:272–279
Vaishnavi, S., Shobana, M., Sabitha, R., & Karthik, S. (2021). Agricultural crop recommendations based on productivity and season. In: 2021 7th international conference on advanced computing and communication systems (ICACCS), pp 883–886
Wang D, Liang Y, Xu D, Feng X, Guan R (2018) A content-based recommender system for computer science publications. Knowl-Based Syst 157:1–9
Xiong H, Cao Z, Lu H, Madec S, Liu L, Shen C (2019) TasselNetv2: in-field counting of wheat spikes with context-augmented local regression networks. Plant Methods 15(1):1–14
Zema DA, Nicotra A, Mateos L, Zimbone SM (2018) Improvement of the irrigation performance in water users associations integrating data envelopment analysis and multi-regression models. Agric Water Manag 205:38–49
Zhang C, Yue P, Di L, Wu Z (2018) Automatic identification of center pivot irrigation systems from landsat images using convolutional neural networks. Agriculture 8(10):147